Recent Developments in Infant Nutrition

Tenth Nutricia Symposium

Recent Developments in Infant Nutrition

Scheveningen, 29 November – 2 December 1995

Edited by

J.G. BINDELS, A.C. GOEDHART and H.K.A. VISSER

KLUWER ACADEMIC PUBLISHERS
DORDRECHT / BOSTON / LONDON

Distributors

for the United States and Canada: Kluwer Academic Publishers, PO Box 358, Accord Station, Hingham, MA 02018-0358, USA
for all other countries: Kluwer Academic Publishers Group, Distribution Center, PO Box 322, 3300 AH Dordrecht, The Netherlands

A catalogue record for this book is available from the British Library

ISBN 0-7923-8707-4

Published in the United Kingdom by Kluwer Academic Publishers, PO Box 55, Lancaster, UK.

Kluwer Academic Publishers BV incorporates the publishing programmes of D. Reidel, Martinus Nijhoff, Dr W. Junk and MTP Press.

Typeset by Lasertext Ltd, Stretford, Manchester.

Printed and bound in Great Britain by Hartnolls Ltd., Bodmin, Cornwall.

Contents

Preface

After 17 years, the Nutricia Symposium returned to its home grounds in The Netherlands, where the first five Symposia were held. The objective of the 10th Nutricia Symposium was to bring together a limited number of opinion leaders and key researchers in selected topics of infant nutrition to discuss the current state of the art based on original contributions and reviews. The discussion sessions after the papers were taped and edited and may give additional information and views. As a result of time constraints, the discussions on a few papers had to be cut short, or, unfortunately, could not take place at all. Nevertheless, we strongly recommend reading the discussion sections, such as the panel discussion on the requirements for LCPUFA for term and/or preterm infants.

During the preparation of the Symposium the sad news that Professor J.H.P. Jonxis died prompted the organisers to dedicate one special lecture to the memory of the principal organiser of the first four Nutricia Symposia. This lecture entitled "What determines the production and composition of breast milk?" was given by Dr. Kathleen Motil.

Acknowledgement should be made to the co-organizers and co-chairmen Professor Sir David Hull, Professor Pieter Sauer and Dr. Terence Stephenson for their contribution. Finally, without mentioning names, the editors wish to thank a number of persons within the Nutricia organization without whom the Symposium and its Proceedings would not have been possible.

May 1996 **The Editors**

List of participants

C. Agostoni, Milan, Italy
D.J.P. Barker, Southampton, UK
F.C. Battaglia, Denver, CO, USA
H. Berger, Leiden, The Netherlands
M. Bertrand-Van Caillie, Antwerp, Belgium
H. Böhles, Frankfurt, Germany
I. Booth, Birmingham, UK
P. Brandtzaeg, Oslo, Norway
D. Brasseur, Brussels, Belgium
N.C. de Bruin, Rotterdam, The Netherlands
H. Büller, Amsterdam, The Netherlands
J.P. Buts, Brussels, Belgium
S. Cadranel, Brussels, Belgium
S.E. Carlson, Memphis, TE, USA
V.P. Carnielli, Rotterdam, The Netherlands
P. Clarke, London, UK
O.M. Conneely, Houston, TX, USA
R. Cooke, Newcastle, UK
P.R. Dallman, San Francisco, CF, USA
Sir David Hull, Nottingham, UK
G. Davidson, Adelaide, Australia
D. Davies, Cardiff, UK
J.F. Desjeux, Paris, France
M. Deutz, Maastricht, The Netherlands
G. Donzelli, Florence, Italy
R. Eeckels, Leuven, Belgium
M. Elia, Cambridge, UK
C. Fabris, Turin, Italy
J.S. Forsyth, Dundee, UK
M.K. Gabr, Cairo, Egypt
R.A. Gibson, Adelaide, Australia
O.L. Giuffré, Palermo, Italy
D.L. Hachey, Houston, TX, USA
M. Hamosh, Washington DC, USA
J. Hautvast, Wageningen, The Netherlands
O. Hernell, Umeå, Sweden
H. Heymans, Groningen, The Netherlands
G. Hornstra, Maastricht, The Netherlands
S. Hudault, Jouy-en-Josas, France
J. Janda, Prague, Czech Republic
J. Karlberg, Hong Kong
C. Kattamis, Athens, Greece
B. Koletzko, Munich, Germany
J.G. Koppe, Amsterdam, The Netherlands
D. Kyle, Columbia, MD, USA
H. Lafeber, Amsterdam, The Netherlands
D.S. Lewis, Ames, IO, USA
H. Loeb, Brussels, Belgium
M. Makrides, Adelaide, Australia
A. Marini, Milan, Italy
T. Mazurczak, Warsaw, Poland
R. Morley, Cambridge, UK
K.J. Motil, Houston TX, USA
H. Murtada, Damascus, Syria
J. Nuyens, Leiden, The Netherlands
A. Okken, Utrecht, The Netherlands
P.B. Pencharz, Toronto, Ontario, Canada
J. Pietrzyk, Krakow, Poland
A. Plaut, Boston, MA, USA
S.O. Poppit, Cambridge, UK
H.D. Pusponegoro, Jakarta, Indonesia
S. Reisbick, Portland, OR, USA
J. Rigo, Liège, Belgium
M. Rowland, Cambridge, UK
J. Salazar de Sousa, Lisbon, Portugal
P.J.J. Sauer, Rotterdam, The Netherlands
J. Socha, Warsaw, Poland
T.J. Stephenson, Nottingham, UK
I. Sudigbia, Semarang, Indonesia
D. Turck, Lille, France
J. Taminiau, Amsterdam, The Netherlands
R. Uauy, Santiago, Chile
Y. Vandenplas, Brussels, Belgium
H.J. Verkade, Groningen, The Netherlands
L. Vega-Franco, Mexico City, Mexico
H.K.A. Visser, Rotterdam, The Netherlands
B. Wharton, London, UK
J.H.P. Wilson, Rotterdam, The Netherlands
K. Zwiauer, St. Pölten, Austria

List of authors, chairpersons and editors

C. Agostoni, Clinica Pediatrica V, Ospedale S. Paolo, Via A. di Rudini, 8, I-20142 Milano, Italy.
Co-authors: S. Trojan, R. Bellù, E. Riva, D. Luotti and M. Giovannini.

D.J.P. Barker, MRC Environmental Epidemiology Unit, Southampton General Hospital, Southampton SO16 6YD, UK.

F.C. Battaglia, Perinatal Medicine, Department of Pediatrics, University of Colorado School of Medicine, 4200 East Ninth Avenue, Box B-199, Denver, CO 80262, USA

J.G. Bindels, Nutricia Research, PO Box 1, 2700 MA Zoetermeer, The Netherlands.

P. Brandtzaeg, LIIPAT Institute of Pathology, University of Oslo, The National Hospital, N-0027 Oslo 1, Norway.

N.C. de Bruin, Sophia Children's Hospital, Dr. Molewaterplein 60, 3015 GJ Rotterdam, The Netherlands. *Current address*: Stichting Slingedael, Dreischorstraat 85, 3086 PP Rotterdam, The Netherlands.
Co-authors: H.J. Degenhart and H.K.A. Visser.

S.E. Carlson, Department of Pediatrics and Obstetrics/Gynecology, Newborn Center, University of Tennessee, 853 Jefferson Avenue, Memphis, TN 38163, USA.

V.P. Carnielli, Sophia Children's Hospital, Dr. Molewaterplein 60, 3015 GJ Rotterdam, The Netherlands. *Current address*: Dipartimento di Pediatria, Universita' degli Studi di Padova, Via Giustiniani 3, I-35128 Padova, Italy.
Co-author: P.J.J. Sauer.

O.M. Conneely, Cell Biology Dept., Baylor College of Medicine, One Baylor Plaza, Houston, TX 77030, USA.
Co-authors: P.P. Ward, X. Zhou and S. Wagner.

P.R. Dallman, 2201 Ninth Avenue, San Francisco, CF 94116, USA.

M. Elia, Dunn Clinical Nutrition Centre, Hills Road, Cambridge CB2 2DH, UK.

J.S. Forsyth, Directorate of Paediatrics, Ninewells Hospital and Medical Schools, Dundee DD1 9SY, UK.
Co-author: P. Willatts.

A.C. Goedhart, Nutricia Research, PO Box 1, 2700 MA Zoetermeer, The Netherlands.

D.L. Hachey, USDA/ARS Children's Nutrition Research Center, Department of Pediatrics, Baylor College of Medicine, 1100 Bates Street, Houston, TX 77030, USA.
Co-authors: W.G. Pond and W.W. Wong.

G. Hornstra, Department of Human Biology, Limburg University, PO Box 616, 6200 MD Maastricht, The Netherlands.
Co-authors: M.D.M. Al, A.C. van Houwelingen and M.M.H.P. Foreman-van Drongelen.

S. Hudault, Unité d'Ecologie et de Physiologie du Système Digestif, Institut National de la Recherche Agronomique, Bat 440, CRJ INRA, F-78352 Jouy-en-Josas, France.

Sir David Hull, Department of Child Health, Queen's Medical Centre, University of Nottingham, Nottingham NG7 2UH, UK.

J. Karlberg, Department of Paediatrics, Queen Mary Hospital, University of Hong Kong, Hong Kong.
Co-author: K. Albertsson-Wikland.

B. Koletzko, Kinderpoliklinik Klinikum der Universität München, Pettenkoferstraße 8a, D-80336 München, Germany.
Co-authors: T. Desci, H. Demmelmair and T. Sauerwald.

D.S. Lewis, Department of Food Science and Human Nutrition, 1127 LaBaron Addition, Iowa State University, Ames, IO 50011, USA.

M. Makrides, Department of Paediatrics & Child Health, Flinders Medical Centre, Bedford Park (Adelaide), South Australia 5042, Australia.
Co-author: R.A. Gibson.

K.J. Motil, USDA/ARS Children's Nutrition Research Center, Department of Pediatrics, Baylor College of Medicine, 1100 Bates Street, Houston, TX 77030, USA.

P.B. Pencharz, Division of Gastroenterology and Nutrition, Hospital for Sick Children, 555 University Avenue, Toronto, Ontario, Canada, M5G 1X8.
Co-authors: J.D. House, L.J. Wykes and R.O. Ball.

S.D. Poppitt, Dunn Clinical Nutrition Centre, Hill's Road, Cambridge CB2 2DH, UK.

S. Reisbick, Department of Medicine, L465, The Oregon Health Sciences University, 3181 S.W. Sam Jackson Park Road, Portland, OR 97201-3098, USA.
Co-authors: M. Neuringer and W.E. Connor.

J. Rigo, Department of Neonatology, Hôpital de la Citadelle, B-4000 Liège, Belgium.
Co-authors: K. Nyamugabo, M. De Curtis and J. Senterre.

P.J.J. Sauer, Sophia Children's Hospital, Dr. Molewaterplein 60, 3015 GJ Rotterdam, The Netherlands.

T.J. Stephenson, Department of Child Health, Queen's Medical Centre, University of Nottingham, Nottingham NG7 2UH, UK.

R. Uauy, Institute of Nutrition INTA, University of Chile, Casilla 138-11, Santiago, Chile.
Co-authors: P. Peirano, P. Mena, D. Hoffman, D. Birch and E. Birch.

H.K.A. Visser, Ghisebrecht Bokellaan 29, 3054 CA Rotterdam, The Netherlands.

B. Wharton, British Nutrition Foundation, High Holborn House, 52–54 High Holborn, London WC1V 6RQ, UK.

Nutricia Symposia

I **The Adaptation of the Newborn Infant to Extra-uterine Life**
edited by J.H.P. Jonxis, H.K.A. Visser and J.A. Troelstra
1964, xvi+332 pp., ISBN 90-207-0027-8

II **Aspects of Praematurity and Dysmaturity**
edited by J.H.P. Jonxis, H.K.A. Visser and J.A. Troelstra
1968, xii+356 pp., ISBN 90-207-0028-6

III **Metabolic Processes in the Foetus and Newborn Infant**
edited by J.H.P. Jonxis, H.K.A. Visser and J.A. Troelstra
1971, xii+308 pp., ISBN 90-207-0261-0

IV **Therapeutic Aspects of Nutrition**
edited by J.H.P. Jonxis, H.K.A. Visser and J.A. Troelstra
1973, xii+344 pp., ISBN 90-207-0419-2

V **Nutrition and Metabolism of the Fetus and Infant**
edited by H.K.A. Visser
1979, xvi+420 pp., ISBN 90-247-2202-0

VI **Child Health in the Tropics**
edited by R.E. Eeckels, O. Ramsone-Kuti and C.C. Kroonenberg
1985, xvi+348 pp., ISBN 0-89838-719-1

VII **Topics and Controversies in Enteral Nutrition**
edited by D.B.A. Silk
1986, *Gut* Vol. 27, Supplement 1, pp. 1 – 122 (ISSN 0017-5749)
(journal of the *British Society of Gastroenterology*)

VIII **Child Nutrition in South East Asia**
edited by H.K.A. Visser and J.G. Bindels
1990, xvi+284 pp., ISBN 0-7923-0690-2

IX **Developments in Enteral Nutrition**
edited by M. Elia
1993, *Clinical Nutrition* Vol. 12, Supplement 1, pp. 1 – 118 (ISSN 0261-5614)
(official journal of the *European Society of Parenteral and Enteral Nutrition*)

X **Recent Developments in Infant Nutrition**
edited by J.G. Bindels, A.C. Goedhart and H.K.A. Visser
1996, xiv+400 pp., ISBN 0-7923-8707-4

Session 1
Maternal nutrition and its effects on the fetus and newborn

1 Fetal undernutrition and disease in adult disease

D.J.P. BARKER

ABBREVIATIONS

CHD coronary heart disease
PEPCK phosphoenol pyruvate carboxykinase

INTRODUCTION

During embryonic life, that is during the 8 weeks after conception, the basic form of the human baby is laid down in miniature. The 5-week-old embryo does not contain a description of the person to whom it will give rise. Rather it contains in its genes a generative programme for making a person, a programme that has been likened to a recipe (1). As development proceeds the destiny of cells becomes determined by their surroundings, by the position they come to occupy in the body, by the signals they receive from neighbouring cells and hence by the genes which become activated.

FETAL GROWTH

The body does not increase greatly in size during embryonic life but in the fetal period, from 9 weeks after conception onwards, there is a phase of rapid growth which continues until after birth. Growth does not simply expand the miniature human being but, through differences in growth rates of different parts of the body, it moulds the baby's form. The main feature of fetal growth is cell division. Different tissues of the body grow during periods of rapid cell division, so called 'critical' periods (2). The timing of these critical periods differs for different tissues: the kidney, for example, goes through a critical period in the weeks immediately before birth. Growth depends on nutrients and oxygen, and the main adaptation of the fetus to lack of these is a slowing of the rate of cell division, particularly in those tissues which are undergoing 'critical periods' at the time. Cell division slows either as a direct effect of undernutrition on the cell or as a result of altered concentrations of growth factors or hormones, of which insulin and growth hormone are particularly important.

Disproportionate growth can occur in utero because different tissues have critical periods of growth at different times. The diversity of size and form of babies born after normal pregnancies is remarkable. Studies of the birth weights of relatives have led to the conclusion that variation in size at birth is essentially determined by the intrauterine environment rather than the fetal genome (3,4). Widdowson and McCance (2,5) were among the first to show that brief periods of undernutrition may permanently

J.G. Bindels, A.C. Goedhart and H.K.A. Visser (eds), Recent Developments in Infant Nutrition, 3–14.

reduce the numbers of cells in particular organs. This is one of the mechanisms by which undernutrition may permanently change or 'programme' the body (6). Other lasting 'memories' of undernutrition include change in the distribution of cell types, in hormonal feedback, in metabolic activity and in organ structure, each of which will be discussed further.

It is not in question that the human body can be programmed by undernutrition. Rickets has for a long while served as a demonstration that undernutrition at a critical stage of early life leads to persisting changes in structure. What is new is the realization that some of the body's 'memories' of early undernutrition become translated into pathology and thereby determine disease in later life. This is perhaps unsurprising, given the numerous experiments on animals which have shown that undernutrition in utero leads to persisting changes in blood pressure, cholesterol metabolism, insulin response to glucose, and in a range of other metabolic, endocrine and immune functions known to be important in human disease (6–8).

FETAL GROWTH AND CORONARY HEART DISEASE

The main focus for research into coronary heart disease (CHD), the most common cause of death in the Western world, has been the lifestyles of men and women. Inappropriate behaviours, such as a high fat diet, cigarette smoking and obesity, have been implicated. Adult lifestyles, however, fail to explain much about the geography of the disease, its trends over time, and why one person dies from the disease while another does not.

In the search for a new model for coronary heart disease an important clue, suggesting that it might originate in utero, came from studies of death rates among babies in Britain during the early years of the century (9), when death during infancy was remarkably common. In 1917 the Bishop of London remarked 'while nine soldiers died every hour in 1915, twelve babies died every hour, so that it was much more dangerous to be a baby than a soldier'. The usual certified cause of death in newborn babies was low birth weight. Death rates in the newborn differed considerably between one part of the country and another, being highest in some of the northern industrial towns and the poorer rural areas in the north and west. We now know that this geographical pattern in death rates closely resembles today's large variations in death rates from CHD, variations which form one aspect of the continuing north–south divide in health. A conclusion suggested by this observation was that low rates of growth before birth are linked to the development of CHD in adult life. The suggestion that events in childhood influence the pathogenesis of CHD was not new. A focus on intrauterine life, however, offered a new point of departure for research.

SMALL SIZE AT BIRTH AND LATER DISEASE

Early epidemiological studies were based on the simple strategy of examining in middle and late life individual men and women whose size at birth was recorded. The records on which these studies were based came to light as a result of the Medical Research Council's systematic search of the archives and records offices of Britain –

***Table 1* Death rates from CHD among 15 726 men and women according to birth weight**

Birth weight (pounds (kg))	*Standardized mortality ratio*	*No. of deaths*
≤5.5 (2.50)	100	57
6.5 (2.95)	81	137
7.5 (3.41)	80	298
8.5 (3.86)	74	289
9.5 (4.31)	55	103
>9.5 (4.31)	65	57
Total74	941	

***Table 2* Prevalence of non-insulin-dependent diabetes and impaired glucose tolerance in men aged 59–70 years**

Birth weight (pounds (kg))	*Total no. of men*	*% with impaired glucose tolerance or diabetes*	*Odds ratio adjusted for body mass index* (95% confidence interval)
≤5.5 (2.50)	20	40	6.6 (1.5–28)
6.5 (2.95)	47	34	4.8 (1.3–17)
7.5 (3.41)	104	31	4.6 (1.4–16)
8.5 (3.86)	117	22	2.6 (0.8–8.9)
9.5 (4.31)	54	13	1.4 (0.3–5.6)
>9.5 (4.31)	28	14	1.0
Total370	25		

a search which led to the discovery of three important collections of birth records, in Hertfordshire, Preston and Sheffield. In Hertfordshire the Lady Inspector of Midwives from 1905 onwards, Margaret Burnside, had recruited an army of trained nurses to attend women in childbirth and to advise mothers on how to keep their babies healthy. Throughout the county when women had their babies they were attended by a midwife. The baby was weighed at birth and again at one year. These weights were recorded in ledgers. We traced 15 726 men and women born in Hertfordshire during 1911–1930 from birth to the present day (10,11). Death rates from CHD among these individuals fell progressively between those who weighed less than 5.5 pounds at birth and those who weighed 9.5 pounds (Table 1). Another study, in Sheffield, showed that babies who were small because they failed to grow, rather than those who were small because they were born prematurely, were at increased risk of coronary heart disease. By examining men and women who still live in Hertfordshire we showed that these trends in coronary heart disease with birth weight are paralleled by similar trends in hypertension and diabetes. Table 2 shows the steep fall in prevalence of adult-onset diabetes, or its preclinical state of impaired glucose tolerance, between men who were small and men who were large at birth (12). This finding has been confirmed in both men and women in two other studies in Britain, in two studies in the USA and in a study in Sweden (13).

One response to such findings is to argue that people who were exposed to an adverse environment in utero and failed to grow continue to be exposed to an adverse environment in childhood and adult life, and it is this later adverse environment that produces the effects attributed to programming in utero. There is, however, little evidence to support this argument. Rather, associations between birth weight and later disease are found in each social group, and are independent of adult behaviours such as smoking and becoming obese (11). Variations in disease rates associated with birth weight are large and highly statistically significant. They are being replicated in different populations around the world, and are supported by animal experiments. It is reasonable to conclude that influences that lead to low growth rates in utero also determine later coronary heart disease. Adult lifestyle does, however, add to these intrauterine effects. The highest prevalences of non-insulin-dependent diabetes and impaired glucose tolerance, for example, are seen in people who were small at birth but obese as adults (12). Around the world communities with high prevalences of diabetes generally conform to this pattern: they are peoples such as the Ethiopian Jews air-lifted to Israel, or Indian people who migrate to Britain, whose fetal growth was poor but who subsequently became affluent and obese in their adult life. We know something of why people with low growth rates in utero are unable to withstand the stress of becoming obese as adults. There is some evidence that poor fetal growth results in a reduced number of pancreatic β-cells and hence a reduced capacity to make insulin (14). There is stronger evidence that they became resistant to the effects of insulin.

DISPROPORTIONATE SIZE AT BIRTH

In the early years of the century a new record form came into use in maternity hospitals in Britain. Known as the Queen Charlotte form it required measurements of the newborn baby's head circumference and length at birth as well as its weight, measurements which allow thin and short babies to be identified. The Hertfordshire study showed that low birth weight was associated with the so-called insulin-resistance syndrome (Table 3), a common disorder in adult life in which impaired glucose tolerance, raised blood pressure and disturbed lipid metabolism coincide in the same patient. Biochemically, the syndrome is characterized by raised serum insulin concentrations and it leads to coronary heart disease. Studies in Preston, where the Queen Charlotte form was used, showed that it is specifically thinness at birth, indicated by a low ponderal index (birth weight/length^3), that is associated with resistance to insulin and its associated disorders in later life (15).

Muscle growth in utero

In fetal life insulin has a key role in stimulating cell division. The thin newborn baby lacks skeletal muscle, as well as fat. It is thought that at some point in mid–late gestation the thin neonate becomes undernourished, and in response its muscles

Table 3 Prevalence of the insulin resistance syndrome in men aged 59–70 years according to birth weight

Birth weight (pounds (kg))	*Total no. of men*	*% with syndrome X*	*Odds ratio adjusted for body mass index* (95% confidence interval)
≤5.5 (2.50)	20	30	18 (2.6–118)
6.5 (2.95)	54	19	8.4 (1.5–49)
7.5 (3.41)	114	17	8.5 (1.5–46)
8.5 (3.86)	123	12	4.9 (0.9–27)
9.5 (4.31)	64	6	2.2 (0.3–14)
>9.5 (4.31)	32	6	1.0
Total	407	14	

become resistant to insulin. Growth of muscle is therefore sacrificed but the brain, which does not require insulin to utilize glucose, is spared. Persisting resistance to insulin through childhood into adult life impairs the response to a glucose challenge, because muscle is the main peripheral site of insulin action. Studies using magnetic resonance spectroscopy show that adults who were thin at birth have reduced rates of glycolysis in their muscles (16). It is unclear how this is linked to insulin resistance, but further research into muscle metabolism of people who were thin at birth may lead to a new understanding of insulin resistance.

Liver growth in utero

Studies in Sheffield, where the Queen Charlotte form was also used, show that neonates who have a short body in relation to the size of their heads, although within the normal range of birth weight, have persisting disturbances of cholesterol metabolism and blood coagulation. Disproportion in body length relative to head size is thought to result from undernutrition in late gestation. The fetus uses an adaptive response present in mammals and diverts oxygenated blood away from the trunk to sustain the brain. One of the organs whose growth is prejudiced by this effect is the liver, two of whose functions, regulation of cholesterol and of blood clotting, seem to be permanently perturbed. Disturbances of cholesterol metabolism and blood clotting are both important features of coronary heart disease.

A small abdominal circumference at birth, reflecting small liver size, predicts raised serum low density lipoprotein cholesterol concentrations, a major coronary disease risk factor (17). The difference in serum cholesterol concentrations between people who had a small abdominal circumference at birth and those who had a large abdominal circumference (Table 4) is equivalent to around 30% difference in risk of coronary heart disease. The findings for plasma fibrinogen concentrations, a measure of blood coagulability, are of similar size (18).

In keeping with these associations a small abdominal circumference at birth is also associated with raised death dates from CHD (17). This, however, is only seen in

Table 4 Mean serum cholesterol concentrations according to abdominal circumference at birth in men and women aged 50–53 years

Abdominal circumference (inches (cm))	*Total no. of people*	*Total cholesterol* (mmol/l)	*Low density lipoprotein cholesterol* (mmol/l)
≤11.5 (29.2)	53	6.7	4.5
12.0 (30.5)	43	6.9	4.6
12.5 (31.8)	31	6.8	4.4
13.0 (33.0)	45	6.2	4.0
>13.0 (33.0)	45	6.1	4.0
Total	217	6.5	4.3

babies of below average weight. In large babies the trend is reversed: it is the large baby with the large abdominal circumference who is at increased risk. This kind of baby is known to result from a pregnancy in which the mother develops diabetes. The fetus is exposed to abnormally high concentrations of glucose and is therefore, in a sense, overnourished. In these babies the abdomen enlarges rapidly in late gestation. Accelerated as well as reduced liver growth in late gestation seem to be linked to later coronary heart disease.

Experiments on rats have shown how undernutrition may permanently alter the way in which the liver metabolizes glucose (19). A low protein diet in gestation was followed by permanent changes in the activity of two enzymes one of which, phosphoenolpyruvate carboxykinase (PEPCK), is part of the process by which glucose is synthesized while the other, glucokinase, acts in the breakdown of glucose. A low protein diet in gestation permanently changes the balance of activity in favour of synthesis, even if it is followed by normal protein intake after birth. A low protein diet after birth has no effect. PEPCK is largely made in the cells around the portal veins of the liver, while glucokinase is made in the cells around the hepatic vein. An interpretation of these findings, therefore, is that undernutrition re-programmes the zonal structure of the liver, enhancing the development of periportal cells in relation to the perivenous cells. The findings are of particular interest because they show clearly that undernutrition after birth has no effect and because the two enzymes are not normally synthesized until after birth, suggesting that their synthesis can be altered before the genes are transcribed.

GROWTH IN INFANCY

In late gestation, rates of cell division fall and growth slows. After birth growth mainly consists of the development and enlargement of existing cells rather than addition of new cells (2,5). Babies who are short at birth, with reduced abdominal circumferences, tend to grow slowly after birth. Low rates of infant weight gain are highly predictive of coronary heart disease among men (11,20). In Hertfordshire, men who were small

Figure 1 Mortality from CHD in 8175 men born during 1911–1930 according to weight at one year

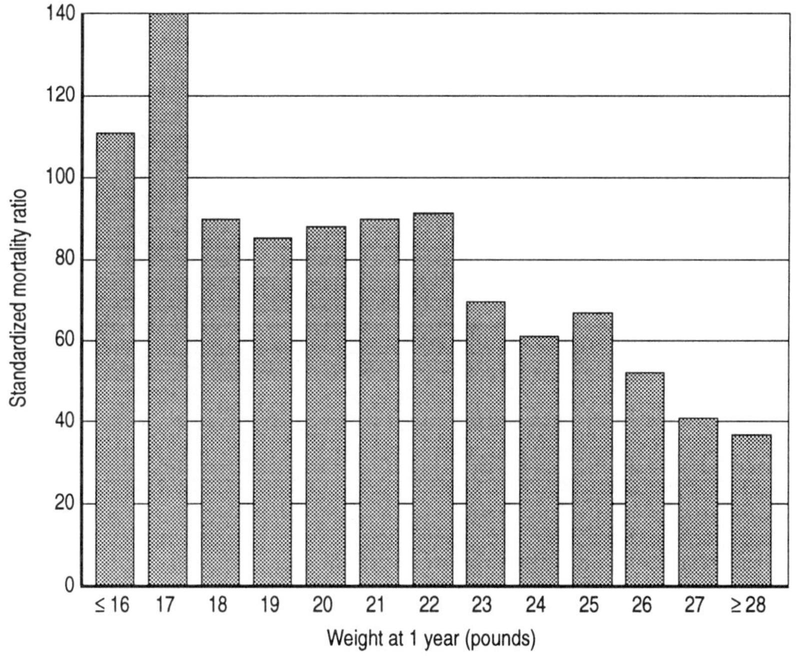

at one year were three times more likely to develop or die from coronary heart disease than those who were large, an association which did not depend on the way in which the infants were fed (Figure 1). Low weight gain during infancy is also followed by hypertrophy of the left ventricle in adult life, which predicts CHD independently of systemic blood pressure (21). One possible explanation of this is that, in the short baby, the structure of the heart is permanently changed by the adaptive responses occurring before birth. Redistribution of blood flow in favour of the brain increases left ventricular blood flow and peripheral resistance and may therefore lead to muscular hypertrophy. Another possible explanation, for which there is only limited evidence, is that short babies are resistant to growth hormone, which takes over control of growth from insulin in late fetal life. Resistance to growth hormone is associated with high circulating concentrations of the hormone. Observations on patients with pituitary tumours producing growth hormone have shown that high concentrations cause cardiac enlargement, atheroma in the vessels and death from CHD. We have proposed that long term consequences of programmed patterns of hormone release, or of tissue sensitivity to hormones, are important disease mechanisms underlying not only CHD and diabetes but also hormonally related cancers (22,23).

Table 5 Mean systolic pressure in men and women aged 60–71 years according to birth weight

Birth weight (pounds (kg))	*Systolic blood pressure* (mmHg; adjusted for sex)
≤5.5 (2.50)	168 (*n*=54)
6.5 (2.95)	165 (*n*=174)
7.5 (3.41)	165 (*n*=403)
8.5 (3.86)	164 (*n*=342)
9.5 (4.31)	160 (*n*=183)
>9.5 (4.31)	163 (*n*=72)
Total	164 (*n*=1228)
SD	25

The thin and short babies are two forms of disproportionate baby, whose growth was restricted at different stages of gestation and in whom different tissues, muscle and liver, were sacrificed to sustain the brain. Perhaps the origins of CHD partly lie in the large size of the human brain, in comparison with that of other mammals. Adaptive responses that protect the brain do so at exaggerated costs to other tissues.

PROPORTIONATE SMALL SIZE AT BIRTH

Some babies are proportionately small; their length, head size and weight are reduced in proportion to each other. Such babies are thought to have established a slow trajectory of growth in early gestation which they were able to sustain throughout late gestation and thereby avoided becoming disproportionate. Early slowing of the growth trajectory is a major adaptation to undernutrition because it reduces the subsequent demand for nutrients (5). We know little about what sets the early growth trajectory in humans or when in embryonic or early fetal life it is set. It may be hormonally controlled. Recent experiments in sheep have shown that administration of progesterone to ewes after mating resets the trajectory upwards and leads to the birth of giant lambs (24).

As yet we know little about proportionately small babies except that they develop raised blood pressure. Persisting elevation of blood pressure seems to follow interference with growth at any stage of gestation, since it is seen in people who were small, thin or short babies. Table 5 shows the progressive fall in blood pressure with increasing birth weight among men and women in Hertfordshire (25). There have been similar findings in 19 other studies of men, women and children. The mechanisms underlying these associations between reduced fetal growth and raised blood pressure may depend on the stage of gestation when they are initiated. One possible mechanism is changes in vascular structure, with persisting loss of elasticity. People who were small at birth have reduced compliance in their large arteries as adults (26). This will increase pulse pressure. Another mechanism may be the effects of glucocorticoid hormones (27,28). In animals modest glucocorticoid excess retards intrauterine growth

Table 6 Mean systolic blood pressure (mmHg) of men and women aged 46–54 years, born after 38 completed weeks of gestation, according to placental weight and birth weight

	Placental weight (pounds (kg))				
Birth weight (pounds (kg))	*≤1.0 (0.45)*	*1.25 (0.57)*	*1.5 (0.68)*	*>1.5 (0.68)*	*All*
≤6.5 (2.95)	149 (*n*=24)	152 (*n*=46)	151 (*n*=18)	167 (*n*=6)	152 (*n*=94)
7.5 (3.41)	139 (*n*=16)	148 (*n*=63)	146 (*n*=35)	159 (*n*=23)	148 (*n*=137)
>7.5 (3.41)	131 (*n*=3)	143 (*n*=23)	148 (*n*=30)	153 (*n*=40)	149 (*n*=96)
Total	144 (*n*=43)	148 (*n*=132)	148 (*n*=83)	156 (*n*=69)	149 (*n*=327)

and programmes raised blood pressure. Such an excess may occur either as a result of feto-placental stress or due to deficiency in the normal placental enzyme barrier which protects the fetus from its mother's glucocorticoids. An enzyme deficiency of this kind can be produced experimentally by undernourishing pregnant rats, which also causes persisting elevation of the offspring's blood pressure (29).

THE PLACENTA

At an early stage of development an embryo comprises two groups of cells, the inner and outer cell masses. The group on the outside does not give rise to any structures in the embryo itself but develops into the placenta: it is from the inner cell mass that we have our origins. Experiments in animals suggest that the distribution of cells between the two groups is influenced by nutrition and hormones. In the ewe undernutrition in early pregnancy leads to placental enlargement, thought to be an adaptation to allow extraction of more nutrients (30). This will only occur, however, if the ewe was well nourished before mating – one of many pointers to the importance to the fetus of the mother's nutritional plane before conception (24). There is evidence that placental enlargement may also be an adaptive response in man. Ultrasound studies in humans show that at around 18 weeks fetuses of a given size already have a range of placental volumes (31), and mothers who are anaemic, who exercise heavily in pregnancy or who live at altitude have babies with large placentas. Our observations suggest that expansion of the placenta is another fetal adaptation that exacts a long term price. The blood pressures of a group of men and women in Preston were measured and related to their birth weights and placental weight (Table 6) (32). As expected, blood pressures fell with increasing birth weight. At any birth weight, however, pressures rose as placental weight increased, so that the highest pressures were in people who, in fetal life, allocated a greater proportion of their resources to placental development rather than to their own growth. Our studies have shown that placental enlargement is followed in adult life not only by elevated blood pressure, but by impaired glucose tolerance, disordered blood coagulation and death from coronary heart disease. Placental enlargement, therefore, seems to be a general marker of fetal undernutrition and its consequences, rather than a specific marker of

Table 7 Death rates from coronary heart disease among 3108 men according to the ratio of placental weight to birth weight

Placental weight to birth weight ratio	*Standardized mortality ratio*	*No. of deaths*
≤15.0	107	57
17.5	92	77
19.0	78	46
21.5	103	78
>21.5	130	83
Total	99	341

later hypertension. Table 7 shows the raised death rates from coronary heart disease associated with a high ratio of placental weight to birth weight among a group of men born in Sheffield. The distribution of death rates is U-shaped, both high and low ratios of placental weight to birth weight being associated with raised rates.

INTER-GENERATIONAL EFFECTS

During the winter of 1944 there was widespread famine in western Holland which ended abruptly after 7 months when Holland was liberated. A group of girl babies conceived during the famine but born after liberation attained normal height as adults but when they had babies in their turn, the babies were small (33). It seems that fetal experience had permanently constrained the girls' ability to deliver nutrients to their own fetuses. Other evidence also suggests that a constraint on nutrient delivery to the fetus is established during the mother's fetal life, though the nature of this constraint is unknown (34,35). It may be a result of changes in the structure of the mother's ovaries or uterus, which are laid down in fetal life.

TIME TRENDS OF CORONARY HEART DISEASE

CHD was rare in Britain at the turn of the century. Having risen rapidly to become the commonest cause of death, it is now declining. Such rapid changes cannot be genetically determined; neither, it seems, can they be explained by coincident alterations in the lifestyles of adults. Could they result from changes during development? An important difference between fetal growth in the Third World and in the western world is that proportionate growth retardation is common in the former while disproportionate growth retardation prevails in the latter. Perhaps CHD occurs in populations at a stage of nutrition between chronic maternal malnutrition with early down-regulation of fetal growth and nutrition at a plane that allows adequate fetal nutrition through gestation. This intermediate phase of nutrition is characterized by disproportionate growth retardation. Studies currently in progress in India and China should clarify this.

TWINS

Because twins are growth retarded it is sometimes suggested that they must have an increased risk of CHD. Twins are, however, heterogeneous, a mixture of proportionately and disproportionately small babies (36). Depending on which predominates, a group of twins may have low or high rates of CHD. Studies of the long term effects of retarded growth due to twinning are, however, of interest apart from this issue, and a number of such studies are in progress.

CONCLUSION

The search for the environmental causes of coronary heart disease has hitherto focused on a 'destructive' model in which influences acting in adult life, such as smoking and obesity, hasten ageing processes – the formation of atheroma, a rise in blood pressure and loss of the ability to metabolize glucose. The observations described in this paper, however, suggest that coronary heart disease may originate during fetal development. This points to a new focus for research. We know surprisingly little about how the nutrition of human mothers influences that of their fetuses: much more is known about this in domestic animals. We need to discover how maternal nutrition, body composition and hormonal profile interact to establish the growth trajectory of a fetus in early gestation, and to sustain its growth thereafter.

It seems that many, if not all, human fetuses have to adapt to a limited availability of nutrients and oxygen. These adaptations programme the body's physiology, metabolism and structure. We now need to know more about them: what they are, what induces them, how they leave a lasting mark upon the body and how they give rise to the diseases of later life.

ACKNOWLEDGEMENT

This paper is based on a lecture first given at the Royal Society, London, 1995, and published in the Proceedings of the Royal Society of London, series B, 1995; 262: 37–43.

REFERENCES

1. Wolpert L 1991 The Triumph of the Embryo. Oxford University Press, Oxford
2. Widdowson EM, McCance RA 1975 A review: new thoughts on growth. Pediatr Res 9:154–156
3. Penrose LS 1954 Some recent trends in human genetics. Caryologia 6 (suppl):521–529
4. Morton NE 1955 The inheritance of human birthweight. Ann Hum Genet 20:123–134
5. McCance RA, Widdowson EM 1974 The determinants of growth and form. Proc R Soc Lond 185:1–17
6. Lucas A 1991 Programming by early nutrition in man. In: Bock GR, Whelan J (eds) The Childhood Environment and Adult Disease. Wiley, Chichester, pp. 38–55
7. Barraclough CA 1961 Production of anovulatory, sterile rats by single injections of testosterone propionate. Endocrinology 68:62–67
8. Barker DJP 1994 Mothers, Babies and Disease in Later Life. BMJ Publishing Group, London
9. Barker DJP, Osmond C 1986 Infant mortality, childhood nutrition, and ischaemic heart disease in England and Wales. Lancet i:1077–1081

10. Barker DJP, Winter PD, Osmond C, Margetts B, Simmonds SJ 1989 Weight in infancy and death from ischaemic heart disease. Lancet ii:577–580
11. Osmond C, Barker DJP, Winter PD, Fall CHD, Simmonds SJ 1993 Early growth and death from cardiovascular disease in women. Br Med J 307:1519–1524
12. Hales CN, Barker DJP, Clark PMS, Cox LJ, Fall C, Osmond C, Winter PD 1991 Fetal and infant growth and impaired glucose tolerance at age 64. Br Med J 303:1019–1022
13. Lithell HO, McKeigue PM, Berglund L, Mashen R, Lithell U-B, Leon DA 1996 Relation of size at birth to non-insulin dependent diabetes and insulin concentrations in men aged 50–60 years. Br Med J 312:406–410
14. Hale CN, Barker DJP 1992 Type 2 (non-insulin-dependent) diabetes mellitus: the thrifty phenotype hypothesis. Diabetologia 35:595–601
15. Phillips DIW, Barker DJP, Hales CN, Hirst S, Osmond C 1994 Thinness at birth and insulin resistance in adult life. Diabetologia 37:150–154
16. Taylor DJ, Thompson CH, Kemp GJ, Barnes PRJ, Sanderson A, Radda GK, Phillips DIW 1996 A relationship between impaired fetal growth and reduced muscle glycolysis revealed by ^{31}P magnetic resonance spectroscopy. Diabetologia. In press
17. Barker DJP, Martyn CN, Osmond C, Wield GA 1995 Abnormal liver growth in utero and death from coronary heart disease. Br Med J 310:703–704.
18. Martyn CN, Meade TW, Stirling Y, Barker DJP 1995 Plasma concentrations of fibrinogen and factor VII in adult life and their relation to intra-uterine growth. Br J Haematol 89:142–146.
19. Desai M, Crowther N, Lucas A, Hales CN 1994 Programming of hepatic metabolism by low protein diet during early life. Diabetic Med 11:537–591
20. Fall CHD, Vijayakumar M, Barker DJP, Osmond C, Duggleby S 1995 Weight in infancy and prevalence of coronary heart disease in adult life. Br Med J 310:17–19
21. Vijayakumar M, Fall CHD, Osmond C, Barker DJP 1995 Birth weight, weight at one year and left ventricular mass in adult life. Br Heart J 73:363–367
22. Barker DJP, Winter PD, Osmond C, Phillips DIW, Sultan HY 1995 Weight gain in infancy and cancer of the ovary. Lancet 345:1087–1088
23. Barker DJP, Gluckman PD, Godfrey KM, Harding JE, Owens JA, Robinson JS 1993 Fetal nutrition and cardiovascular disease in adult life. Lancet 341:938–941
24. Robinson JS, Owens JA, DeBarro T, Lok F, Chidzanja S 1994 Maternal nutrition and fetal growth. In: Ward RHT, Smith SK, Donnai D (eds). Early Fetal Growth and Development, RCOG Press, London
25. Law CM, de Swiet M, Osmond C, Fayers PM, Barker DJP, Cruddas AM, Fall CHD 1993 Initiation of hypertension in utero and its amplification throughout life. Br Med J 306:24–27
26. Martyn CN, Barker DJP, Jespersen S, Greenwald S, Osmond C, Berry CL 1995 Growth in utero, adult blood pressure and arterial compliance. Br Heart J 73:116–121
27. Edwards CRW, Benediktsson R, Lindsay RS, Seckl JR 1993 Dysfunction of placental glucocorticoid barrier: link between fetal environment and adult hypertension? Lancet 341:355–357
28. Seckl JR 1994 Glucocorticoids and small babies. Q J Med 87:259–262
29. Langley SC, Jackson AA 1994 Increased systolic blood pressure in adult rats induced by fetal exposure to maternal low protein diets. Clin Sci 86:217–222
30. McCrabb GJ, Egan AR, Hosking BJ 1992 Maternal undernutrition during mid-pregnancy in sheep: variable effects on placental growth. J Agric Sci 118:127–132
31. Wheeler T, Sollero C, Alderman S, Landen J, Anthony F, Osmond C 1994 Relation between maternal haemoglobin and placental hormone concentrations in early pregnancy. Lancet 343: 511–513
32. Barker DJP, Bull AR, Osmond C, Simmonds SJ 1990 Fetal and placental size and risk of hypertension in adult life. Br Med J 301:259–262
33. Lumey LH 1992 Decreased birthweights in infants after maternal in utero exposure to the Dutch famine of 1944–45. Paediatr Perinatal Epidemiol 6:240–253
34. Ounsted M, Ounsted C 1966 Maternal regulation of intra-uterine growth. Nature 212:995–997
35. Gluckman P, Harding J 1992 The regulation of fetal growth. In: Hernandez M, Argente J (eds) Human Growth: Basic and Clinical Aspects. Elsevier, Amsterdam, pp. 253–259
36. Leveno KJ, Santos-Ramos R, Duenhoelter JH, Reisch JS, Whalley PJ 1979 Sonar cephalometry in twins: a table of biparietal diameters for normal twin fetuses and a comparison with singletons. Am J Obstet Gynecol 135:727–730

2 Nutrition of pregnant mothers in developing countries

S.D. POPPITT

ABBREVIATIONS

BMI	body mass index (weight/height^2)
BMR	basal metabolic rate
HDP	hypertensive disorders of pregnancy
IUGR	intrauterine growth retardation
LBW	low birth weight
LDC	less developed country

INTRODUCTION

In an attempt to attain delivery of a full-term, well-developed and healthy infant considerable attention has focused on nutritional status, weight gain and dietary intake of the pregnant mother. While achieving optimal outcomes in developed countries is within reach, less developed countries (LDC) remain slow to eradicate problems of maternal and infant morbidity and mortality. Pregnancy commits the mother to a considerable investment of resources which must be met by a combination of increased dietary intake and mobilization of body stores. Since fertility tends to be suppressed only under conditions of severe malnutrition, in many countries conception and development of the fetus commonly proceed under conditions of marginal nourishment. Undernutrition during pregnancy is certainly one of the major causes of both maternal and infant mortality in LDC. Approximately 500 000 pregnant women in LDC die each year, mainly as a result of haemorrhage, sepsis, hypertensive disorders of pregnancy (HDP), obstructed labour and anaemia (1, Fig. 1). In many cases both obstructed labour and anaemia can be attributed directly to problems of undernutrition, as may HDP. Considerable evidence also suggests that maternal diet and nutritional status can significantly affect birth outcome. Twenty million low birth weight (LBW) infants are born worldwide (1) and maternal undernutrition is commonly cited as a cause for retarded fetal growth, malformations of the central nervous system such as spina bifida and hydrocephalus, and prematurity. Both premature and small-for-date babies are at considerable increased risk during their early years of life.

NUTRITIONAL ANAEMIA

Iron deficiency and its consequent anaemia is recognized as the most prevalent nutrient deficiency worldwide (1). Nutritional anaemia is responsible for up to 20% of maternal

J.G. Bindels, A.C. Goedhart and H.K.A. Visser (eds), Recent Developments in Infant Nutrition, 15–35.

Figure 1 Major causes of maternal mortality during pregnancy. OL, obstructed labour; HDP, hypertensive disorders of pregnancy (compiled from Ref. 1)

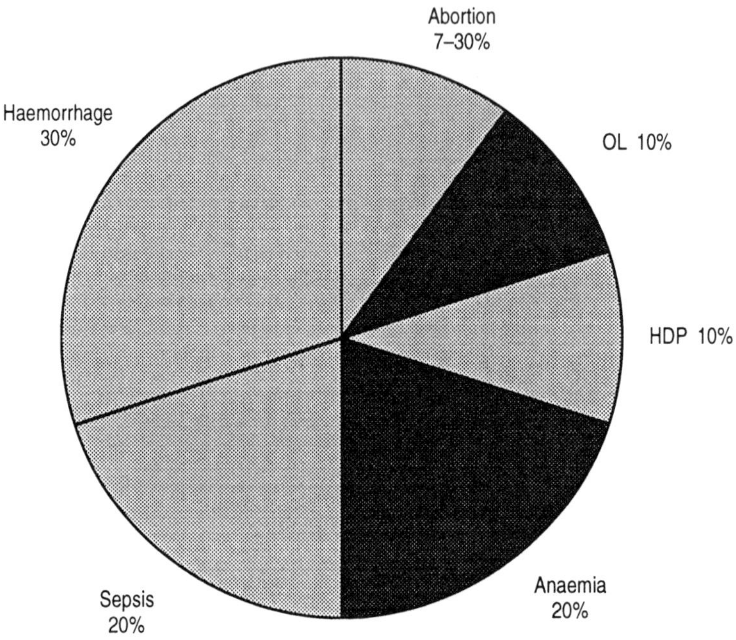

Table 1 Estimated prevalence of anaemia and numbers of pregnant women affected in developed and less developed countries (data from WHO 1991)

	Pregnant women	
Region	*%*	*millions*
Africa	63	11.3
Eastern Asia	20	0.5
Southern Asia	65	27.1
Latin America	30	3.0
Developing countries	59	41.7
Developed countries	14	2.0

deaths during pregnancy in LDC, affecting 60–70% of women at some stage during gestation (1, Table 1). Severe nutritional anaemia may lead to death via infection as a result of a lowered resistance threshold, or from heart failure or shock. Blood loss is poorly tolerated and the risk of death from pre- or post-partum haemorrhage is increased, as are anaesthetic and operative risks (2). It must also be borne in mind, however, that the increase in the prevalence of anaemia which is often seen between the first and third trimesters may in part be artifactual, a result of the physiologically normal expansion of the plasma volume. Red cell mass also increases throughout pregnancy but at a quite different rate, and this may often lead to a decline in haemoglobin concentration in the first two trimesters, a plateau early in the third and a further increase late in the third trimester. It may be quite difficult to determine which women are truly anaemic and which are simply testing positive as a result of haemodilution.

Iron deficiency is certainly the most common cause of anaemia in the first 24 weeks of gestation (3), although factors such as increased placental weight and a high placenta to birth weight ratio are also linked to maternal anaemia. While women can complete pregnancy without becoming iron deficient, this is only possible with an abundant and highly absorbable source of iron. The diet must be rich in haem iron, animal protein and ascorbic acid and low in inhibitors of iron absorption. Oral supplementation with iron, folate and/or vitamin B_{12} has been successful in many, although not all, intervention studies, leading to a significant increase in haemoglobin (4–6). Where iron deficiency is the cause, therapeutic trials have shown that doses of iron as low as 30–65 mg are as effective as higher doses in improving outcome (7–9). A modest dose of iron is preferable since it may inhibit the absorption of zinc in particular (10).

Anaemia during pregnancy may be harmful to the fetus, although the data are not conclusive. For example, in a study of 54000 pregnancies in the UK the risk of LBW, preterm birth and perinatal mortality was greater when the haemoglobin concentration fell into the anaemic range (11). There is little evidence that the haemoglobin concentration of newborn infants from anaemic mothers may also be low, but studies have shown a relationship between maternal anaemia at full term and LBW (12–14), prematurity and perinatal mortality (15,16). Certainly in countries where such deficiencies are widespread (see Table 1) combined supplementation of the haematinics is effective and routine administration during the second and third trimesters is justified.

OBSTRUCTED LABOUR

Obstructed labour may be a result of nutritional insufficiency when the obstruction arises from cephalopelvic disproportion, and may be responsible for 5–10% of maternal deaths in LDC (see Fig. 1). Such disproportion between the mother and the neonate may arise from a number of factors including immaturity of the mother during adolescent pregnancy, delayed bone growth or physical stunting as a result of chronic undernutrition in childhood and adolescence. Many factors may affect bone growth and peak bone mass, including genetics, sex, diet, exercise and hormones. Delayed bone growth is certainly common in many LDC, where teenage marriage may be the custom and obstructed labour is common (17). Studies in The Gambia have shown females aged between 8 and 25 to have lower bone mineral content than their western counterparts, not achieving maximal bone mineral content or maximal growth in stature until their mid-twenties (18, Fig. 2) by which time they may have completed several pregnancies (19).

Figure 2 Bone mineral content and density of rural Gambian girls is lower than their age-matched western counterparts (compiled from Ref. 18)

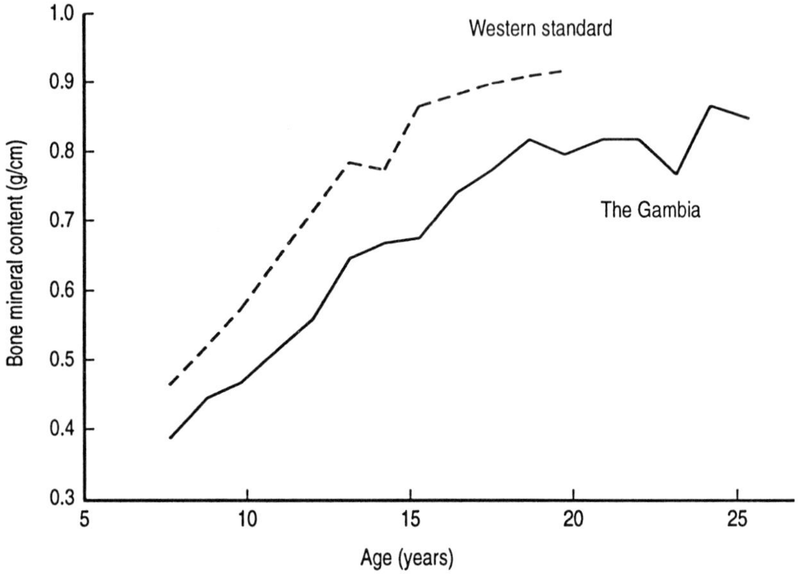

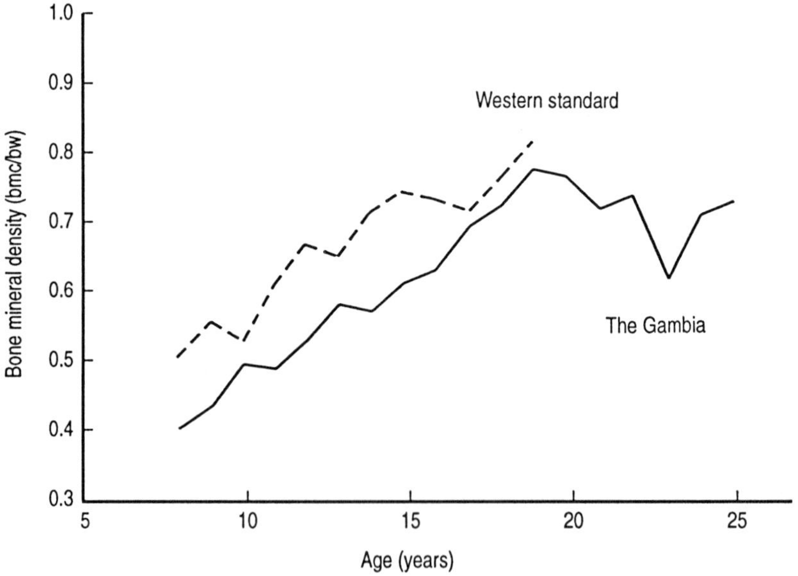

HYPERTENSIVE DISORDERS OF PREGNANCY (HDP)

HDP can be identified from a number of clinical conditions including pre-eclampsia (high blood pressure with proteinuria), dependent oedema and gestational hypertension (elevated blood pressure alone). HDP increases the risk of placental abruption, maternal renal failure and cerebral haemorrhage. On current evidence it is difficult to define the precise role of nutrition in HDP, although it has been suggested that most women who develop the condition have below average total weight gains (20) and that low early gains may be a marker of subsequent hypertensive disorders (21). Other studies have shown that large weight gains during the second and third trimesters increase the risks of HDP and pre-eclamptic toxaemias in primiparous western women (21–24). This is difficult to interpret since an increase in body water, characteristic of oedema, necessarily results in an increase in body weight irrespective of whether fat or lean mass also increase.

VITAMIN AND MINERAL DEFICIENCIES

An insufficient supply of vitamins and minerals such as zinc, magnesium, calcium and vitamin D may also have detrimental effects on both maternal and infant mortality. Zinc is extensively involved in cell replication and differentiation, and in particular nucleic acid and protein metabolism (see 25 for review), and as such is essential for reproduction. Numerous studies in LDC show that a diet deficient in zinc (<20 mg/day) is common (26–30), although there have been few reported cases of severe human zinc deficiency, potentially leading to an increase in congenital defects and LBW babies (31–35). Women with low plasma zinc concentrations may be at greater risk of prolonged labour, atonic bleeding and post-date delivery (36). Supplementation with zinc (10–20 mg/day) may reduce rather than eliminate this fall in blood levels (37–39), although it has been shown to improve the outcome of pregnancy in western women (40,41).

Magnesium is required for the release of parathyroid hormone (PTH), which exerts its action on the intestine, bone and kidney. Studies, albeit in developed countries, have suggested that supplementation during pregnancy is associated with lower frequencies of growth retardation and pre-eclampsia (42).

Adequate calcium is essential for fetal skeletal growth. Changes in calcium regulating hormones, which control intestinal absorption, renal reabsorption and bone turnover of calcium, ensure the transfer of calcium from mother to fetus while maintaining maternal serum and bone calcium levels. Inadequate prenatal transfer from the mother, possibly in part a result of low dietary intakes, may reduce bone growth or even cause bone loss from the fetus. No serious consequences of low calcium intake in pregnancy have been documented, although several studies in LDC have shown it to be associated with an increase in blood pressure leading to gestational hypertension (43,44).

Since vitamin D is required for the maintenance of a stable calcium concentration, hypovitaminosis D of the mother may result in a deficiency in the fetus (45,46). Since most vitamin D is synthesized from a precursor in the skin after exposure to ultraviolet light from the sun, only in LDC where sunlight levels are low are conditions such as neonatal hypocalcaemia and tetany, tooth enamel hypoplasia and maternal osteomalacia likely to occur (47–49). There is some evidence to suggest that women in northern latitudes should receive vitamin D supplementation during winter months, although care must be taken as excessive vitamin D can lead to hyperabsorption of calcium, hypercalcaemia and calcification of the soft tissues (50–52). Few LDC are situated in areas of low sunlight.

ENERGY BALANCE

One of the most important sequelae of undernutrition in LDC is the inability to maintain energy balance throughout pregnancy, a consequence of insufficient food supply. Over the course of a 40 week pregnancy extra energy is required to cover the costs of tissue deposition comprising the products of conception (fetus, placenta and amniotic fluid) and the maternal tissues (uterus, breasts and blood), the costs of maternal fat deposition plus the costs of maintaining these extra tissues (Fig. 3) (53). In many species, such as the rat, the failure to increase food intake during gestation in response to these energy demands has extreme consequences such as abortion, resorption of the fetuses or consumption of the neonates by the mother at birth. In humans, however, once the act of conception has occurred numerous factors act to protect the fetus and ensure a successful pregnancy, even under marginal and sometimes severely inadequate nutritional conditions. Human gestation is unusually slow (280 days) compared with other similar sized primates, in which a 55 kg female would generally give birth after only 180 days. While this is often assumed to be an adaptation to allow growth and development of a large brain (54), it also ensures that the extra energy costs of pregnancy are relatively small on a day-to-day basis. Where the diet cannot cover these costs body fat may be used as a short or medium term buffer, even in relatively thin women from LDC. Humans as a species tend to be fat (Fig. 4) with an almost 10-fold greater percentage of body fat than other primates and with a marked sexual dimorphism between males and females, amounting to almost twice as much body fat in women.

Much of the success of human pregnancy appears to lie in the high degree of plasticity of metabolic response between women in widely diverse nutritional settings. Ten recent longitudinal studies carried out in seven different countries (England (55), Scotland (56), The Netherlands (57,58), Sweden (59), The Gambia (60,61), The Philippines (62), Thailand (63); for review see Ref. 64) have shown that the response to pregnancy differs widely between well-nourished and undernourished women. Each of these studies measured weight gain, fat deposition and basal metabolic rate (BMR) in the pre-pregnant state and then followed the women longitudinally every 6 weeks until term (except in The Philippines and Thailand, where measurements began in the first trimester of pregnancy). From these studies it was possible to establish that women from LDC tend to gain less weight, deposit less fat and have lower maintenance costs than their affluent, well-nourished counterparts (Table 2). Figure 5 shows the average changes in BMR relative to the pre-pregnant baseline. If the studies are separated into developed (left hand panel) and less developed countries (right hand panel), in

Figure 3 During pregnancy energy is required for the production of new tissue (top panel) plus the maintenance of these structures (bottom panel)

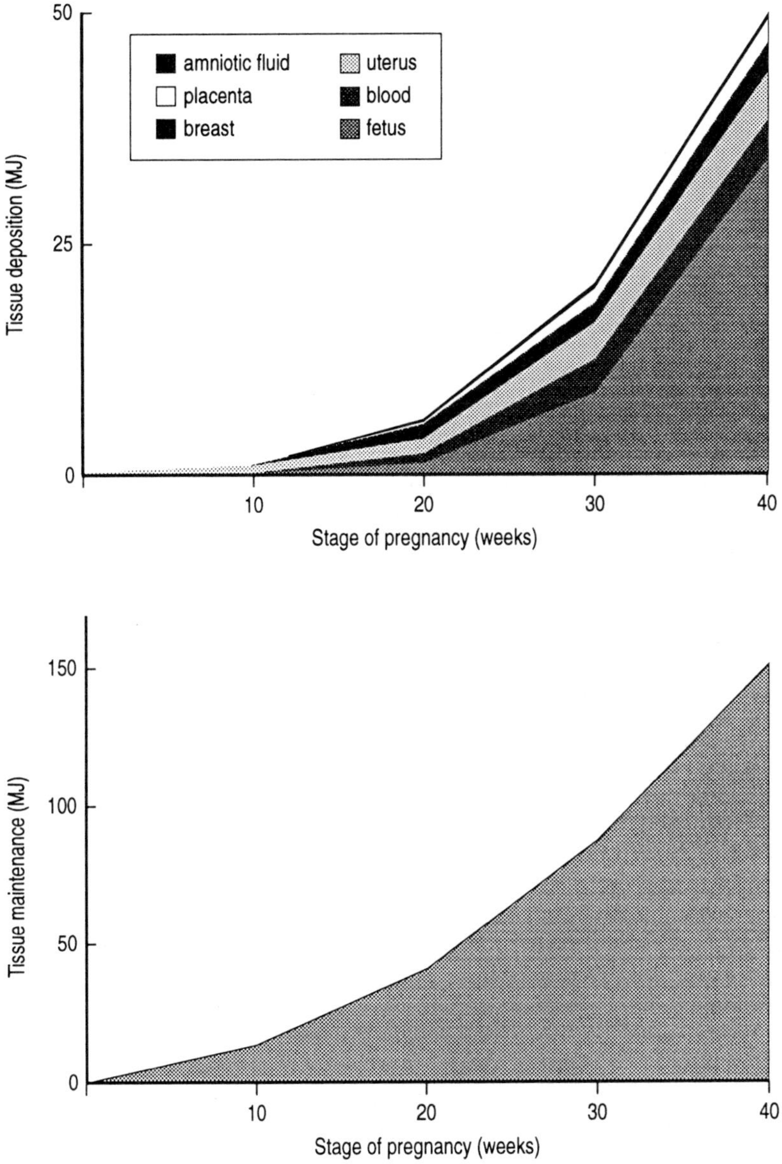

***Figure 4** The relatively high proportion of body fat in humans may act as a short-term buffer against periods of famine during gestation*

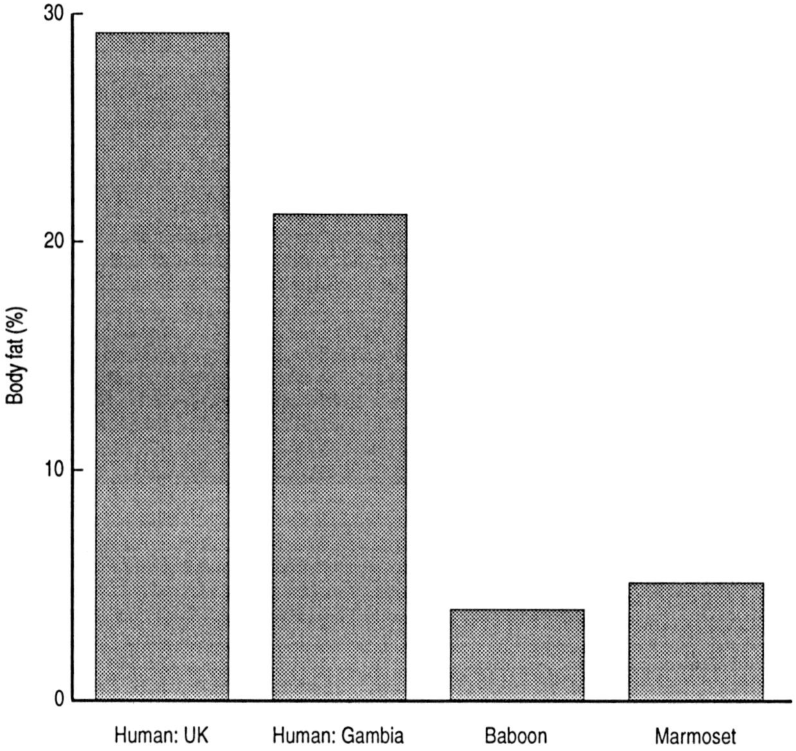

accordance with both economic and nutritional factors (e.g. maternal weight: developed countries, 61.0±0.9 kg; LDC, 49.4±1.3 kg), there is a clear difference in metabolic response. Well-nourished women show an immediate 'energy profligate' increase in BMR, whilst marginally nourished women show an immediate 'energy-sparing' decrease which offsets the later increase.

The energy required for pregnancy may be up to 5-fold lower in women from LDC. For example the total costs of pregnancy, calculated as the additive costs of new tissue accretion in the fetus (estimated *pro rata* on the basis of maternal weight from the average value of 43 MJ) (65), plus maternal fat deposition and maintenance expenditure, varied between 523 MJ in Swedish women to −30 MJ in unsupplemented Gambian women (Fig. 6). The average cost in LDC was well below the current WHO international assumptions of 336 MJ (66). Several good predictors of the metabolic costs of pregnancy may help to explain the mechanisms responsible for these wide differences (64). While maternal height, reflecting both early and chronic nutritional

Table 2 Pre-gestational and pregnant nutritional status in recent longitudinal studies of affluent and less developed countries

Countries	*Reference*	*Height* (m)	*Weight* (kg)	*Fat mass* (kg)	*Weight gain* (kg)	*Fat gain* (kg)	*Birth weight* (kg)
Developed							
England	55	1.64	61.7	18.6	13.2	3.4	3.77
Scotland	56	1.62	57.3	14.9	11.7	2.3	3.37
Netherlands I	57	1.69	62.5	17.7	11.6	2.0	3.46
Netherlands II	58	1.69	62.6	17.3	11.7	2.1	3.52
Sweden	59	1.65	61.0	17.2	13.6	5.8	3.56
Less developed							
Gambia I (S)	60	1.57	51.2	9.9	7.9	2.0	2.99
Gambia I (U)	60	1.58	51.6	11.0	6.4	−0.5	2.96
Gambia II	61	1.57	52.0	12.2	8.7	2.3	3.02
Philippines	62	1.51	44.5	11.3	8.4	1.3	2.89
Thailand	63	1.52	47.6	11.4	8.9	1.3	2.98

S, supplemented; U, unsupplemented

***Figure 5** Longitudinal changes in basal metabolic rate (BMR) during pregnancy in studies from developed [left hand panel, ■ England (55), ▲ Scotland (56), □ ○ Netherlands (57,58), ● Sweden (59)] and less developed countries [right hand panel, △ ○ □ The Gambia (60,61), ● The Philippines (62), ■ Thailand (63)]. NPNL, non-pregnant non-lactating. Reproduced with permission from Ref. 64*

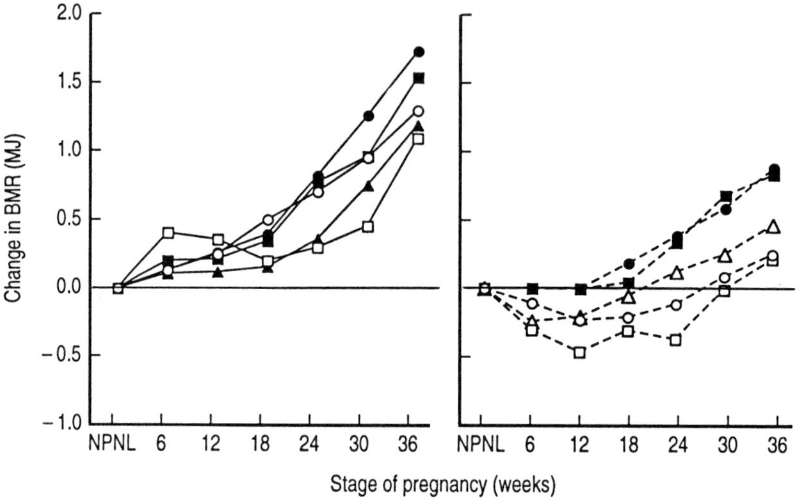

status, is not well correlated with metabolic pregnancy outcome, both maternal pre-pregnant weight and fatness (Fig. 7, left hand panel) are significantly correlated with maintenance costs (weight: $r=0.67, p<0.05$; fatness: $r=0.84, p<0.01$) and total costs (weight: $r=0.67$, $p<0.05$; fatness: $r=0.80$, $p<0.01$). This has led to speculation that a mechanism exists by which the maternal pre-pregnant nutritional status may be monitored, leading to considerable changes in metabolism, particularly during early pregnancy. Interestingly, the process of 'energy-sparing' is not limited only to women under nutritional stress in LDC but also occurs in women from affluent countries who are of low body weight and fatness (Fig. 8, England, unpublished data; The Netherlands, Spaaij personal communication) and as such merely represents one end of a wide continuum onto which women from all backgrounds and nutritional status can be placed.

The costs of pregnancy, however, are also highly correlated with the amount of weight gained during pregnancy (maintenance: $r=0.92$, $p<0.001$; fat deposition: $r=0.81$, $p<0.01$; total costs: $r=0.94$, $p<0.001$; see Fig. 7, right hand panel). Pre-pregnant fatness and weight gain are themselves highly correlated ($r=0.88, p<0.001$) making it difficult to determine which of the two is dominant in the regulation of the metabolic outcome of pregnancy. It is entirely possible that pre-pregnant fatness merely represents a proxy measure of nutritional conditions at conception which continue unchanged throughout pregnancy, the decrease in maintenance costs in

***Figure 6** The total costs of pregnancy (bottom panel; U, unsupplemented; S, supplemented) comprise the energy deposited as fat (top panel), the fetus and associated structures [= 43 MJ, not shown (65)] and the extra energy maintaining these structures (middle panel). Reproduced with permission from Ref. 64*

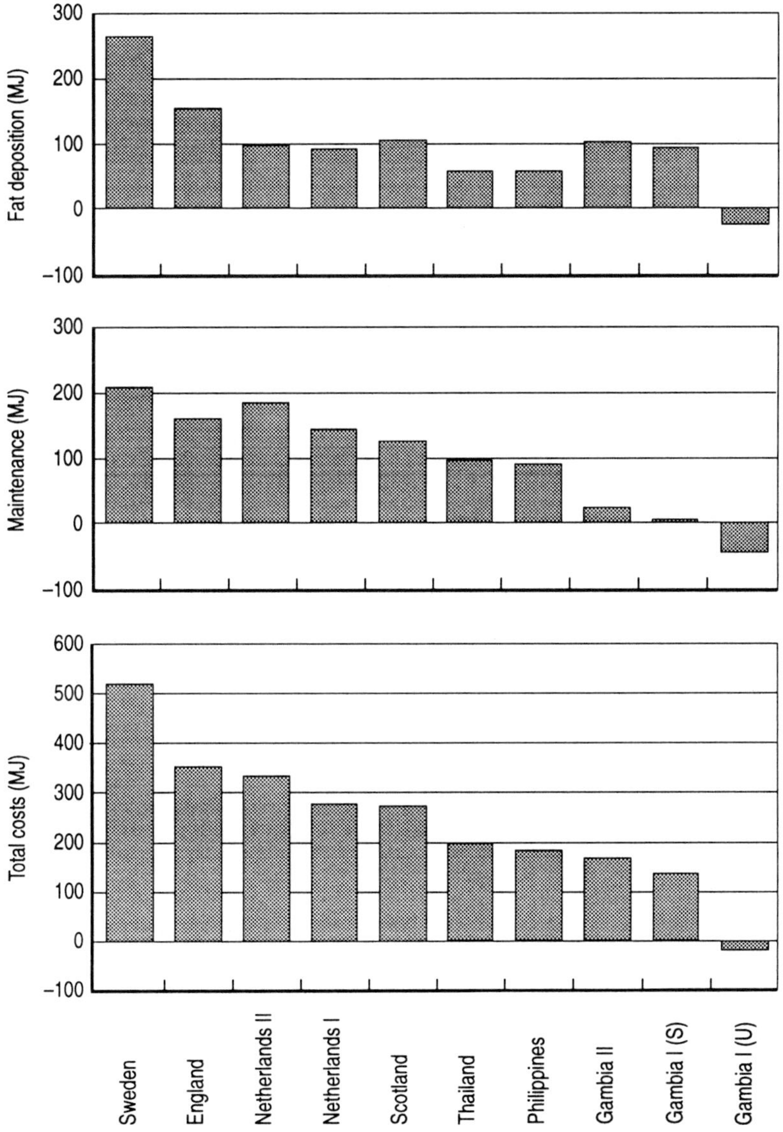

***Figure 7** Correlations between maintenance, fat deposition and total costs of pregnancy with pre-pregnant fatness and pregnancy weight gain [○ England (55), □ Scotland (56), ▲▼ Netherlands (57,58), ● Sweden (59), ★◆◇ The Gambia (60,61), ■ The Philippines (62), △ Thailand (63)]. Reproduced with permission from Ref. 64*

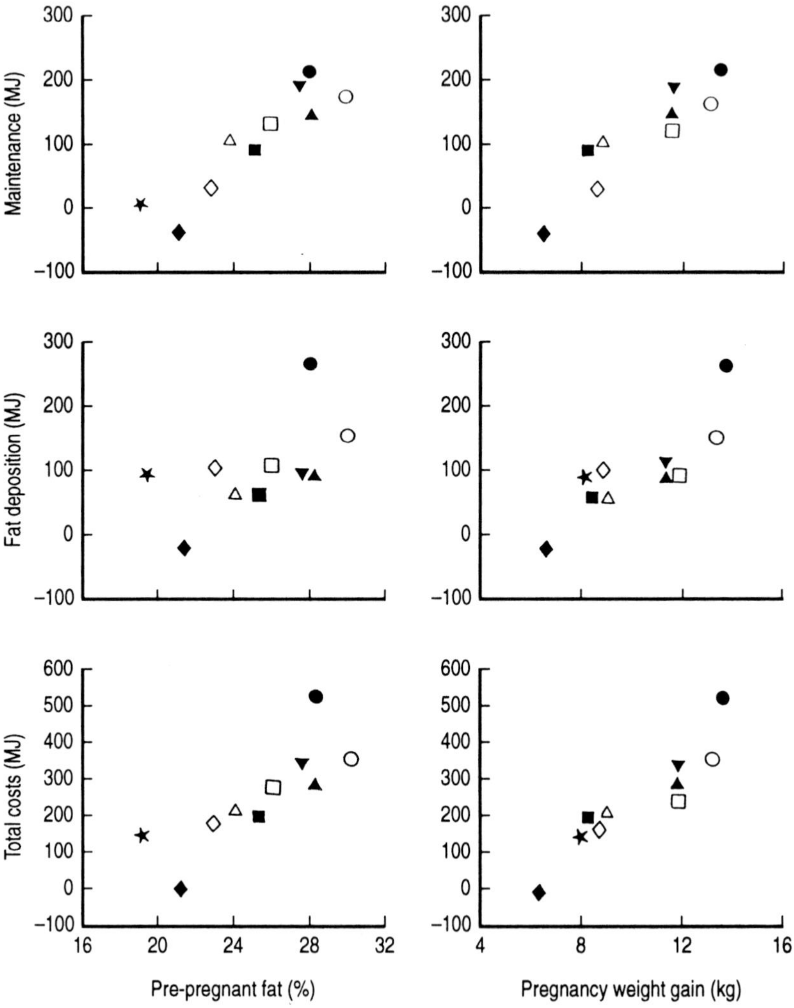

'energy-sparing' women simply being a response to a low energy intake in the first trimester rather than an autoregulatory mechanism directed by pre-pregnant fatness *per se*.

Irrespective of whether it is the balance of intake and expenditure during pregnancy,

Figure 8 The longitudinal change in basal metabolic rate (BMR) throughout pregnancy in individual affluent English women. NPNL, non-pregnant, non-lactating. A proportion of women of low body weight and fatness show a rapid early decrease in BMR

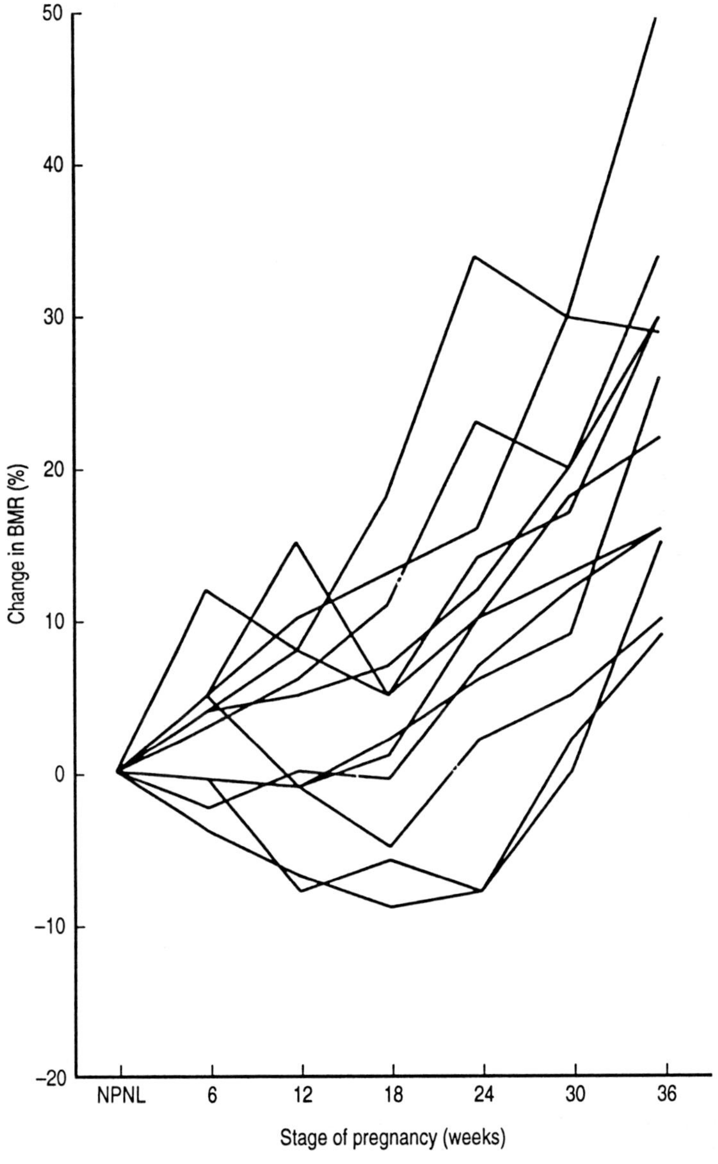

existing energy stores, or a combination of the two that is acting as effector, it is clear that the plasticity in energy metabolism is an effective mechanism for matching energy requirements to energy availability. This must not, however, be grounds for condoning nutritional hardship. The response to pregnancy is unable fully to protect fetal growth under situations of severe nutritional hardship, as demonstrated by the large numbers of LBW infants born in LDC (1).

INFANT MORTALITY

On the basis of simple biological criteria, human reproduction is extremely successful, clearly shown by the continuing growth of the world population. However, it is also true that 20 million LBW infants (<2500 g) are born in LDC every year (1). The consequences of LBW include increased risks of neonatal and postnatal mortality (67–70) and morbidity (71,72), and abnormal growth and development (73–75). There are numerous potential nutritional risk factors for LBW in LDC, including low maternal pre-pregnant weight, short stature, undernutrition during gestation and anaemia. Adequate weight gain during pregnancy is certainly crucial to both maternal and infant outcome. Women from LDC who are of low pre-pregnant weight need to gain more weight than their fatter counterparts to achieve comparable birth outcomes (Fig. 9) (76). Women with a BMI < 19.8 are recommended to gain 12.5–18 kg, but average weight gain in women from LDC is only 5–9 kg, far below recommendations. While there are many other important factors, including adolescent pregnancy, infections such as malaria, inadequate birth spacing, hard physical exercise during the final trimester, preterm delivery and twin or multiple birth, maternal undernutrition appears to be an important contributor to the problem of intrauterine growth retardation (IUGR). Infants born at 40 weeks to marginally nourished Gambian mothers are, after correction for differences in stature, effects of infection and parity, 415 g lighter than their well-nourished British counterparts (77). There is also a highly significant seasonal fluctuation in birth weight in The Gambia, paralleling the pattern of food availability. During periods of food shortage there is both a decrease in birth weight of 200–300 g and an increase from 12.5% to 23.7% in the proportion of LBW infants born (Table 3). These changes must be due to genetic or environmental factors and it seems highly likely that maternal nutrition during pregnancy plays an important role.

MATERNAL ENERGY SUPPLEMENTATION

If undernutrition is a causal factor then provision of a dietary supplement for the mother should remove the nutritional inhibition that has been placed upon the fetus and result in larger, more robust babies. Increasing infant birth weight via maternal supplementation is certainly preferable to providing supplementary food for small, poorly developed IUGR infants during their first months of life. Many maternal supplementation trials have been performed worldwide in both developed and less developed countries, yet the results have failed to provide a consensus of opinion as to when supplementation may be effective (Table 4). This may in part be a consequence of inadequacies of the individual studies, including supplementation of women with

***Figure 9** Women of low pre-pregnant weight must gain more weight than their fatter counterparts during pregnancy to achieve comparable birth outcomes (compiled from Ref. 76)*

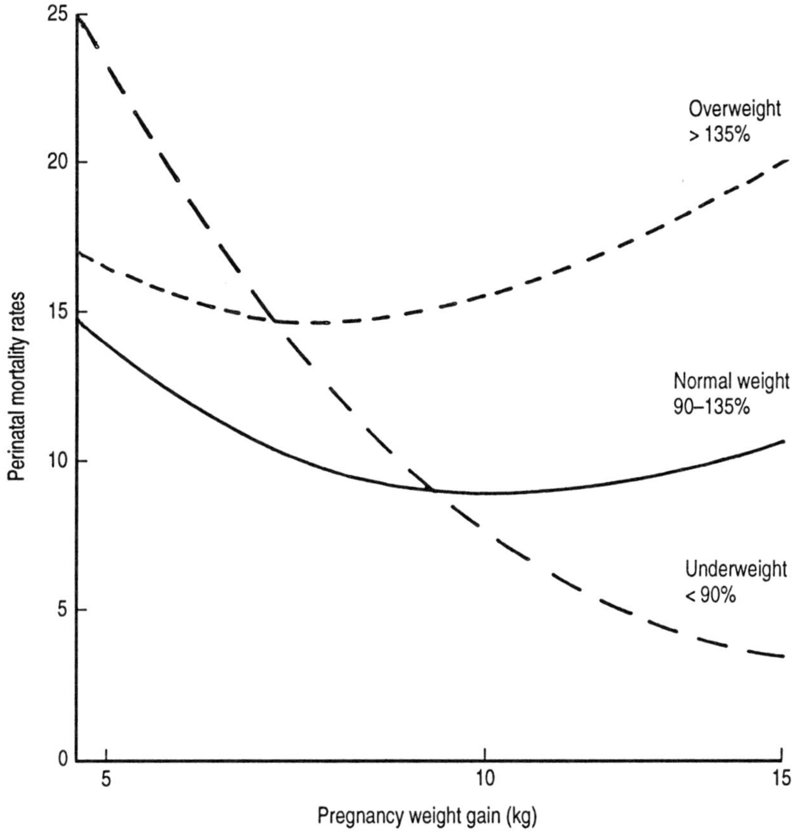

little evidence of nutritional hardship, insufficient energy content within the supplement and poor compliance of the participating women. Several studies in LDC, however, have demonstrated that supplementation of the mother during pregnancy does reduce the number of LBW infants born. In The Gambia between 1974 and 1980 an energy dense and nutritionally fortified supplement of locally grown ground-nut biscuits was given to women soon after conception was established, on 6 days of each week (19). The supplement provided a net increase in intake (supplement minus decrease in home food intake) of 1.8 MJ per day. The supplementation continued in a single village for a period of 4 years and the results were compared with retrospectively collected data. During the wet season (when food is less readily available, the physical costs of farming

Table 3 Birth weight in rural Gambia declines during the nutritionally deprived 'hungry' season relative to the more affluent 'harvest' season. Babies are commonly born smaller than their counterparts in developed countries such as Scotland

Country	*Measurement*	*Outcome*
The Gambia	Harvest season	
	Birth weight	2944 g
	Adjusted birth weight*	2959 g
	% LBW (<2500 g)	12.5%
	% small-for-dates	25.5%
	Hungry season	
	Birth weight	2808 g
	Adjusted birth weight*	2810 g
	% LBW (<2500 g)	23.7%
	% small-for-dates	43.0%
Scotland	Birth weight	3500 g

*Adjusted for sex, month of birth and parity; LBW, low birth weight

Table 4 Prenatal supplementation studies in less developed countries

		Increase in birth weight		
Country	*Reference*	*n*	*g*	*p*
The Gambia	19	197	200	<0.001
The Gambia	78	406	173	<0.05
Guatemala	79	405	111	<0.05
Chile	80	526	300	<0.01
Mexico	81	80	180	<0.05
Colombia	82	207	51	NS
Taiwan	83	108	56	NS
Indonesia	84	753	100	NS

high and the mothers nutritionally stressed), the supplement reduced the prevalence of LBW infants from 23.7% to 7.5%. This was a consequence of an increase of birth weight from 2810 g to 3010 g, close to that of the nutritionally 'affluent' dry season (presupplementation birth weight=2959 g, postsupplement birth weight=2972 g), which remained largely unaffected. When extended to the rather more realistic setting of the wider community (27 villages, with a total of approximately 406 live births, randomized to either supplementation or control), with women supplemented with 2.1 MJ/day from the fifth month of pregnancy to delivery, there was a significant increase in birth weight of 173 g after adjustment for sex, gestational age, month of birth and maternal weight ($p<0.05$, Table 5).

When a programme of energy supplementation was implemented in Guatemala, Central America (79) the percentage of LBW infants in 357 pregnancies also decreased,

Table 5 The effect on infant birth weight of maternal energy supplementation during the second half of pregnancy in 27 rural Gambian villages

	Control	*Intervention*	*Difference*	*p*
Dry season				
Birth weight (g)	2907	2952	45	NS
Length (cm)	49.2	49.2	0	NS
Head circumference (cm)	33.5	33.6	0.1	NS
Wet season				
Birth weight (g)	2853	2973	120*	<0.05
Length (cm)	49.4	49.5	0.1	NS
Head circumference (cm)	33.2	33.6	0.4	NS

*173 g after adjustment for sex, gestational age, month of birth and maternal weight

from 19% to 9% of infants, and birthweight increased significantly by 111 g ($p<0.05$, see Table 4). After controlling for maternal home diet, height, head circumference, parity, gestational age, duration of disease during pregnancy and socioeconomic status the authors concluded that supplementation during pregnancy was responsible for the observed increase in birth weight, although this study has been criticized for abandoning the randomized control design part-way through the study. In both Chile and Mexico, birth weight also increased following maternal prenatal supplementation: supplemented mothers gave birth to infants whose average birth weight was increased by 300 g and 180 g respectively (80,81). Several other studies, however, have found supplementation to be relatively ineffective in preventing IUGR in marginally nourished women. A well-designed, randomized trial in Colombia, in which mothers were supplemented during the final trimester of pregnancy, reported an increase in birth weight of only 51 g (82). This represented a significant effect only in male fetuses and, in addition, it was concluded that the supplement only had an effect on the thinner women within the population. In Taiwan 114 women were supplemented both before and during pregnancy (83). Although there were some significant subgroup effects, overall there were no significant effects on mean birth weight, length or head circumference of either male or female infants. In a more recent study in Indonesia (84) two levels of energy supplementation were tested in randomized, controlled trials in the last trimester of pregnancy. There was no difference between the level of supplementation, yet mean birth weight increased by close to 100 g (2835 g *vs* 2933 g) and the rate of LBW decreased from 12.2% to 9.5%. These changes were, however, found to be non-significant, and the authors concluded that energy supplementation had little effect on mean birth weight.

Despite the disparity in results it has been suggested (77) that prenatal supplementation may indeed have a beneficial effect, although under specific conditions. Improvements in birth weight and percentage of LBW infants born may only be achieved in populations where there is a genuine nutritional deficit across the population (or across a selected subgroup or at certain times of year), where birth weights are lower than predicted on the basis of maternal size and where the supplementary energy is well targeted and delivered. Under such conditions an increase in birth weight of

100–300 g may be expected, the consequences of which may well be improved morbidity and mortality during the subsequent years of infancy. Certainly using the WHO worldwide aggregated statistics (85) an increase in birth weight of up to 300 g would be predicted to reduce infant mortality by almost 40%.

ACKNOWLEDGEMENTS

The preparation of this manuscript was supported by Bristol Myers-Squibb-Mead Johnson, USA.

REFERENCES

1. WHO 1991 Maternal anthropometry for prediction of pregnancy outcomes: Memorandum from a USAID/WHO/PAHO/Mothercare meeting. Bull WHO 69:523–532
2. Flemming AF 1989 Tropical obstetrics and gynaecology. I. Anaemia in pregnancy in tropical Africa. Trans R Soc Trop Med Hyg 83:441–448
3. Puolakka J, Janne O, Pakarinen A, Vihko R 1980 Serum ferritin in the diagnosis of anemia during pregnancy. Acta Obstet Gynecol Scand Suppl 95:57–63
4. Izak G, Levy S, Rachmilewitz M, Grossowicz N 1973 The effect of iron and folic acid on combined iron and folate deficiency anaemia: the results of a clinical trial. Scand J Haematol 11:236–240
5. Aung-Than-Batu, Thane-Toe, Hla-Pe, Khin-Kyi-Nyunt 1976 A prophylactic trial of iron and folic acid supplements in pregnant Burmese women. Israel J Med Sci 12:1410–1415
6. Sood SK, Ramachandran K, Mathur M 1975 WHO sponsored collaborative studies on nutritional anaemia in India. I. The effects of supplemental oral iron administration to pregnant women. Q J Med 44:241–258
7. Chanarin I, Rothman D 1971 Further observations on the relation between iron and folate status in pregnancy. Br Med J 2:81–84
8. Taylor DJ, Matter C, McDougall N, Lind T 1982 Effect of iron supplementation on serum ferritin levels during and after pregnancy. Br J Obstet Gynaecol 89:1011–1017
9. Dawson EB, McGanity WJ 1987 Protection of maternal iron stores in pregnancy. J Reprod Med 32:478–487
10. Sandstrom B, Davidsson L, Cederblad A, Lönnerdal B 1985 Oral iron, dietary ligands and zinc absorption. J Nutr 115:411–414
11. Murphy JF, O'Riordan J, Newcome RG, Coles EC, Pearson JF 1986 Relation of haemoglobin levels in first and second trimesters to outcome of pregnancy. Lancet i:992–995
12. Klein L 1962 Premature birth and maternal prenatal anaemia. Am J Obstet Gynecol 83:588–590
13. MacGregor MW 1963 Maternal anaemia as a factor in prematurity and perinatal mortality. Scott Med J 8:134–140
14. Leiberman E, Ryan KJ, Monson RR, Schoenbaum SC 1987 Risk factors accounting for racial differences in the rate of premature birth. New Engl J Med 317:743–748
15. Murray MJ, Murray AB, Murray NJ, Murray MB 1978 The effect of iron status of Nigerian mothers on that of their infants at birth and 6 months and on the concentration of Fe in breast milk. Br J Nutr 39:627–630
16. Agrawal RMD, Tripathi AM, Agrawal KN 1983 Cord blood haemoglobin, iron and ferritin status in maternal anaemia. Acta Paediatr Scand 72:545–548
17. Harrison KA 1985 Child-bearing, health and social priorities: a survey of 22774 consecutive hospital births in Zaria, Northern Nigeria. Br J Obstet Gynaecol 92:1–119
18. Lo CW, Jarjue LM, Poppitt SD, Cole TJ, Neer R, Prentice A 1990 Delayed development of bone mass in West African adolescents. In: Christiansen C, Overgaard K (eds) Osteoporosis 90 (Third international symposium on osteoporosis). Copenhagen, Denmark
19. Prentice AM, Cole TJ, Foord FA, Lamb WH, Whitehead RG 1987 Increased birthweight after prenatal dietary supplementation of rural African women. Am J Clin Nutr 46:912–925
20. Chesley LC 1976 Blood pressure, edema and proteinurea in pregnancy. I. Historical developments. Prog Clin Bio Res 7:19–66

21. Tompkins WT, Wiehl DG, Mitchell RM 1955 The underweight patient as an increased obstetric hazard. Am J Obstet Gynecol 69:114–123
22. Thomson AM, Billewicz WZ 1957 Clinical significance of weight trends during pregnancy. Br Med J 1:243–247
23. Naeye RL 1982 Maternal nutrition and pregnancy outcome. In: Dobbing J (ed) Maternal nutrition in pregnancy: Eating for two? Academic Press, London, pp 89–111
24. Shepard MJ, Hellenbrand KG, Bracken MB 1986 Proportional weight gain and complications of pregnancy, labor and delivery in healthy women of normal prepregnant stature. Am J Obstet Gynecol 155:947–954
25. Hambidge KM, Casey CE, Krebs NF 1986 Zinc. In: Mertz W (ed) Trace elements in human and animal nutrition. Academic Press, Orlando, FL, pp 1–137
26. Hunt IF, Murphy NJ, Shroads J, Smith JC 1979 Dietary zinc intake of low-income pregnant women of Mexican descent. Am J Clin Nutr 32:1511–1518
27. Hunt IF, Murphy NJ, Cleaver AE 1983 Zinc supplementation during pregnancy: zinc concentration of serum and hair from low-income women of Mexican descent. Am J Clin Nutr 37: 573–582
28. Hunt IF, Murphy NJ, Martner-Hewes PM, Faraji B, Swenstid ME, Reynolds RD 1987 Zinc, vitamin B6, and other nutrients in pregnant women attending prenatal clinics in Mexico. Am J Clin Nutr 46:563–569
29. Abraham R 1982 Trace element intake by Asians during pregnancy. Proc Nutr Soc 41:261–265
30. Turnlund JR, King JC, Wahbeh CJ 1983 Zinc status and pregnancy outcome of pregnant Lebanese women. Nutr Res 3:309–315
31. Hambidge KM, Nelder KH, Walravens PA 1975 Zinc, acrodermatitis enteropathica, and congenital malformations. Lancet i:577–578
32. Prema K 1980 Predictive value of serum copper and and zinc in normal and abnormal pregnancy. Indian J Med Res 71:554–560
33. Meadows NJ, Ruse W, Smith MF 1981 Zinc and small babies. Lancet ii:1135–1137
34. Meadows NJ, Grainger SL, Ruse W 1983 Oral iron and the bioavailability of zinc. Br Med J 287:1013–1014
35. Simmer K, Thomson RPH 1985 Zinc in the fetus and newborn. Acta Paediatr Scand 74 (Suppl 319):158–163
36. Jameson S 1976 Effects of zinc deficiency in human reproduction. Acta Med Scand Suppl 593: 1–89
37. Hunt IF, Murphy NJ, Cleaver AE 1983 Zinc supplementation during pregnancy: zinc concentration of serum and hair from low-income women of Mexican descent. Am J Clin Nutr 37: 572–582
38. Hunt IF, Murphy NJ, Cleaver AE 1984 Zinc supplementation during pregnancy: effect on selected blood constituents and on progress and outcome of pregnancy in low-income women of Mexican descent. Am J Clin Nutr 40:508–521
39. Breskin MW, Worthington-Roberts BS, Knopp RH 1983 First trimester serum zinc concentration in human pregnancy. Am J Clin Nutr 39:943–953
40. Jameson S 1982 Zinc status and pregnancy outcome in humans. In: Prasad AS, Dreosti EI, Hetzel BS (eds) Clinical Applications of Recent Advances in Zinc Metabolism. Alan R Liss, New York, pp 39–52
41. Hambidge KM, Krebs NF, Jacobs MA, Favier A, Guyette L, Ikle DN 1983 Zinc nutritional status during pregnancy: a longitudinal study. Am J Clin Nutr 37:429–442
42. Spatling L, Spatling G 1988 Magnesium supplementation in pregnancy. A double blind study. Br J Obstet Gynaecol 95:120–125
43. Lopez-Jaramillo P, Narvaez M, Yepez R 1987 Effect of calcium supplementation on the vascular sensitivity to angiotensin II in pregnant women. Am J Obstet Gynecol 156:261–262
44. Belizan JM, Villar J, Repke J 1988 The relationship between calcium intake and pregnancy-induced hypertension: up-to-date evidence. Am J Obstet Gynecol 158:898–902
45. Neer RM 1979 Calcium and inorganic phosphate homeostasis. In: DeGroot LU, Cahill GF, Odell WD (eds) Endocrinology. Grune & Stratton, New York, pp 669–692
46. Glorieux FH, Delvin EE, Salle BL 1988 Calcium and vitamin D status during pregnancy. In: Berger H (ed) Vitamins and Minerals in Pregnancy and Lactation (Nestle Nutrition Workshop Series, vol 16). Raven Press, New York

47. Paunier L, Lacourt G, Pilloud P, Schlaeppi P, Sizonenko PC 1978 25-Hydroxyvitamin D and calcium levels in maternal, cord and infant serum in relation to maternal vitamin D intake. Helv Paediatr Acta 33:95–103
48. Cockburn F, Belton NR, Purvis RJ, Giles MM, Brown JK, Turner TL, Wilkinson EM, Forfar JO, Barrie WJM, McKay GS, Pocock SJ 1980 Maternal vitamin D intake and mineral metabolism in mothers and their newborn infants. Br Med J 281:11–14
49. Brooke OG, Brown IRF, Bone CDM, Carter ND, Cleeve HJW, Maxwell JD, Robinson VP, Winder SM 1980 Vitamin D supplements in pregnant Asian women: effects of calcium status and fetal growth. Br Med J 280:751–754
50. Kuoppala T, Tuimala R, Parviainen M, Koskinen T, Ala-Houhala M 1986 Serum levels of vitamin D metabolites, calcium, phosphorus, magnesium and alkaline phosphatase in Finnish women throughout pregnancy and in cord serum at delivery. Hum Nutr: Clin Nutr 40C:287–293
51. Nehama H, Weintroub S, Eisenberg Z, Birger A, Milbauer B, Weisman Y 1987 Seasonal variations in paired maternal-newborn serum 25-hydroxyvitamin D and 24,25-dihydroxyvitamin D concentrations in Israel. Israel J Med Sci 23:274–277
52. Verity CM, Burman D, Beadle PC, Holton JB, Morris A 1981 Seasonal changes in perinatal vitamin D metabolism: maternal and cord blood biochemistry in normal pregnancies. Arch Dis Child 56:943–948
53. Hytten F, Chamberlain G 1980 Clinical Physiology in Obstetrics. Blackwell Scientific Publications, Oxford
54. Martin RD 1980 Adaptation and body size in primates. Z Morphol Anthrop 71:115–124
55. Goldberg GR, Prentice AM, Coward WA, Davies HL, Murgatroyd PR, Wensing C, Black AE, Harding M, Sawyer M 1993 Longitudinal assessment of energy expenditure in pregnancy by the doubly-labelled water method. Am J Clin Nutr 57:494–505
56. Durnin JVGA, McKillop FM, Grant S, Fitzgerald G 1987 Energy requirements of pregnancy in Scotland. Lancet ii:897–900
57. van Raaij JMA, Vermaat-Miedema SH, Schonk CM, Peek MEM, Hautvast JGAJ 1987 Energy requirements of pregnancy in The Netherlands. Lancet ii:953–955
58. Spaaij CJK 1993 The efficiency of energy metabolism during pregnancy and lactation in well-nourished Dutch women. University of Wageningen, The Netherlands: PhD Thesis
59. Forsum E, Sadurskis A, Wager J 1988 Resting metabolic rate and body composition of healthy Swedish women during pregnancy. Am J Clin Nutr 47:942–947
60. Lawrence M, Lawrence F, Coward WA, Cole TJ, Whitehead RG 1987 Energy requirements of pregnancy in the Gambia. Lancet ii:1072–1076
61. Poppitt SD, Prentice AM, Jequier E, Schutz Y, Whitehead RG 1993 Evidence of energy-sparing in Gambian women during pregnancy: a longitudinal study using whole-body calorimetry. Am J Clin Nutr 57:353–364
62. Tuazon MAG, van Raaij JMA, Hautvast JGAJ, Barba CVC 1987 Energy requirements of pregnancy in The Philippines. Lancet ii:1129–1131
63. Thongprasert K, Tanphaichitre V, Valyasevi A, Kittigool J, Durnin JVGA 1987 Energy requirements of pregnancy in rural Thailand. Lancet ii:1010–1012
64. Poppitt SD, Prentice AM, Goldberg GR, Whitehead RG 194 Energy-sparing strategies to protect human fetal growth. Am J Obstet Gynecol 171:118–125
65. Hytten FE 1991Weight gain in pregnancy. In: Hytten FE, Chamberlain G (eds) Clinical Physiology in Obstetrics. Blackwell Scientific Publications, Oxford
66. FAO/WHO/UNU. Report of a Joint Expert Consultation: Energy and Protein Requirements. WHO. 1985. WHO Tech Rep Series No 724. WHO, Geneva
67. Chase HC 1972 A comparison of Neonatal Mortality from Two Cohort Studies. US Health Services and Mental Health Administration, Washington DC
68. Lubchenko LO, Searls DT, Brazie JV 1972 Neonatal mortality rate: relationship to birth weight and gestational age. Fetal Neonatal Med 81:814–822
69. Mata LJ 1978 The children of Santa Maria Cauque. A Prospective Field Study of Health and Growth. MIT Press, Cambridge, MA
70. Ashworth A, Feacham RG 1985 Interventions for the control of diarrhoea diseases among young children: prevention of low birth weight. Bull WHO 63:165–184
71. Taffel S 1980 Factors associated with low birth weight. In: Vital and Health Statistics. Government Printing Office, Washington DC

72. McCormick MC 1985 The contribution of low birthweight to infant mortality and childhood morbidity. New Engl J Med 312:82–90
73. Babson SG 1970 Growth of low-birth-infants. J Paediatr 77:11–18
74. Ounsted M, Taylor ME 1971 The postnatal growth of children who were small-for-dates or large-for-dates at birth. Dev Med Child Neurol 13:421–434
75. Garn SM 1985 Relationship between birth weight and subsequent weight gain. Am J Clin Nutr 42:57–60
76. Naeye RL 1979 Weight gain and the outcome of pregnancy. Am J Obstet Gynecol 135:3–9
77. Prentice AM 1991 Can maternal dietary supplements help in preventing infant malnutrition? Acta Paediatr Scand 80 (Suppl 374):1–11
78. Ceesay SM, Saidykhan S, Prentice AM, Cole TJ, Day KC, Rowland MGM, Whitehead RG 1993 Effect of birth weight on a community based supplementation programme for pregnant Gambian women: first year results. Proc Nutr Soc 52:77A
79. Lechtig A, Habicht JP, Delgado H, Klein RE, Yarborough C, Martorell R 1975 Effect of food supplementation during pregnancy on birthweight. Paediatrics 56:508–520
80. Mardonnes-Santander F, Rosso P, Salivas J, Ahumada E, Vial I, Steckel A 1986 Effect of two supplements on birth weight. In: Taylor TG, Jenkins NK (eds) Proc 13th Int Congr Nutr, Brighton, UK. John Libbey, London
81. Chavez A 1978 Effects of malnutrition on body morphology. In: Sterky G, Mellander I (eds) Birthweight Distribution – an Indicator of Social Development. SAREC, Uppsala, pp 18–20
82. Herrera MG, Mora JO, Paredes BD, Wagner M 1980 Maternal weight/height and the effect of food supplementation during pregnancy and lactation. In: Aebi H, Whitehead RG (eds) Maternal Nutrition During Pregnancy and Lactation. Hans Huber, Bern, pp 252–263
83. Adair LS, Pollitt E 1985 Outcome of maternal nutritional supplementation: a comprehensive view of the Bacon Chow study. Am J Clin Nutr 41:948–978
84. Kardjati SR, Kusin JA, de With C 1988 Energy supplementation in the last trimester of pregnancy in East Java, Indonesia. I. Effect on birthweight. Br J Obstet Gynaecol 95:783–794
85. Ashwood A, Feacham RG 1985 Interventions for the control of diarrhoeal diseases among young children: prevention of low birth weight. Bull WHO 63:165 –184

3 What determines the production and composition of breast milk?

K.J. MOTIL

INTRODUCTION

The regulation of milk production and composition in the human is an intricate and complex process that integrates structure with function in the presence of hormonal, behavioural, and nutritional factors. While some of these factors directly influence milk production and/or composition, other can be inferred only from indirect evidence. This review will highlight four major determinants of milk production and composition: the anatomical factors of the breast at the level of both gross and molecular structure; the hormonal factors, including reproductive and metabolic hormones; maternal behavioural factors as they relate to infant feeding patterns; and nutritional factors as they relate to the maternal diet and metabolic adaptation to the lactational process. While all of these factors are essential for optimal lactational performance, this review will focus primarily on the nutritional factors that influence the outcome of lactation, i.e. milk production and composition.

ANATOMICAL FACTORS

The mammary gland itself is essential to the lactational performance of the nursing mother. The mammary gland is composed of 15–20 lobes of glandular tissue per breast (1) each of which is subdivided into many lobules, and each lobule contains a large number of alveoli. Each alveolus contains many secretory cells surrounded by myoepithelial cells and a network of arterial and venous capillaries. The alveolus is the functional unit of the mammary gland. Milk that is produced by the secretory cells is emptied into the lumen of the alveolus, which in turn empties into a series of small ductules, larger ducts, and lactiferous sinuses prior to its removal by the infant.

The development of the mammary gland takes place to the greatest extent during adolescence and the reproductive phase of the adult woman's life cycle (2). Prior to puberty, limited ductular and alveolar proliferation and differentiation occur under the influence of a number of hormones produced by the hypothalamic–pituitary axis, including growth hormone, prolactin, adrenocorticotrophic hormone and thyroid stimulating hormone. During puberty, ductular maturation is the primary event of mammary gland development. Under the influence of oestrogens, the ducts elongate, branch, and canalize the alveoli. During pregnancy, alveolar proliferation is the major developmental event of the mammary gland, occurring under the influence of placental and ovarian hormones such as human placental lactogen and progesterone, respectively.

J.G. Bindels, A.C. Goedhart and H.K.A. Visser (eds), Recent Developments in Infant Nutrition, 36–50.

Ductular maturation, characterized by further elongation, branching and canalization of alveoli, also accelerates under the influence of these placental and ovarian hormones.

Interruption of the development of the mammary gland for any reason leads to lactational failure by virtue of the absence of the structures of the mammary gland that are essential for milk production. Examples of primary lactation failure associated with insufficient glandular tissues within the breast have been documented (3). In these cases, inadequate milk production resulted, despite the presence of an appropriate hormonal milieu and adequate nursing stimuli.

The molecular structure of the mammary gland also plays a role in the production and composition of breast milk in the human (4). The α-lactalbumin gene regulates the synthesis of the whey protein, α-lactalbumin, which in turn regulates the synthesis of milk lactose through its role as a component of the lactose synthase complex (5). Lactose is a major osmotic component of milk and determines milk volume by regulating water influx. Mice that carry the deletion of the α-lactalbumin gene are α-lactalbumin deficient, produce reduced amounts of thickened milk containing little or no lactose, and cannot sustain their offspring. If the human α-lactalbumin gene replaces the deficient gene in the mouse, these animals lactate normally, providing evidence that the molecular structure of the mammary gland is an essential determinant of milk production and composition.

HORMONAL FACTORS

The hormonal milieu of the lactating woman is a major determinant of lactogenesis, i.e., the initiation of milk production, and galactopoiesis, the maintenance of milk production. Prolactin and oxytocin are central to the initiation of milk production. Prolactin release is triggered at parturition, when human placental lactogen and progesterone levels fall (6), and stimulates the production and subsequent secretion of milk into the alveolus. At the same time, the suck stimulus of the infant triggers oxytocin release, which in turn results in milk ejection from the breast (7).

The role of these hormones, particularly prolactin, in the maintenance of milk production is less clear. The secretion of prolactin generally is proportional to the amount of sucking by the infant, and hence, the amount of milk produced by the mother (8). In support of this observation, administration of bromocriptine, an inhibitor of prolactin secretion (9), inhibits milk production in women. Low milk yields in postpartum women have been associated with decreased prolactin secretion (8), and administration of metoclopramide in women with lactational insufficiency increases milk secretion (10). On the other hand, prolactin may have no relationship with the amount of milk produced (11): in late lactation, high milk yields are sustained even in the presence of low plasma prolactin levels (12). Although the mechanisms by which prolactin stimulates milk production are not understood completely, the hormone clearly has diverse roles in the production of human milk. Prolactin binds to cell surface receptors within the mammary gland and initiates a number of intracellular events that lead to successful milk protein synthesis and secretion. Prolactin is responsible for the synthesis of milk specific proteins such as the caseins and α-lactalbumin. In this regard, prolactin regulates milk protein synthesis via mechanisms that involve transcriptional (13), translational (14) and post-translational (15) events. As an example, casein-specific mRNA accumulates intracellularly in the presence of prolactin because the hormone is associated with increased rates of transcription and decreased rates of degradation of mRNA (13).

Other metabolic hormones, such as insulin, cortisol, thyroxine and triiodothyronine, play a supporting role in the production of milk by modulating the growth and development of the mammary gland and/or the disposal of available nutrients for milk synthesis and secretion (16–18). Glucocorticoids, such as cortisol, and the thyroid hormones exert a permissive effect on the activity of the mammary gland by potentiating the effects of more specific lactogenic factors. Insulin, on the other hand, may have a more direct role in the partitioning, mobilization, and utilization of nutrients within the mammary gland for milk production.

MATERNAL BEHAVIOURAL FACTORS

The behavioural patterns of the nursing mother, as they relate to infant feeding, influence her lactational performance. The frequency of nursing is a major determinant of the adequacy of milk production, as measured by the duration of exclusive breastfeeding and the continuation of partial breast-feeding for prolonged periods of time. Recent studies examined the adequacy of maternal milk production in terms of infant weight gain, the frequency of nursing, and the duration of breast-feeding in Chilean infants (19). When the frequency of nursing at 1 month postpartum was seven feeds per day or more, the mothers were more likely to sustain a longer duration of exclusive breast-feeding than when the nursing frequency was less than seven feeds per day. Furthermore, the probability that their infants would continue to breast-feed for at least 1 year, was 30% higher for infants who breast-fed seven times or more per day during the first 6 months postpartum and four times or more per day during the second 6 months of life than for those infants who breast-fed less frequently.

NUTRITIONAL FACTORS

Our own studies have focused on the role of maternal nutrition, with emphasis on protein, as a determinant of milk production and composition. The importance of nutrition in relation to lactational performance is not a new concept. Early medical textbooks provided lengthy instructions to mothers about the importance of the diet during lactation. The following excerpt is taken from a textbook published in 1920 (20): "*The Diet for the Nursing Mother*... the mother may largely follow her own inclinations as to what she eats... If there be but scanty quantity of milk it is an indication that there should be a more generous diet... plenty of fresh milk, eggs, fresh vegetables, ripe fruit, and other plain simple food are required." This excerpt illustrates the point that lactational performance historically has been associated with good nutrition, i.e. the adequacy of dietary nutrient intakes. Furthermore, in the context of adequate nutrition, the emphasis was placed on foods that were generous in proteins of good quality.

Figure 1 Dietary regulation of milk production and composition: theoretical considerations

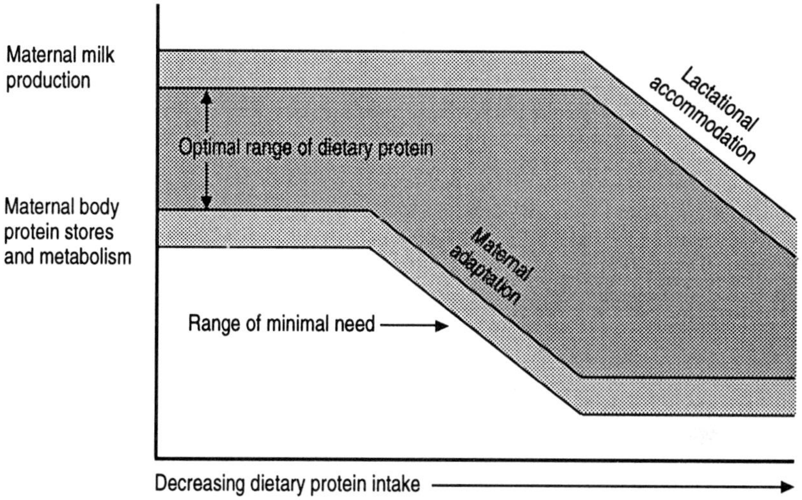

Our current concept of the dietary regulation of milk production and composition, using protein as an example, is illustrated in Figure 1. Maternal milk production is represented by the upper channel, maternal body protein stores and metabolism are represented by the lower channel, and dietary protein intake in decreasing amounts is represented by the x-axis. At some optimal range of dietary protein intake, maternal milk production and maternal body protein metabolism proceed at an optimal rate coincident with maternal health and well-being. As dietary protein intake decreases, the mother must adapt by altering her own body protein metabolic responses in order to sustain her milk production. As dietary protein intake decreases to a level where the mother's protein metabolic responses cannot adapt further, the mother's lactational performance diminishes to accommodate her body's own metabolic needs in the presence of a diminishing protein supply.

It is difficult to understand why nutrition in general, and protein metabolism, specifically, should be pivotal to lactational performance, particularly when maternal milk production appears to be an inconsequential physiological and metabolic process from the perspective of the lactational performance of the mother. This point can be illustrated by estimating the contribution that milk protein production makes to both maternal body protein stores and body protein metabolism. In general, the typical lactating American woman weighs approximately 65 kg, her lean body mass composes approximately 65% of her body weight, and protein composes approximately 20% of her lean body mass (21).With respect to her lactational performance, the lactating woman produces 750 g of milk per day which contains 9 g protein per litre of milk. If the mother's body protein stores provided all of the nitrogen needed for

***Figure 2** Contribution of milk protein synthesis to whole body protein turnover in lactating women*

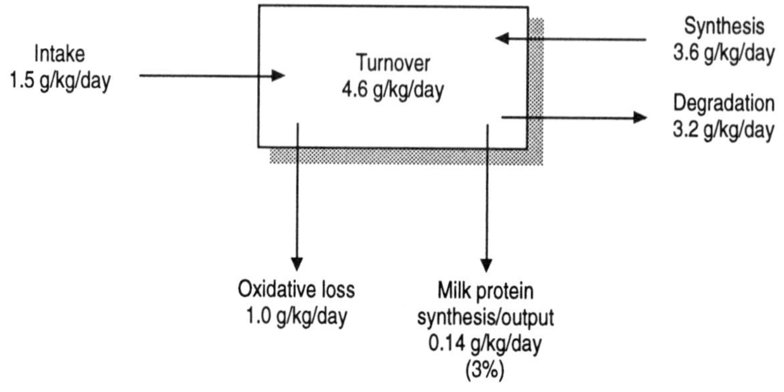

milk protein synthesis, milk protein production would require less than 0.1% of maternal body protein stores.

As shown in Figure 2, the contribution that milk protein synthesis makes to whole body protein turnover in the lactating mother is equally small. Using stable isotope tracers, we have estimated that milk protein output averages 0.14 g/kg/day or approximately 3% of total body protein turnover (22). Thus, from the perspective of body protein metabolism, milk production theoretically should have minimal metabolic and physiological impact on the mother.

To reconcile our construct of the dietary regulation of milk production and composition with these observations, we have assumed that: (1) lactation imposes a significant metabolic stress on the mother that is out of proportion to the net cost of milk production, and (2) the adequacy of the maternal diet is inherently related to the success or failure of the mother's adaptational responses to the lactational process. The discussion that follows provides supportive evidence that lactation imposes a metabolic stress on the mother that is highly responsive to dietary perturbation, that maternal diet directly influences milk production and composition, and that the integration of the maternal diet, adaptive metabolic responses, and milk productivity are modulated by the hormonal milieu associated with the lactational process.

Maternal diet and metabolic adaptation to lactation

Available evidence suggests that lactation imposes a metabolic stress on the nursing mother because of alterations in protein metabolic responses that occur postpartum. We recently studied lactating women 1, 5 and 12 months postpartum and nulliparous women in the postabsorptive state using stable isotope techniques (22). As shown in Figure 3, rates of protein synthesis and degradation tended to increase as lactation

Figure 3 Protein synthesis, degradation, and leucine oxidation in lactating women 1 (■), 5 (▒) and 12 months postpartum (■) and in nulliparous women (□)

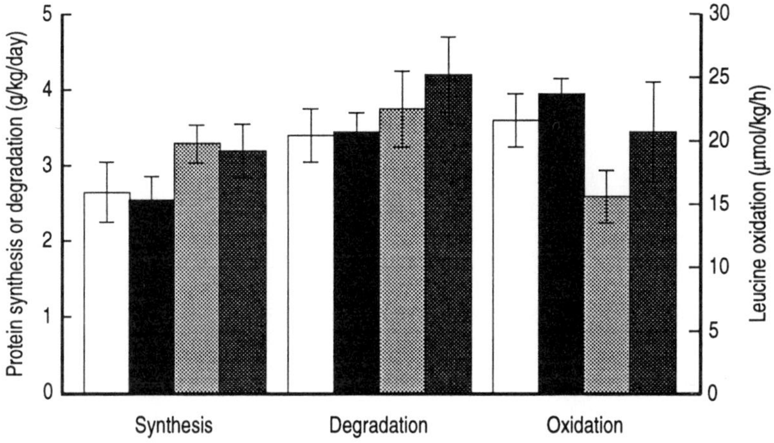

progressed, even though weaning was well under way at 12 months postpartum. Furthermore, rates of protein synthesis and degradation in later lactation tended to be greater in the nursing mothers than in the nulliparous women. These combined observations imply that the lactational event imposes a pronounced metabolic effect on the lactating woman which intensifies with prolonged breast-feeding. In addition, amino acid ([1-^{13}C]leucine) oxidation, another representative marker of body protein metabolism, was lower in the lactating women 5 months postpartum than at 1 and 12 months postpartum, a time of maximum milk production in terms of the duration of lactation and daily milk output. This reduction of amino acid oxidation presumably occurred because of the mother's metabolic need to conserve available nutrients for either milk production or the maintenance of her own body stores.

The maternal diet has a major role in the adaptive responses to lactation. One of the earliest physiological responses to lactation is a change in appetite, as reflected by increases in maternal dietary intakes. Three-day food records repeatedly demonstrate that lactating women in the free-living setting consume dietary protein in amounts that exceed those of their non-lactating counterparts (22–24) (Fig. 4). Moreover, lactating women consume dietary protein in amounts that exceed the theoretical estimate of their obligatory protein nitrogen losses in milk by approximately 50% (25). The mechanisms that account for the heightened maternal appetite during lactation have not been well studied. It has been postulated that alterations in the hormonal profiles of prolactin, oxytocin, cortisol, progesterone and/or oestrogens account for alterations in appetite, although no one hormone appears to be responsible for this change in eating behaviour (26).

Figure 4 Actual dietary protein intakes and theoretical protein needs of lactating, non-lactating postpartum and nulliparous women

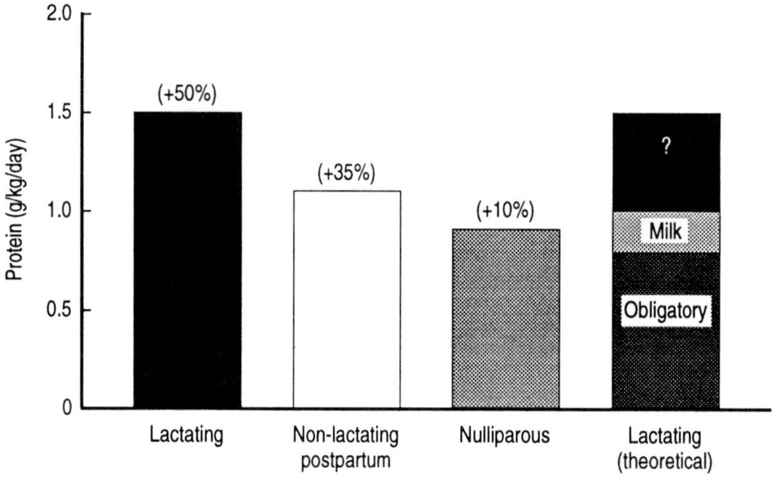

It is clear that both acute and chronic dietary perturbations are associated with adaptive responses of body protein metabolism in lactating women. In our early studies we measured the nitrogen balances of lactating and non-lactating women who were given dietary protein intakes of 1.5 to 1.0 g/kg/day for 7–10 days (22,23) (Fig. 5). A protein intake of 1.5 g/kg/day or its nitrogen equivalent of 240 mg/kg/day represented the typical dietary intake of the American woman, whereas a protein intake of 1.0 g/kg/day or its nitrogen equivalent of 160 mg/kg/day represented the theoretical estimate that meets body protein needs (25). Nitrogen balance was significantly lower in the lactating than in the non-lactating women at both levels of dietary protein. Furthermore, nitrogen balance was significantly lower in the group of lactating women who consumed protein at a level of 1.0 g/kg/day than in those who consumed protein at a level of 1.5 g/kg/day. The variability in nitrogen retention was considerably reduced in lactating women who consumed the lower amount of protein compared with those consuming the higher amount of protein. All of these observations taken together suggest that lactation imposes a metabolic stress on the mother, and that diet clearly influences the adaptive responses of the woman to lactation.

Nitrogen balances also have been measured in marginally nourished lactating women from India (27,28), Africa (29), and Mexico (30), as well as in well-nourished lactating American women (22,23,31). These studies indicate that not only is there a direct relationship between nitrogen retention and dietary protein intake, but also that this relationship is influenced by the nutritional status of lactating women. As shown in Figure 6, nitrogen equilibrium in marginally nourished women is achieved at a

***Figure 5** Nitrogen retention and intake of lactating (■), non-lactating postpartum (■) and nulliparous (▩) women. Values are mean ± SD. *$p < 0.05$: lactating group vs. non-lactating postpartum and nulliparous groups at each level of nitrogen (240, 160 mg/kg/day) intake. ⁺$p < 0.05$: lactating groups who consumed 240 vs. 160 mg N/kg/day*

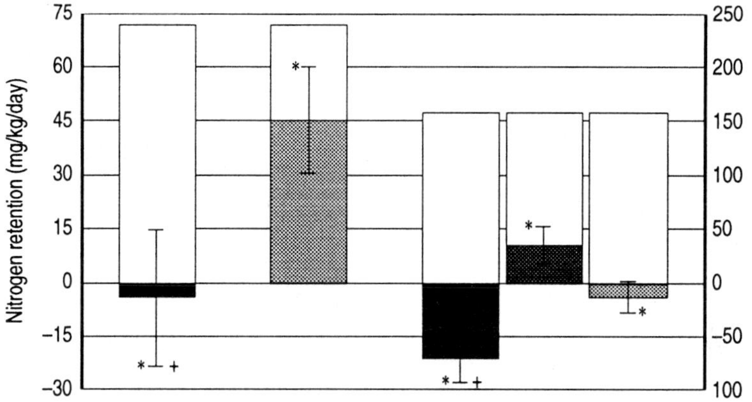

***Figure 6** Nitrogen balances of marginally nourished and well-nourished lactating women (numbers in parentheses represent number of subjects)*

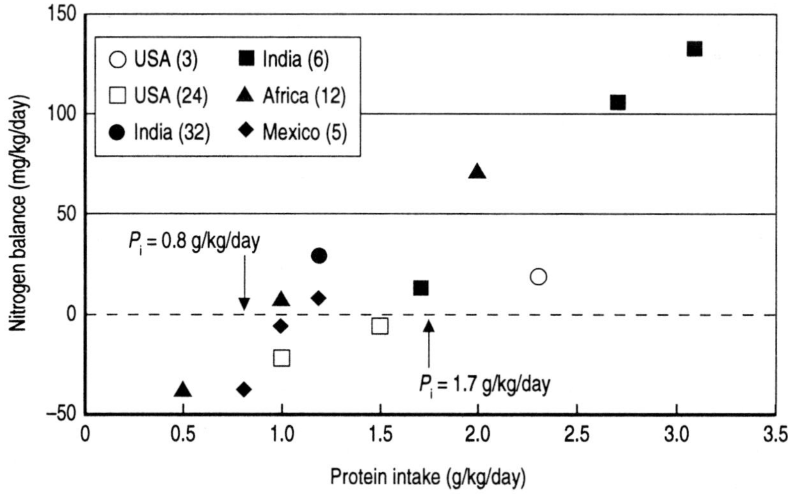

dietary protein intake of 0.8 g/kg/day, whereas in the well-nourished women, nitrogen equilibrium occurs at dietary protein intakes of 1.7 g/kg/day. The implications of these observations are that marginally nourished women adapt to lactation in a different manner and are more efficient at conserving dietary protein than their well-nourished counterparts.

If well-nourished lactating women are in negative nitrogen balance at dietary protein intakes that theoretically meet their metabolic needs, the question arises as to whether they are sacrificing their lean body mass to support the metabolic needs of milk production. We measured longitudinal changes in body composition over the first year postpartum in free-living lactating and non-lactating women, using whole body potassium (^{40}K) counting techniques (21). Our studies show that both lactating and non-lactating postpartum women lose approximately 4% of their body weight over the first 6–12 months after giving birth. Nevertheless, lean body mass is preserved in both groups of women, implying that weight loss is due entirely to loss of fat. It is noteworthy, however, that the lean body mass of the lactating group is preserved, presumably because these women augmented their dietary protein intake to a level of 1.5 g/kg/day whereas the non-lactating women preserved their lean body mass at dietary protein intakes of 1.0 g/kg/day.

When dietary protein intakes are reduced, what are the metabolic mechanisms that allow the well-nourished mother to adapt? In one of our more recent studies, lactating and non-lactating women were placed on protein diets of 1.5, 1.0, and 0.4 g/kg/day (32), representing quantitatively a typical American diet for adults, a diet that theoretically meets the lactating mother's metabolic needs, and an inadequate protein diet for both groups of women, respectively (25). At the end of each dietary period, the women were given an oral bolus dose of the stable isotope, [1-^{13}C]leucine, followed by the administration at hourly intervals of small protein meals. Amino acid ([1-^{13}C]leucine) oxidation, measured by the rate of ^{13}C appearance in expired CO_2, was reduced commensurately with the reduction of dietary protein to a greater extent in the lactating than in the non-lactating women, implying that oxidative losses are a sensitive indicator of the adequacy of the protein content of the diet and that dietary protein is conserved to meet the metabolic needs of the mother for milk production or for the maintenance of her own body stores. As shown in Figure 7, rates of body protein synthesis and degradation were lower in lactating women than in their non-lactating counterparts at all levels of dietary protein intake. Moreover, there was an abrupt fall in the rates of protein degradation at protein intakes between 1.5 and 1.0 g/kg/day which was sustained between levels of 1.0 and 0.4 g/kg/day. Despite lower rates of protein synthesis and degradation, net protein retention, i.e. the difference between these two measures of body protein metabolism, was consistently greater at all levels of dietary protein in the lactating than in the non-lactating group because of the relatively greater reduction in rates of body protein degradation in the lactating group.

When the contribution of maternal milk protein synthesis and output is examined in relation to net protein retention, milk protein composes approximately one-third of the protein retained by the mother at protein intakes of 1.5 and 1.0 g/kg/day. However, at the inadequate level of dietary protein intake, approximately 93% of the protein retained

Figure 7 Whole body protein degradation (upper panel) and synthesis (lower panel) in the fed state in lactating (●) and non-lactating (■) postpartum women. Values are mean ± SE

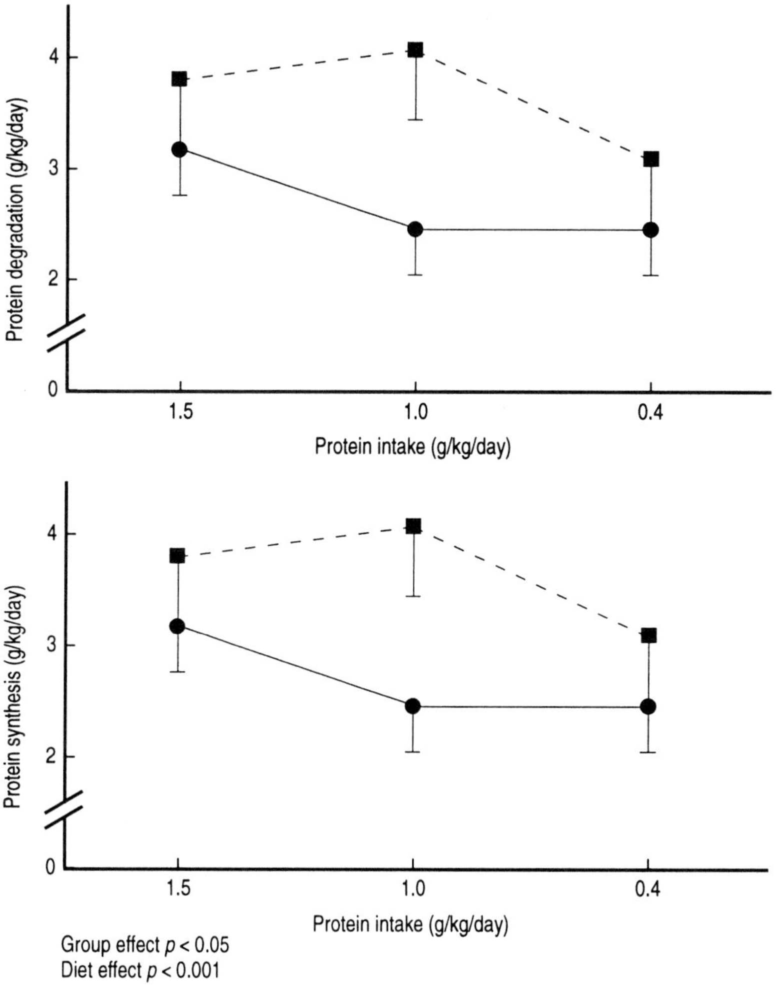

by the lactating woman is milk protein. This imbalance in the partitioning of net protein retention between maternal needs and milk production is important insofar as the availability of dietary nutrients may be channelled into milk production at the expense of maternal body protein stores.

Figure 8 Relationship between maternal milk production and dietary protein (upper panel) and energy (lower panel) intakes

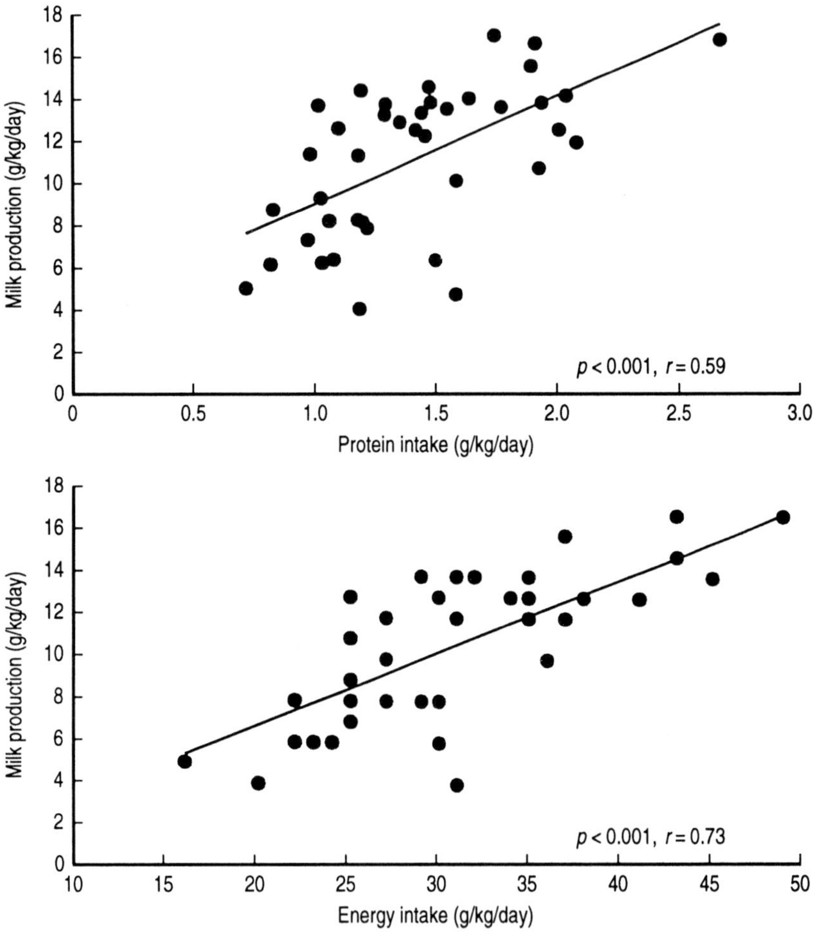

Maternal diet and lactational productivity

The maternal diet directly influences lactational performance, although the major impact appears to be on milk production as opposed to milk composition. We measured the amount of milk produced daily by free-living lactating women, using the test weighing procedure, and their dietary protein and energy intakes, using the 3-day food record technique (21,33). There was a direct relationship between the amount of milk produced by the lactating women and their dietary protein and energy intakes (Fig. 8). In

Figure 9 Relationship between milk protein and non-protein constituents and dietary protein intake in lactating women

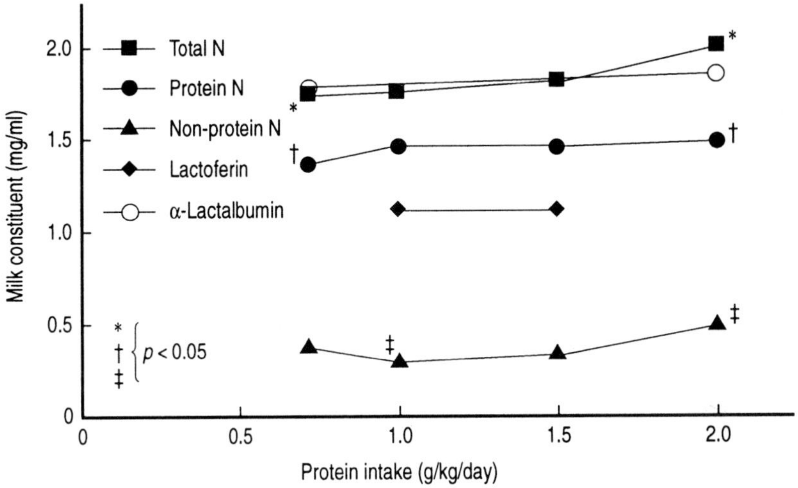

a similar manner, an anecdotal report of a malnourished woman and her infant who was failing to thrive because of insufficient milk production demonstrated that this process can be reversed merely by feeding the mother an adequate diet (34).

Although the protein fraction of human milk, as a whole, is conserved across a broad range of dietary protein intakes, some constituents are altered, although the magnitude of the change is relatively small. For example, as shown in Figure 9, lactoferrin and α-lactalbumin concentrations remain relatively constant across a range of protein intakes from 0.6 to 2.0 g/kg/day (35,36). In contrast, milk total nitrogen, particularly the non-protein nitrogen fraction, falls in response to a reduction of dietary protein intake (35,36). Although these alterations in the protein fraction of milk may have a concurrent regulatory role with respect to milk protein synthesis (37), the effect of these changes in milk composition on the mother or her infant is quantitatively insignificant.

Hormonal integration of maternal diet, metabolic adaptation and lactational productivity

Overall, these observations suggest that, in the human, dietary protein is partitioned away from maternal body protein stores and channelled into the mammary gland for milk production. The homeorhetic mechanisms that regulate the allocation of dietary protein to milk production or the maintenance of tissue protein mass of lactating women are presumably of endocrine origin (16). As an example, the modulating effect

Figure 10 Hormonal modulation of nitrogen balance and hepatic protein synthesis in lactating women 1, 5 and 12 months postpartum

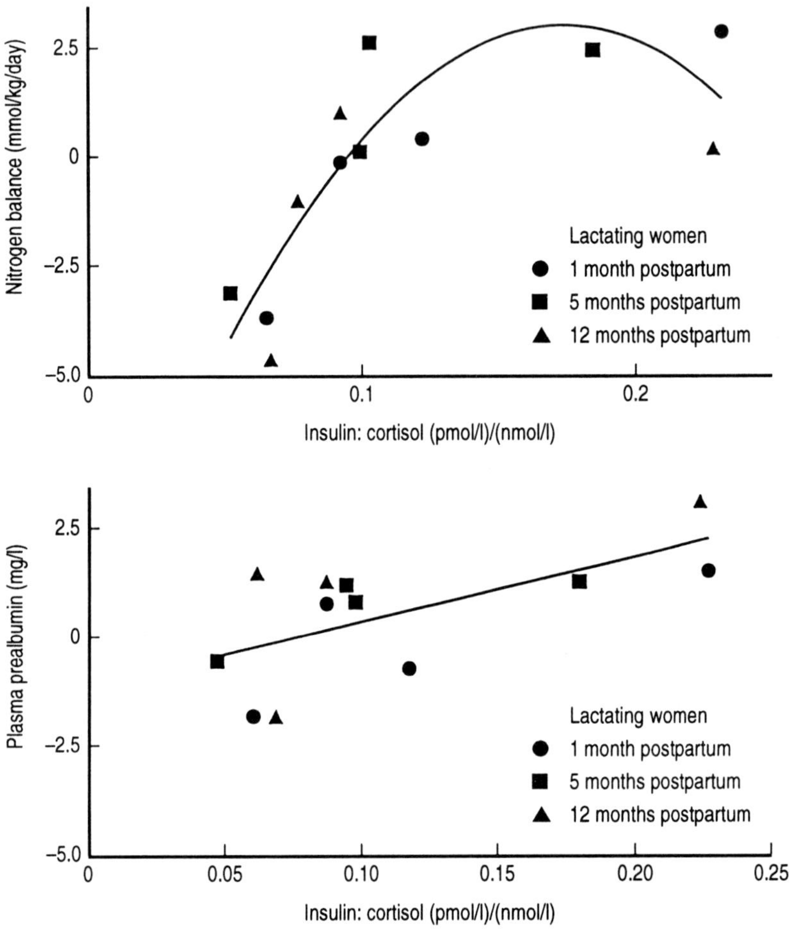

of postabsorptive insulin:cortisol ratios in lactating women 1, 5, and 12 months postpartum on nitrogen balance and hepatic protein synthesis, as measured by plasma prealbumin levels, is illustrated in Figure 10 (38). In both instances, the hormonal milieu shows a direct relationship with the protein metabolic status of the mother. The direct association between plasma thyroxine levels and the amount of milk produced by these women implies that metabolic hormones also have a modulating effect on their milk productivity (38). These observations suggest that maternal milk production and the metabolic adaptation to lactation are integrated, highly regulated, metabolic processes. We presume that these processes reflect the balance between supply and demand for dietary substrate for maternal body protein metabolism and milk production.

ACKNOWLEDGEMENTS

This work is a publication of the USDA/ARS Children's Nutrition Research Center, Department of Pediatrics, Baylor College of Medicine, Houston, TX and has been funded with federal funds from the U.S. Department of Agricultural Research Service, under Cooperative Agreement No. 38-6250-1-003. The contents of this publication do not necessarily reflect the views or policies of the U.S. Department of Agriculture, nor does the mention of trade names, commercial products, or organizations imply endorsement by the United States Government. The author thanks those individuals who have contributed on an ongoing basis to the research efforts described herein, including B. Kertz, C. M. Montandon, S. Seributra, S. Katz, A. Opekun, M. Thotathuchery, S. Vadiya, W. W. Wong, F. Jahoor, D. L. Hachey, T. A. Davis, P. J. Reeds, E. O. Smith, K. Fraley and I. Tapper.

REFERENCES

1. Vorherr H 1974 The Breast: Morphology, Physiology, and Lactation. Academic Press, New York
2. Neville MC 1983 Regulation of mammary development and lactation. In: Neville MD, Neifert MD (eds) Physiology, Nutrition, and Breast-feeding. Plenum Press, New York pp 103–117
3. Neifert MR, Seacat JM, Jobe WE 1985 Lactation failure due to insufficient glandular development of the breast. Pediatrics 76:823–828
4. Stacy A, Schnieke A, Kerr M, Scott A, McKee C, Cottingham I, Binas B, Wilde C, Colman A 1995 Lactation is disrupted by α-lactalbumin deficiency and can be restored by human α-lactalbumin gene replacement in mice. Proc Natl Acad Sci USA 92:2835–2839
5. Mepham TB 1987 Physiology of Lactation. Open University Press, Philadelphia pp 80–82
6. Ostrom KM 1990 A review of the hormone prolactin during lactation. Prog Food Nutr Sci 14: 1–43
7. McNeilly AS, Robinson IC, Houston MJ, Howie PW 1983 Release of oxytocin and prolactin in response to sucking. Br Med J 286:257–259
8. Aono T, Shioji T, Shoda T, Kurachi K 1977 The initiation of human lactation and prolactin response to sucking. J Clin Endocrinol Metab 44:1101–1106
9. Brun del Re R, Del Pozo E, De Grandi P, Friesen H, Hinselmann M, Wyss H 1973 Prolactin inhibition and suppression of puerperal lactation by a Br-ergocriptine (CB154). A comparison with estrogen. Obstet Gynecol 41:884–890
10. Kauppila A, Kivinen S, Ylikorkala O 1981 A dose response relation between improved lactation and metoclopramide. Lancet i:1175–1177
11. Howie PW, McNeilly AS, McArdle T, Smart L, Houston M 1980 The relationship between suckling-induced prolactin response and lactogenesis. J Clin Endocrinol Metab 50:670–673
12. Hennart P, Delogne-Desnoeck J, Vis H, Robyn C 1981 Serum levels of prolactin and milk production in women during a lactation period of thirty months. Clin Endocrinol 14:349–353
13. Guyette WA, Matusik RJ, Rosen JM 1979 Prolactin-mediated transcriptional and post-transcriptional control of casein gene expression. Cell 17:1013–1023
14. Teyssot B, Houdebine L-M 1981 Induction of casein synthesis by prolactin and inhibition by progesterone in the pseudopregnant rabbit treated by colchicine without any simultaneous variations of casein mRNA concentration. Eur J Biochem 117:563–568
15. Ollivier-Bousquet M 1978 Early effects of prolactin on lactating rabbit mammary gland. Cell Tissue Res 187:25–43
16. Hart IC, Bines JA, Morant SV, Ridley JL 1978 Endocrine control of energy metabolism in the

cow: comparison of the levels of hormones (prolactin, growth hormone, insulin, and thyroxine) and metabolites in the plasma of high- and low-yielding cattle at various stages of lactation. J Endocrinol 77:333–345

17. Girard J, Burnol A-F, Leturque A, Ferre P 1987 Glucose homeostasis in pregnancy and lactation. Biochem Soc Trans 15:1028–1030
18. Thatcher WW, Tucker HA 1970 Lactational performance of rats injected with oxytocin, cortisol-21-acetate, prolactin, and growth hormone during prolonged lactation. Endocrinology 86: 237–240
19. Diaz S, Herreros C, Aravena R, Reyes MV, Casado M, Schiappacasse V 1995 Breast-feeding duration and growth of fully breast-fed infants in a poor urban Chilean population. Am J Clin Nutr 62:371–376
20. Scholl BF, McFarland-Sharp A, Miller FE 1920 Library of Health. Complete Guide to Prevention and Cure of Disease. Historical Publishing Co, Philadelphia, p 1189
21. Motil KJ, Sheng HP, Montandon CM, Wong WW, Ellis K 1992 Partitioning of dietary protein into lean body mass in lactating and nonlactating women. FASEB J 6:A1082
22. Motil KJ, Montandon CM, Hachey DL, Boutton TW, Klein PD, Garza C 1989 Whole-body protein metabolism in lactating and nonlactating women. J Appl Physiol 66:370–376
23. Motil KJ, Montandon CM, Thotathuchery M, Garza C 1990 Dietary protein and nitrogen balance in lactating and nonlactating women. Am J Clin Nutr 51:378–384
24. Motil KJ, Montandon CM, Garza C 1990 Basal and postprandial metabolic rates in lactating and nonlactating women. Am J Clin Nutr 52:610–615
25. Subcommittee on the Tenth Edition of the RDAs, Food and Nutrition Board 1989 Recommended Dietary Allowances (10th edn). National Research Council, Washington, DC pp 52–77
26. Rosso P 1988 Regulation of food intake during pregnancy and lactation. Curr Concepts Nutr 16:91–99
27. Agarwal KN, Bhatia BD, Agarwal DK, Sharma D, Prakash M, Sharma ML 1986 Urinary nitrogen and glucose excretion in pregnant and lactating women. Indian Pediatr 23:331–337
28. Nageswara Rao C, Narasinga Rao BS 1974 Nitrogen balance in pregnancy and lactation in women whose protein intake is marginal. Indian J Med Res 62:1619–1626
29. Edozien JC, Rahim Khan MA, Waslien CI 1976 Human protein deficiency: results of a Nigerian village study. J Nutr 106:312–328
30. DeSantiago S, Villalpando S, Ortiz N, Alanso L, Ramirez I 1995 Protein requirements of marginally nourished lactating women. Am J Clin Nutr 62:364–370
31. Hunscher HA, Donelson AE, Nims B, Kenyon F, Macy IG 1933 Metabolism of women during the reproductive cycle. V. Nitrogen utilization. J Biol Chem 99:507–520
32. Motil KJ, Davis TA, Montandon CM, Wong WW, Klein PD, Reeds PJ 1996 Whole body protein turnover in the fed state is reduced in response to dietary protein restriction in lactating women. Am J Clin Nutr (in press)
33. Motil KJ, Montandon CM, Hachey DL, Boutton TW, Klein PD, Garza C 1989 Relationships among lactation performance, maternal diet, and body protein metabolism in humans. Eur J Clin Nutr 43:681–691
34. Sosa R, Klaus M, Urrutia JJ 1976 Feed the nursing mother, thereby the infant. J Pediatr 88: 668–670
35. Motil KJ, Thotathuchery M, Bahar A, Montandon CM 1995 Marginal dietary protein restriction reduced nonprotein nitrogen, but not protein nitrogen, components of human milk. J Am Coll Nutr 14:184–191
36. Forsum E, Lonnerdahl B 1980 Effect of protein intake on protein and nitrogen composition of breast milk. Am J Clin Nutr 33:1809–1803
37. Meijer GAL, van der Meulen J, van Vuuren AM 1993 Glutamine is a potentially limiting amino acid for milk production in dairy cows: a hypothesis. Metabolism 42:358–364
38. Motil KJ, Thotathuchery M, Montandon CM, Hachey DL, Bouton TW, Klein PD, Garza C 1994 Insulin, cortisol and thyroid hormones modulate maternal protein status and milk production and composition in humans. J Nutr 124:1248–1257

4 Essential fatty acids, pregnancy and pregnancy outcome

G. HORNSTRA, M.D.M. AL, A.C. VAN HOUWELINGEN and M.M.H.P. FOREMAN-VAN DRONGELEN

ABBREVIATIONS

AA	arachidonic acid
CA	cervonic acid (=DHA)
CADI	cervonic acid deficiency indicator
CASI	cervonic acid sufficiency index
DHA	docosahexaenoic acid
EFA	essential fatty acid
EPA	eicosapentaenoic acid (=timnodonic acid)
LCPUFA	long-chain polyunsaturated fatty acids
PIH	pregnancy-induced hypertension

ESSENTIAL FATTY ACIDS AND THEIR DERIVATIVES

Essential fatty acids (EFA) are essential for human health but cannot be synthesized by man: they have to be consumed with the diet. EFA are important structural and functional membrane components and also the ultimate precursors of the prostanoids (prostaglandins and thromboxanes), a group of local hormone-like substances with important bioregulatory functions (1). Fetal development is associated with a high EFA requirement, and for its EFA supply the developing fetus depends on the availability of maternal EFA. There are two families of essential fatty acids, the $n-6$ and the $n-3$ family. The essential nature of the $n-6$ family has been recognized for decades (2), but that of the $n-3$ family has been a matter of debate for some time. However, at present there is no doubt that $n-3$ fatty acids are essential for reproductive, brain and visual functions (3). The 'parent EFA' are linoleic acid (18:2 $n-6$), and α-linolenic acid (18:3 $n-3$), which are present in seed oils and green leaf vegetables. Both parent EFA can be desaturated and elongated in the human body to a series of longer chain, more unsaturated 'derived EFA' (Fig. 1). The most important EFA derived from linoleic acid are dihomo-γ-linolenic acid (20:3 $n-6$) and arachidonic acid (20:4 $n-6$, AA). These fatty acids are the precursors of the prostanoids of the 1- and 2-series, respectively. In addition, AA is an important structural fatty acid in the brain. The most important EFA derived from the $n-3$ series are timnodonic acid (also called eicosapentaenoic acid, EPA, 20:5 $n-3$) and cervonic acid (CA, also called docosahexaenoic acid, DHA, 22:6 $n-3$). Timnodonic acid is the precursor of the prostanoids of the 3-series, whereas CA, together with AA the major polyenoic fatty acid in brain

J.G. Bindels, A.C. Goedhart and H.K.A. Visser (eds), Recent Developments in Infant Nutrition, 51–63.

***Figure 1** Metabolic interconversion of the parent essential fatty acids linoleic acid, (18:2 n−6) and α-linolenic acid, (18:3 n−3), into their longer-chain, more-unsaturated derived EFA. D: desaturase; E: elongase. If insufficient EFA are available, the body starts to synthesize 'surrogate fatty acids' by desaturating and elongating non-essential oleic acid, (18:1 n−9), to mead acid, (20:3 n−9). If the cervonic acid status is too low, the desaturation of adrenic acid, (22:4 n−6) is promoted, leading to higher amounts of osbond acid (22:5 n−6)*

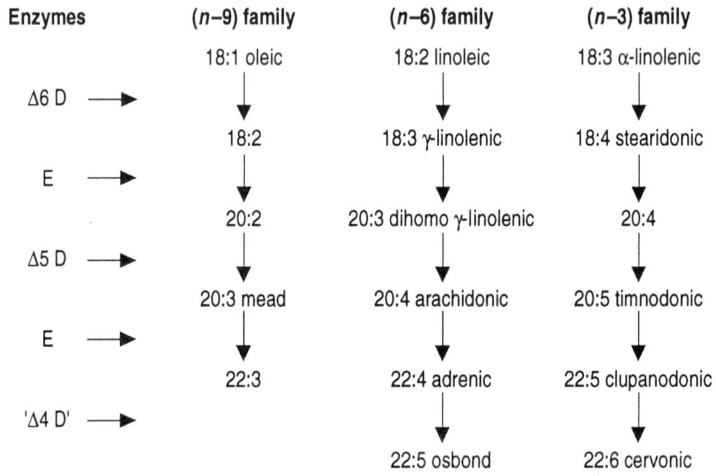

and retina, can be converted by a lipoxygenase to form hydroxy fatty acids with biological activity. The other EFA, parent as well as derived ones, are also good lipoxygenase substrates. Many of the resulting hydroxy fatty acids have significant biological functions (1,4). The derived EFA containing three or more double bonds and 20 or more carbon atoms are called the long-chain polyunsaturated fatty acids (LCPUFA).

EFA STATUS, EFA DEFICIENCY, AND EFA DEFICIENCY MARKERS

If the dietary intake of EFA is too low to meet the requirement of the body, the EFA body stores will be used as an additional supply. Considerable amounts of linoleic acid are stored in adipose tissue and can readily be mobilized. However, adipose tissue contains hardly any α-linolenic acid and LCPUFA. Since the amount of adipose tissue is rather limited in neonates, and in preterm infants in particular, an adequate postnatal EFA intake is very important to guarantee an adequate EFA status. The developing fetus depends on the mother for an adequate EFA supply. The diet of a pregnant woman should therefore contain sufficient EFA to cover not only her own requirement, but that of her baby as well.

In the assessment of the EFA status of an individual, the total amount of the various EFA in blood plasma is a useful indicator. Since, in general, phospholipids have a slower turnover than free fatty acids, triglycerides and cholesterol esters, the EFA profile of plasma or red blood cell phospholipids is considered suitable to document the overall EFA status of a given individual (5). It should be realized, however, that the plasma EFA supply does not necessarily meet the EFA requirements of the various tissues, nor does the presence of EFA in blood guarantee their proper use by cells and tissues. Additional 'EFA status markers' are also required to assess the EFA status of a given individual accurately. In general, if insufficient EFA are available to meet the EFA requirements, the body starts to synthesize 'surrogate fatty acids', which are barely detectable under normal conditions. The best known EFA status marker is mead acid, 20:3 $n-9$ (see Fig. 1). This fatty acid is hardly present under normal conditions, but its synthesis is promoted if insufficient parent EFA (linoleic and α-linolenic acids) are available to cover the need for derived EFA. Since the derived EFA inhibit mead acid synthesis, the presence of this latter fatty acid indicates a general shortage of all parent and derived EFA. Another suitable indicator of a general EFA shortage is the 'EFA status index', which is the ratio between all essential fatty acids (total amount of all the $n-3$ and $n-6$ fatty acids) and all non-essential unsaturated fatty acids (the sum of all $n-7$ and $n-9$ fatty acids). The higher this index, the better the EFA status.

Finally, if there is a functional shortage of cervonic acid, the body starts to synthesize the most comparable ($n-6$) LCPUFA, osbond acid (22:5 $n-6$, Ref. 6). Therefore, under steady-state conditions, the ratio between cervonic acid and osbond acid, the cervonic acid sufficiency index (CASI), is a reliable indicator of the functional CA status. Since the synthesis of osbond acid also depends on the availability of its precursor, adrenic acid (22:4 $n-6$, see Fig. 1), an increased ratio between osbond and adrenic acids is also a reliable marker of the CA status. This ratio is called the cervonic acid deficiency index (CADI) and increases with lower CA status.

EFA STATUS IN TERM AND PRETERM INFANTS AND DURING FETAL DEVELOPMENT

Fatty acid profile measurements revealed that the EFA status of the walls of the umbilical vein is higher than that of the umbilical arteries. In addition, the EFA status of umbilical arterial walls is lower than that of adult arterial tissue[6]. The umbilical vein is the supplying blood vessel to the fetus, whereas the umbilical arteries carry the blood away from the fetus, back to the placenta. Although direct comparisons with other fetal tissues have not yet been performed, and certain tissues may be preferred sites of EFA/LCPUFA uptake, the EFA status of umbilical venous and arterial walls probably reflects the EFA status of 'upstream' and 'downstream' fetal tissue, respectively. Consequently, the typical fatty acid profiles of umbilical veins and arteries indicate that the EFA status of the developing fetus is relatively low, and is lower in 'downstream' than in 'upstream' areas.

Preterm infants have an EFA status significantly lower than that of full term neonates (8). However, since the EFA status in cord blood of preterm infants at birth is no lower than that in cord blood obtained by fetal blood sampling at a comparable gestational age (Fig. 2), the low EFA status of preterm infants is probably a physiological situation and not a pathological condition. These comparative studies also demonstrate

***Figure 2** EFA status index, $\{\Sigma(n-6) + \Sigma(n-3)\}/\{\Sigma(n-9) + \Sigma(n-7)\}$, measured in plasma phospholipids isolated from blood obtained by fetal blood sampling of ongoing pregnancies (fetal ■), or from cord blood, immediately after delivery (neonatal ▩). Fetal samples at term were obtained by cord blood sampling after caesarean section, but before clamping the cord. n: number of samples. Data are mean ± SEM. Multiple regression analysis revealed that the fetal EFA status index during ongoing pregnancies is significantly (p = 0.04) lower than that after birth at a comparable gestational age*

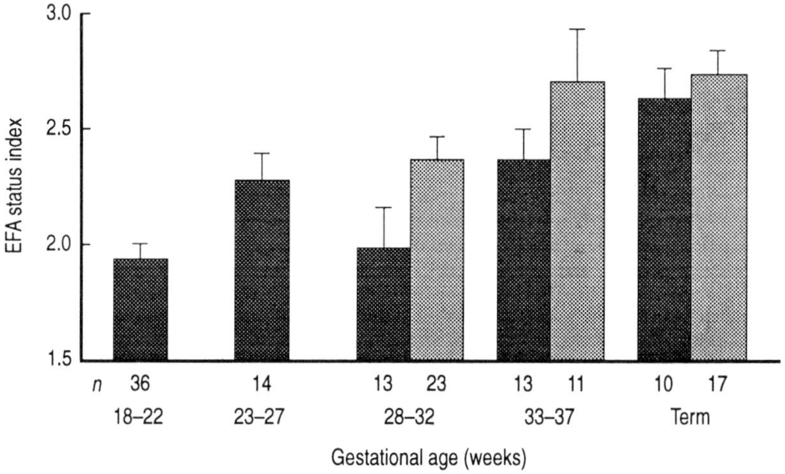

that the EFA status of the fetus is not stable during its development, but changes with the gestational age in a fatty-acid-specific way. Thus, the linoleic acid content of fetal tissue strongly decreases during early gestation (9), after which it increases slightly (as measured in umbilical plasma) during the second and third trimesters (Fig. 3). Arachidonic acid levels, however, slowly decrease as pregnancy progresses, whereas cervonic acid levels rise strongly during the last 2 months of gestation (Fig. 3). This implies that a proper interpretation of the fetal EFA status during its development needs correction for gestational age.

The 'upstream' EFA status of twins and triplets at birth, as indicated by the fatty acid profile of umbilical vein walls, is lower than that of newborn singletons (37), whereas the EFA status of their mothers does not differ at delivery (Fig. 4). The same holds for the CA status. This suggests that the neonatal FA status may be limited by the transplacental EFA supply.

MATERNAL EFA STATUS DURING PREGNANCY AND AT DELIVERY AND ITS CORRELATION WITH NEONATAL VALUES

In a prospective, longitudinal study (10), we demonstrated that the general maternal EFA status progressively declines during pregnancy, as indicated by a reduction

***Figure 3** The relative amounts (% of total fatty acids) of linoleic acid (LA, ●), arachidonic acid (AA, ○) and cervonic acid (CA, ■) in plasma phospholipids isolated from blood obtained by fetal blood sampling of ongoing pregnancies. Data from this cross-sectional study are given as mean ± SEM (shown only if permitted by symbol dimensions). n: number of samples. From multiple regression analyses it appeared that the time courses observed for the various fatty acids were significant: linoleic acid: $p < 0.05$; arachidonic acid: $p < 0.0001$; cervonic acid: $p < 0.0001$*

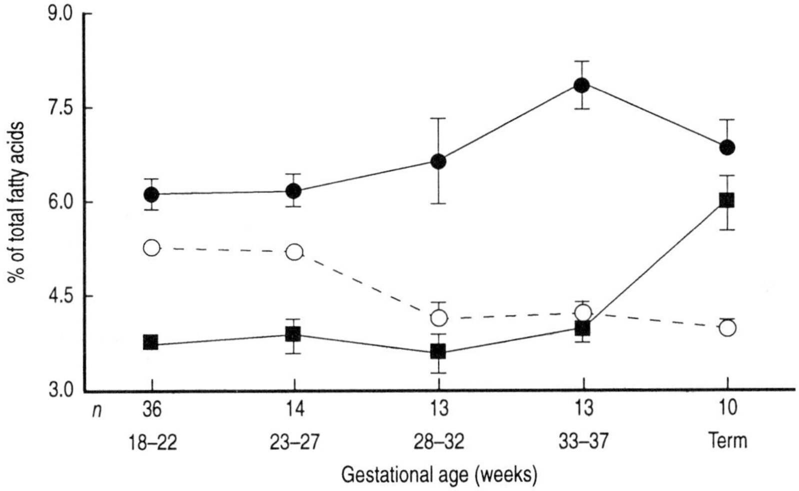

of the EFA status index. The same is observed for the cervonic acid status. About 6 months after delivery cervonic acid levels remain lower than during early pregnancy, indicating that normalization of the maternal CA status after delivery is rather slow. In a cross-sectional study, the CA level of maternal plasma phospholipids was significantly lower in multiparous than in primiparous women, at all stages of pregnancy. In addition, infants born to multiparous women appeared to have significantly lower CA levels than infants born to primiparous women (Fig. 5). Although these results need to be confirmed in a longitudinal study, they nonetheless indicate that pregnancy is associated with a high demand for EFA, and for CA in particular, and that today's maternal diet seems unable to meet this demand fully.

The same prospective, longitudinal study (10) confirmed earlier observations (11) that the EFA status of the neonate is strongly correlated to that of its mother. Since the EFA status of the mother appears to be determined by her EFA intake (12), it is highly likely that the neonatal EFA status can be altered by nutritional intervention during pregnancy.

***Figure 4** (a) EFA status index, $\{\Sigma(n-6)+\Sigma(n-3)\}/\{\Sigma(n-9)+\Sigma(n-7)\}$ and (b) cervonic acid deficiency index (22:5n−6/22:4n−6, CADI) in umbilical plasma phospholipids (cord) and maternal plasma phospholipids (mother) of singletons and multiples. Numbers in parentheses indicate number of samples from infants/mothers, respectively. Results are given as mean ± SEM. Differences between neonates are partly due to differences in gestational age; in the statistical analyses results were corrected for these differences*

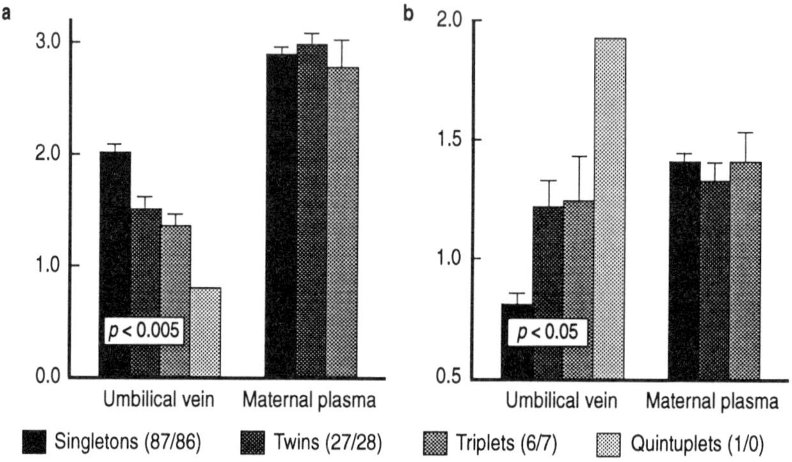

EFFECTS OF MATERNAL EFA SUPPLEMENTATION DURING PREGNANCY ON MATERNAL AND NEONATAL EFA STATUS

Two studies have suggested that an increase in neonatal EFA status can be achieved via the maternal diet. Supplementation of pregnant women with 2.7 g fish oil per day from the 31st week of pregnancy significantly increased their $n-3$ status, but lowered the amount of $n-6$ fatty acids in their plasma phospholipids. Compared with a control group, fish oil supplementation of pregnant women produced a significantly higher $(n-3)$ LCPUFA status of the cord plasma phospholipids of their newborns. This was associated with a comparable reduction of the amounts of $(n-6)$ LCPUFA. The fatty acid composition of phospholipids isolated from umbilical cord vessel walls was less pronounced: although levels of $(n-3)$ LCPUFA were significantly increased the decrease in total $(n-6)$ fatty acids was not statistically significant (13). These less pronounced effects could be due to a slower turnover of tissue than of plasma phospholipids, in combination with the relatively short period of supplementation.

When pregnant women were given linoleic acid rich food products (providing an additional 10 g of linoleic acid per day at the expense of saturated fatty acids) from the 20th week of pregnancy, the overall EFA status of maternal plasma phospholipids remained relatively stable, whereas in non-supplemented control women the EFA status continued to decrease until delivery. In addition, a slightly higher neonatal

***Figure 5** Amounts of cervonic acid (% of total fatty acids) in maternal plasma (a) and umbilical cord vessel phospholipids (b) of primiparous and multiparous women and their neonates. Numbers in parentheses refer to average parity number and number of mother/child pairs per group. Numbers within bars refer to p-values of comparisons between both groups*

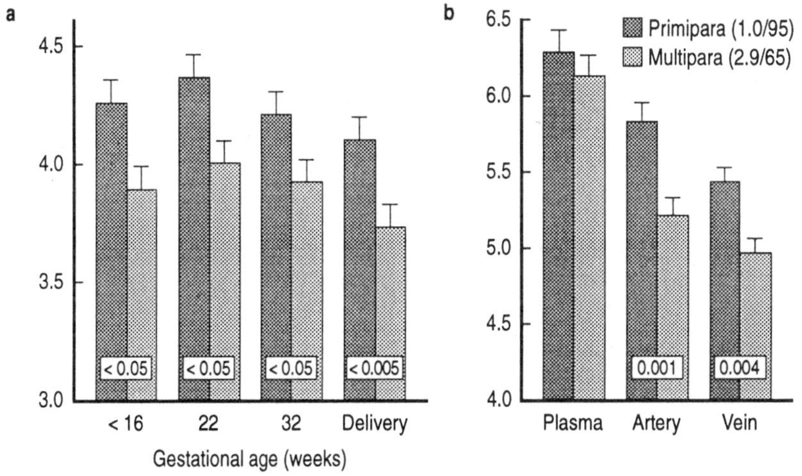

($n-6$) LCPUFA status was induced, although the neonatal ($n-3$) LCPUFA status was significantly lower (14). Therefore, if it is felt necessary to increase the neonatal EFA status, this would require maternal supplementation with a combination of both $n-6$ and $n-3$ EFA. Further studies are in progress to determine the optimal composition of such a dietary supplement.

RELATIONSHIP BETWEEN MATERNAL EFA INTAKE AND SOME MEASURES OF PREGNANCY OUTCOME

A few years ago, we observed that Inuit women living in coastal settlements were 2.6 times less likely to develop PIH than were Inuit women living in inland communities, where caribou rather than fish is the major source of animal fat. The lower risk of PIH was associated with a higher $n-3$ status of plasma phospholipids (15). These findings indicate that the consumption of $n-3$ LCPUFA may protect against PIH and that administration of $n-3$ LCPUFA to pregnant women could prevent the development of PIH. However, we also demonstrated that Dutch women suffering from PIH have higher amounts of $n-6$ and $n-3$ LCPUFA in their plasma phospholipids at delivery (16). This could theoretically imply that LCPUFA are involved in the aetiology of PIH, in which case supplementation with $n-3$ fatty acids would be contraindicated. To investigate a possible relationship between EFA status and PIH development, venous blood samples of pregnant women were collected three times during pregnancy and immediately after delivery. Each woman who was shown to have suffered from PIH was matched with three control women with an uncomplicated pregnancy and the EFA status of the plasma of these 'sets' was then determined, together with that of the phospholipids isolated from umbilical plasma and cord vessel walls of their neonates. The results of this study confirmed that, compared with uncomplicated pregnancy, PIH is associated with higher amounts of $n-6$ and $n-3$ LCPUFA in maternal plasma phospholipids at delivery. However these differences were not observed before the clinical manifestation of the disease. The altered EFA status in PIH is, therefore, a late phenomenon and is unlikely to have contributed to the pathogenesis of PIH (17). The higher maternal LCPUFA status observed at delivery after PIH was associated with lower amounts of their precursors, the parent EFA linoleic and α-linolenic acids. On the basis of this finding, we suggest that PIH is associated with a more active desaturation and elongation of parent into derived EFA, possibly to compensate for compromised placental transport. This putative compensatory mechanism appears to be successful, since the LCPUFA status of infants born after PIH was slightly better than that of infants born after uncomplicated pregnancy (17).

In the course of this prospective, longitudinal study, dietary intake data were collected by means of the dietary history technique and from repeated food frequency questionnaires; some overall pregnancy outcome data were also collected at delivery and by backwards multiple regression analysis, the relationship between nutrient intake and pregnancy outcome was evaluated after correction for maternal body weight, height, age, smoking habits, energy intake, education, parity, gestational diabetes, PIH, infant sex and gestational age. Because of multiple comparisons, the required p-value for significance was set at 0.005. A positive relationship was observed between the maternal intake of polyunsaturated fatty acids without linoleic acid and length of the newborn (38). In addition, a negative relationship was found between maternal linoleic acid intake and head circumference shortly after birth. Such a negative relationship was also observed between the amount of linoleic acid in umbilical plasma phospholipids and head circumference (10). No significant ($p>0.005$) correlations were observed for any of the dietary EFA with birth weight, placental weight/birth weight, and ponderal index (weight/length^3).

COMPARISON OF MATERNAL AND NEONATAL EFA STATUS BETWEEN SOME EUROPEAN COUNTRIES

To compare the EFA status during pregnancy in The Netherlands with that in other European countries, plasma from pregnant women and their neonates was collected in Finland (Dr Manninen, Tampere), England (Dr K. Godfrey, Southampton) and Hungary (Dr M. Antal, Budapest). It appeared that differences in maternal EFA status parameters between countries are often much more pronounced than the differences occurring during pregnancy within a given country (see Fig. 6 for the CA status). This finding casts some doubt on the functional implications of the observed decrease in EFA status during pregnancy and of the relatively low EFA status at birth. Since marked differences with respect to pregnancy outcome and infant development have

***Figure 6** Cervonic acid status (CASI; 22:6 n−3/22:5 n−6) in maternal plasma phospholipids during pregnancy and after delivery (a), and in cord plasma phospholipids (b) in different European countries. F: Finland (Helsinki), n = 50; GB: Great Britain (Southampton), n = 43; N: The Netherlands (Maastricht), n = 47; H: Hungary (Budapest), n = 48. Results are given as mean ± SEM (shown only if permitted by symbol dimensions). Regression lines fitted on the basis of mean values*

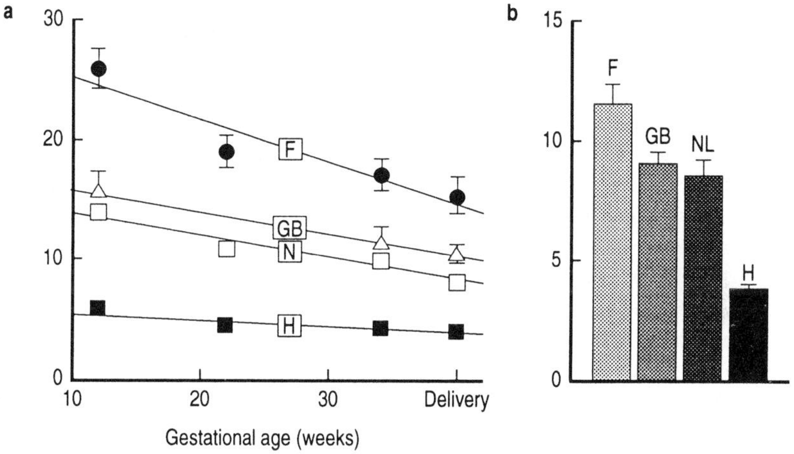

not been described between these countries, low values, as seen in Hungary, may still be functionally adequate.

POSSIBLE IMPLICATIONS OF THESE FINDINGS FOR NUTRITION DURING PREGNANCY

Dietary composition and genetic constitution in man

Although the western diet is constantly criticized for its high amount of fat at the expense of complex carbohydrates in general and fibre in particular, its essential fatty acid content has long been considered reasonably adequate. However, there are indications that in earlier times, when man lived as a hunter–gatherer, the diet was much richer in $n-6$ and $n-3$ LCPUFA obtained from structural fat in animal brain and from fish (18). Under these conditions there was no pressing need for a powerful enzyme system to desaturate and elongate parent EFA, because the resulting LCPUFA were readily available from the daily diet; consequently, this system remained rather inefficient (19). Due to fundamental changes in food production (the introduction of agriculture and animal domestication) and processing, however, major changes in the dietary patterns of man have occurred over the last millennia, leading to the consumption of large amounts of depot fat, relatively poor in LCPUFA (18). This would require a more active system to desaturate and elongate parent EFA, but environmentally driven changes in the human genetic constitution are known to occur very slowly only. The developments described above are likely to have resulted in a situation where the human fatty acid desaturation and elongation system no longer matches the dietary EFA composition, potentially leading to a relative LCPUFA shortage. This may be of particular importance under conditions where the LCPUFA requirement is high, e.g. during pregnancy and fetal development and during lactation and neonatal development.

Should the EFA content of the maternal diet be increased?

From our results summarized above, it may be felt necessary to increase the dietary EFA and/or LCPUFA intake of pregnant women in order to prevent a fall in EFA levels during pregnancy and to optimize the fetal EFA status. This latter may be of particular importance for infants born preterm, because these infants have a significantly lower EFA status than term neonates (8). In addition, their EFA status drops considerably during the first postnatal weeks, even when given breast milk (20), whereas during intrauterine life the fetal EFA status increases considerably during the same post-conceptional period (Fig. 2). Consequently, during the growth spurt of the brain, less LCPUFA is available for infants born preterm than for the intrauterine fetus of comparable postconceptional age. In preterm infants, the arachidonic and cervonic acid levels at birth are positively related to these LCPUFA levels at the expected date of delivery (21). A high intrauterine LCPUFA status would therefore be of particular benefit to a fetus born preterm, since a higher EFA status would result in a higher postnatal EFA status. Since CA (20) and AA (22,23) have been shown to be associated with increased prenatal and neonatal growth, our present observations indicate that enhancement of the fetal LCPUFA status may promote fetal and early neonatal development and may, thereby, improve the 'starting condition' of these infants and, thereby, their prognosis.

We have not yet investigated whether observed changes in the biochemical EFA status have functional consequences. Data from small mammals and primates, however, show that EFA in early nutrition have potentially important long term effects on behaviour and learning (24,25). Moreover, it has been demonstrated (26) that in rats 0.6 energy% of $n-3$ fatty acids is insufficient for optimum nervous system functions. Studies in primates show that the insufficient availability of $n-3$ fatty acids at a critical stage in retinal development results in a functional impairment of the retina (26,27). Comparable results have been obtained in newborn infants (28–31). Bjerve et al (32) found that the CA level in plasma phospholipids of 21 low-birth-weight infants at the corrected age of 1 year correlated with Bayley's psychomotor and mental development indices, and Carlson et al. (33) demonstrated that supplementation of infant formula with cervonic acid for 48 weeks results in a significant improvement of the Bayley score in infants born preterm. See also Chapters 10–13 of this volume.

Human breast milk contains small but significant amounts of AA and CA, which are not present in the usual formulas. Preterm infants, when given breast milk for at least 4 weeks have, on average, a significantly higher IQ at 7 years of age than comparable neonates reared on formula only (34), and breast-feeding of term neonates for at least 3 weeks is associated with significantly fewer neurological abnormalities at 9 years of age than bottle-feeding (35). These results were corrected for all possible confounders and both research groups suggest that the LCPUFA present in breast milk may have been responsible, at least in part, for the differences observed. However, fatty acid measurements to support this view were not performed.

Recently, Agostoni et al. (36) demonstrated that feeding human milk or an LCPUFA-fortified formula to term infants for 4 months is associated with a significantly higher neurodevelopmental quotient and with higher amounts of arachidonic and cervonic acids in plasma and erythrocyte phospholipids than feeding a 'classical' formula not containing additional LCPUFA. This study, also included in this volume as Chapter 14, clearly indicates that even in term infants, which have a higher EFA status at birth than preterm infants (8), the postnatal intake of additional LCPUFA promotes neurodevelopment. Whether the maternal EFA status during pregnancy and lactation affects infant development and prognosis is not yet known. We therefore intend to follow the children born during the course of our studies, in order to investigate whether their EFA status at birth and during early infancy is associated with their later neurological condition and their cognitive and psychomotor development.

ACKNOWLEDGEMENTS

The assistance of Hasibe Aydeniz, Anita Badart-Smook, Suzie Otto, Marianne Simonis, and Evelijn Zeydner (Department of Human Biology, Limburg University, Maastricht, The Netherlands) and of André de Jong and Taco Zijlstra (Analytical Biochemical Laboratory, Assen, The Netherlands) is gratefully acknowledged. Financial support for these studies was obtained from Nutricia Research, Zoetermeer, The Netherlands.

REFERENCES

1. Hornstra G 1982 Dietary fats, prostanoids and arterial thrombosis. Chapter 2: Biochemical physiology of dietary fats. Martinus Nijhoff Publishers, pp 15–29
2. Aaes-Jørgensen E 1961 Essential fatty acids. Physiol Rev 41:1–51
3. Bjerve KS 1991 Omega 3 fatty acid deficiency in man: implications for the requirement of alpha-linolenic acid and long-chain $\omega3$ fatty acids. World Rev Nutr Diet 66:133–142
4. Salem N, Ward GR 1993 Are $\omega3$ fatty acids essential nutrients for mammals? World Rev Nutr Diet 72:128–147
5. Hornstra G 1992 Essential fatty acids, pregnancy, and pregnancy complications: a roundtable discussion. In: Sinclair A, Gibson R (eds) Essential Fatty Acids and Eicosanoids. American Oil Chemists' Society, Champaign, IL pp 177–182
6. Neuringer M, Connor WE, Lin DS, Barstad L, Luck S 1986 Biochemical and functional effects of prenatal and postnatal $\omega3$ fatty acid deficiency on retina and brain in rhesus monkeys. Proc Natl Acad Sci USA 83:4021–4025
7. Hornstra G, van Houwelingen AC, Simonis M, Gerrard JM 1989 Fatty acid composition of umbilical arteries and veins: possible implications for the fetal EFA status. Lipids 24:511–517
8. Foreman-van Drongelen MMHP, Al MDM, van Houwelingen AC, Blanco CE, Hornstra G 1995 Comparison between the essential fatty acid status of preterm and full-term infants, measured in umbilical vessels. Early Hum Dev 42:241–251
9. van Houwelingen AC, Puls J, Hornstra G 1992 Essential fatty acid status during early human development. Early Hum Dev 31:97–111.

10. Al MDM, van Houwelingen AC, Kester ADM, Hasaart THM, de Jong AEP, Hornstra G 1995 Maternal essential fatty acid patterns during normal pregnancy and their relationship with the neonatal essential fatty acid status. Br J Nutr 74:55–68
11. Al MDM, Hornstra G, van der Schouw YT, Bulstra-Ramakers MTEW, Huisjes HJ 1990 Biochemical EFA status of mothers and their neonates after normal pregnancy. Early Hum Dev 24:239–248
12. Al MDM, Badart-Smook A, van Houwelingen AC, Hasaart THM, Hornstra G 1996 Fat intake of women during normal pregnancy: relationship with maternal and neonatal essential fatty acid status. J Am Coll Nutr 15:49–55
13. van Houwelingen AC, Sørensen JD, Hornstra G, Simonis MMG, Boris J, Olsen SF, Secher NJ 1995 Essential fatty acid status in neonates after fish-oil supplementation during late pregnancy. Br J Nutr 74:723–731
14. Al MDM, van Houwelingen AC, Badart-Smook A, Hornstra G 1995 Some aspects of neonatal essential fatty acid status are altered by linoleic acid supplementation of women during pregnancy. J Nutr 125:2822–2830
15. Popeski D, Ebbeling LR, Brown PB, Hornstra G, Gerrard JM 1991 Blood pressure during pregnancy in Canadian Inuit: community differences related to diet. Can Med Assoc J 145: 445–454
16. van der Schouw YT, Al MDM, Hornstra G, Bulstra-Ramakers MTEW, Huisjes HJ 1991 Fatty acid composition of serum lipids of mothers and their babies after normal and hypertensive pregnancies. Prost Leukotr Essential Fatty Acids 44:247–252
17. Al MDM, van Houwelingen AO, Badart-Smook A, Hasaart TH, Roumen FJ, Hornstra G 1995 The essential fatty acid status of mother and child in pregnancy-induced hypertension: A prospective longitudinal study. Am J Obstet Gynecol 172:1605–1614
18. Eaton SB, Konner N 1985 Paleolithic nutrition: a consideration of its nature and current implications. New Engl J Med 312:283–289
19. Chambaz J, Ravel D, Manier M-C, Pepin D, Mulliez N, Béréziat G 1985 Essential fatty acid interconversion in the human fetal liver. Biol Neonate 47:136–140
20. Foreman-van Drongelen MMHP, van Houwelingen AC, Kester ADM, de Jong AEP, Blanco CE, Hasaart THM, Hornstra G 1995 Long-chain polyene status of preterm infants with regard to the fatty acid composition of their diet: comparison between absolute and relative fatty acid levels in plasma and erythrocyte phospholipids. Br J Nutr 73:405–422.
21. Foreman-van Drongelen MMHP, Hornstra G, van Houwelingen AC, Kester ADM, Hasaart THM, Blanco CE 1995 Long chain polyunsaturated fatty acids in preterm infants: status at birth and its influence on postnatal levels. J Pediatr 126:611–618
22. Koletzko B, Braun M 1991 Arachidonic acid and early human growth: is there a relation? Ann Nutr Metab 35:128–131
23. Carlson SE, Werkman SH, Peeples JM, Cooke RJ, Tolley EA 1993 Arachidonic acid status correlates with first year growth in preterm infants. Proc Natl Acad Sci USA 90:1073–1077
24. Lucas A 1991 Programming by early nutrition in man. Ciba Found Symp 157:38–55
25. Wainwright PE 1992 Do essential fatty acids play a role in brain and behavioral development? Neurosci Biobehav Rev 16:193–205
26. Okuyama H 1992 Essentiality of α-linolenic acid for the nervous system functions in rats and mice: Food-behavior link. In: Sinclair A, Gibson R (eds) Essential Fatty Acids and Eicosanoids. American Oil Chemists' Society, Champaign, IL pp 153–155
27. Neuringer M, Connor WE, van Petten C, Barstad L 1984 Dietary omega-3 fatty acid deficiency and visual loss in infant rhesus monkeys. J Clin Invest 73:272–276
28. Innis SM 1991 Essential fatty acids in growth and development. Prog Lipid Res 30:39–103
29. Uauy R, Birch E, Birch D, Peirano P 1992 Visual and brain function measurements in studies of n-3 fatty acid requirements of infants. J Pediatr 120:168–180
30. Birch DG, Birch EE, Hoffman DR, Uauy R 1992 Dietary essential fatty acid supply and visual acuity development. Invest Ophthalmol Vis Sci 33:3242–3253
31. Makrides M, Simmer K, Goggin M, Gibson RA 1993 Erythrocyte docosahexaenoic acid correlates with visual response of healthy term infants. Pediatr Res 33:425–427
32. Bjerve KS, Brubakk AM, Fougner KJ, Johnson H, Midthell K, Vik T 1993 Omega-3 fatty acids: essential fatty acids with important biological effects, and serum phospholipid fatty acids as markers of dietary ω-3 fatty acid intake. Am J Clin Nutr 57:801S–806S

33. Carlson SE, Wilson WE 1994 Docosahexaenoic acid supplementation of preterm infants: effect of the 12-month Bayley development. Pediatr Res 34:20a
34. Lucas A, Morley R, Cole TJ, Lister G, Leeson-Payne C 1992 Breast milk and subsequent intelligence quotient in children born preterm. Lancet 339:261–264
35. Lanting CI, Fidler V, Huisman M, Touwen BCL, Boersma ER 1994 Neurological differences between 9-year-old children fed breast-milk or formula-milk as babies. Lancet 344:1319–1322
36. Agostoni C, Trojan S, Bellù R, Riva E, Giovannini M 1995 Neurodevelopmental quotient in healthy term infants at 4 months and feeding practice: the role of long-chain polyunsaturated fatty acids. Pediatr Res 38:262–266
37. Foreman-van Drongelen MMHP, Zeijdner EE, van Houwelingen AC, Kester ADM, Al MDM, Hasaart THM, Hornstra G. Essential fatty acid status measured in umbilical vessel walls of infants born after a multiple pregnancy. Early Hum Devel. Accepted for publication
38. Badart-Smook A, van Houwelingen AC, Al MDM, Kester ADM, Hornstra G. Nutrient intakes of Dutch women during pregnancy and relationship with pregnancy outcome. Manuscript submitted

5 Experimental approaches to improve intrauterine growth retardation

F.C. BATTAGLIA

ABBREVIATIONS

FGR	fetal growth restriction
GDR	glucose disposal rate
MPE	molar percentage enrichment
PI	pulsatility index

INTRODUCTION

The topic of potential therapy for pregnancies complicated by fetal growth restriction (FGR) is a timely one considered from several viewpoints. First, advances in ultrasound imaging and better normative data for the growth rate in fetal biparietal diameter, abdominal circumference and other measurements of bone growth mean that obstetricians are now able to diagnose FGR much earlier in gestation. If the data reported by Wolf using three-dimensional imaging of placental volume are confirmed (1,2), diagnosis will be possible even earlier in gestation. Second, clinical mortality and morbidity data have shown that the fetuses recognized as having FGR relatively early in gestation (i.e. <30 weeks) are the group to which therapeutic intervention should be directed. Fetuses in whom the diagnosis can be made early in gestation are the most severely affected and have a high neonatal as well as fetal mortality (3–5). If therapeutic approaches were available early in gestation, such pregnancies might be extended to a gestation length where survival rates are excellent (>34 weeks).

Early recognition of FGR has another theoretical advantage in that therapeutic strategies can be induced at a stage when placental growth and development are still not complete.

In this paper I should like to discuss two different, but related, areas of research that bear on the topic of FGR and its therapy; one addresses studies of an animal model of severe FGR and the other topic is directed at reviewing some of our current clinical research in pregnancies complicated by FGR.

HEAT STRESS MODEL OF FGR

This model was developed at the University of Colorado by Dr Alan Bell, Dr Randall Wilkening and Dr Giacomo Meschia (6). The protocol involves exposing pregnant

J.G. Bindels, A.C. Goedhart and H.K.A. Visser (eds), Recent Developments in Infant Nutrition, 64–72.

sheep to 40°C and 50% humidity for 13 h each day, the remaining 11 h being spent at normal room temperature. Sheep exposed to this temperature cycle from the 35th to the 100th day of gestation show striking fetal and placental growth restriction. Using this model, we can study the effects upon placental transport of nutrients to the fetus as a consequence of the small placenta.

Glucose transport

Placental glucose transport expressed as mg/min is reduced consistent with the small size of the placenta, but when expressed per gram of placental tissue fetal glucose uptake is normal, suggesting that there may be little adaptation of placental glucose transporters to reduced placental size (7). Fetal adaptation to the smaller placenta results in a lowered fetal arterial glucose concentration and increased maternal–fetal glucose concentration gradient. This increases placental glucose transport, partially correcting the reduced glucose transport. As we shall discuss later, a lower fetal glucose concentration and increased maternal–fetal glucose gradient is also found in clinical studies of FGR pregnancies.

Amino acid studies

We have carried out a number of studies of amino acid fluxes and net umbilical uptake in lambs with FGR (8,9). I shall concentrate on the data obtained for two essential amino acids, leucine and threonine. There is a decreased transplacental flux of leucine from the maternal to the fetal circulation (8). Unlike glucose transport, this is true even if the data are expressed per gram of placenta. Thus, in this model of FGR, the fetus is compromised both by a reduced placental size and by a decreased placental efficiency for the transport of leucine. The growth-retarded fetus maintains significantly lower leucine concentrations than normal fetuses. The advantage of this derives from the fact that both fetal leucine oxidation and the back flux of leucine from fetal plasma into the placenta are a function of leucine concentration. Thus, a lower fetal leucine concentration redirects a larger fraction of leucine entry from the placenta into the fetus towards fetal protein synthesis and away from oxidation and back flux.

Studies with stable isotope-labelled threonine have led to similar conclusions. Transplacental threonine flux from the maternal circulation to the fetal circulation is reduced both in absolute terms and when expressed per gram of placenta. As with leucine, its oxidation rate and back flux are also reduced. Data obtained by Lane et al. in a rat model of FGR showed that the neutral amino acid transporter is down-regulated, consistent with the flux data in FGR fetal lambs (9). This specific impact on amino acid transport within the FGR placenta also appears to be a characteristic of the FGR human placenta.

CLINICAL STUDIES OF FGR PREGNANCIES

Fetal surveillance

Changes in technology, driven largely by improvements in computer software which reduce background noise, are leading to the introduction of a vast array of new

Table 1 Classification of FGR pregnancies

	Velocimetry	*Fetal heart rate*	*% acid-base complications**
Group 1	Normal	Normal	0
Group 2	P.I. ↑	Normal	17
Group 3	P.I. ↑	Abnormal	63

See Ref. 10 for details of velocimetry (umbilical artery blood flow) and fetal heart rate coding
* = pregnancies in which cordocentesis sample showed evidence of acidosis and/or lacticacidaemia
P.I. = pulsatility index (systolic velocity minus diastolic velocity divided by mean velocity)

information for obstetricians. The problem is to define the role, if any, of new techniques and to evaluate the interpretation of such data. As a first step to assessing clinical severity and to defining which groups of FGR fetuses are most in need of therapeutic intervention we proposed subdividing FGR pregnancies into three groups based upon clinical information which is readily available (10) (Table 1). All FGR pregnancies, recognized by ultrasound exams, are subdivided into these three groups based upon velocimetry (umbilical artery blood flow) and fetal heart rate data. We can then ask whether other biochemical or physiological parameters correlate with this definition of clinical severity. Table 1 demonstrates that for acid–base data collected at the time of cordocentesis the frequency of acidosis and lacticacidaemia increased from group 1 to group 3 (10).

Carbohydrate studies in FGR pregnancies

The relationship of conceptus mass to maternal utilization rate

It has been a commonly held assumption that the metabolic demand upon the pregnant woman increases in multiple pregnancies, reflecting the increased nutrient requirements of the larger conceptus mass. However, this hypothesis has never been tested. By selecting pregnancies with FGR, normal fetal growth and multiple pregnancies we were able to study patients whose conceptus mass ranged from 0.9 kg to 7.4 kg (11). There was a highly significant relationship among conceptus mass (W), maternal glucose concentration (G) and maternal plasma glucose disposal rate (GDR) ($p<0.01$). A multiple linear regression analysis yielded the following regression, which is presented graphically in Figure 1.

$$\mathrm{GDR} = -0.157 + 0.178W + 0.932(G) \pm 0.24$$

From this analysis one can estimate the glucose utilization rate of the human conceptus as between 10 and 15 mg/kg/min.

Maternal–fetal glucose relationships in FGR pregnancies

In a series of studies which utilized cordocentesis data, it was possible to demonstrate that the glucose concentration difference across the placenta increased in normal

Figure 1 Impact of conceptus mass and fasting glucose concentration on maternal glucose disposal rate (GDR) in singleton (○), twin (●) and triplet (■) pregnancies. Taken from Ref. 11

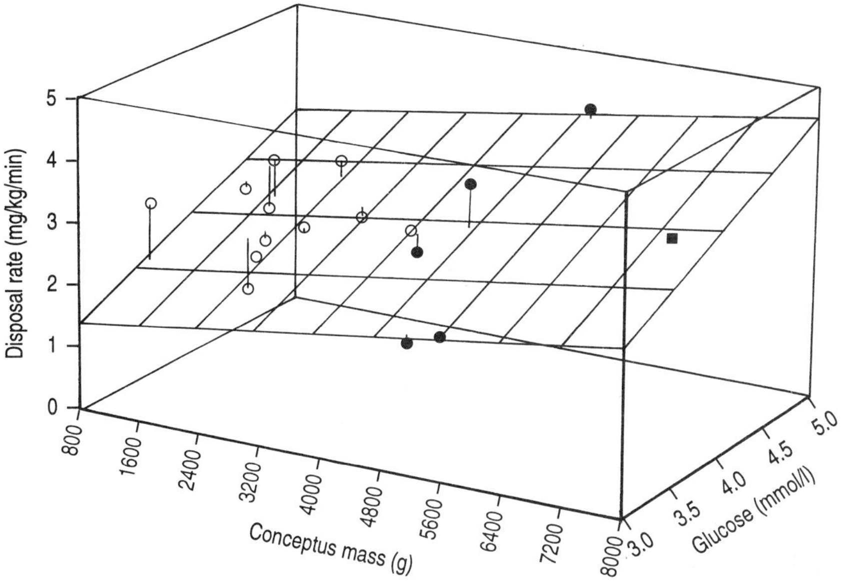

pregnancies as gestation advanced (12,13). This was also observed in studies of normal ovine pregnancy (7). It would appear that this is a general mechanism to achieve an increased fetal glucose delivery across the placenta to meet the demands of the increasing fetal mass. This is in addition to the large increase in placental glucose transport capacity during gestation (14), presumably reflecting up-regulation of glucose transporters within the placenta. More recently, the maternal–fetal glucose concentration difference has been compared for FGR pregnancies by subgroups (Table 1). The concentration difference is significantly elevated from group 1 through group 3 (13).

Fetal gluconeogenesis in FGR pregnancies

It has been clearly established that the normal ovine fetus has no appreciable rate of gluconeogenesis (15). This is readily understandable: fetal gluconeogenesis would be self-defeating since it would reduce the delivery of glucose from the maternal to the fetal circulation by decreasing the transplacental gradient. However, in animals, as umbilical glucose delivery decreases, fetal glucose production increases (15). In order to determine whether FGR fetuses show any evidence of glucose production, we infused stable isotopically labelled glucose into the maternal circulation until a steady

***Figure 2** Maternal (▽) and fetal (▲) enrichments in a single study. The fetal sample was collected 63 min after the start of the maternal infusion. Taken from Ref. 16*

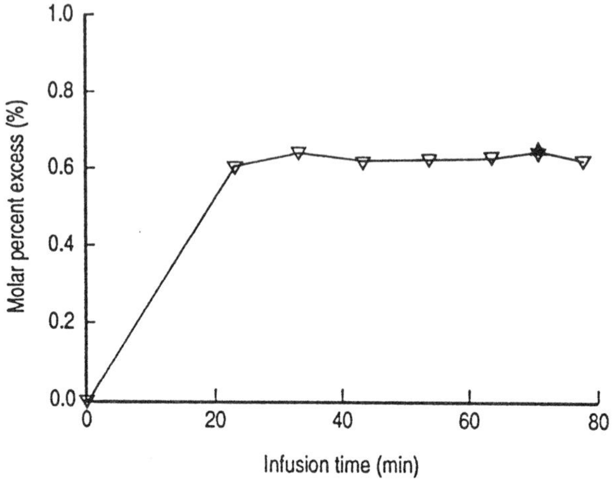

***Figure 3** Fetal and maternal enrichment (MPE) plotted against an identity line. Regression analysis of fetal MPE vs maternal MPE was as follows: fetal MPE = −0.03 + 1.06 maternal MPE. $R^2 = 0.98$, $p < 0.001$. This was not significantly different from the identity line. Taken from Ref. 16*

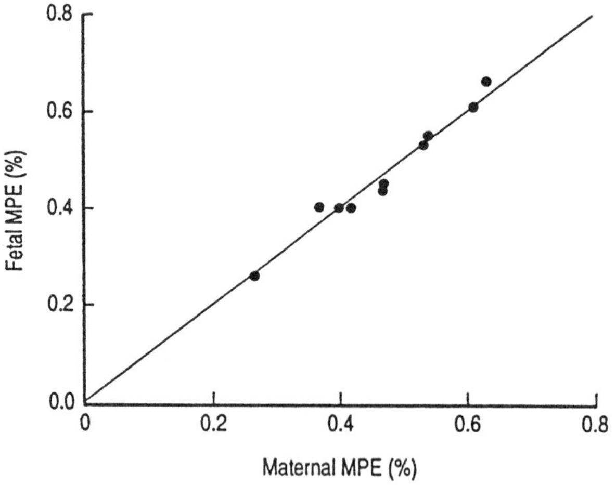

state enrichment was obtained in maternal plasma (Fig. 2). This was performed in 10 women with FGR pregnancies who were scheduled for cordocentesis after an overnight fast (16). No evidence of dilution of fetal plasma glucose enrichment was found in any of these pregnancies (Fig. 3).

***Figure 4** Umbilical venous plasma concentration of leucine, isoleucine and valine (μmol/l) in growth retarded fetuses who underwent transabdominal cord sampling (■) compared with midterm cordocentesis data (▩) and data collected at caesarean section near term (▤). See also Ref. 18*

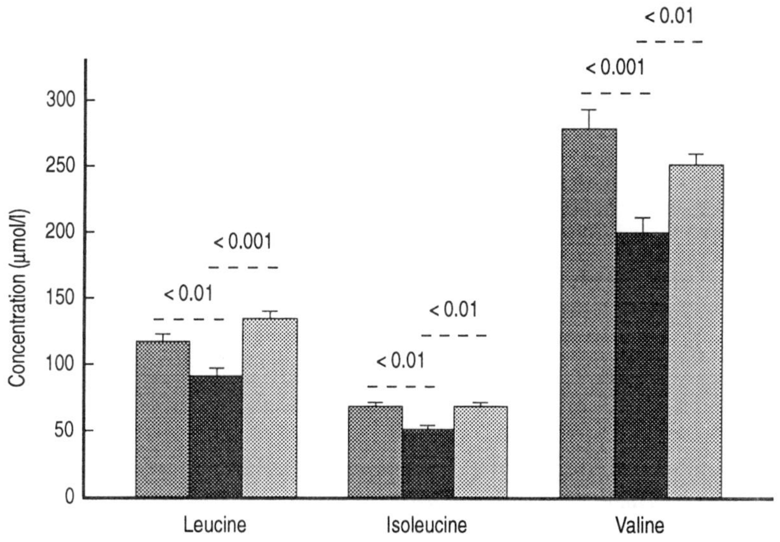

Amino acid studies in FGR pregnancies

One of the first studies completed examined the umbilical veno–arterial differences for plasma amino acid concentrations (17). Since this involved sampling both the vein and the artery the studies could only be conducted at caesarean section. In order to minimize variability due solely to blood flow changes, the data for each amino acid were expressed per μmol O_2 content, i.e. μmol amino acid/μmol O_2. This study showed a marked decrease both in amino acid concentrations and in their uptake for FGR pregnancies compared to controls.

In a later study (18) cordocentesis data were used to reflect fetal amino acid concentrations in a relatively undisturbed state and to determine whether fetal amino acid concentrations were reduced in FGR pregnancies even many weeks before delivery. Figure 4 summarizes some of these data showing that FGR fetal concentrations were reduced whether they were compared to mid-term cordocentesis data from normal pregnancies or data near term collected at caesarean section.

More recently, we have begun to use stable isotopically labelled amino acids to investigate in vivo transport characteristics of the human placenta (19). Thus far the data have been collected only from normal pregnancies. The protocol for these studies is shown in Figure 5. A bolus of L[1-^{13}C]leucine and L[1-^{13}C]glycine is given

Figure 5 ***A bolus of L-[1-^{13}C]leucine is injected into the peripheral circulation of the mother (point B). Maternal 'arterialized' samples are taken at 2 min intervals and a fetal cordocentesis sample is obtained at point F***

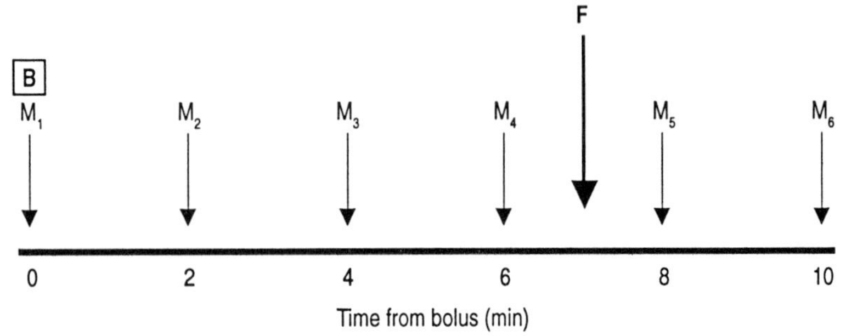

Figure 6 ***Fetal/maternal enrichment ratio for leucine (■) and glycine (●) vs time of fetal sampling. Non-linear regression lines and their 95% confidence limits are: leucine $(F/M)^{MPE} = 0.94t/(60.6 + t) \pm 1.91$; glycine $(F/M)^{MPE} = 0.23t/(153.4 + t) \pm 0.036$. These curves were significantly different ($p < 0.0001$, Wald test). Taken from Ref. 19***

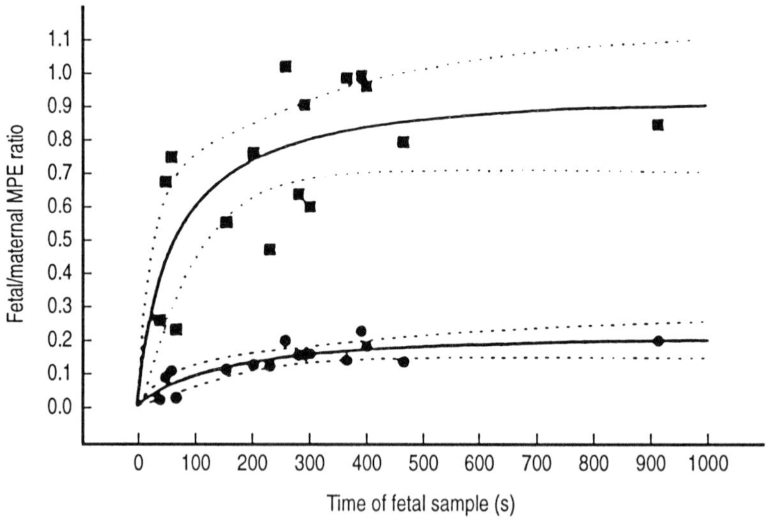

into the peripheral circulation of the mother, and maternal 'arterialized' samples are obtained at 2 min intervals. Cordocentesis is carried out for clinical indications some time during the 10 min window following the bolus of tracers. The fetal sample and maternal samples are analysed to determine both molar percentage enrichments

(MPE) and leucine, glycine concentrations. Figure 6 presents the fetal/maternal MPE ratio for a series of patients in which the fetal sample was obtained at varying times during the 10 min window. The leucine fetal/maternal MPE ratio was significantly ($p<0.001$) higher than that of glycine. These studies are currently being extended to other amino acids and to pregnancies complicated by FGR.

Given the data reviewed above, the animal studies presented earlier and the reports from several centres describing reduced amino acid transport across microvesicles of the human placenta (20,21), a reasonable hypothesis is that amino acid delivery to the FGR fetus is reduced, both because of a reduced placental growth and because of a change within the placental membrane, possibly a down-regulation of amino acid transporters. In any case, the large body of anecdotal information in the clinical literature (22–29) supports the hypothesis that despite these changes within the placenta it may still be possible to increase the placental delivery of amino acids by a sustained increase of maternal concentrations. To do so safely for all essential amino acids may require adjustments in the molar ratios of these amino acids in the solutions currently available for clinical intravenous use. Therapeutic intervention may also include efforts directed at improving placental perfusion, both uterine and umbilical. This may include the use of vasodilators or of chronic administration of supplemental oxygen. Hopefully, as fetal surveillance techniques improve, obstetric management can avoid the risk of increasing fetal mortality by attempting to prolong gestation during therapeutic management.

ACKNOWLEDGEMENTS

I wish to acknowledge the contributions of my colleagues. Dr Wilkening and Dr Meschia developed the model of fetal growth restriction produced by environmental heat and humidity. They were involved, along with post doctoral fellows listed in the literature cited, in all of the basic studies I have presented. The clinical studies were carried out in the Obstetrics-Gynecology department of Dr Giorgio Pardi's at the Hospital San Paolo in Milan. Dr Marconi and Dr Cetin were involved in most of the biochemical and tracer studies, Dr Ferrazzi in the ultrasound and velocimetry studies and Dr Buscaglia in the cordocentesis studies.

REFERENCES

1. Wolf H, Oosting H, Treffers P 1989 Second-trimester placental volume measurement by ultrasound: Prediction of fetal outcome. Am J Obstet Gynecol 160:121–126
2. Wolf H, Oosting H, Treffers P 1989 A longitudinal study of the relationship between placental and fetal growth as measured by ultrasonography. Am J Obstet Gynecol 161:1140–1145
3. Ounsted M, Moar V, Scott WA 1981 Perinatal morbidity and mortality in small for dates babies: the relative importance of some maternal factors. Early Hum Dev 5:367–375
4. Amon E, Sibai BM, Anderson GD, Mabie WC 1987 Obstetric variables predicting survival of the immature newborn (less than or equal to 1000 gm): a five-year experience at a single perinatal center. Am J Obstet Gynecol 156:1380–1389
5. Schauseil-Zipf U, Hamm W, Stenzel B, Bolte A, Gladtke E 1989 Severe intra-uterine growth retardation: obstetrical management and follow up studies in children born between 1970 and 1985. Eur J Obstet Gynecol Reprod Biol 30:1–9

6. Bell AW, Wilkening RB, Meschia G 1987 Some aspects of placental function in chronically heat-stressed ewes. J Dev Physiol 9:17–29
7. Thureen PJ, Trembler KA, Meschia G, Makowski EL, Wilkening RB 1992 Placental glucose transport in heat-induced fetal growth retardation. Am J Physiol 263:R578–R585
8. Ross J, Fennessey P, Wilkening R, Battaglia F, Meschia G 1996 Placental transport and fetal utilization of leucine in a model of fetal growth retardation. Am J Physiol 270:E491–E503
9. Anderson A. Threonine metabolism in control and IUGR fetal sheep. In preparation
10. Pardi G, Cetin I, Marconi AM, Lanfranchi A, Ferrazzi E, Bozetti P, Buscaglia M, Battaglia FC 1993 Diagnostic value of blood sampling in fetuses with growth retardation. N Engl J Med 328:692–696
11. Marconi AM, Davoli E, Cetin I, Lanfranchi A, Zerbe G, Fanelli R, Fennessey PV, Pardi G, Battaglia FC 1993 The impact of conceptus mass upon glucose disposal rate in pregnant women. Am J Physiol 264:E514–E518
12. Bozzetti P, Ferrari MM, Marconi AM, Ferrazzi E, Pardi G, Makowski EL, Battaglia FC 1988 The relationship of maternal and fetal glucose concentrations in the human from mid gestation until term. Metabolism 37:358–363
13. Marconi AM, Paolini C, Cetin I, Perugino G, Buscaglia M, Battaglia FC, Pardi G 1996 The impact of gestational age and of intrauterine growth upon the maternal–fetal glucose concentrations difference. Obstet Gynecol (in press)
14. Molina RD, Meschia G, Battaglia FC, Hay WW 1991 Gestational maturation of placental glucose transfer capacity in sheep. Am J Physiol 261:R697–R704
15. Hay WW Jr, Sparks JW, Wilkening RB, Battaglia FC, Meschia G 1984 Fetal glucose uptake and utilization as functions of maternal glucose concentration. Am J Physiol 246:E237–E242
16. Marconi AM, Cetin I, Davoli E, Baggiani AM, Fanelli R, Fennessey PV, Battaglia FC, Pardi G 1993 An evaluation of fetal glucogenesis in intrauterine growth retarded pregnancies: steady state fetal and maternal enrichments of plasma glucose at cordocentesis. Metabolism 42: 860–864
17. Cetin I, Marconi AM, Bozzetti P, Sereni LP, Corbetta C, Pardi G, Battaglia FC 1988 Umbilical amino acid concentrations in appropriate and small for gestational age infants: a biochemical difference present in utero. Am J Obstet Gynecol 158:120–126
18. Cetin I, Corbetta C, Sereni LP, Marconi AM, Bozzetti P, Pardi G, Battaglia FC 1990 Umbilical amino acid concentrations in normal and growth retarded fetuses sampled in utero by cordocentesis. Am J Obstet Gynecol 162:253–261
19. Cetin I, Marconi AM, Baggiani AM, Buscaglia M, Pardi G, Fennessey PV, Battaglia FC 1995 In vivo placental transport of glycine and leucine in human pregnancies. Pediatr Res 37:571–575
20. Dicke JM, Henderson G 1988 Placental amino acid uptake in normal and complicated pregnancies. Am J Med Sci 295:223–227
21. Mahendran D, Donnai P, Glazier JD, D'Souza SW, Boyd R, Sibley C 1993 Amino acid (system A) transporter activity in microvillous membrane vesicles from the placentas of appropriately and small for gestational age babies. Pediatr Res 34:661–665
22. Rivera-Alsina ME, Saldana LR, Stringer CA 1984 Fetal growth sustained by parenteral nutrition in pregnancy. Obstet Gynecol 64:138–141
23. Beischer NA, O'Sullivan EF 1972 The effect of rest and intravenous infusion of hypertonic dextrose on subnormal estriol excretion in pregnancy. Am J Obstet Gynecol 113:771–778
24. Webb GA 1980 The use of hyperalimentation and chemotherapy in pregnancy: a case report. Am J Obstet Gynecol 137:263–266
25. Hatjis CG, Meis PJ 1985 Total parenteral nutrition in pregnancy. Obstet Gynecol 66:585–589
26. Main AN, Shenkin A, Black WP, Russel RI 1981 Intravenous feeding to sustain pregnancy in a patient with Crohn's disease. Br Med J 283:1221–1222
27. Smith C, Rufleth P, Phelan JP, Nelson KJ 1981 Long term enteral hyperalimentation in the pregnant woman with insulin-dependent diabetes: a report of two cases. Am J Obstet Gynecol 141:180–183
28. Lavin JP, Gimmon Z, Miodovnik M, von Meyenfeldt M, Fischer JT 1982 Total parenteral nutrition in a pregnant insulin-requiring diabetic. Obstet Gynecol 59:660–664
29. Tresadern JC, Falcone GF, Turnberg LA, Irving MH 1983 Successful completed pregnancy in a patient maintained on home parenteral nutrition. Br Med J 286:602–603

Discussion Session 1

CHAIR: P.J.J. SAUER

CHAPTER 1

Karlberg Could I ask you about the postnatal growth? Is the catch up growth or failure to catch up important for the outcome later in life?

Barker In Hertfordshire the highest rate of coronary heart disease occurred in men who were small at birth and then did not catch up but remained small. In women it is the reverse: it is highest in women who were small at birth and then caught up. I think ideas about catch up growth in some way being harmful long term are very important and we will see a lot more in the coming 3–4 years.

Battaglia In your talk you discussed the fact that these relationships held if you took into account gestational age. But nowhere in the talk did you present the data on gestational age. We have had a somewhat different view of intrauterine growth retardation (IUGR), and that is that proportionate IUGR is simply the baby you recognize as IUGR very early in gestation, before fat depositional carcass growth is predominant. Disproportionate IUGR is simply the baby that is recognized later in gestation. For me, it would be really interesting to see data that look at this not just in birth weight but birth weight versus gestational age.

Barker I entirely accept that. Partly I do not mention it because in Hertfordshire as opposed to the other two places, Sheffield and Preston, we do not know duration of gestation. The earlier studies did not take on board what you have said, but the later studies have made it clear that it is the small disproportionate term baby.

Okken How did you exclude in your studies the role of postnatal feeding, either formula or breast-feeding, or feeding practice. My mother told me that when I was born I was not given any feed for 24–48 h, as that was common practice in Holland. Is this kind of thing being recorded, and how could you exclude this?

Barker The data we have on these men and women in Hertfordshire said whether they were exclusively breast-fed, whether they were breast- and bottle-fed or whether they were exclusively bottle-fed, and also whether they were weaned at one year. That is all we know. That does not answer the point you made. But we see that the relationship between weight at one year and later coronary heart disease is present in each feeding group. Over and above that is this curious effect of prolonged breast-feeding which took place in 20% of the people, because we are going back 60–70 years.

Sauer May I add one question to that? Do you know the weight of the children that were breast-fed for more than a year. Were they lighter than the supplemented group?

J.G. Bindels, A.C. Goedhart and H.K.A. Visser (eds), Recent Developments in Infant Nutrition, 73–78.

Barker Yes, the babies that were breast-fed for more than a year were somewhat lighter, but not much. They also had fewer teeth, which might be related. The midwives had recorded how many teeth they had at one year.

Sauer Was there a relationship with outcome when you just take that subgroup?

Barker Was their higher expectation related to the fact that they are a bit smaller? No, not at all.

Wharton It is a point made to Dr Barker, but the others may wish to comment. I wonder whether it might be safer at this stage instead of saying fetal undernutrition to be a little less specific and say fetal growth retardation. I think that is what you have shown. Many people, if they say fetal undernutrition, take it that that therefore means maternal undernutrition and I think we are a long way away yet from that particular step.

Barker Yes, but I would have thought that the lecture from Dr Battaglia we just heard tells us that what goes into the mother does not go into the fetus. I am scrupulous to avoid maternal undernutrition. I am just talking about the fetus.

Wharton Would it not be better then to call it fetal growth retardation?

Barker I can conceive that a fetus may make adaptations which allow it to proceed with its growth successfully, but these adaptations might have a long term cardiovascular cost. I suggest that there is a possibility that an infant might be born without growth retardation but having staged a series of successful adaptations which, like many fetal adaptations, have long term costs.

CHAPTER 2

Pencharz Some years ago I was involved in the Montreal study. This achieved a difference of about 120 g in birth weight by intervening in poor women, which made a difference of about 50–60% in low birth weight and neonatal mortality. The important thing with respect to targeting, if you look at your Gambian women, is that you still only have a mean birth weight of 3 kg, and I think you should be aiming for 3.5 kg. If you look at the Dutch famine in 1944, the decrease of birth weight was about 300 g. The same was found in the famine of Leningrad. If the mother is well nourished at the beginning of pregnancy she does not really need supplements. The baby will weigh about 3.5 kg if the mother is undernourished or underweight during pregnancy. I've certainly been convinced one needs to intervene. But the mother has to be undernourished to start with, and you have to make an adequate intervention. I think the intervention that you are able to achieve in the Gambia is still far too small.

Poppitt I am often approached with this kind of question. Certainly the ideal situation is that everybody produces babies of about 3.5 kg. It is a little more difficult perhaps to agree that this is our goal in the developing countries, simply for reasons I mentioned at the beginning. Consider, for example, obstructed labour. A situation that in the West may be completely straightforward may require operative intervention, and that opportunity is not available in many of the developing countries. In the past we have been criticized by people saying that we are producing superbabies and that the mothers will have problems in giving birth. I think 3.5 kg is a wonderful aim but I am not really sure that through our supplementation that is what we should be aiming for.

Battaglia You bring up a couple of interesting points. Some 30 years ago I published a study on teenage pregnancies in Baltimore and we found that though there was a small pelvis in a 13-year-old child there was no functional cephalo-pelvic disproportion because these girls tended to produce smaller babies that would be delivered a few weeks premature. Recent studies in the USA basically confirm that finding so I am not sure that is the reason why you are having operative problems in functional cephalo-pelvic disproportion in teenage pregnancy. In your data you never addressed the issue of gestational length. Did length of gestation change in the supplemented group?

Poppitt It did not change significantly, no. You are suggesting that perhaps a lot of the low birth weight babies are potentially preterm, and the only reason that their size increases is because you are prolonging gestation for several weeks.

Comment from the audience A problem is that many African women do not know their length of gestation.

CHAPTER 4

Donzelli Do you think that we can improve fetal growth by giving essential fatty acids to fetuses with intrauterine growth retardation?

Hornstra This is a very difficult question to address. We did see a positive relation between the content of docosahexaenoic acid in the umbilical arteries of preterm babies and three measures of fetal growth: birth weight, head circumference and birth length. From that you could suggest that if you would have been successful in increasing the cervonic acid status of these fetuses, they might have been bigger, larger, and that may have improved their prognosis. Whether or not this also would be successful in growth retarded babies I do not know.

Gibson I have great trouble with two concepts that are the basis of your work. One is the concept of using the word 'fetal' when you are talking about your cord tissue. We also find low levels of essential fatty acids in the cord tissue but not in the infant itself, of those infants that we've been able to follow. You tend to use all that as one expression and I feel a little bit uncomfortable about that.

Hornstra If I say fetal, then I mean tissue from the umbilical cord. There are two transportation routes in the cord, there is an afferent blood vessel and an efferent blood vessel. If you want to relate what happens in the fetus to these two routes of transportation, I think the only possibility is to take the cord tissue content. I agree that it is not necessarily so that what you find in the umbilical cord is a reflection of everywhere else in the fetal body. What we do see is that at the entrance, so at the venous side, the essential fatty acid status of the tissue is always better, biochemically speaking, than at the arterial side, so where the blood leaves the fetal body. I agree, it doesn't tell you directly about the other tissues of the fetus, but it is the best we have.

Sauer Did you ever measure whether there is a difference between veins and arteries, in other than umbilical veins and arteries in babies that had just died?

Hornstra No. We used surgical material and there you see hardly any difference between artery and vein.

Gibson We have looked at veins in babies that have died and found no fatty acids, which brings me to the marker fatty acids themselves. Mead acid for example, you have claimed as a marker of $n-6$ fatty acid deficiency, but in fact mead acid can appear as a consequence of other things, other than low intake of linoleic acid. Populations in Australia consume large amounts of olive oil, in other words have a kind of Mediterranean type diet. We find levels of EFA or mead acid as high as 1% in phospholipids and the fatty acids in the plasma, so you have to correct for dietary intake of monounsaturated fat. We also find that the level of docosahexaenoic acid (DHA) in plasma depends on dietary intake. We feel that that should be corrected for. So, while it is appealing to look at things as changes in gestational age and showing changes in DHA or other marker fatty acids, unless there is accurate dietary information, I think it is a pretty long shot.

Hornstra In the long term longitudinal study, we collected data throughout pregnancy and we were able to demonstrate that as pregnancy progresses there is hardly any major difference in the intake of dietary lipid. At the 22nd week of pregnancy we obtained a dietary history from more than 400 women. These data were correlated with pregnancy outcome and maternal and neonatal EFA status. The negative relationship between linoleic acid and cervonic acid was one of our most pronounced findings. We fully agree with you about the danger for overinterpretation of LCPUFA levels if information about the dietary situation is lacking.

Carnielli I want to confirm what Dr Gibson said, that our data in neonatal brain do not confirm this relationship of 22:5 $n-6$ and DHA. In some situations they seem to be both depressed.

Hornstra It is quite possible that there is a preferential uptake of the building bricks, as Crawford calls them, for the brain during fetal development. That may be the reason why changes in the brain comparable to those in the umbilical artery are not seen. This is another reason why we should no longer rely on biochemical data only. We have to start with functional studies and we now have at least 1200 babies in our databank for which we know exactly at least the essential fatty acid status of their umbilical plasma and umbilical vein and arterial wall. What we need to know of course is whether or not these biochemical data are in one way or another related to their performance later.

Comment from the audience Of course also in relation to their later feeding.

CHAPTER 5

Marini You depicted the intrauterine growth-retarded fetus as a hibernating animal, reducing the metabolism in order to survive. This is probably due to the reduced oxygen supply to the fetus. You state that we have to increase the transfer of nutrients through the placenta to the fetus, and you have demonstrated that this can be done. In case of a chronic treatment of the fetus, you have also to increase the oxygen supplied to the fetus. Otherwise the situation will be worse, because it will increase its acidic metabolism.

Battaglia Two things. First of all, the growth retarded fetus has an oxygen consumption per kilo that is normal, although obviously its absolute oxygen consumption is less. It is therefore not hypometabolic in that respect. I agree with you that if you have a small placenta and then begin to increase nutrient transport, which has an impact on fetal growth, you have to have better surveillance techniques to enable identification of a baby who may be developing problems. That is why in our protocol we have a lot of research directed at improved fetal surveillance. I think you need that, to feel that you can go ahead safely.

Marini Do you think that there is some way to improve oxygenation of the fetus? There is some anecdotal experience of given hyperbaric oxygen to the mother, claiming positive results.

Battaglia Whether you would need to do anything like that depends on several things. First of all can we produce effective vasodilators? Several other groups are looking at this. Better vasodilators mean that increased oxygen does not have to be given to the mother. Second, you are making the presumption that it is not possible simply to extract more oxygen coming in with a slightly lower PO_2 and by that means have a higher coefficient of extraction across the placenta. To do it safely, it still comes down to having good surveillance techniques.

Pencharz I want to address the issue of maternal blood volume expansion and how much that relates to the circulation of the placenta. In the Montreal diet expenditure study we looked at the relationship between a fall in maternal haemoglobin as a reflection of plasma volume expansion and outcome expressed as birth weight. We found that the greater the fall in haemoglobin, the greater the birth weights. We did look at red cell indices so we were not looking at iron deficiency. Modern imaging techniques allow flow through the umbilical vessels to be monitored. Have you any measure of the degree of expansion of the mother's blood volume?

Battaglia The imaging of the uterine flow in humans is very difficult, although colour Doppler is making it more and more feasible. That is why all of the clinical data tend to be on the umbilical side of the circulation. I cannot answer your question, and I have seen no data like that.

Plaut I was very interested in finding that this efficiency of the placenta to transport amino acids is related to the placental weight per gram of placenta. Glucose transport is related to placental overall weight, however. How do you deal with this entire issue when you want to supplement? Do you adjust the amount of supplement in order to cope with this?

Battaglia We are not convinced that glucose restriction is what restricts fetal growth. Even in the presence of a reduced supply of glucose, which is not sufficient to drive gluconeogenesis, it is possible to achieve a pretty good increase in fetal growth if you can get enough amino acids across. Obviously you are hoping that it also triggers a change in the endocrine setting in the fetus. We don't know that, but you would expect it. We have not started the right type of intervention strategy in these growth-retarded animals because we first wanted to develop good fetal growth curves with ultrasound. Techniques requiring insertion of wires to measure length are not adequate for this kind of research. We have now published fetal growth curves by ultrasound in breeding sheep and we can pick up growth retardation by 60 days. I think we are in a position now to be implementing an intervention strategy in those animals.

Sauer Did you look at insulin levels or other hormones?

Battaglia Yes, we have recently begun to collect data on that. You get into the issue of how frequently you have to sample, how you pick up changes in pulsatile secretion. One of our technicians looked at the maternal cord temperature when the animals go into the chamber. She pointed out that if you put six animals into this chamber only three or four would develop growth retardation and two would never show any growth retardation or any placental growth retardation. 24h after they were in the chamber, she could tell us which ones would be growth retarded. If the sheep did lower the body temperature, they had no growth retardation. An environmental effect is not enough: you must have an increase in maternal temperature to produce growth retardation.

Session 2
Infant nutrition and body composition

6 Body composition in the first year of life

N.C. DE BRUIN, H.J. DEGENHART and H.K.A. VISSER

ABBREVIATIONS

FFM	fat-free mass
TBF	total body fat
TBW	total body water
TOBEC	total-body electrical conductivity

INTRODUCTION

The two-compartment model is still the most widely used model for the study of body composition in infants. The model assumes that the body consists of fat and fat-free mass (FFM). Body composition data are important for adequate follow-up and monitoring of nutritional status and quality of growth, the nutritional adequacy of specific preterm and term infant diets, severity of malnutrition, and the progress of recovery from malnutrition, particularly in fast-growing preterm and term young infants.

For infants no 'gold standard' method for assessing body composition exists: most methods that yield accurate values in adults and children are not applicable to infants. Methods used to assess body composition in infants have to be non-invasive, non-radioactive and without the need for active cooperation of the subject. Methods such as bioelectrical impedance (1), dual energy X-ray absorptiometry (2,3), absorption and desorption of xenon (4) and NMR (5) are either inaccurate or still in an experimental phase for use in infants. Near-infrared interactance yields erroneous values for the thickness of the subcutaneous fat layer (6). At present total body fat (TBF) in infants is measured mainly by isotope dilution (using ^{2}H- or ^{18}O-enriched water) (7,8), or by various anthropometric methods. However, calculation of FFM from total body water (TBW) estimations by isotope dilution is complicated by the (individual and age-related) changes in hydration of the FFM. This body compartment changes rapidly in the growing infant, from a FFM hydration of approximately 80% in the neonate to 73.2% in children and adults.

Recently, measurement of total-body electrical conductivity (TOBEC) has emerged as an accurate, precise and reproducible method for the estimation of FFM and TBF in infants. All reported TOBEC data for TBF of full-term infants throughout the first year of life (9–11) are in excellent agreement with reference data on TBF, calculated from the combination of total body water (TBW), total body potassium and skinfold thickness measurements (12). In addition, Fiorotto showed that when the changes in

J.G. Bindels, A.C. Goedhart and H.K.A. Visser (eds), Recent Developments in Infant Nutrition, 81–97.

the actual amount of TBF present in intrauterine life (measured by carcass analysis of human fetuses) and extrauterine life (measured by TOBEC during the first 4 months of life) are plotted against age, the lines of the extra- and intrauterine periods coincide (with equal slopes) around the time of birth (13). Physiological changes in hydration of the FFM during the process of FFM maturation in early life are accounted for by the calibration procedure (14). The method has been found to be resistant to changes in extracellular fluid volume (15), so physiological changes in FFM hydration will not seriously affect TOBEC outcome (16). TOBEC is rapid, safe and easy to perform, and suitable for measurement of large numbers of infants. The instrument has been commercially available since 1989. At present TOBEC is one of the most reliable methods of estimating infant body composition, but is not widely used, due to the costs of the instrument (approximately $45 000), and the fact that the instrument is large, difficult to move and therefore not suitable for field studies. However, its good reproducibility, precision and accuracy justify the use of TOBEC in infancy for nutritional studies and as a reference method.

Calibration of TOBEC instrument

Calibration of TOBEC was performed using data obtained from minipig carcasses. Minipigs were chosen because their body geometry and FFM maturation (in terms of intra- and extracellular electrolyte concentrations) closely matches that of human infants. The calibration was performed in two centres (Houston TX, USA and Rotterdam, The Netherlands). Instrument stability, accuracy and precision were determined, and a universal calibration equation was calculated (17), which will be summarized in the present paper. To determine whether this calibration curve is valid for use throughout the entire first year of life (FFM hydration and body geometry both change with age in human infants as well as in minipiglets) FFM was estimated concomitantly by TOBEC and 18-oxygen dilution in 149 healthy term infants aged 1–12 months.

Centile standards for TBF and FFM

No data have been published on the magnitude of the biological scatter of body fat and FFM in the growing infant, although an average indication of fat and FFM growth in boys and girls has been calculated from 40-K measurements and TBW measurements using the falling drop method (12). We recently calculated centile standards for body fat, and FFM against age, length and weight in boys and girls from cross-sectional data of 423 healthy, term infants aged 2–54 weeks of age (18). An outline of these standards will be given in the present paper.

Validation of anthropometry against TOBEC

Because the TOBEC instrument is heavy and immobile, it is not suitable for field studies and bedside measurements. In these latter circumstances the use of anthropometry for determination of body composition and hence nutritional status might still be important. Anthropometry is cheap, easy to use, but has been found to be inaccurate in untrained hands (interobserver variation). Due to lack of a valid reference method, traditional anthropometric prediction equations and indexes for the prediction of body fat have not yet been well validated in infants. We performed concomitant TOBEC and anthropometry in 435 healthy term infants (total length, limb lengths, body weight, circumferences of trunk/head and limbs, and skinfold thickness at five sites: biceps, triceps, subscapular, suprailiac and above the quadriceps muscle). Using these measurements we calculated TBF with the widely used method described for infants by Dauncey et al. (19) and calculated its predictive value to estimate TBF (10). Dauncey et al. assumed all fat to lie subcutaneously in cylinders lying around the limbs and trunk. The fat volume thus could be calculated from the lengths and circumferences of trunk and limbs and the thickness of the fat layer as derived from skinfold thickness.

Traditional anthropometric indexes used in adults and children (as Quetelet's index, upper-arm circumference, upper-arm muscle area/fat area etc.) are also often applied in infants. We calculated these traditional indexes and estimated their predictive value for TBF and FFM by regressing these parameters against TOBEC-derived TBF and FFM. New anthropometric prediction equations and indexes were also calculated (20) and are summarized here.

SUBJECTS AND METHODS

Subjects

Animals

For the calibration procedure the Dutch laboratory used 12 miniature piglets of the Göttingen strain ranging in age from 7 to 99 days and weighing 1.03–10.10 kg. The Houston laboratory studied 26 miniature piglets of the Hanford strain ranging in age from 7 to 33 days and weighing 1.87–5.53 kg. All piglets were fasted (but provided with water) for at least 6 h before TOBEC measurements were made under anaesthesia (17). On completion of the TOBEC measurement the pigs were killed with an overdose of anaesthetic and weighed. The analytical procedures for carcass analysis have been reported in detail elsewhere (17). The animal protocols were reviewed and approved by the respective institutional review boards.

Human infants

Comparison of FFM by TOBEC and 18-oxygen dilution by linear regression was derived from 149 measurements on healthy, term infants aged 1–12 months (21). In cooperation with the local child health clinics of the Rotterdam Home Care Foundation a random sample of 2000 infants (living in the Rotterdam metropolitan area and aged 0–12 months) was drawn from their data base. A total of 601 parent–infant couples responded. All healthy term, Caucasian infants born vaginally to healthy mothers ($n = 423$) were used for the calculation of centile standards (18). From the total recruited population all infants who were measured by the main observer ($n=435$) were enrolled in the study to compare TOBEC and anthropometry (20). A second sample of healthy term infants from this population ($n=110$) used for validation of the newly calculated prediction equations were measured by a second observer. Written informed consent was obtained from all parents. The study protocol was approved by the ethical review boards of the Erasmus University/University Hospital Rotterdam and the Rotterdam Home Care Foundation.

Anthropometry

All infants were weighed naked on an electronic baby scale (Instru Vaaka Oy, Finland) to the nearest 1 g (0–3 kg body weight), 2 g (3–6 kg) or 5 g (6–10 kg). Recumbent crown–heel length and crown–rump length was measured to the nearest millimetre on a length board. Body circumferences and limb lengths were measured with a standard plastic measuring tape (1 cm wide), circumferences to the nearest millimetre, and limb lengths to the nearest 0.5 cm. Circumferences measured were: head (fronto-occipital), chest (at the level of the nipples), abdomen (recumbent, at the level of the largest cross-sectional area), mid-upper arm, mid-thigh and calf. Limb lengths measured were: upper arm (inferior border of acromion to tip of olecranon) and lower arm (tip of olecranon to tip of the lateral styloid). All skinfold thickness (SFT) measurements were performed in duplicate with one calibrated Harpenden caliper at the following five sites: biceps, triceps, subscapular, suprailiac and quadriceps femoris. The caliper was applied while keeping the skinfold between thumb and forefinger. Special care was taken to ensure that the caliper remained applied until the measurement had been fully stabilized. All anthropometric measurements were performed (before each TOBEC measurement) on the left side of the body according to Lohman et al (22).

TOBEC

Instrument specification and safety, measurement procedure and FFM calculation from raw TOBEC data were described earlier (10,14,16,23). TBF and FFM were calculated from raw TOBEC data using a (theoretically deduced) transformed TOBEC value (T#) which is the square root from the product of the TOBEC value and the subject's conductive length (14). Measurements were carried out with the instrument set in 'peak' mode.

Animals

Measurements were made on anaesthetized pigs. The measurement procedure has been described in detail previously (17). The piglets were centred on the subject carrier in a lateral recumbent position (on their side), which is geometrically consistent with human infants placed in a recumbent position (on their back), with respect to their oval shape on cross-sectional view.

Human infants

Since body temperature affects TOBEC outcome (16), no infants with apparent or anamnestic fever were measured. Infants were undressed and carefully swaddled in a large blanket to prevent cooling and to ensure geometric homogeneity between infants with respect to the introduction of the conductive mass into the electromagnetic field. Care was taken that limbs were not flexed and did not touch each other or the trunk. Infants were placed on their back on the sled of the instrument. A pacifier was allowed when necessary. One TOBEC reading took approximately 10 s. A complete TOBEC measurement consisted of 10 reliable 10-s readings which were averaged for calculation of FFM_{TOBEC}. If the infant urinated it was swaddled again in a dry blanket and remeasured. Movement or crying during a reading was also a reason for remeasuring the infant. Infants were not fed for at least 2 h preceding the measurement.

Statistical methods

Comparison of FFM by TOBEC and 18-oxygen dilution from combined longitudinal data was assessed using the linear regression method as described by Draper and Smith (24).

The statistical procedure we used for establishing centile standards was that suggested by Altman (25). P50 values were fitted as polynomials by entering first to fourth powers of the independent variable into a stepwise linear regression. Altman then suggests modelling of absolute residuals, also by stepwise linear regression and then calculation of each centile by applying the appropriate Z-value.

The 'All Possible Subsets Regression' module of the BMDP Statistical Software package (1990) (26) was used to calculate the best combinations of variables to predict TOBEC-derived TBF, percentage body fat (TBF%) and FFM. For each subset (containing 1–5 independent variables) the best to tenth-best combination was calculated by the program. First, the best to tenth-best single predictor variable ($k=1$) of these 24 variables was found by the program. Next, the best to tenth-best combinations of two predictor variables ($k=2$) were determined. The same was performed for $k=3$, $k=4$ and $k=5$. In all cases TOBEC-derived body composition estimates were entered as the dependent variables. With the best scoring variables from the 'All Possible Subsets Regressions' new anthropometric indexes were calculated, using a heuristic approach. An anthropometric index was defined as a simple combination of anthropometric variables (analogous to e.g. Ponderal or Quetelet's index). The optimal exponent of each variable in an index was calculated by nonlinear regression.

In general, percentage explained variability (i.e. the square of the correlation coefficient (r^2) of the linear regression) was used as a measure for the predictive value. An effect was assumed to be statistically significant if $p<0.05$. Unless stated otherwise data are expressed as mean (SD).

***Table 1** Coefficients for linear regression of FFM versus the transformed TOBEC value T#, derived from measurements on infant miniature piglets*

Equation	*n*	*Intercept*	*Slope*	*SEE* (kg)	r^2
(1) Rotterdam, The Netherlands	12	−0.0047	0.0264	0.082	0.999
(2) Houston TX, USA	26	0.0261	0.0258	0.078	0.991
(3) Pooled data from (1) and (2)	38	−0.0213	0.0264	0.077	0.997

***Figure 1** Regression line (equation 3, Table 1) of FFM determined by chemical analysis against the transformed TOBEC value T# (which is √{raw TOBEC value × conductive length}), showing the 95% confidence and prediction intervals. ■, values for individual piglets (from ref. 17)*

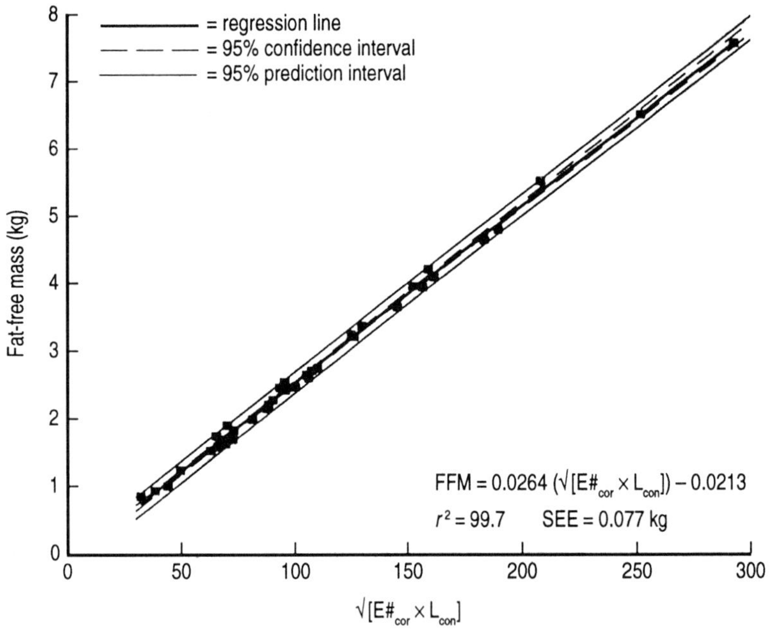

RESULTS

Calibration of TOBEC instruments

Table 1 shows the coefficients of the linear regressions of FFM by carcass analysis against the transformed TOBEC value T#. Because the calibration curves from Houston and Rotterdam (equations 1 and 2) were not significantly different, all data

***Figure 2** FFM (kg) by TOBEC against FFM (kg) by 18-oxygen dilution for boys (A) and girls (B). Dotted line represents the calculated regression equation*

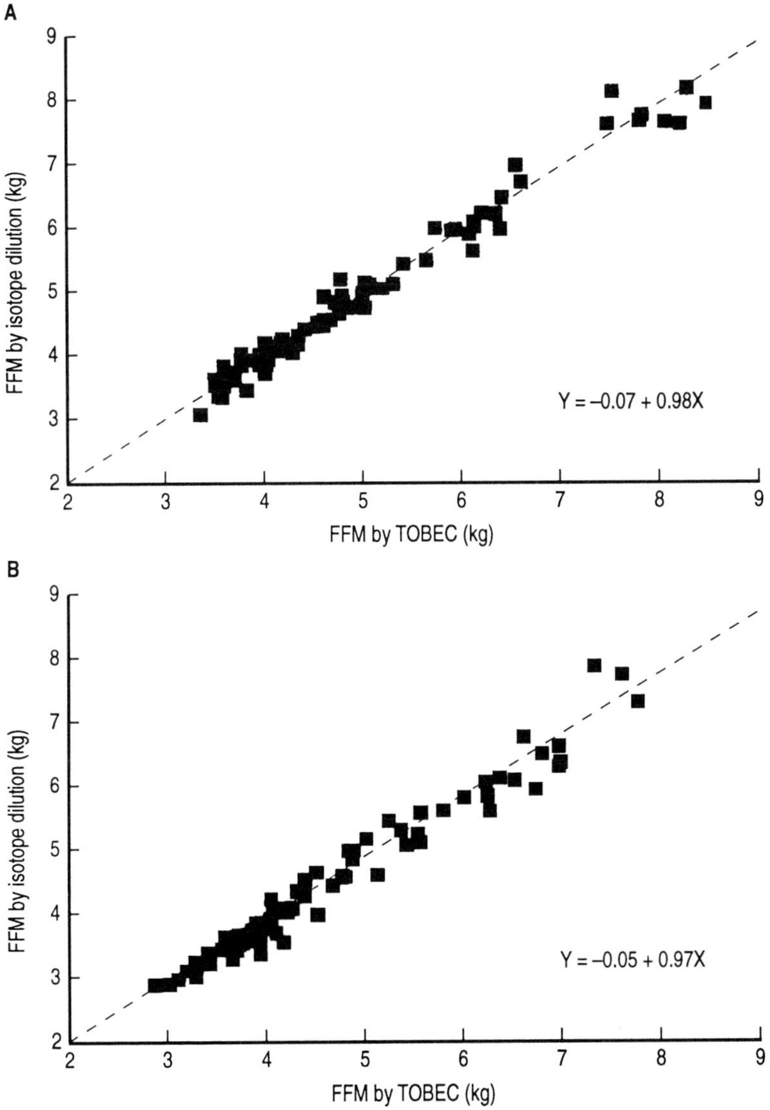

were pooled and a general calibration equation was calculated (equation 3). Figure 1 is a graphical representation of the general calibration equation. Precision for an individual measurement was 0.154 kg of FFM (i.e. the 95% prediction interval of the

Figure 3 Centile charts of percentage body fat (measured by TOBEC) against age in months

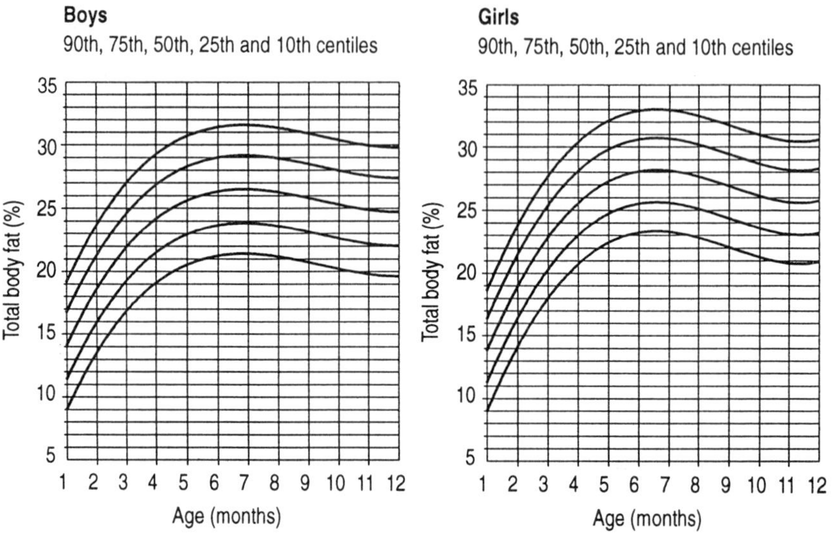

minipig calibration curve), which is consistent with an uncertainty in the FFM estimate of <5% in infants above ~3 kg FFM. The SD did not increase with body weight (17). Intra-measurement variation was <0.5% and long-term instrument drift, measured over a 2 year period, was 0.5% as measured with a cylindrical reference phantom with known conductivity index and supplied by the manufacturer (17).

The linear regression of FFM by TOBEC and 18-oxygen dilution is shown in Figure 2. In boys and girls the regression was strictly linear throughout the entire first year of life and not significantly different from the line of identity.

Centile standards for TBF and FFM

Figure 3 is an example of the centile charts of percentage body fat against age in months. Percentage TBF increased steeply during the first half year of life, and slowly declined beyond this age. Figure 4 is an example of the raw data points from which the centiles were calculated. The dotted line indicates the reference data as calculated by Fomon et al. (12), which are in good agreement with our experimental data. The algorithms (for TBF, TBF(%) and FFM *vs* age, weight and length) on which the centile standards are based are given in Appendix A. These algorithms allow calculation of Standard Deviation Scores (SDS), and also personal printing of the centiles and plotting of individual data on the centiles.

***Figure 4** Individual data points and 10th, 25th, 50th, 75th and 90th centiles of percentage body fat plotted against age in boys (A) and girls (B). Dotted lines = reference data of Fomon et al. (12)*

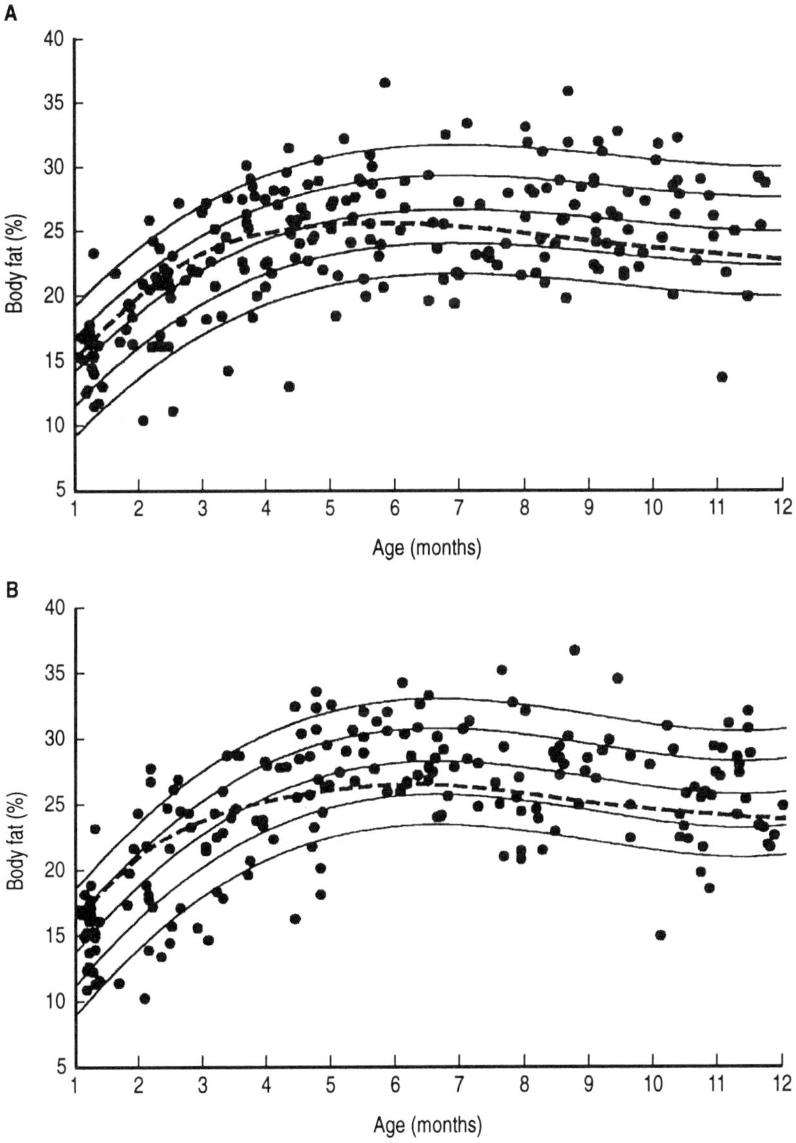

***Figure 5** Body fat by total-body electrical conductivity (TOBEC) plotted against body fat calculated by the original equation of Dauncey et al. (18). Dotted line = regression line: $Y = 0.49 + 0.69X$ ($r^2 = 0.61$). Solid line = line of identity*

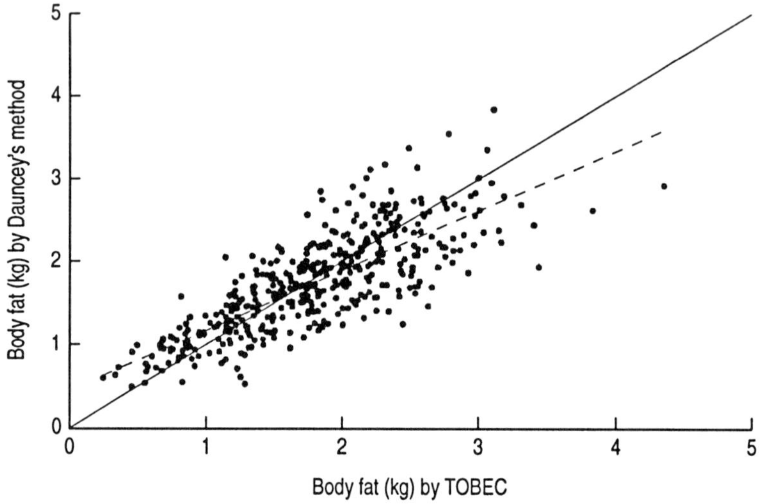

Validation of anthropometry against TOBEC

Figure 5 shows TBF (g) by Dauncey's method plotted against TBF_{TOBEC}. TBF is overestimated by Dauncey's method in the first months of life (as was also observed by Kabir et al. (6)) and slightly underestimated in later infancy. The enormous scatter, indicated by an SD of ±300–500 g, which also increases with age, is impressive; for the TOBEC method the error (95% prediction interval of the minipig calibration curve) is only approximately 150 g (20).

Table 2 shows that most anthropometric parameters poorly predict infant body composition, except weight by length and calf circumference for TBF_{TOBEC} and total length and body weight for FFM_{TOBEC}. Appendix B summarizes the best scoring anthropometric prediction equations (calculated by 'All Possible Subsets Regression'. New simple indexes (analogous to eg. Quetelet's index) were calculated using an heuristic approach (20). For TBF the best scoring indexes were: (weight×calf circumference/length), $r^2 = 0.87$ and (weight×calf circumference×sum of 3 skinfolds/length), $r^2 = 0.89$. For FFM the best scoring index was: $\sqrt{(\text{weight} \times \text{length})}$ ($r^2 = 0.95$). The coefficients for calculation of TBF and FFM in kg from these indexes have been described elsewhere (20). As observed from Table 2 and the new indexes, calf circumference is apparently one of the major single anthropometric predictors of infant body fat.

DISCUSSION

Body composition data give a better insight into nutritional status and quality of growth than body weight alone. Now that TOBEC has been established as a safe, simple and accurate method of assessing infant body composition, nutritional status in infants can be assessed easily on multiple occasions. The method has been thoroughly calibrated independently in two centres against data obtained from analysis of minipig carcasses, which show a change in body geometry and FFM–electrolyte maturation comparable to human infants. Moreover, regression of FFM derived from TOBEC and 18-oxygen dilution (methods which are based on widely divergent methodological principles) was not significantly different throughout the entire first year of life. The method has not been thoroughly tested in sick infants. A study measuring TOBEC in non-human objects (16) showed that errors arising from changes in temperature and major deviations from the normal cross-sectional area of the subject will disturb TOBEC-derived body composition estimates: care must therefore be taken in infants with fever. Due to tissue autolysis and body temperature decline, TOBEC measurements made after death are susceptible to considerable error (16). Most physiological changes in FFM electrolyte concentrations will not disturb the TOBEC outcome (15,16), although significant changes in bicarbonate levels may affect the TOBEC signal (e.g. in the case of diabetic ketoacidosis and renal failure). Ions bound in a crystalline structure (such as in bone) will not elicit a significant TOBEC signal, nor do deviations in plasma levels of protein and amino acids affect TOBEC. We conclude that TOBEC is currently the most accurate method for safe and convenient determination of infant body composition.

Reference centiles help to identify the severity of nutritional and growth disorders in preterm and term infants and to monitor and evaluate the adequacy of therapeutic interventions during this important period of rapid growth. We suggest that the present centiles are valid for monitoring nutritional status. Children at either extreme of the centile curves might be regarded as possibly at risk for obesity or undernutrition, although the numbers of infants were insufficient to accurately predict more explicit extremes as e.g. P97 and P3 centiles. Further research should disclose the relationship between infants at either extremes of the centile curves and a possible associated risk of future health hazards.

Upper arm circumference is widely used as a simple proxy for body fatness, and is also used in infants (26,27). As described in the present paper, upper arm circumference in infants is very poorly correlated with TBF and FFM. Skinfold measurements and Quetelet's index have also been found to be poorly correlated with TBF (28–31). It can be concluded, therefore, that in infants these traditional local anthropometric measurements, used in children and adults as a proxy for total body composition, poorly reflect the actual total energy and protein stores of the infant's body. Centile charts of these variables (32–35), therefore, are of limited value in infants. Measurement of total body composition, represented by TBF and FFM, would provide better estimates of nutritional reserves than regional anthropometric measurements and may provide a more accurate assessment of nutritional status. For infants, quantitation of TBF has been performed traditionally using the anthropometric algorithm described

***Table 2** Relationship between TOBEC-derived body composition estimates and traditional and single anthropometric variables (derived from the main study population,* n *= 435)*

	r^2		
Independent variable	*TBF*[a]	*TBF%*[b]	*FFM*[c]
Ratio weight/length (kg/m)	**0.84**	0.43	**0.85**
Quetelet's index (kg/m^2)	0.59	0.44	0.38
Ponderal index (kg/m^3)	0.10	0.03	0.11
Weight	**0.81**	0.33	**0.93**
Age	0.60	0.19	0.73
Upper arm anthropometry			
Area	0.65	0.37	0.47
Muscle circumference	0.40	0.16	0.46
Muscle area	0.40	0.15	0.35
Fat area	0.45	0.38	0.20
Muscle ratio	0.02	0.02	0.12
Fat ratio	0.09	0.08	0.03
Ratio of arm/head circumference	0.25	0.26	0.07
Skinfold thickness			
Biceps	0.30	0.26	0.17
Triceps	0.21	0.25	0.05
Subscapular	0.15	0.20	0.05
Suprailiac	0.13	0.24	0.08
Quadriceps	0.57	0.49	0.22
Sum of 3 skinfolds	0.52	0.52	0.16
Sum of 5 skinfolds	0.51	0.53	0.13
Circumference			
Head	0.58	0.18	**0.84**
Chest	0.65	0.25	**0.85**
Mid-thigh	0.75	0.44	0.64
Calf	**0.83**	0.52	0.66
Mid-upper arm	0.62	0.36	0.53
Length			
Total	0.63	0.22	**0.89**
Crown–rump	0.63	0.24	**0.81**
Upper arm	0.50	0.20	0.64
Lower arm	0.50	0.19	0.66

[a]TBF and independent variables both log-transformed
[b]TBF% and independent variables not transformed
[c]FFM log-transformed, independent variables not transformed
$r^2 > 0.80$ have been printed **bold**

by Dauncey et al. (18). This paper summarized our validation of this method, which showed a moderate accuracy and a very poor precision of Dauncey's method (10). We conclude that Dauncey's method is not suitable for measurement of TBF in an individual patient due to the large error.

The foregoing validations do not inevitably mean that anthropometric measurements are obsolete. Because the vast majority of anthropometric parameters and indexes appeared to be invalid for infants, we calculated new predictive equations for TBF and FFM (20). This was done by searching for the best combinations of two or three anthropometric parameters and the accessory prediction equations for TBF and FFM (data given as Appendix B). For TBF no improvement in the predictive value (r^2 and SEE) was seen with inclusion of more than three parameters and for FFM with inclusion of more than two parameters. Depending upon the available anthropometric data the appropriate equations can be chosen. When no accurate body composition method (e.g. TOBEC or isotope dilution) is available, these new anthropometry-based prediction equations are more accurate for the assessment of nutritional status in infants than upper-arm anthropometry of skinfold thickness, and are suitable for screening purposes in conjunction with the present centile standards. However, anthropometric methods are still less precise than TOBEC or isotope dilution, and caution should be exercised when using these data for accurate individual total body composition estimations.

The new indexes and prediction equations were validated in a second population of slightly older children by a second well-trained observer (20). While the precision of the various equations was equal to that in the first population, it appears that equations incorporating skinfold thickness show the largest deviation in accuracy from TBF_{TOBEC} and FFM_{TOBEC}. We therefore estimated the interobserver variation separately for all anthropometric parameters (20). For skinfold thickness the estimated interobserver variation was ~10% (at all five locations), for upper and lower arm length ~6%, and for circumferences of calf, arm, thigh and chest ~2%. No interobserver difference was found for head circumference and total length. We conclude that the indexes and prediction equations with calf circumference, weight and length are least prone to interobserver variation, and should be used in combination with the TBF and FFM centile standards.

CONCLUSIONS

TOBEC is a suitable method for rapid, safe and accurate measurements of infant body composition. The method has been thoroughly calibrated independently in two centres against data obtained from carcass analysis of minipigs, which show a change in body geometry and FFM–electrolyte maturation comparable to human infants. Moreover, regression of FFM derived from TOBEC and 18-oxygen dilution (methods based on widely divergent methodological principles) was not significantly different throughout the entire first year of life. Traditional anthropometric prediction equations and indexes were validated against TOBEC and found to be invalid for use in infants, except from weight by length and calf circumference (for TBF) and length and weight (for FFM). Centile standards were assessed for TBF and FFM *vs* age, weight and length, while new, infant-specific anthropometric prediction equations and anthropometric indexes were calculated which can be used in conjunction with these centile standards.

ACKNOWLEDGEMENTS

We gratefully acknowledge financial support from the Sophia Foundation for Medical Research, Nutricia Research, Procter & Gamble Division Holland, Trustfonds Foundation of the Erasmus University Rotterdam, University Hospital Rotterdam and Foundation 'De Drie Lichten' in the Netherlands. We extend our appreciation to the parents and infants who gave their time to take part in this study.

REFERENCES

1. Mayfield SR, Uauy R, Waidelich D 1991 Body composition of low-birth-weight infants determined by using bioelectrical resistance and reactance. Am J Clin Nutr 54:296–303
2. Van Loan MD, Mayclin PL 1992 Body composition assessment: dual-energy X-ray absorptiometry (DEXA) compared to reference methods. Eur J Clin Nutr 46:125–130
3. Jebb SA, Goldberg GR, Elia M 1992 DEXA measurements of fat in relation to depth and adiposity. Clinical Nutrition 11:145
4. Mettau JW, Degenhart HJ, Visser HKA 1977 Measurement of total body fat in newborns and infants by absorption and desorption of nonradioactive xenon. Pediatr Res 11:1097–1101
5. Lewis DS, Rollwiitz WL, Bertrand HA, Masoro EJ 1986 Use of NMR for the measurement of total body water and estimation of body fat. J Appl Physiol 60:836–840
6. Kabir N, Forsum E 1993 Estimation of total body fat and subcutaneous adipose tissue in full-term infants less than 3 months old. Pediatr Res 34:448–454
7. Schoeller DA, Van Santen E, Petersen DW, Dietz W, Jaspan J, Klein PD 1980 Total body water measurement in humans with ^{18}O and ^{2}H labeled water. Am J Clin Nutr 33:2686–2693
8. Trowbridge FL, Graham GG, Wong WW, Mellits ED, Rabold JD, Lee LS, Cabrera MP, Klein PD 1984 Body water measurements in premature and older infants using $H_2{}^{18}O$ isotopic determinations. Pediatr Res 18:524–527
9. De Bruin NC, Van Velthoven CAM, Brugman R, Degenhart HJ, Visser HKA 1994 Measuring body fat in infancy: anthropometry versus total body electrical conductivity (TOBEC). Pediatr Res 35:268
10. De Bruin NC, Van Velthoven CAM, Stijnen T, Juttmann RE, Degenhart HJ, Visser HKA 1995 Quantitative assessment of infant body fat by anthropometry and TOBEC. Am J Clin Nutr 61: 279–286
11. Fiorotto ML, Cochran WJ, Klish WJ 1987 Fat-free mass and total body water of infants estimated from total body electrical conductivity measurements. Pediatr Res 22:417–421
12. Fomon SJ, Haschke F, Ziegler EE, Nelson SE 1982 Body composition of reference children from birth to age 10 years. Am J Clin Nutr 35:1169–1175
13. Fiorotto ML 1991 Measurements of total body electrical conductivity for the estimation of fat and fat-free mass. In: Whitehead RG, Prentice A (eds) New Techniques in Nutritional Research. Academic Press, San Diego CA, pp 281–301
14. Fiorotto ML, Cochran WJ, Funk RC, Sheng HP, Klish WJ 1987 Total body electrical conductivity measurements: effects of body composition and geometry. Am J Physiol 252:R794–R800
15. Cochran WJ, Fiorotto ML, Sheng HP, Klish WJ 1989 Reliability of fat-free mass estimates derived from total-body electrical conductivity measurements as influenced by changes in extracellular fluid volume. Am J Clin Nutr 49:29–32
16. DeBruin NC, Luijendijk IHT, Visser HKA, Degenhart HJ 1994 Effect of alterations in physical and chemical characteristics on TOBEC-derived body composition estimates: validation with non-human models. Phys Med Biol 39:1143–1156
17. Fiorotto ML, De Bruin NC, Brans YW, Degenhart HJ, Visser HKA 1995 Total body electrical conductivity measurements: an evaluation of current instrumentation for infants. Pediatr Res 37:94–100
18. De Bruin NC, Van Velthoven CAM, De Ridder M, Stijnen T, Juttmann RE, Degenhart HJ, Visser HKA Standards for total body fat and fat-free mass in infants. Arch Dis Child (in press)
19. Dauncey MJ, Gandy G, Gairdner D 1977 Assessment of total body fat in infancy from skinfold thickness measurements. Arch Dis Child 52:223–227

20. De Bruin NC, Van Velthoven CAM, De Ridder M, Stijnen T, Juttmann RE, Degenhart HJ, Visser HKA 1995 Body fat and fat-free mass in infants: new and classic anthropometric indexes and prediction equations compared with total-body electrical conductivity. Am J Clin Nutr 61: 1195–1205
21. De Bruin NC, Westerterp KR, Degenhart HJ, Visser HKA 1995 Measurement of fat-free mass in infants. Pediatr Res 38:411–417
22. Lohman TG, Roche AF, Martorell R 1988 Anthropometric Standardization Reference Manual. Human Kinetic Books, Champaign, IL
23. EM-SCAN Inc. 1989 Operator's Manual. Springfield, IL
24. Draper N, Smith H 1981 Applied Regression Analysis, 2nd edn. J Wiley & Sons, New York, Chap 1
25. Altman DG 1993 Construction of age-related reference centiles using absolute residuals. Stat Med 12:917–924
26. Jelliffe EFP, Jelliffe DB 1969 The arm circumference as a public health index of protein-calorie malnutrition of early childhood. J Trop Pediatr 15:179–192
27. Gurney JM, Jelliffe DB 1973 Arm anthropometry in nutritional assessment: nomogram for rapid calculation of muscle circumference and cross-sectional muscle over fat areas. Am J Clin Nutr 26:912–915
28. Davies PSW, Lucas A 1989 Quetelet's index as a measure of body fatness in young infants. Early Hum Dev 20:135–141
29. Davies PSW, Lucas A 1990 The prediction of total body fatness in early infancy. Early Hum Dev 21:193–198
30. Kabir N, Forsum E 1993 Estimation of total body fat and subcutaneous adipose tissue in full-term infants less than 3 months old. Pediatr Res 34:448–454
31. Sheng HP, Muthappa PB, Wong WW, Schanler RJ 1993 Pitfalls of body fat assessments in premature infants by anthropometry. Biol Neonate 64:279–286
32. Tanner JM, Whitehouse RH 1975 Revised standards for triceps and subscapular skinfolds in British children. Arch Dis Child 50:142–145
33. Oakley JR, Parsons RJ, Whitelaw AGL 1977 Standards for skinfold thickness in British newborn infants. Arch Dis Child 52:287–290
34. Sann L, Durand M, Picard J, Lasne Y, Bethenod M 1988 Arm fat and muscle areas in infancy. Arch Dis Child 63:256–260
35. Frisancho AR 1990 Anthropometric standards for the assessment of growth and nutritional status. University of Michigan Press, Ann Arbor, MI

Appendix A ***Regression coefficient of the 50th centile polynomials[a] and of the regression with their residual SD***

Independent variable	Dependent variable: *TBF* (kg)	*TBF* (%)	*FFM* (kg)
Boys			
Age (months)	$\ln Y = -0.9638 + 0.5665X - 0.0595X^2 + 0.0021X^3$	$Y = 8.2231 + 6.5941X - 0.7565X^2 + 0.0268X^3$	$Y = 3.4314 + 0.4627X - 0.00903X^2$
	$r^2 = 0.72$	$r^2 = 0.45$	$r^2 = 0.83$
	$SD = 0.26040$	$SD = 3.96160$	$SD = 0.3497 + 0.0033X + 0.0033X^2$
Length (cm)	$Y = -6.3515 + 0.1334X - 3.723 \times 10^{-8}X^4$	$Y = -171.0996 + 5.4210X - 0.0372X^2$	$Y = 1.6835 + 1.2563 \times 10^{-5}X^3$
	$r^2 = 0.77$	$r^2 = 0.42$	$r^2 = 0.94$
	$SD = -0.6597 = 0.0158X$	$SD = 4.08511$	$SD = -0.4862 + 0.0119X$
Weight (kg)	$Y = -1.6332 + 0.5752X - 0.0143X^2$	$Y = -44.3129 + 19.9832X - 1.6061X^2 + 0.0031X^4$	$Y = 1.6190 + 0.4290X + 0.0140X^2$
	$r^2 = 0.89$	$r^2 = 0.59$	$r^2 = 0.96$
	$SD = -0.0711 + 0.0462X$	$SD = 2.2214 + 0.1792X$	$SD = 0.2285 - 0.0478X + 0.0068X^2$
Girls			
Age (months)	$Y = -0.0527 + 0.5325X - 0.0341X^2 + 4.3767 \times 10^{-5}X^4$	$Y = 7.3883 + 7.0665X - 0.6505X^2 + 0.001314X^4$	$Y = 3.1353 + 0.4500X - 0.0150X^2 + 3.40 \times 10^{-5}X^4$
	$r^2 = 0.76$	$r^2 = 0.55$	$r^2 = 0.87$
	$SD = 0.1973 + 0.0230X$	$SD = 3.76954$	$SD = 0.2693 + 0.0253X$
Length (cm)	$Y = -8.6848 + 0.1840X - 8.3826 \times 10^{-8}X^4$	$Y = -139.9504 + 3.5352X - 2.33 \times 10^{-4}X^3$	$Y = 1.2680 + 1.7481 \times 10^{-5}X^3 - 6.4164 \times 10^{-8}X^4$
	$r^2 = 0.75$	$r^2 = 0.49$	$r^2 = 0.95$
	$SD = -0.1681 + 0.0084X$	$SD = 3.97150$	$SD = 0.8091 - 0.0286X + 3.089 \times 10^{-4}X^2$
Weight (kg)	$Y = -1.3754 + 0.4829X - 0.4829X - 6.7071 \times 10^{-5}X^4$	$Y = -42.5733 + 20.3583X - 1.6595X^2 + 0.003259X^4$	$Y = 1.3754 + 0.5171X + 6.7071 \times 10^{-5}X^4$
	$r^2 = 0.90$	$r^2 = 0.59$	$r^2 = 0.96$
	$SD = -0.0148 + 0.0380X$	$SD = 3.56802$	$SD = -0.0148 + 0.0380X$

[a]All regressions with non-constant residual SD were obtained by weighted linear stepwise regression, using $1/SD^2$ as weighting factor

***Appendix B** Results of 'All Possible Subsets Regression'[a] and regression coefficients of the algorithm $y=A_0+A_1x+A_2x^2+A_3x^3$*

				Regression coefficients			
Eq.	r^2	*SD*	*Independent variables*	A_0	A_1	A_2	A_3
Dependent variable : ln(body fat [kg])							
2a	0.86	0.153	ln(calf circ.), ln (weight)	−6.180	1.800	0.8140	
2b	0.86	0.158	ln(calf circ.), ln(sum of 5 skinfolds)	−8.9995	2.7389	0.4623	
2c	0.85	0.159	ln(calf circ.), ln(sum of 3 skinfolds)	−8.8219	2.7431	0.4511	
2d	0.84	0.164	ln(calf circ.), ln(age)	−7.0276	2.7139	0.1520	
3a	0.90	0.131	ln(calf circ.), ln(weight), ln(sum of 5 skinfolds)	−6.3032	1.1052	0.8876	0.5149
3b	0.90	0.134	ln(calf circ.), ln(weight), ln(sum of 3 skinfolds)	−6.1506	1.1453	0.8722	0.4951
3c	0.89	0.136	ln(calf circ.), ln(weight), ln(suprailiac skinfold)	−5.4117	1.2223	1.0348	0.2587
3d	0.88	0.143	ln(calf circ.), ln(sum of 5 skinfolds), ln(age)	−6.9185	1.9585	0.5430	0.1949
Dependent variable : body fat (%)							
2a	0.65	2.93	calf circ., sum of 5 skinfolds	−6.8645	1.1384	0.2586	
2b	0.62	3.04	calf circ., suprailiac skinfold	−8.9557	1.6423	0.8990	
2c	0.62	3.07	thigh circ., sum of 5 skinfolds	−3.6903	0.6664	0.2908	
2d	0.55	3.31	calf circ., gender[b]	−10.7796	1.9030	1.8624	
3a	0.67	2.84	calf circ., sum of 5 skinfolds, chest circ.	0.4664	1.6515	0.2576	−0.3757
3b	0.67	2.86	calf circ., sum of 5 skinfolds, gender[b]	−9.4029	1.2128	0.2435	1.3216
3c	0.65	2.94	calf circ., triceps skinfold, suprailiac skinfold	−9.4356	1.4948	0.7818	0.3671
3d	0.65	2.95	calf circ., suprailiac skinfold, gender[b]	−11.8208	1.6912	0.8542	1.5495
Dependent variable : ln(fat-free mass [kg])							
2a	0.95	0.045	weight, total length	0.3571	0.080	0.0110	
2b	0.95	0.045	weight, sum of 5 skinfolds	0.8987	0.1252	−0.0029	
2c	0.95	0.045	weight, head circ.	0.2185	0.0904	0.0187	
2d	0.91	0.058	chest circ., head circ.	−1.1872	0.0313	0.0355	
3a	0.95	0.042	weight, total length, gender[b]	0.3969	0.0757	0.0116	−0.0337
3b	0.95	0.042	weight, total length, sum of 5 skinfolds	0.5309	0.0948	0.0083	−0.0022
3c	0.95	0.042	weight, total length, head circ.	0.0291	0.0689	0.0084	0.0136
3d	0.92	0.056	head circ., chest circ., thigh circ.	−1.0989	0.0333	0.0264	0.0097

[a]In the 'All Possible Subsets Regressions' we used the adjusted r^2 instead of the normally used r^2 as selection criterion: $r^2_{adjusted}=r^2-\{[k(1-r^2]/(n-k-1)\}$, where k is the number of independent variables in the equation. Use of adjusted r^2 allows comparison between two subsets with different numbers of predictors

[b]gender: male = 1, female = 2

7 Bone mineralization during the first year of life

J. RIGO, K. NYAMUGABO, M. DE CURTIS and J. SENTERRE

ABBREVIATIONS

BMC	bone mineral content
BMD	bone mineral density
BW	body weight
DXA	dual X-ray absorptiometry
VLBW	very low birth weight

INTRODUCTION

The goal of feeding regimens for low birth weight (LBW) infants is to obtain a prompt postnatal resumption of growth to a rate approximating that seen during intrauterine growth and to follow after term a growth rate similar to that of breast-fed term infants (1). Human intrauterine and postnatal growth, and body composition, have therefore been studied extensively in order to obtain reference values (2–8). Among the various body compartments, bone mineralization and calcium retention have received considerable attention and sophisticated techniques have been developed in order to define the most appropriate diet for preterm and term infants.

For many years, body composition reference values at different gestational ages have been obtained from analyses of stillborn infants and neonates who died soon after birth. First values were reported in 1877, and around 169 infants have been analysed to date (8). The chemical reference values of whole body Ca content were recently reconfirmed by Ellis et al. (9,10) using gamma neutron activation in 19 preterm and term infants who died in the first 2 weeks of life (Fig. 1). Ca content was shown to be linearly related to birth weight, increasing from 10 g to 30 g as birth weight rises from 1500 g to 3500 g. From those data, it was calculated that intrauterine Ca accretion represents about 140 mg/kg body weight (BW) per day or 0.9% of BW gain (~15 g/kg per day). There are no reference values for whole body calcium content during infancy. Fomon et al. (11) studied body composition of reference children during infancy. They assumed that the concentration of osseous minerals in fat-free mass was constant from birth to 12 months and suggested that bone mineral content (BMC) represented around 92 g at 3500 g, 185 g at 8000 g and 233 g at 10 000 g.

Development of non-invasive techniques is necessary to permit collection of much of the needed information on body composition in healthy subjects during infancy.

J.G. Bindels, A.C. Goedhart and H.K.A. Visser (eds), Recent Developments in Infant Nutrition, 98–111.

***Figure 1** Whole body calcium content related to birth weight. Comparison with values obtained by chemical carcass analyses (●) and by neutron activation (▲). Data from Refs 7 and 10*

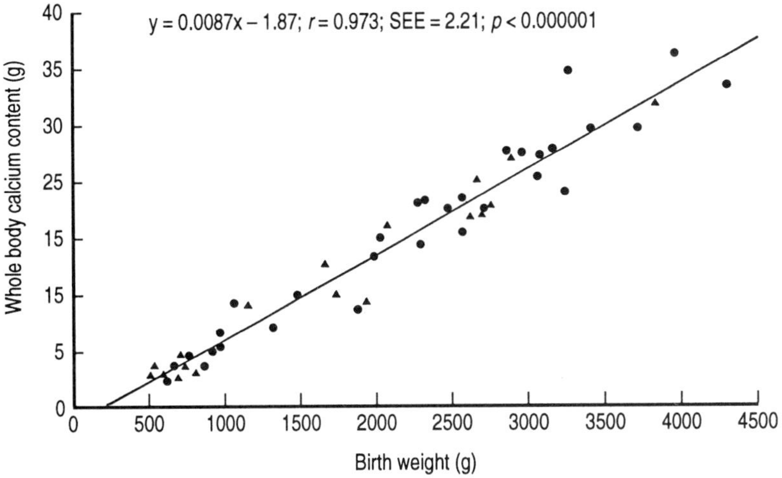

Photon absorptiometry for the study of bone mineralization represents an example of an advanced technique that is of promise for use in the clinical setting. Dual X-ray absorptiometry (DXA) allows more precise estimates of whole body bone mineral content and is a novel step in determining whole body composition during development. Validation studies of dual X-ray absorptiometry for body composition have been performed in small piglets and term neonates (12–15). In a recent study using a new software program (QDR 2000 infants whole body software 5.64, 1993, Hologic Inc., Waltham, MA, USA), reproducibility of DXA measurements was 0.09% for BW and 1.95% for BMC (16). BMC estimated by DXA was significantly correlated with whole body ash weight ($r=0.955$) and whole body chemical calcium content ($r=0.991$) determined in 13 piglets with weights ranging from 1470 to 5500 g (Fig. 2). Equations for the conversion of BMC into ash and of BMC into calcium were calculated from these data and used to determine the precision of DXA measurements as ±10.99% for ash and ±4.44% for calcium content (Fig. 2). Since radiation exposure during whole body scanning of an infant is low (3 μSV or 0.3 mRem), we concluded that DXA is a non-invasive, safe and accurate method for evaluating bone mineral and calcium content in human newborn babies.

The aim of the present study was to evaluate the use of DXA for the determination of intrauterine and postnatal reference values of whole body composition and calcium content.

***Figure 2** Correlation between BMC measured by DXA and calcium content measured in small piglets (n = 13) (a). Estimate of the precision of BMC measured by DXA; measurements difference between the measured DXA values and the reference values as a percentage of the measured values in the piglets (b)*

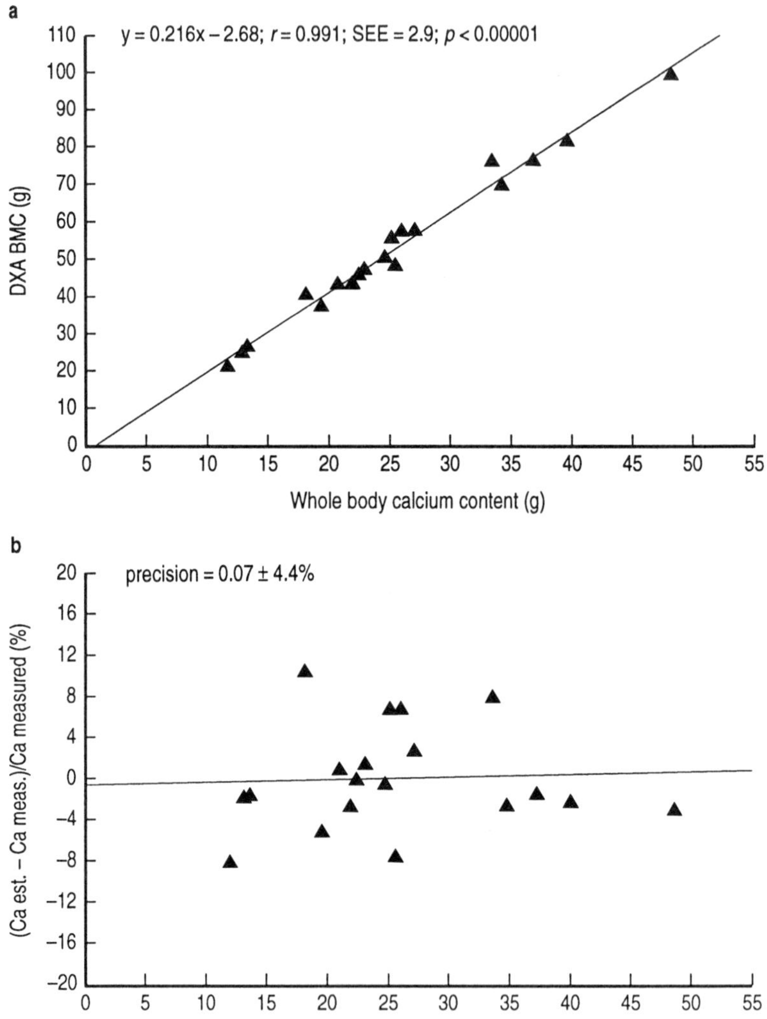

Table 1 Characteristics of infants studied (mean ± SD)

	Preterm infants (n=53)	Term infants (n=53)	Term infants (n=36)	Term infants (n=24)
Gender (F:M)	28:25	30:23	19:17	18:6
At birth				
Weight (g)	1908 ± 434	3243 ± 426	3080 ± 424	3298 ± 386
Length (cm)	42.9 ± 2.6	49.5 ± 2.0	49.0 ± 2.2	50.3 ± 1.1
Gestational age (weeks)	33.3 ± 2.4	39.0 ± 1.3	40.0 ± 1.4	39.6 ± 0.7
At DXA analysis				
Weight (g)	1856 ± 407	3144 ± 437	3825 ± 629	7476 ± 1282
Length (cm)	43.5 ± 2.6	49.9 ± 2.0	52.5 ± 2.9	65.1 ± 4.3
Age (days)	8 ± 3	4 ± 2	30 ± 14	187 ± 77.4

MATERIALS AND METHODS

DXA measurements

To determine BMC reference values, DXA measurements were performed in 106 healthy preterm and term infants at birth and in 60 infants at 1 – 12 months of age. Body composition was measured using a new software program (QDR 2000 infant whole body software 5.64, 1993, Hologic Inc, Waltham, MA, USA). The principle of DXA has been previously described. Infants were placed supine on an infant table pad without an interposing paper sheet. They were naked and wrapped in a paper sheet in order to avoid movement during the measurement and to obtain abduction of legs. No sedation was given. The table pad, which was specifically designed for whole body measurements in infants, filters the low energy beam to improve system linearity in small subjects and to reduce radiation dose (15). Whole body composition data analysis allows the measurements of body weight (g), bone mineral content (BMC) (g), bone area (cm^2), bone mineral density (BMD; g/cm^2), fat mass (g), lean body mass (g) and fat-free mass (g).

To determine intrauterine reference values, 53 preterm infants with a birth weight ranging from 1100 to 3360 g and a gestational age from 27 to 36 weeks and 53 term infants with a birth weight ranging from 2280 to 4210 g and a gestational age from 37 to 41 weeks were evaluated soon after birth (Table 1). Time from birth to analysis ranged from 1 to 14 days (mean 8 days) in the preterm infants and 1 to 7 days (mean 4 days) in the term infants. The delay was related to clinical conditions, particularly in the preterm infants, and/or to availability of the DXA apparatus. Since BMC does not change significantly during this short period, data obtained were considered as intrauterine values and analysed according to birth weight, birth length and gestational age at birth. From these data, intrauterine reference values in centiles were calculated for BMC and BMD.

In order to obtain reference values during the first year of life, the results obtained in term infants at birth (n=53) were combined with the results obtained in 60 healthy term infants analysed during the first year of life, between 1 week and 2 months ($n=36$) and between 2 and 12 months ($n=24$; Table 1). These data were analysed according to body weight and length and postnatal age at the time of the study. Due to the small number of infants presently available centile values were not calculated.

Data in Table 1 are given as mean ± SD. The relationships between body weight, gestational age, length, fat-free mass and BMC or BMD were analysed using a PC statistical software program (Statistica version 5.0, 1995). This study was approved by the human ethics committee of the University of Liège. Informed parental consent was obtained.

RESULTS

Intrauterine reference values

At birth BMC is positively linearly related to birth weight (Fig. 3). In addition there is a square relationship between BMC and birth length and a cubic relationship with gestational age. BMC is about 20 g in a preterm infant of 1500 g, whereas it reaches 60 g in a term infant of 3500 g. Bone mineral density (Fig. 4) is positively related to birth weight and birth length and there is a square relationship between BMC and gestational age. There was a 60% increase in BMD with an increase in intrauterine weight from 1500 g to 3000 g.

From those results, intrauterine reference curves of BMC related to birth weight and length were calculated in centiles (Fig. 5). From those data, it appears that in a 1500 g preterm infant appropriate BMC values range from 17 to 26 g whereas in a 3500 g term infant the normal value is 50–70 g.

Bone mineralization during the first year of life

Between birth and the end of the first year of life, BMC was positively linearly related to body weight (Fig. 6). By contrast, there was a cubic relationship with body length. In a 3500 g term infant BMC was 60 g, reaching 150 g in an infant of 8000 g. BMD was also linearly related to body weight (Fig. 7), whereas there was a square relationship with body length. BMD increased from 195 mg/cm^2 in a 3500 g term infant to 255 mg/cm^2 at 8000 g.

DISCUSSION

Assessment of body composition is one of the best indices for evaluating the adequacy of growth and nutrition of infants (17). The fetus during the last trimester of gestation and the infant during the first months of life undergo important changes in body composition including fat deposition, muscle mass accretion and bone mineralization. Preterm infants partly or completely miss this period of mineral accretion and the postnatal calcium retention appears limited to 50–90 mg/kg/day, leading to a relative bone mineral depletion or osteopenia of prematurity (18,19). Postnatal rickets and incidence of bone fracture are increased in very low birth weight (VLBW) infants

***Figure 3** Correlation between BMC measured by DXA and birth weight (a), birth length (b) and gestational age (c) in 106 preterm and term newborn infants*

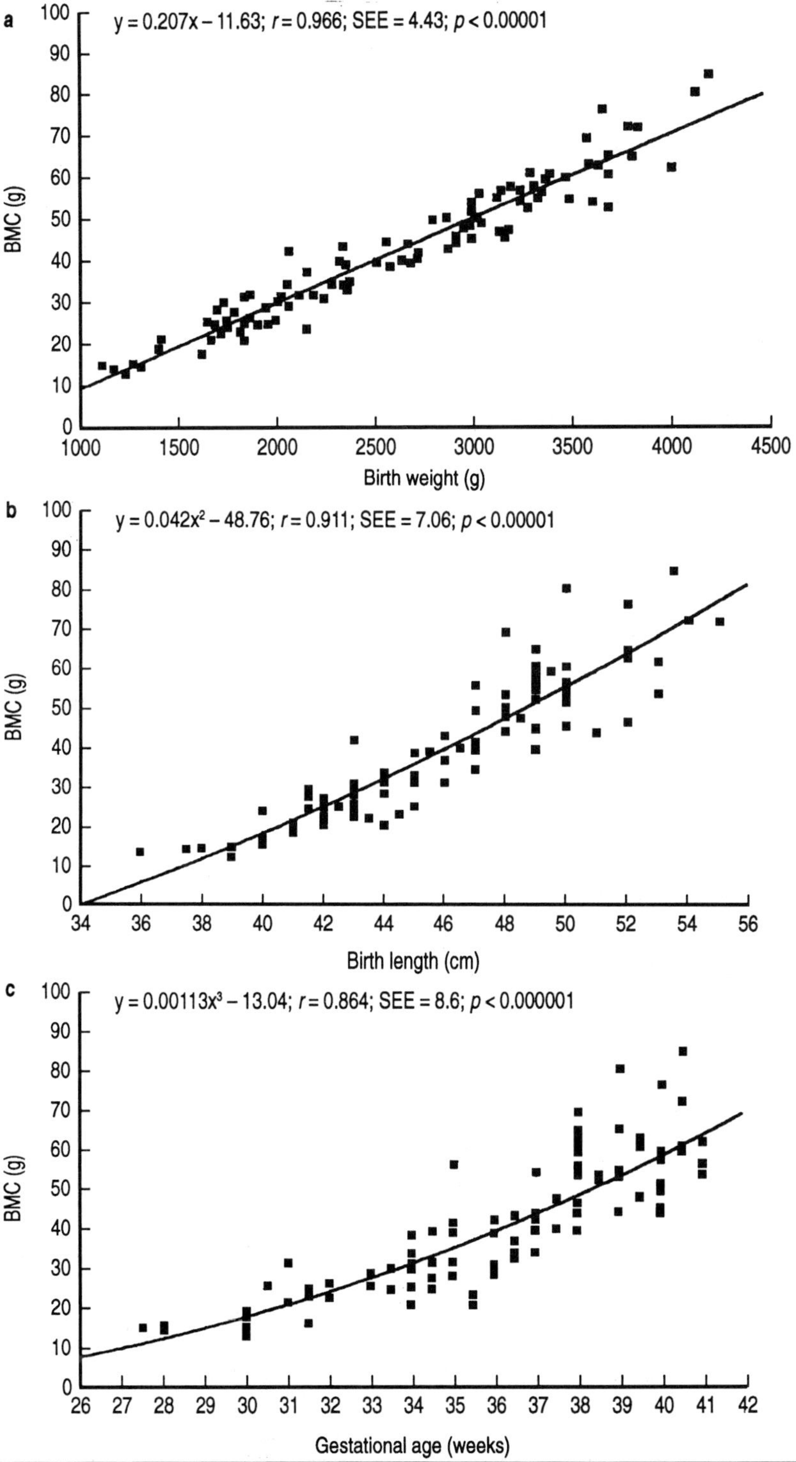

Figure 4 Correlation between BMD measured by DXA and birth weight (a), birth length (b) and gestational age (c) in 106 preterm and term newborn infants

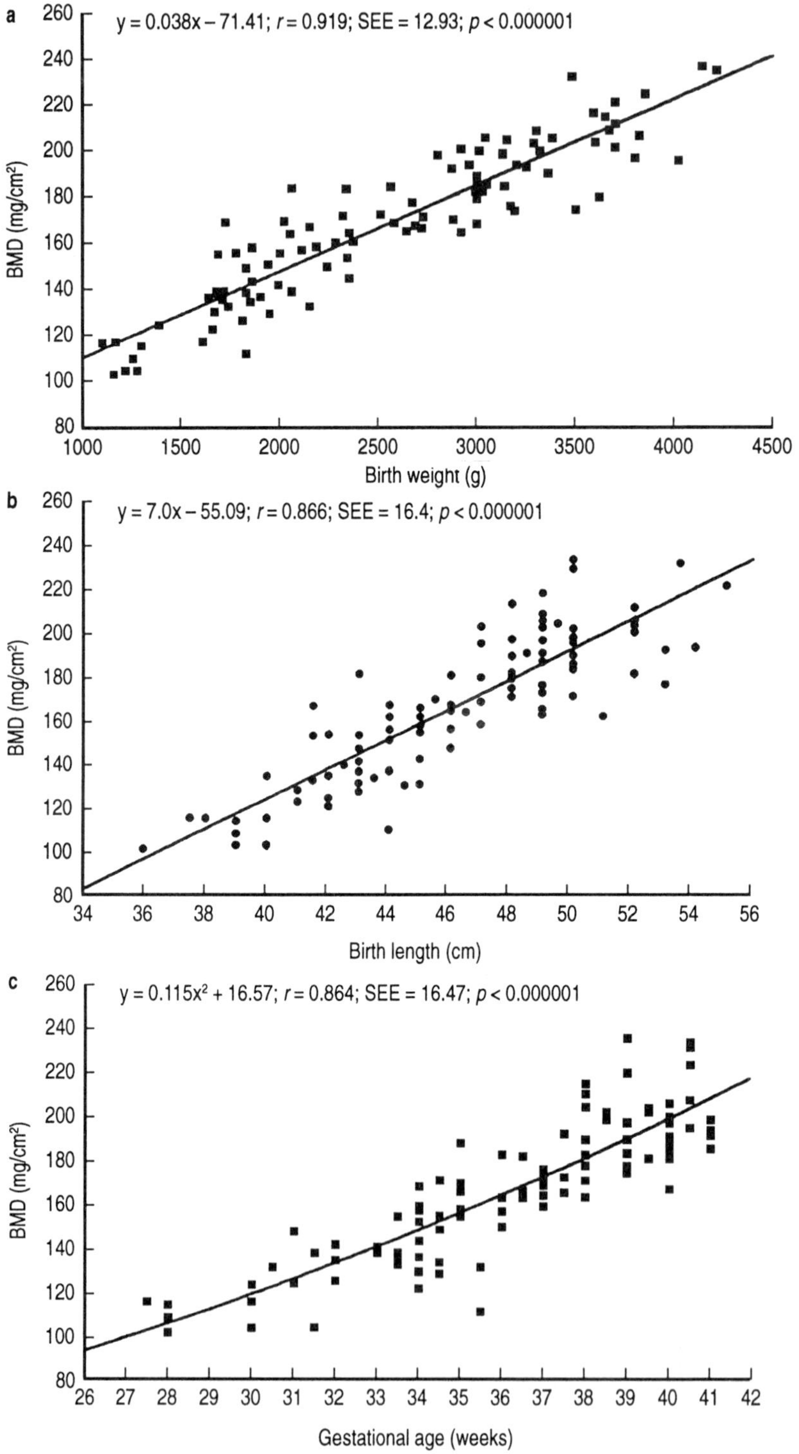

Figure 5 Reference curves of BMC related to body weight (a) and body length (b). Centile values were calculated from the data obtained in 106 preterm and term newborn infants

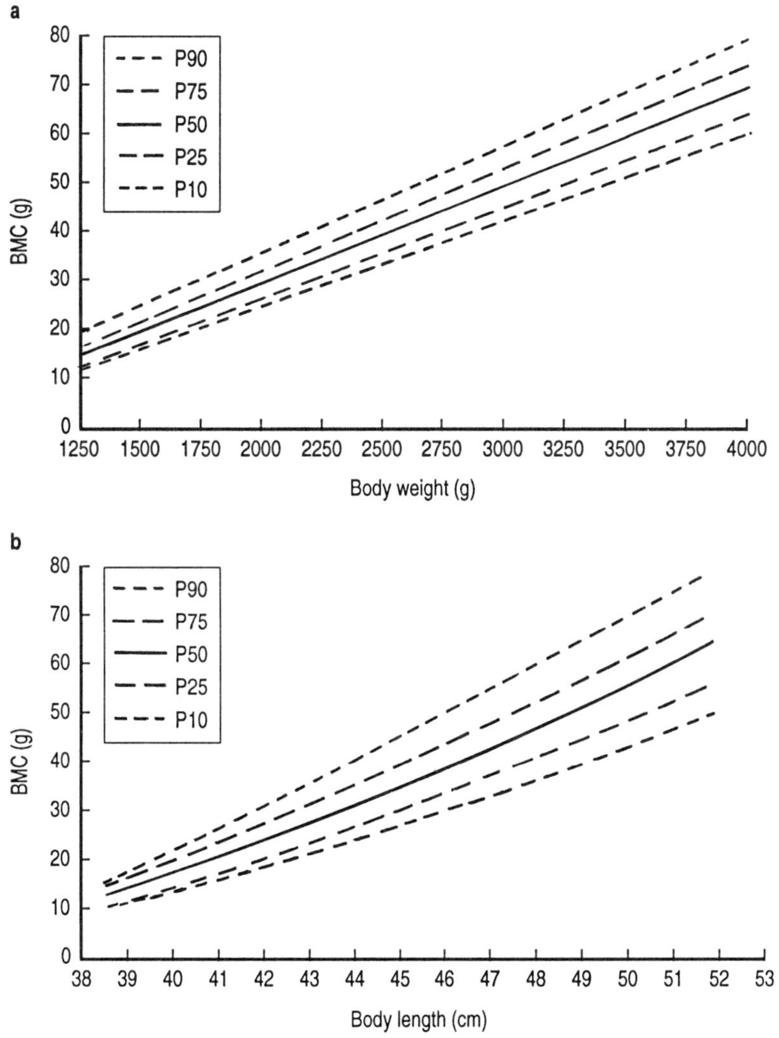

receiving long term parenteral nutrition, and VLBW infants with bronchopulmonary dysplasia and/or those given corticosteroids or diuretics. In older infants, other pathological conditions and medications may affect bone mineralization. Practical, non-invasive and safe methods are therefore needed to evaluate and follow BMC in preterm and term infants, and during infancy.

Figure 6 Correlation between BMC measured by DXA and body weight (a) and body length (b) in 113 infants measured during the first year of life

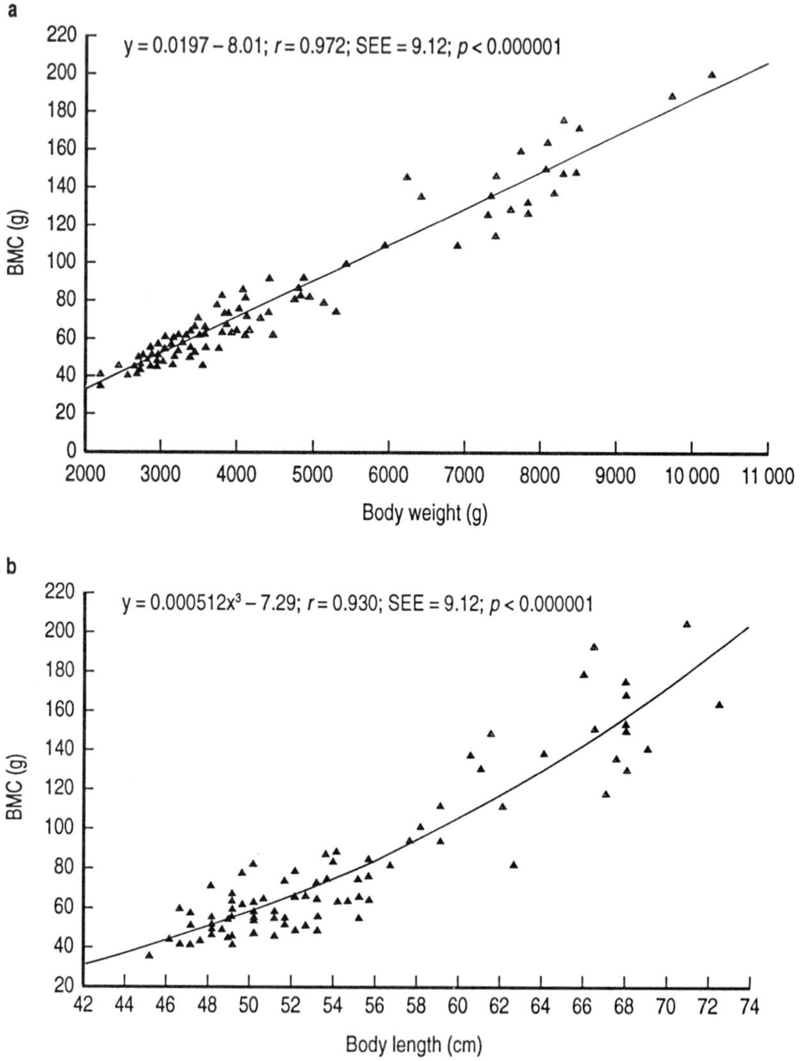

Recently DXA was suggested as the preferred technique for the evaluation of whole body composition: BMC, BMD, fat mass and lean body mass. The use of a special infant whole body programme with an infant pad improves the linearity of the system in small subjects and reduces the radiation dose (15). Unlike fat mass, BMC

Figure 7 Correlation between BMD measured by DXA and body weight (a) and body length (b) in 113 infants measured during the first year of life

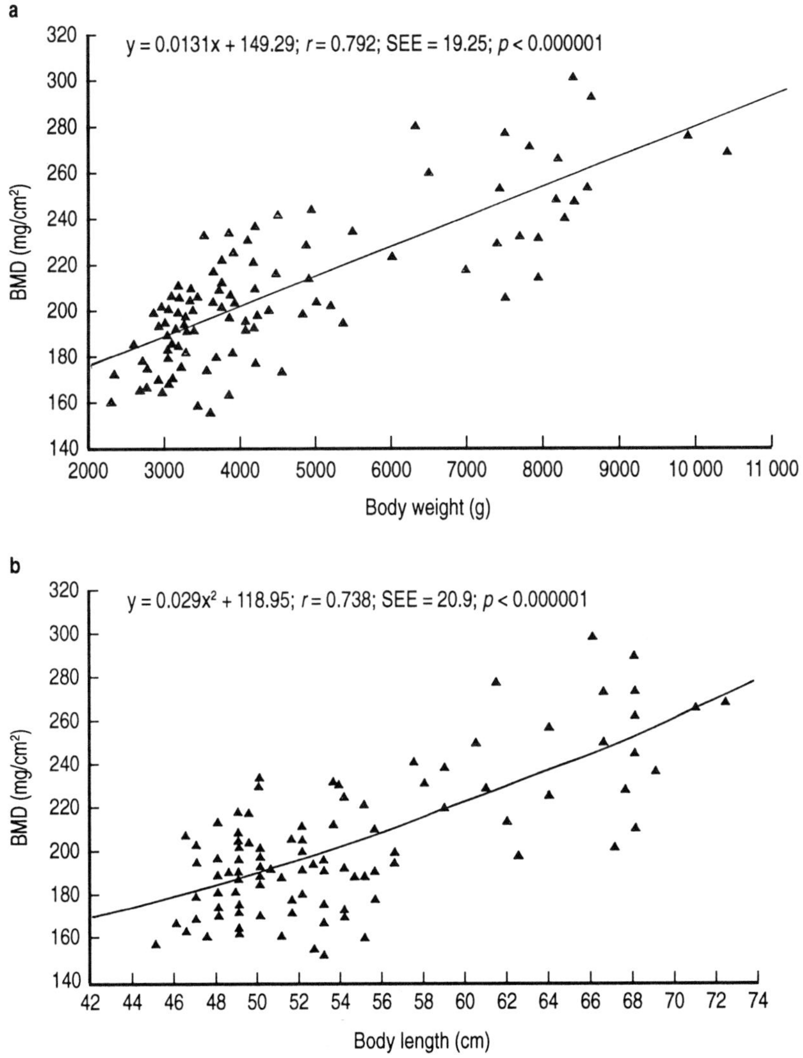

is not directly accessible to chemical analysis. In an experimental validation study in newborn piglets (16), DXA measurement of BMC was highly significantly correlated with ash weight and whole body calcium content. In this study, DXA measurement of BMC predicted whole body calcium content (±4.4%) better than ash weight

***Figure 8** Correlation between whole body calcium content estimated from BMC measured by DXA (□) and birth weight in 106 preterm and term newborn infants. Comparison with the reference values reported by Widdowson (7, ■) and by Ellis (10, ▲)*

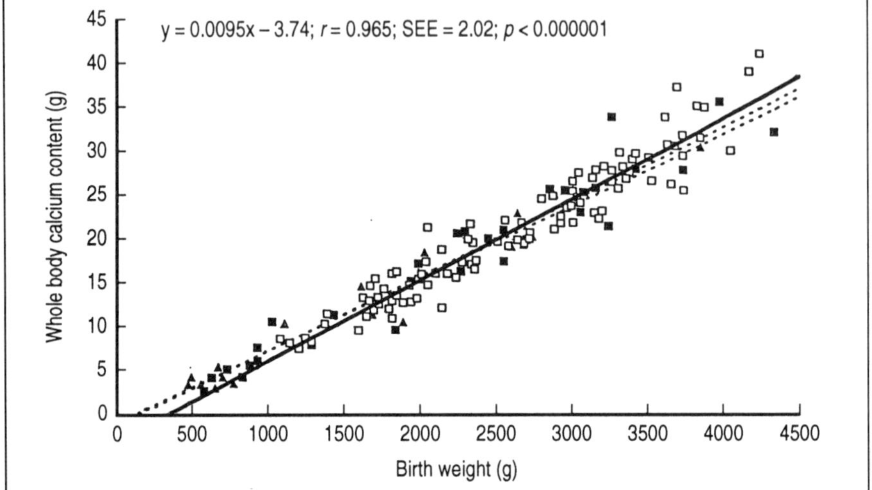

(10.99%). In the present study, 'intrauterine' BMC was linearly correlated with birth weight whereas a polynomial relationship was observed for birth length ($length^2$) and gestational age (GA^3). Similar observations were reported for Ca content determined by carcass analysis. As in adults, the BMC to calcium ratio during infancy is not yet defined. In piglets with body weight of 1500–5500 g, we found that the ratio of BMC to calcium was expressed by the equation $y = 0.456x + 1.56$. Applying this conversion equation to our BMC intrauterine reference values, a relationship between whole body calcium content and body weight was obtained and compared with the whole body chemical calcium carcass analysis reported by Widdowson (7) and to the values recently determined by Ellis et al. (10) from postmortem examination of newborn infants using the neutron activation technique. As shown in Figure 8, there is an excellent agreement between the in vivo (DXA) and postmortem (chemical and neutron activation) Ca determination. According to this relationship, whole body calcium content represents 10 g in preterm infants of 1500 g and 30 g in a term infant of 3500 g.

After term, BMC increases linearly with body weight and there is a polynomial relationship between BMC and body length ($length^3$). The rate of BMC accretion in relation to body weight appears to be constant during the last trimester of gestation and during the first year of life (±20 g/kg BW gain; Fig. 9). The whole body calcium content during the first year of life has never been evaluated by chemical analysis (17). Nevertheless, the percentage of calcium content in the fat-free human femur has been shown to be relatively constant from 24 weeks of gestational age to the end of

Figure 9 Correlation between BMC measured by DXA, whole body calcium content and body weight in 166 preterm and term infants measured at birth and during the first year of life

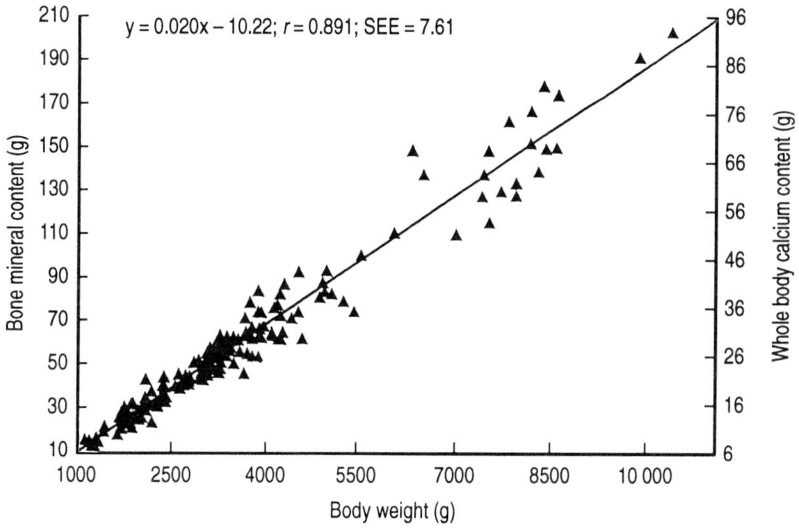

first year of life (20). Therefore, the conversion equation was also applied to the BMC values obtained in infants during the first year. As shown in Figure 9, whole body calcium content increases from 30 g to 70 g as body weight increases from 3500 to 8000 g. Such an increase represents a mean calcium retention of 25 mg/kg/day. This value is close to that generally accepted as the calcium requirement in infancy during the first 6 months of life (21).

Comparison of the present data with those previously reported in newborn infants is difficult because of the difference in populations studied, in apparatus, in data acquisition and in data analysis (14,22).

Bone mineral density represents the ratio between BMC (g) and bone area determined during the whole body scanning. The area measured is highly dependent on the position of the infant during the DXA scan, and all infants were therefore similarly positioned, supine with the legs in abduction and the head on lateral side in order to assure area reproducibility. Bone area is highly significantly correlated with body weight ($r=0.996$). BMD increases linearly with body weight during the last trimester of gestation and during the first year of life, whereas there is a polynomial relationship between BMD and length in these two periods. In relation to body weight, the increase in BMD appears significantly higher during the last trimester of gestation (37.6 ± 1.6 mg/cm^2/kg BW gain) than during the first year of life (13.4 ± 1.0 mg/cm^2/kg BW gain; Fig. 10). The greater increase in bone mineral density

Figure 10 ***Correlation between BMD and body weight in preterm (▲) and term (□) infants at birth (n = 106) and in 113 infants measured during the first year of life (■)***

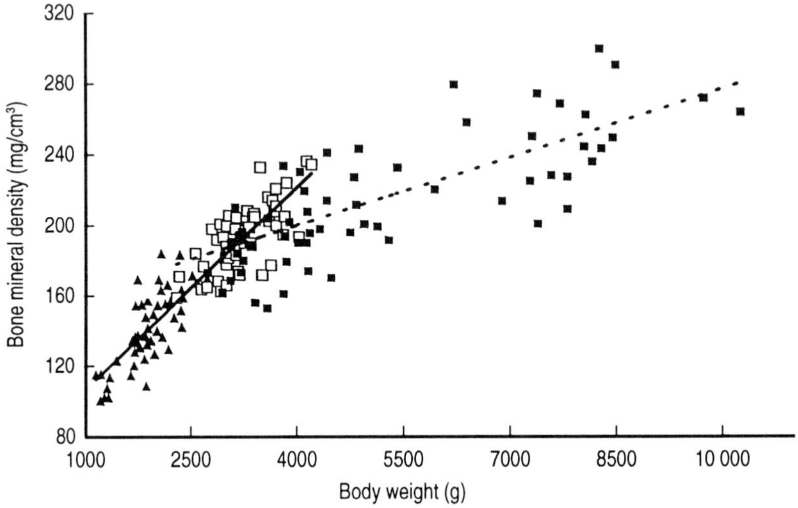

during the last trimester of gestation confirms the fact that bone mineralization is relatively more important than bone extension in length or width during fetal life than thereafter. However, the slow increase in BMD observed during the first year of life could also reflect the progressive reduction in bone water content reported during the same period (20).

In conclusion, dual X-ray absorptiometry appears a useful, safe and sensitive technique for the evaluation and the follow-up of BMC in preterm and term infants. In addition, our study suggests that whole body calcium content can be estimated accurately from BMC values. New studies need to be performed to determine the sensitivity of DXA measurement for the estimation of calcium retention from two successive DXA scans.

ACKNOWLEDGEMENT

This work was supported in part by a Bristol-Myers Squibb/Mead Johnson unrestricted grant.

REFERENCES

1. Barness LA, Mauer AM, Andersen AS, Dallman PR and American Academy of Pediatrics Committee on Nutrition 1977 Nutrition needs of low-birth-weight infants. Pediatrics 60:519–530

2. Lubchenco LO, Hansman C, Dressler M, Boyd E 1963 Intrauterine growth is estimated from liveborn birth-weight data on 24 to 42 weeks of gestation. Pediatrics 32:793–800
3. Lubchenco LO, Searls DT, Brazie JV 1972 Neonatal mortality rate: relationship to birth weight and gestational age. J Pediatr 81:814–822
4. Babson SG, Behrman RE, Lessel R 1970 Fetal growth. Liveborn birth weights for gestational age of white middle class infants. Pediatrics 45:937–944
5. Usher R, McLean F 1969 Intrauterine growth of liveborn Caucasian infants at sea level: standards obtained from measurements in 7 dimensions of infants born between 25 and 44 weeks of gestation. J Pediatr 74:901–910
6. Ziegler EE, O'Donnel AM, Nelson SE, Fomon SJ 1976 Body composition of the reference fetus. Growth 40:329–341
7. Widdowson EM 1982 Importance of nutrition in development, with special reference to feeding low-birth-weight infants. In Sauls HS, Bachhuber WL, Lewis LA (eds.): Nutritional Goals for Low-Birth-Weight Infants (Proceedings of the Second Ross Clinical Research Conference, Oct. 2–3, 1980). Ross Laboratories, Columbus, OH, pp 4–11
8. Sparks JW 1984 Human intrauterine growth and nutrient accretion. Semin Perinatol 8:74–93
9. Ellis KJ, Shypailo RJ, Schanler RJ, Langston C 1993 Body elemental composition of the neonate: new reference data. Am J Hum Biol 5:323–330
10. Ellis KJ, Shypailo RJ, Pratt JA, Pond WG 1994 Accuracy of dual-energy X-ray absorptiometry for body-composition measurements in children. Am J Clin Nutr 60:660–665
11. Fomon SJ, Haschke F, Ziegler EE, Nelson SE 1982 Body composition of reference children from birth to age 10 years. Am J Clin Nutr 35:1169–1175
12. Braillon PM, Salle BL, Brunet J, Glorieux FH, Delmas PB, Meunier PJ 1992 Dual energy X-ray absorptiometry measurement of bone mineral content in newborns: validation of technique. Pediatr Res 32:77–80
13. Salle BL, Braillon PM, Glorieux FH, Brunet J, Cavero E, Meunier PJ 1992 Lumbar bone mineral content measured by dual energy X-ray absorptiometry in newborns and infants. Acta Paediatr 81:953–958
14. Brunton JA, Bayley HS, Atkinson SA 1993 Validation and application of dual X-ray absorptiometry to measure bone mass and body composition in small infants. Am J Clin Nutr 58:839–845
15. Koo WWK, Massom LR, Walters J 1995 Validation of accuracy and precision of dual energy X-ray absorptiometry for infants. J Bone Mineral Res 10:1111–1115
16. Picaud JC, Rigo J, Nyamugabo K, Millet J, Senterre J 1996 Evaluation of dual energy X-ray absorptiometry (DXA) for body composition assessment in 1500 to 5500 g piglets and term human neonates. Am J Clin Nutr 63:157–163
17. Friis-Hansen B 1971 Body composition during growth. In vivo measurements and biochemical data correlated to differential anatomical growth. Pediatrics 47(Suppl 2):264–274
18. Bronner F, Salle BL, Putet G, Rigo J, Senterre J 1992 Net calcium absorption in premature infants results of 103 metabolic balance studies. Am J Clin Nutr 56:1037–1044
19. Senterre J 1991 Osteopenia versus rickets in premature infants. In: Glorieux FH (ed) Rickets. Nestlé Nutrition Workshop Series, 21, Nestec Ltd., Vevey/Raven Press Ltd., New York, pp 145–154
20. Dickerson JWT 1962 Changes in the composition of the human femur during growth. Biochem J 82:56–61
21. Kahn B, Straub CP, Robbins PJ, Wellman HN, Seltzer RA, Telles NC 1969 Intake and excretion of calcium and phosphorus by infants; calcium retention and model. Pediatrics 43, Suppl. 4, part II:668–686
22. Venkataraman PS, Ahluwallia BW 1992 Total bone mineral content and body composition by X-ray densitometry in newborns. Pediatrics 90:767–770

8 Nutrition and linear growth in childhood

J. KARLBERG and K. ALBERTSSON-WIKLAND

ABBREVIATIONS

BMI	body mass index
H	height
SDS	standard deviation score
W	weight

INTRODUCTION

Postnatal linear growth can be described in terms of three distinct phases: the nutrition-dependent infancy phase, the growth-hormone-dependent childhood phase and puberty, with the additional growth during adolescence which is driven by sex steroids (1). The main objective of this work was to evaluate the importance of nutrition in the linear growth seen during childhood, i.e. from 2 years of age to the onset of puberty.

It is generally believed that over-nutrition accelerates linear growth, just as under-nutrition retards it. Most studies on over-nutrition and linear growth were made two to three decades ago in obese children referred to paediatric clinics (2–7). These studies seem to support the hypothesis that over-nutrition accelerates linear growth or that obese children forced to lose weight have a deceleration in height growth. Some of these studies also imply that obese children are usually tall in childhood, but due to an earlier pubertal onset they end up close to or just below the normal mean final height. So far, no large population-based study has addressed the question of how obesity/leanness and excessive weight gain/loss are related to height and height gain in childhood.

In order to measure obesity or under-nutrition through body measurement the optimum weight-for-height index should be independent of (not correlated with) height in healthy subjects. Common expressions of weight-for-height such as body mass index (BMI) are, however, correlated with height; for instance, a taller child will have a higher BMI (8–10). Another reason why BMI is not an appropriate estimate of obesity in children is that mean BMI values differ with age: low values in early infancy are followed by an increase and a decrease in childhood and finally an increase in adolescence (9,10). Weight-for-height standards during childhood are usually constructed on the assumption that the weight and height are related equally at different ages, an assumption which is also wrong when relating lean body mass to height (10–12). For these reasons another objective of this study was to develop an age- and

J.G. Bindels, A.C. Goedhart and H.K.A. Visser (eds), Recent Developments in Infant Nutrition, 112–127.

gender-specific weight-for-height index from birth to maturity, which should be independent of height in a healthy population. The method used here is based on the linear regression between weight standard deviation scores (SDS) and height SDS, and has been presented previously for adults (13).

The data presented here to illustrate a new weight-for-height index and the relationship between nutritional status, expressed as an increase or decrease in weight-for-height, and linear growth in childhood were obtained from a longitudinal growth study of 3656 healthy children born at full term between 1973 and 1975 in Göteborg, Sweden (14).

SUBJECTS AND METHODS

The study population was enrolled from 5111 children who were in the last grade of school in 1992 in Göteborg, Sweden (14). Only 97 girls and 77 boys were unwilling to participate, and another 222 girls and 227 boys failed to attend the investigations in school. Of the remaining 4488, 84% were born in Göteborg or in close proximity to the city and 94% were born in 1973 or 1974. Boys and girls were equally represented.

Data were collected between April and November 1992, when the study team visited all the schools in Göteborg. All health records from birth to the last grade in school were obtained and all children were measured (weight and height) in a standardized way with a calibrated Harpenden stadiometer. Information relating to birth and the perinatal period were taken individually from the Swedish Birth Register at the Swedish National Board for Health and Welfare. Standing height was measured by skilled personnel at the Child Health Monitoring Centres between 2 and 6 years of age, and from 7 years of age by trained school nurses. The accuracy of the height measuring scale in school was tested in this study: in 70% of the schools the difference between the results obtained by the two measuring devices was less than 2 mm.

The number of healthy full-term children for whom data were available relating to both birth and the last grade in school was 3656 (1807 girls and 1849 boys); 529 children were excluded since information relating to their birth was lacking and another 303 children did not fulfil the inclusion criteria for reasons such as multiple births, prematurity and/or growth-related disorders (10).

The study was approved by the Ethical Committee of the Medical Faculty, University of Göteborg, and by the Swedish Data Authorities. Written informed consent was obtained from the parents and/or those children who were older than 16 years of age.

Length and weight at birth were both converted into SDS using the current Swedish reference values for size at birth and adjusting for gender and length of gestation (15). Postnatal body measurements were also expressed in SDS using the total population of this study as the reference population. These reference values have been reported elsewhere (16). Growth data were used from the children at birth and at the ages of approximately 3, 6, 9, 12 and 18 months, and annually thereafter until the examination in school at about 18 years of age. Only one body measurement value was used per child and age. All children who had gained less than 0.5 cm in height during the previous year and who had reached the age at peak height velocity during puberty at least 2 years prior to the last examination were considered to have reached final height.

Using this definition, 99.2% of the children had reached final height at the time the study team visited the schools (at about 18 years of age) or at re-examination one year later.

The following common expressions of weight were correlated with height (H); weight (W); weight/height (W/H); weight/height^2 (W/H^2); weight/height^3 (W/H^3); weight$_{SDS}$ (W_{SDS}). An additional expression was taken from the age-specific linear regression between weight SDS and height SDS:

$$\text{Weight}_{SDS} - \beta \times \text{height}_{SDS}$$

where β is the slope between weight SDS and height SDS. This index can be expressed as SDS (WH_{SDS} SDS) using the residual SD around the regression line.

The impact of nutrition on linear growth during childhood was examined by relating the changes in weight-for-height SDS [δWH_{SDS} SDS] at 5–8 years of age to changes in height SDS (δ height SDS) during the same period. All children ($n=91$) in whom WH_{SDS} SDS changed by more than 1.5 units, representing the limits of the 95% confidence interval for $\delta\, WH_{SDS}$ SDS, were characterized in more detail with respect to their growth in height and weight during childhood.

Statistical analysis include bivariate correlation, simple linear regression analysis and Student's *t*-test with an expected mean value of zero.

RESULTS

Weight-for-height index

The results are based on the girls, but similar results were found for the boys. Figure 1 shows the scatter plot, linear regression and correlation coefficient between four different expressions of weight and height in 7-year-old girls. W/H^2 and W/H^3 have the lowest absolute correlation value, so the optimum value for p (W/H^p) at this age would be between 2 and 3. However, there is an age dependent pattern in the correlation coefficients (Fig. 2); W/H^2 has correlation values close to zero in early life and in the adolescent region, while W/H^3 has such values in the prepubertal and pubertal years.

Figure 3 shows the relationship between weight SDS and height SDS for the same sample as shown in Figure 1 and the correlation ($r=0.7$) is similar to that between weight and height (Fig. 1). However, after correcting weight SDS for the slope (β) of the linear regression between weight SDS and height SDS, the correlation coefficient becomes zero, as shown in Figure 3. The slope of the regression line is somewhat age-dependent, but a β value of 0.7 is an approximation over all ages and both sexes. In Appendix A the complete data on the correlation analyses of data sets of ages 0–21 years are given. The correlation coefficients when using $\beta=0.7$ and the optimum or age-specific β value between the corresponding weight-for-height index and height are included in Figure 2.

Body mass index (BMI) and other similar expressions (W/H^2 or W/H^3) are thus all correlated with height. The two alternative, i.e. 'uncorrelated' indices are the age specific W/H^p and the method presented here, i.e. $W_{SDS} - \beta \times H_{SDS}$. The latter method has a practical advantage, offering a rapid method for estimating the individual weight-for-height value; weight SDS and height SDS are read from the individual growth chart and $W_{SDS} - 0.7 \times H_{SDS}$ is computed, which can be standardized into SDS using the residual SD as given in Appendix A, or by using an overall SD approximation of 0.7 SDS. An illustration is given in Figure 4.

Weight-for-height and linear growth in childhood

Table 1 gives the correlation coefficients between height SDS, weight SDS and WH_{SDS} SDS at 5 years of age and the change in these values at 5–8 years of age. As expected, there is a low correlation ($r = -0.03$) at 5 years of age between height and weight-for-height SDS. A low correlation ($r = -0.10$) is also found between the change in height SDS and the change in weight-for-height SDS at 5–8 years of age.

The impact of nutrition on linear growth during childhood was further examined by relating the changes in weight-for-height SDS at 5–8 years of age to changes in height SDS during the same period. This examination was made according to the following three different types of data analysis.

Table 2 includes the δ SDS values for height, weight and WH_{SDS} at 5–8 years of age for different groups of children as defined by the change in WH_{SDS} SDS at this age (integer groups of 0.5 units). A few children with extreme values in both tails of the δ WH_{SDS} SDS distribution were excluded as they did not form any larger group of their own. The mean δ height SDS value was significantly different from zero in four of the eight groups. However, the mean δ height SDS values were relatively small in all four groups (<0.15 SDS) and the change of the mean height SDS also took the opposite direction to the mean δ WH_{SDS} SDS.

All children ($n = 91$) with a change in WH_{SDS} SDS of more than 1.5 units, representing the limits of the 95% confidence interval, were characterized in more detail regarding their growth in height and weight during childhood. Table 3 depicts the height, weight and WH_{SDS} SDS at different ages for two specific groups of children ($n = 34$ for each group); one with a decrease in WH_{SDS} with 1.5–2.0 SDS at 5–8 years of age, and a group with an increase in WH_{SDS} with 1.5–2.0 SDS at the same age. During this period of highly significant change in WH_{SDS} SDS ($p < 0.0001$, *t*-test with an expected mean value of zero) the mean change in height SDS was 0.09 and 0.14 SDS in the two groups, respectively (non-significant in both groups). Also note that the mean height SDS is close to zero for all ages in both groups; during the same period there were major changes in both mean weight SDS and mean WH_{SDS} SDS.

A final analysis was made by observing the individual growth curves of 103 children. All of these children showed a change (increase or decrease) in WH_{SDS} SDS of >1.5 units at 5–8 years of age representing the limits of the 95% confidence interval. The data from 12 children were not used as the number of observations between 2 and 10 years reached a maximum of three in each case. The remaining children displayed eight different growth patterns during childhood (Table 4); an increase or a decrease in WH_{SDS} SDS in combination with an increase, decrease or no change in height SDS. Figures 5–12 show typical growth patterns for all eight groups. The majority (66%, groups 2, 4 and 6; Table 4) had an increase or decrease in WH_{SDS} SDS with no clear change in height SDS, i.e. a change of <0.5 SDS during childhood. In 24% (groups

Figure 1 Scatter-plot, linear regression and correlation coefficient between four different expressions of weight-for-height. The data points represent girls at 7 years of age (n = 1333)

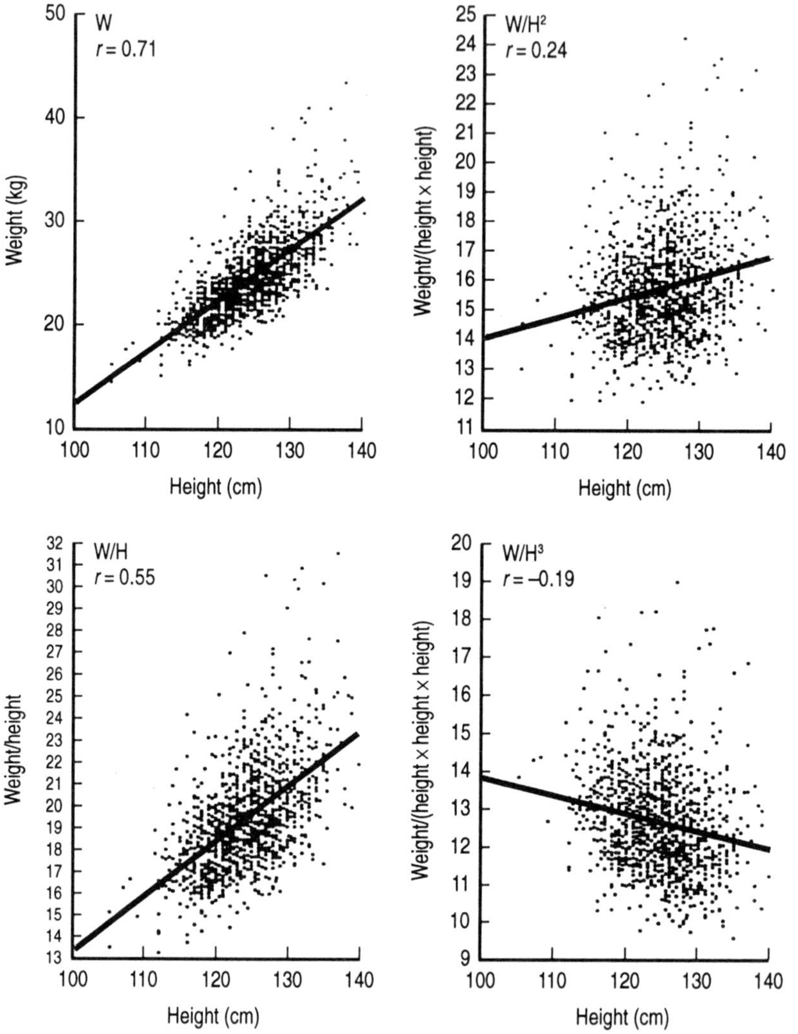

Figure 2 Correlation coefficients during childhood for the different expressions of weight-for-height with height

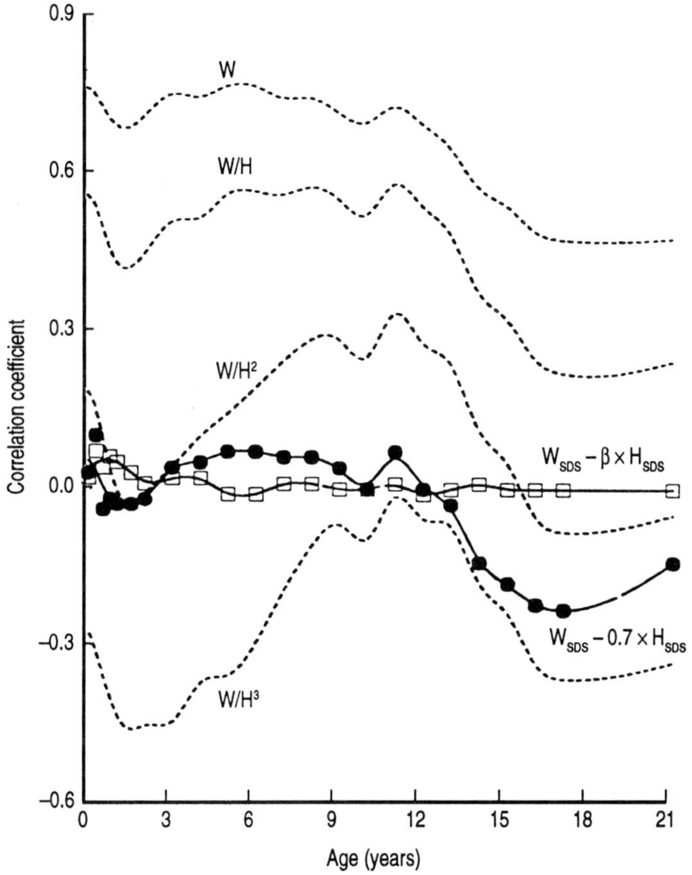

***Figure 3** Scatter-plot, linear regression and correlation coefficient between two expressions of weight SDS and height SDS. The data points represent girls at 7 years of age (n = 1333). The left graph gives the scatter plot between weight SDS and height SDS, and the slope and the residual SD values are given in Appendix A. The right graph includes an expression of weight-for-height, i.e. weight$_{SDS} - \beta \times$height$_{SDS}$ where β is the slope between weight SDS and height SDS as given in Appendix A*

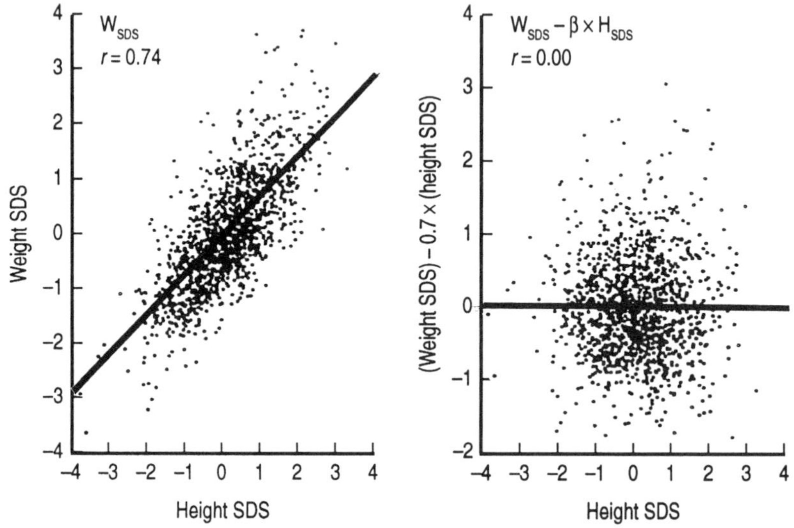

1 and 7) there was a similar change in both WH_{SDS} SDS and height SDS, i.e. increase or decrease during childhood. However, the change in height SDS came after the change in WH_{SDS} SDS in only 50% of the latter children. In 9% the δ SDS was reverse, i.e. an increase or decrease in height SDS when the weight-for-height estimate showed the reverse change. One single case (group 8) showed an increase in height SDS with constant weight SDS leading to a reduction in WH_{SDS} SDS.

DISCUSSION

These results contradict the general belief that over-nutrition accelerates linear growth, just as under-nutrition retards it. Our results show that significant increases or decreases in weight-for-height have no or only marginal effects on the gain in height of normal, healthy children during the prepubertal years.

Most previous studies were performed two to three decades ago, frequently in obese children treated at paediatric clinics (2–7). These studies, often based on a relatively low number of cases, seem to support the hypothesis that over-nutrition accelerates linear growth or that obese children forced to lose weight have a deceleration in

***Figure 4** Estimating weight-for-height SDS from growth charts of weight and height for one girl followed at four points during childhood. The height SDS values estimated from the growth chart (left graph) are 0.2, 0.3, 0.7 and 1.3 and the weight SDS values −0.4, 1.3, 2.4 and 3.4. Based on the approximate equation for WH_{SDS} ($W_{SDS} - 0.7 \times H_{SDS}$) we will obtain values of −0.6, 1.1, 1.8 and 2.5, respectively. After dividing these values by the approximate SD of 0.7 we will have the following WH_{SDS} SDS values: −0.8, 1.6, 2.6 and 3.5 SDS. A quicker way is to plot height SDS against weight SDS in the graph given to the right including the approximate values of the linear regression ($W_{SDS} = 0.7 \times HSDS$) and the residual SD (0.7 SDS)*

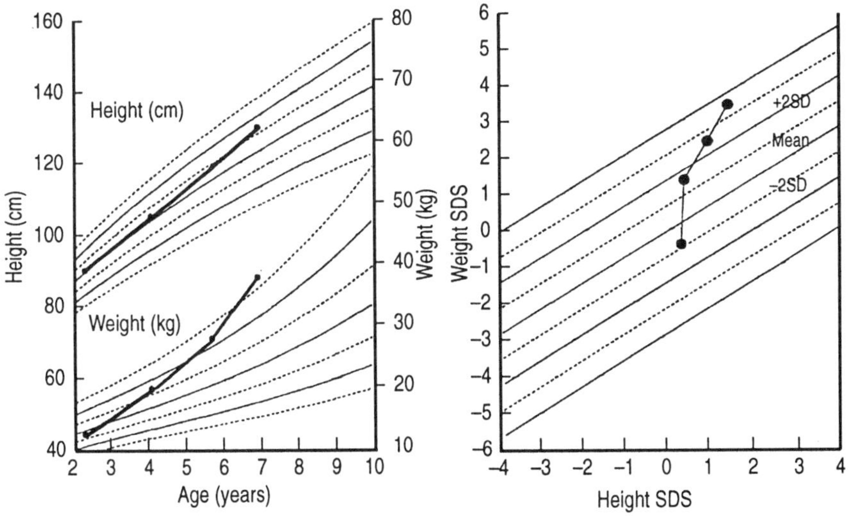

***Table 1** The bivariate correlation (Pearson) coefficient between height SDS, weight SDS and WH_{SDS} SDS at 5 years of age and their change at 5–8 years of age. Both sexes pooled (n = 2958)*

	Measures					
Measures	*1*	*2*	*3*	*4*	*5*	*6*
At 5 years						
1. WH_{SDS} SDS	1.00	−0.03	0.60	−0.37	0.10	−0.26
2. Height SDS		1.00	0.74	0.00	−0.14	−0.13
3. Weight SDS			1.00	−0.23	−0.05	−0.27
Change at 5–8 years						
4. WH_{SDS} SDS				1.00	−0.10	0.80
5. Height SDS					1.00	0.43
6. Weight SDS						1.00

***Table 2** Delta height, weight and WH_{SDS} SDS at 5–8 years of age for different groups of children as defined by the change in WH_{SDS} SDS at 5–8 years (integer groups of 0.5 units). Both sexes pooled. The p value for the t-test with an expected value of zero is also included*

δ WH_{SDS}		δ height SDS			δ weight SDS			δ WH_{SDS} SDS		
Group	n	Mean	SD	p	Mean	SD	p	Mean	SD	p
−1.75	34	0.09	0.54	0.35	−1.05	0.41	0.00001	−1.72	0.15	0.00001
−1.25	136	0.14	0.41	0.0002	−0.69	0.30	0.00001	−1.21	0.14	0.00001
−0.75	488	0.05	0.34	0.002	−0.43	0.27	0.00001	−0.72	0.14	0.00001
−0.25	952	−0.00	0.32	0.69	−0.16	0.26	0.00001	−0.24	0.14	0.00001
0.25	841	−0.06	0.30	0.00001	0.10	0.25	0.00001	0.23	0.14	0.00001
0.75	406	−0.05	0.37	0.01	0.43	0.30	0.00001	0.71	0.14	0.00001
1.25	135	0.03	0.36	0.30	0.82	0.29	0.00001	1.21	0.15	0.00001
1.75	34	0.15	0.50	0.10	1.24	0.39	0.00001	1.71	0.14	0.00001

***Table 3** Height, weight and WH_{SDS} SDS at different ages for two groups of children: (A) with a decrease in WH_{SDS} with 1.5–2.0 SDS at 5–8 years of age, (B) with an increase in WH_{SDS} with 1.5–2.0 SDS at 5–8 years of age. Both sexes pooled. The p value for the t-test with an expected value of zero is also included*

			Height SDS			Weight SDS			WH_{SDS} SDS		
Group	Age (years)	n	Mean	SD	p	Mean	SD	p	Mean	SD	p
A	0.0	34	0.06	1.18	0.75	0.23	1.17	0.26	0.24	1.09	0.26
A	0.5	33	0.22	1.15	0.27	0.20	1.24	0.36	0.06	1.10	0.36
A	1.0	32	−0.02	1.22	0.94	0.54	1.31	0.03	0.69	1.23	0.03
A	2.0	33	0.15	1.23	0.48	0.74	1.29	0.003	0.85	1.06	0.003
A	5.0	34	0.19	1.12	0.32	1.00	1.09	0.00001	1.32	0.95	0.00001
A	8.0	34	0.28	1.18	0.18	−0.05	1.15	0.78	−0.40	0.97	0.78
A	18.0	34	0.29	1.14	0.15	0.01	1.03	0.95	−0.12	1.02	0.95
B	0.0	34	−0.13	0.95	0.42	−0.35	0.92	0.04	−0.37	0.87	0.04
B	0.5	33	0.16	0.89	0.31	0.10	0.79	0.47	−0.00	0.96	0.47
B	1.0	33	0.29	0.91	0.07	−0.19	0.87	0.21	−0.48	0.83	0.21
B	2.0	30	0.16	0.98	0.38	−0.22	0.91	0.21	−0.42	0.83	0.20
B	5.0	34	0.19	0.92	0.25	−0.01	0.92	0.93	−0.24	0.99	0.93
B	8.0	34	0.33	1.06	0.08	1.22	1.09	0.00001	1.46	1.00	0.00001
B	18.0	33	−0.14	0.85	0.36	0.67	1.15	0.003	0.87	1.05	0.003

height growth. It may very well be that such series are biased, for instance by the fact that a tall obese child is more likely to seek advice than a short obese child.

Our data represent a larger healthy Swedish population going through childhood between 1975 and 1983. The results are thus representative for the general normal population. The extremes of obesity and leanness are represented in our data, but not in the same way as in paediatric clinics.

***Table 4** Growth patterns of 103 children with an increase or decrease in WH_{SDS} at 5–8 years of age with ≥ 1.5 SDS. The individual growth patterns are shown in Figs. 5–12. An 'increase' or 'decrease' corresponds to a change of 1.5 SDS for WH_{SDS} and 0.5 SDS for height. Twelve children with three examinations or less during the childhood years were omitted*

Group	*WH_{SDS} SDS*	*Height SDS*	*n*	*%*	*Figure*
1	Increase	Increase	13	14	5
2	Increase	No change	32	35	6
3	Increase	Decrease	2	2	7
4[a]	Inc + dec	No change	10	11	8
5	Decrease	Increase	6	7	9
6	Decrease	No change	18	20	10
7	Decrease	Decrease	9	10	11
8[b]	Decrease	Increase	1	1	12
	Total		91	100	

[a]Increase/decrease followed by decrease/increase in WH_{SDS} SDS
[b]Increase in height SDS, but no change in weight SDS

A number of other methodological issues may influence the outcome of different studies, including the definition of obesity/leanness or change in weight-for-height over time. In this study we have presented a new index for weight-for-height during childhood. This method seems to be a better choice than other such indices, as our index is not correlated with height in normal subjects, and it also offers a simple method of estimating weight-for-height SDS in the clinic. The reason for introducing this method here was merely to avoid the discussion about the influence of the weight-for-height index on the results. Other weight-for-height indices have been criticized for being correlated with height and also because they have not considered the fact that the relationship between weight and height is age dependent (9–12); all of this has been taken into account in the method presented here.

It is important to stress that a weight-for-height index does not necessarily represent the true nutritional status of an individual as it does not give information on single nutritional factors, body composition or bone density (17). For this reason, we cannot state from this study that nutrition has no impact on skeletal growth during childhood or whether it leads to any long-term consequences which persist into adulthood. The main conclusion from this study is simply that gaining or losing excess weight during childhood has no significant impact on height gain in healthy subjects.

ACKNOWLEDGEMENTS

Supported by grants from the Faculty of Medicine, University of Hong Kong, Swedish Medical Research Council (no 6465, 7509), Pharmacia and Upjohn Inc.

***Figure 5** The impact of nutrition on linear growth during childhood based on children with a change in WH_{SDS} SDS (large dot) of > 1.5 units. The two children in this graph represent pattern 1 (Table 4; an increase in WH_{SDS} SDS and an increase in height SDS) which was found in 14%*

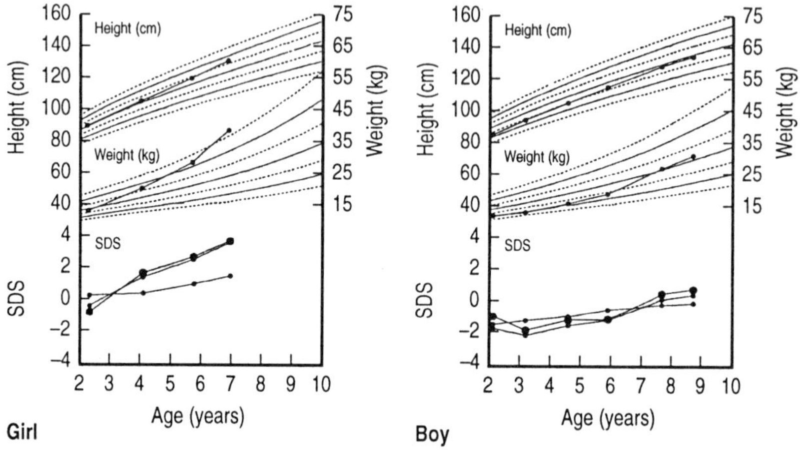

***Figure 6** The impact of nutrition on linear growth during childhood based on children with a change in WH_{SDS} SDS (large dot) of > 1.5 units. The two children in this graph represent pattern 2 (Table 4; an increase in WH_{SDS} SDS and no change in height SDS) which was found in 35%*

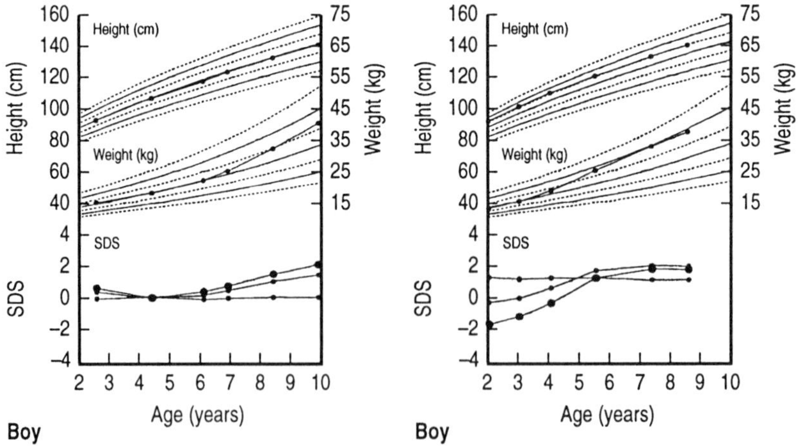

***Figure 7** The impact of nutrition of linear growth during childhood based on children with a change in WH_{SDS} SDS (large dot) with > 1.5 units. The two children in this graph depict pattern 3 (Table 4; an increase in WH_{SDS} SDS and a decrease in height SDS) which was found in 2%*

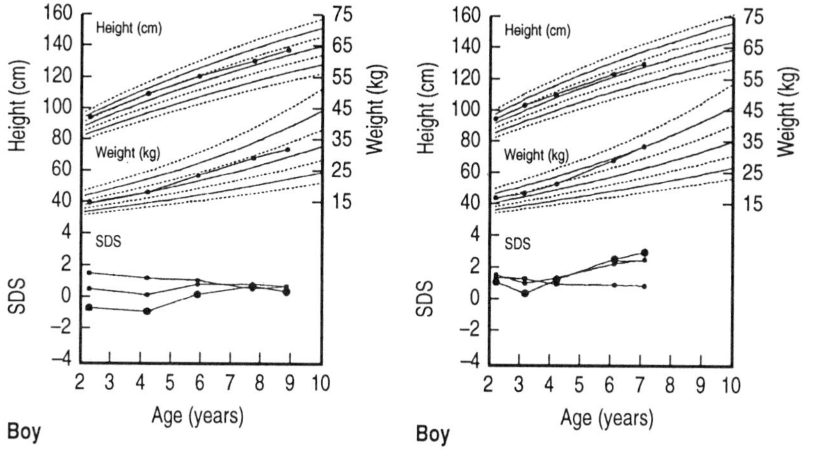

***Figure 8** The impact of nutrition on linear growth during childhood based on children with a change in WH_{SDS} SDS (large dot) of > 1.5 units. The two children in this graph represent pattern 4 (Table 4; an increase and a decrease in WH_{SDS} SDS and no change in height SDS) which was found in 11%*

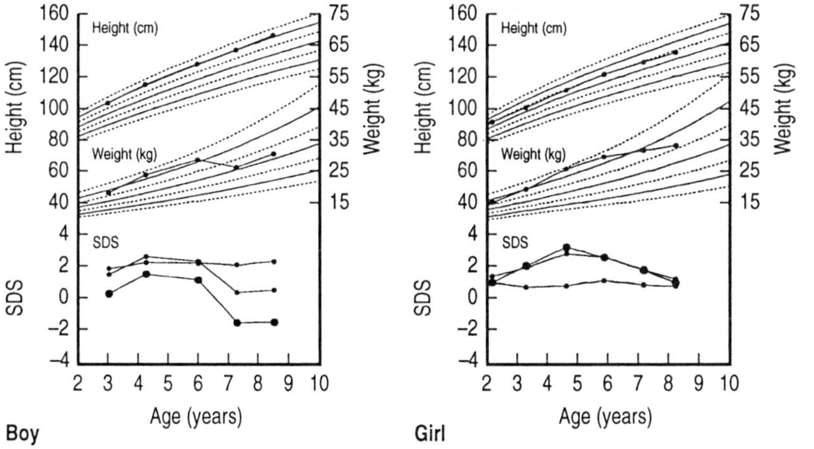

***Figure 9** The impact of nutrition on linear growth during childhood based on children with a change in WH_{SDS} SDS (large dot) of > 1.5 units. The two children in this graph represent pattern 5 (Table 4; a decrease in WH_{SDS} SDS and an increase in height SDS) which was found in 7%*

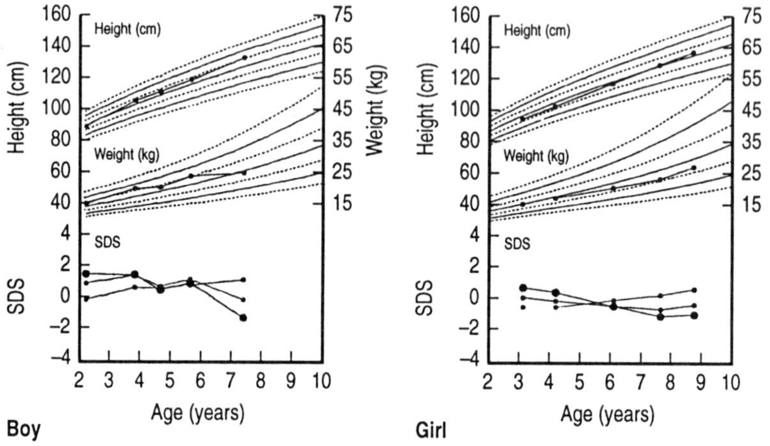

***Figure 10** The impact of nutrition on linear growth during childhood based on children with a change in WH_{SDS} SDS (large dot) of > 1.5 units. The two children in this graph represent pattern 6 (Table 4; a decrease in WH_{SDS} SDS and no change in height SDS) which was found in 20%*

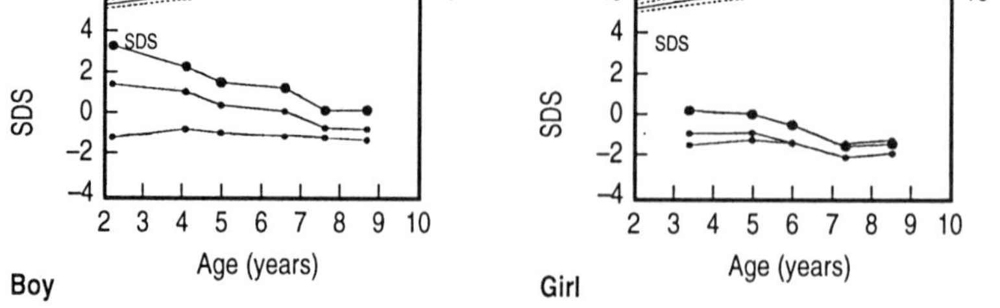

***Figure 11** The impact of nutrition on linear growth during childhood based on children with a change in WH_{SDS} SDS (large dot) of > 1.5 units. The two children in this graph represent pattern 7 (Table 4; a decrease in WH_{SDS} SDS and a decrease in height SDS) which was found in 10%*

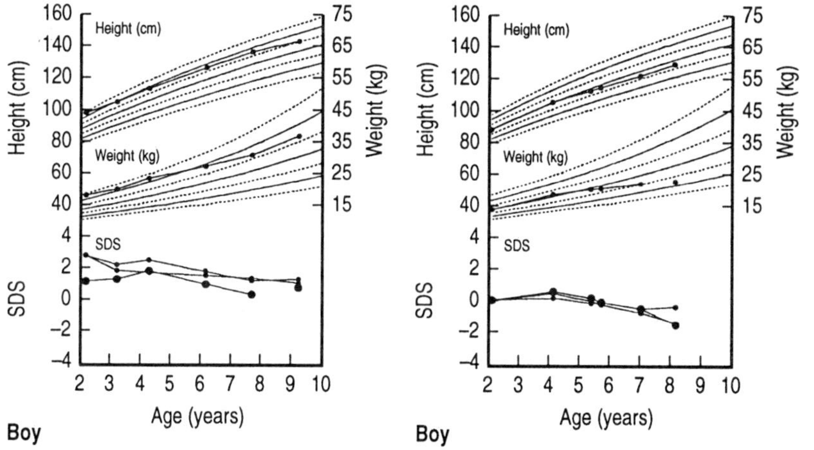

***Figure 12** The impact of nutrition on linear growth during childhood based on children with a change in WH_{SDS} SDS (large dot) of > 1.5 units. The child in this graph represents pattern 8 (Table 4; a decrease in WH_{SDS} SDS and an increase in height SDS) which was found in 1%, i.e. one child*

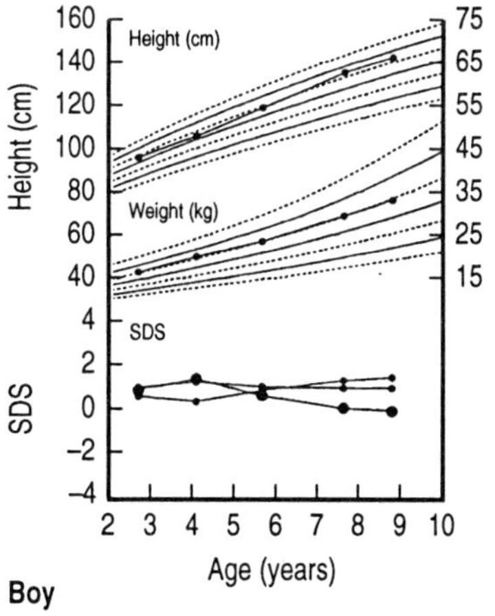

REFERENCES

1. Karlberg J 1989 On the construction of the infancy-childhood-puberty growth standard. Acta Paediatr Scand 356:26–37
2. Lloyd JK, Wolff OH, Whelen WS 1961 Childhood obesity. A long-term study of height and weight. Br Med J 2:145–148
3. Garn SM, Clark DC, Guire KE 1973 Level of fatness and size attainment. Am J Phys Anthropometry 40:447–450
4. Forbes GB 1977 Nutrition and growth. J Pediatr 91:40–42
5. Wilkinson PW, Parkin JM, Pearlson J, Philips PR, Sykes P 1977 Obesity in childhood: a community study in Newcastle Upon Tyne. Lancet i:350–352
6. Poskitt EME, Cole TJ 1977 Do fat babies stay fat? Br Med J 1:7–9
7. Dietz WH Jr, Hartung R 1985 Changes in height velocity during weight reduction. Am J Dis Child 139:705–707
8. Karlberg J, Albertsson-Wikland K 1995 Assessing weight for height. Acta Paediatr Suppl 411:93
9. Fung KP, Lee J, Lau SP, Chow OKW, Wong TW, Davies DW 1990 Properties and clinical implications of body mass indices. Arch Dis Child 65:516–519
10. Cole TJ 1979 A method for assessing age-standardized weight-for-height in children seen cross-sectionally. Ann Hum Biol 3:249–268
11. Poskitt EME 1995 The fat child. In: Brook CED (ed) Clinical Paediatric Endocrinology. Blackwell Science Ltd, Oxford, pp 210–233
12. Forbes GB 1972 Relation of lean body mass to height in children and adolescents. Pediatr Res 6:32–37
13. Karlberg J, Mossberg HO 1991 Weight-for-height standards in adulthood. J Intern Med 229: 303–308
14. Karlberg J, Albertsson-Wikland K 1995 Growth in full-term small-for-gestational-age infants: from birth to final height. Pediatr Res 38:733–739
15. Niklasson A, Ericson A, Fryer JG, Karlberg J, Lawrence C, Karlberg P 1991 An update of the Swedish reference standards for weight, length and head circumference at birth for given gestational age (1977–1981). Acta Paediatr Scand 80:756–762
16. Albertsson-Wikland K, Karlberg J 1994 Natural growth in children born small for gestational age and with and without catch up growth. Acta Paediatr 83:64–70
17. Allen LH 1993 Nutritional influences on linear growth: a general review. Eur J Clin Nutr 48 (suppl):S75–S89

Appendix A **Weight-for-height reference values at different ages (** $Weight_{SDS} = \beta \times Height_{SDS}$ **)**

Age (years)	Girls *n*	Girls Slope (β)	Girls Residual SD	Boys *n*	Boys Slope (β)	Boys Residual SD
0.00	1804	0.70667	0.70206	1857	0.74633	0.69703
0.25	1502	0.72070	0.68544	1534	0.71749	0.68664
0.50	1468	0.63535	0.78165	1484	0.65229	0.75890
0.75	1347	0.63324	0.78801	1358	0.63979	0.77636
1.00	1155	0.63198	0.78713	1179	0.63532	0.79514
1.50	1294	0.64923	0.77696	1323	0.57929	0.81546
2.00	1116	0.67753	0.74816	1106	0.64595	0.76346
3.00	1173	0.71826	0.69615	1168	0.73431	0.68189
4.00	1383	0.72636	0.69699	1387	0.76784	0.68976
5.00	741	0.75848	0.65813	739	0.77462	0.63258
6.00	1189	0.75564	0.65343	1187	0.78784	0.64344
7.00	1333	0.73884	0.68214	1387	0.75173	0.65915
8.00	1054	0.73439	0.68044	1028	0.75241	0.65544
9.00	812	0.72639	0.68123	841	0.74657	0.66006
10.00	878	0.69383	0.72328	840	0.73966	0.69282
11.00	934	0.74659	0.67140	964	0.74265	0.67335
12.00	694	0.70979	0.70852	651	0.73765	0.68990
13.00	1160	0.67521	0.74045	1151	0.76912	0.64483
14.00	840	0.57774	0.81719	781	0.75933	0.66432
15.00	982	0.55270	0.83307	1033	0.72353	0.69852
16.00	693	0.49918	0.86720	770	0.58424	0.81648
17.00	613	0.48794	0.88605	739	0.56076	0.82105
21.00	1791	0.56337	0.87278	1851	0.51049	0.85888

9 Infant feeding and body composition in later life

D.S. LEWIS

ABBREVIATIONS

BMI	body mass index
CHD	coronary heart disease
EGF	epidermal growth factor
IGF-I	insulin growth factor I
ODC	ornithine decarboxylase
TSH	thyroid stimulating hormone

INTRODUCTION

Although obesity is associated with increased mortality, one does not have to be obese to be at increased risk. Body weight in the high end of the 'normal' range increases the risk of coronary heart disease (CHD) (1) and mortality (2,3). A major weight gain ultimately resulting in overweight and obesity typically occurs between 20 and 35 years of age (4). One approach to the prevention of adult obesity is to identify when the trend for overweight and obesity first becomes evident. In a significant proportion of 36-year-old obese adults the trend for development of obesity begins with onset of overweight in childhood (5) (Fig. 1). Critical periods for the development of obesity were recently considered to be the prenatal period, early childhood, and adolescence (6) and the early pre-weaning period during the first year of life (7–9).

Observations in humans have led to inconsistent conclusions regarding whether infant diet influences child, adolescent and adult body composition, yet animal experiments consistently demonstrate long term effects of infant diet on adult body composition or on factors that can influence adult body composition. How the effects of infant under- and overnutrition on adult adiposity in animals apply to humans is not known. Results from animal studies may not apply to humans because of different developmental processes and possibly different nutritional requirements during development. Identifying the mechanisms by which infant nutrition affects adult body fat in animals is likely to provide a better understanding of the potential for infant nutrition to influence adiposity in humans.

RELATIONSHIP BETWEEN INFANT FEEDING AND FATNESS IN HUMANS

There is no evidence for a large effect of nutrition or other factors during infancy on child, adolescent, or adult obesity (Tables 1 and 2) (10–36). With a few exceptions,

J.G. Bindels, A.C. Goedhart and H.K.A. Visser (eds), Recent Developments in Infant Nutrition, 128–147.

***Figure 1** Percentage of obese 36-year-old men and women who were either obese (■; BMI > 29.9) or overweight (□; BMI 25.0–29.9) at earlier ages. Data from Ref. 5 (BMI = weight/height²)*

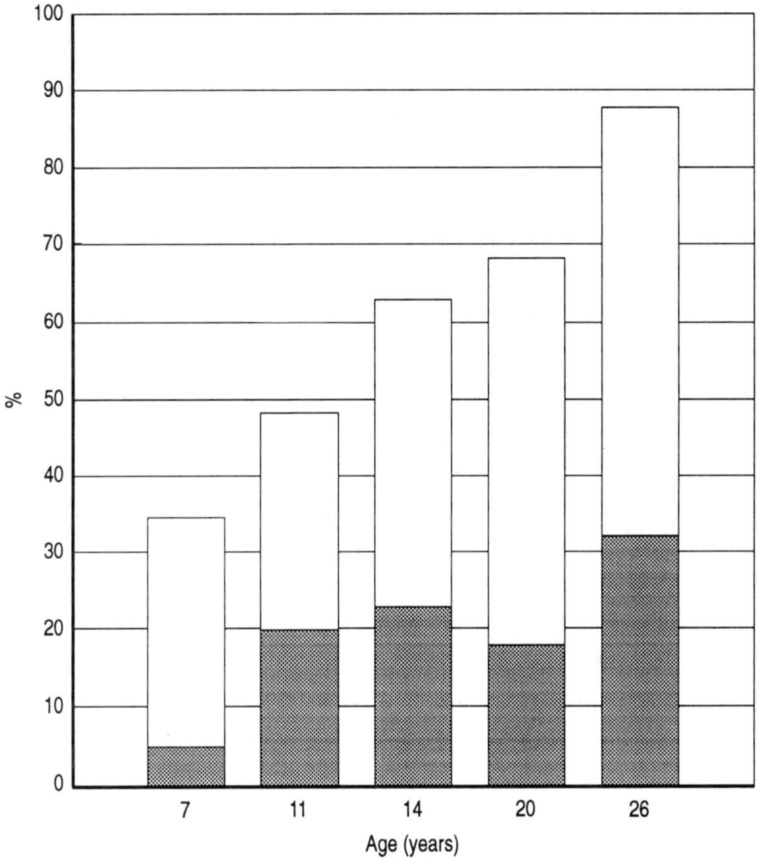

the degree of fatness during childhood appears to be independent of infant nutrition, infant body weight status, or formula *vs* breast-feeding (Table 1). Although infant overnutrition, infant body weight status or formula *vs* breast-feeding are associated with increased adolescent and adult adiposity in many studies (Table 2), the magnitude of the effects is small. The incidence of obesity in 19-year-old men born during a severe food shortage in Holland during World War II (35), presumably reflecting undernutrition, was 0.82% compared with 1.32% in a control area. Obese infants are more likely to become obese later in life (15,33) than are non-obese infants, but most do not develop obesity. The lack of significant associations between infant feeding and

Table 1 Relationship between infant (0–1 years) nutrition and childhood (2–10 years) fatness

Infant nutrition	*Follow-up age (years)*	*Result*	*Reference*	*Country*
High caloric intake	3	Increased BMI (r=0.33)	10	USA
	6	No effect	11	USA
Early introduction of solid food	5	No effect	12	UK
Rapid weight gain	10	No effect	13	Sweden
Infant obesity/ overweight	3	Increased relative weight (r=0.33–0.35)	14	Canada
	5	One of three obese infants remained obese	15	UK
	6–8	Increased weight and skinfold thickness	16	UK
	6	No effect	17	American Somoa
	9	No effect	18	USA
Formula *vs* breast feeding	2–5	No effect	19–21	USA
	4	No effect	22	USA
	5	No effect	23	USA
	4–5	Increased incidence of of obesity (from 2 to 7%)	24	Slovakia
	7	No effect	25,26	New Zealand, UK
	7	Increased skinfold thickness	17	American Somoa
	8	No effect	27	USA
	10	No effect	28	UK

obesity in childhood together with weak associations between infant nutrition and obesity in adult life suggest that the early postnatal period may not be important, or that the degree of nutrition studied was not sufficient to influence the development of obesity.

The available studies are probably not sufficient to warrant rejecting the role of either infant feeding or the preweaning period as factors contributing to fat development in children and adults. The interpretation of many studies depends on the extent to which infant adiposity and weight gain are related to overnutrition because food intake was not measured. There is no convincing evidence that infant overfeeding causes obesity within the first 2 years of life (Table 3). Infant obesity is unique because obese infants spontaneously lose body fat and become normal weight, whereas obese adolescents and adults do not. Rapid fat gain during infancy (55) may be regulated by developmental factors unique to infants, such as enhanced expression of anti-lipolytic alpha-2 adrenergic receptors in infant adipocytes (56). Infant growth rates and adiposity may also be influenced by genetic potential (57) or reduced energy expenditure (58).

The amount of body fat at any age is a reflection of a complex interaction between

Table 2 Relationship between infant (0–1 years) nutrition and adolescent and adult fatness

Infant nutrition	*Follow-up age (years)*	*Result*	*Reference*	*Country*
Early introduction of solid food	11, 15, 20	No effect	26	UK
Rapid weight gain	15	No effect	29	Germany
	20–26	Increased weight : length ratio	30	Germany
Low weight at birth and 1 year	>50	Increased waist : hip ratio	31	UK
Infant obesity/ overweight	9–14	Increased skinfold thickness	32	USA
	20–30	Increased incidence of overweight (36%) compared with lean infants (14%)	33	USA
Formula *vs* breast feeding	12–18	Higher incidence of obesity in formula-fed (27%) than breast-fed (8%)	34	Canada
Exposure to reduced food supply prenatally to 5 months of age	19	Decreased incidence of obesity compared to control area (0.82% *vs* 1.32%)	35	Holland
1–10 months of age	19	No effect	35	Holland
1–12 months of age	27	Reduced weight and obesity : length ratio	36	Germany

genes and fatty acid esterification in adipocytes and oxidation of fatty acids in lean tissues. Most associations between early nutrition and child and adult obesity are confounded by the heterogeneous nature of genetic factors, diet, physical activity, and normal changes in body composition during childhood and adolescence (59). Animal experiments circumvent many of these problems.

INFANT NUTRITION AND ADULT ADIPOSITY EXPERIMENTS IN RODENTS

Altering milk intake by litter size manipulation has been used to investigate the effects of infant overnutrition (small litters) and undernutrition (large litters) on adult body composition (Table 4). Neonatal undernutrition reduces adult body fat, adipocyte number and size. Reduced adipocyte replication is detectable at 10 days of age (71), but most of the differences in fat mass, adipocyte number and size between rats under- and overfed as pups occur after sexual maturation (68,70), Rats overfed as pups gained 3.1 million more epididymal adipocytes and 0.226 μg more triacylglycerol per

Table 3 ***Relationship between infant (0–1 years) nutrition and infant (< 2 years) fatness***

Infant nutrition	*Follow-up age*	*Result*	*Reference*	*Country*
Food intake				
Formula	4 months	Varying energy density 54, 67 100, 133 kcal/dl increased growth, not fatness	37, 38	USA
	4 months	100 kcal/dl increased weight gain (no information on fatness)	39	Italy
Human milk	2½ months	Increased milk intake associated ($r = 0.68$) with fat deposition, not lean tissue accretion	40	Sweden
	1½–5 months	Overweight infants increased breast milk intake	41	USA
	1–2 years	Vigorous sucking/no. feeds per day associated with increased skinfold thickness ($r^2 = 0.13–0.21$)	10	USA
Early introduction of solid food	4 months	No effect	42	USA
	6 months	No effect	43–45	UK, USA, Canada
	1 year	Higher percentage (44.3%) were overweight or obese	46	UK
	1 year	No effect	47	USA
Infant obesity	4–12 months	Energy intake similar between obese and lean infants	48–51	USA, Sweden, Canada
Formula *vs* breast feeding	1½ months	More formula-fed than breast-fed infants were overweight or obese (59% *vs* 19%)	52	UK
	3 months	No effect	53	Germany
	5–6 months	No effect	44	USA
	6 months	Increased weight, skinfold thickness, weight : length ratio	45	Canada
	2 years	Increased adiposity	54	Canada

cell between 5 and 20 weeks compared with underfed pups. Only in rat models where the period of adipocyte replication extends beyond the preweaning period did neonatal undernutrition not have a permanent effect on body fat and adipocyte number (60,62,63,72). If undernutrition was continued until 12 weeks of age adult fat mass and adipocyte number was reduced (62,63).

Whether neonatal overnutrition increases adult body weight, fat mass and adipocyte number depends on the type of rodent studied (Table 4). Infant overnutrition increases

Table 4 Effects of litter size manipulation on fatness later in life in rodents

Nutrition (litter size)	*Follow-up age (weeks)*	*Total*			*Fat depot mass*	*Fat cell*		*Reference*
		Body mass	*Fat mass*	*Protein mass*		*No.*	*Size*	
Undernutrition	26 lean[a]	↓	↓	ND[b]	↓	↓	↔	60
(large *vs* standard)	26 obese[a]	↓	↓	ND	↓	↔	↓	60
	8	↓	↓	↓	ND	ND	ND	61
	3–32	↓	↓	↓	↓	↔	↓	62
	24	↓	ND	ND	↓	↔	↓	63
	24	↓	↓	↔[c] ↓[d]	↓	↓	↓	64
	43	↓	ND	ND	↓	↓	↓	65
Overnutrition	26 lean[a]	↑	↑	ND	↑	↑	↔	60
(small *vs* standard)	26 obese[a]	↑	↔	ND	↔	↑	↓	60
	8	↔	↔	↔	ND	ND	ND	61
	3–32	↑	↑	↑	↔	↔	↔	62
	24	↔	ND	ND	↓	↔	↓	63
	24	↑	↑	↔[c] ↑[d]	↑	↑	↑	64
	15	↔	ND	ND	↔	↔	↔	66
	43	↔	ND	ND	↔	↔	↔	65
Extremes in food intake	23	↓	↓	↓	ND	ND	ND	67
(large *vs* small)	5–20	↓	ND	ND	↓	↓	↓	68
	48	↓	ND	ND	↓	↓	↔	69
	3–43	↓	ND	ND	↓	↓	↓	70

[a]Zucker lean and obese rats; [b]ND not determined; ↑ increase; ↓ decrease; ↔ no significant difference
[c]female; [d]male

adult fat mass, adipocyte number and size in mice (64) and Zucker obese and lean rats (60). Enhanced adipocyte differentiation and increased adipocyte number are detectable in overfed pups at 10 days of age (73). Studies often cited as evidence that infant overfeeding increases adult adipocyte number compared the effects of small and large litters (68,69). However, the difference in adipocyte number is probably due to undernutrition since other experiments with similar rat species found no difference between adult rats reared on small and standard litters on body weight (61,74), total body fat mass (61), and adipocyte cell number and size (65).

The development of formulas for rearing rat pups (75–77) allows the effects of energy intake to be observed independent of growth factors and other compounds in rat milk. Rats reared on artificial formulas appear to grow normally (77,78), but formula feeding is not initiated until 4–6 days and these experiments require 'robust' pups maintained under stringent conditions. Brain weights are reduced, and there are various subtle changes in some metabolic parameters (77).

Overfeeding rat pups in normal (79) or small (66) litters with supplementary feeding of a milk-based formula had no effect on adipocyte number and only slightly increased adipocyte size. Adult rats reared solely on a high fat formula between the ages of 4 and 18 days were 27% heavier and had increased fat depot mass, due mostly to increased adipocyte size, compared with rats reared in a standard litter or rats reared by gastrostomy without the fat supplement (78). Feeding a high carbohydrate formula also increased adiposity after sexual maturation (80).

Potential mechanisms for effects of infant diet on adiposity in rats

The effect of infant over- and undernutrition on adipocyte number is an attractive hypothesis (7,8) to explain the tendency for some obese infants to become obese adults (16,33) and early onset obesity being associated with increased adipocytes (8). Although the effect of infant feeding on adipocyte number in adult obesity has been criticized (81) there is no direct evidence supporting or disproving this hypothesis in humans. Why adipocyte number is influenced by neonatal overfeeding in rodents is not clear.

Litter size manipulation affects more than just food intake in the pups. Manipulation of litter size affects milk composition (82,83), energy, protein and fat intake (82), intake of bioactive compounds present in milk (84), maternal behaviour, pup sucking behaviour, maternal–pup interactions (85,86), learning behaviour in pups (87) and brain weight (88). The long term effects of litter size manipulation on growth, body size and adiposity in adult rats could result from nutritional differences, behaviour differences or a combination of both. Identification of the actual factors accounting for the long term effects of infant nutrition in the rat are critical to understanding their applicability to humans.

Altered polyamine metabolism may mediate the effects of litter size on growth and adipocyte number for several reasons. Polyamines are essential for cell multiplication (89), which increases the proportion of adipose cells in culture (90,91). Intracellular synthesis of polyamines is required for DNA synthesis and mitosis of preadipocytes (92) and for preadipocytes to differentiate into adipocytes (93–96). Inhibition of ornithine decarboxylase (ODC) immediately prior to or during preadipocyte differentiation completely blocks the formation of adipocytes (Fig. 2). Only partial recovery of differentiation was observed if putrescine was added 1, 2 or 3 days after initiation of differentiation was initiated (96). Therefore, depletion of preadipocyte polyamine concentrations can subsequently reduce the number of adipocytes in vitro. These experiments mimic the situation that may occur with pups reared in large and small litters. Food deprivation (97,98), lack of tactile interaction between mother and pup (99), and large litter sizes (100) reduce ODC in pup tissues. ODC is increased in tissues of pups reared in small litter sizes (100). There have been no reports concerning ODC activity in adipose tissue of over- and underfed pups.

Other observations consistent with a role for polyamines in mediating the effect of infant over- and undernutrition in rodents include a high polyamine concentration in milk (rat milk 46 μmol/l; human milk 4.5 μmol/l) (101). Different milk intakes would result in 25% less intake (0.7 μmol/day) in underfed pups and 40% greater intake (1.3 μmol/day) in overfed pups compared with standard fed pups (0.9 μmol/day). These intakes were estimated from milk intakes provided from Fiorotto et al. (82), and milk compositions from Pollack et al. (101). Enterally administered polyamines are absorbed and transported to peripheral tissues (102,103), promote intestinal development and accelerate intestinal sucrase and maltase activities (104,105).

***Figure 2** Role of polyamines on preadipocyte differentiation in vitro. Murine 3T3-L1 preadipocytes were treated 3 days prior to insulin and 1-methyl-3-isobutylxanthine stimulation of differentiation into adipocytes with difluoromethylornithine, an inhibitor of ornithine decarboxylase activity (ODC inhibitor, ●). Control cells were not exposed to the ODC inhibitor (○). Putrescine, the product of ornithine decarboxylase, was added at 0 (▽), 1 (▼) and 3 (□) days following stimulation of differentiation to cultures previously exposed to the ODC inhibitor. The number of fat cell colonies was determined at 12 and 18 days after differentiation was initiated. Data are from Ref. 96*

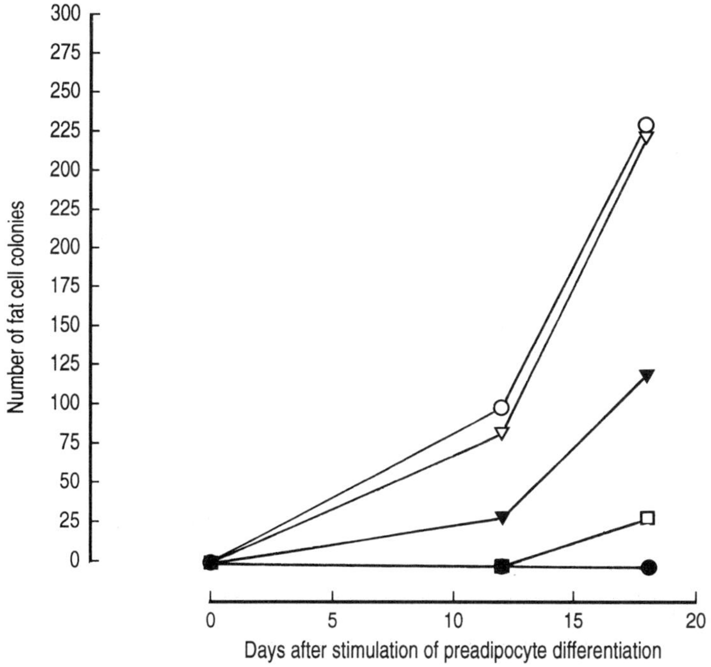

In addition to polyamines, rat milk contains biologically active compounds, including epidermal growth factor (EGF, 3.6–12 μg/l; 106) and insulin growth factor I (IGF-I, 16.2 μg/l; 107), which may affect growth and development of the infant directly by increasing gastrointestinal development, improving digestion and nutrient utilization, and by entering the circulation and interacting with target tissues (84,108). These factors stimulate cell proliferation in vitro (108,109) and EGF administered enterally stimulates somatic, intestinal and hepatic growth and DNA synthesis in rat pups (110–114). IGF-I is required for in vitro preadipocyte differentiation (115), and EGF as a mitogen inhibits differentiation but stimulates the proliferation of preadipocytes (116).

Pups in large litters probably consume considerably less EGF and IGF-I than do standard and overfed pups, assuming milk-borne EGF and IGF-I concentrations are not altered by litter size manipulation. Reduced exposure of intestinal cells to milk-derived growth factors, together with reduced milk polyamines (and perhaps reduced intracellular polyamine synthesis) may contribute to delayed or reduced intestinal growth and development, subsequently reducing overall body growth.

Thyroxine, thyroid stimulating hormone (TSH), and thyroid releasing hormone are present in breast milk (84) and may influence triiodothyronine levels in pups (117). TSH is present in rat milk (35 μU/ml) and daily intake of TSH in suckling pups has been estimated as 1 mU/100 g body weight. Suckling rats given TSH exhibit better growth (118).

It is not known whether the magnitude of the difference in intake of biologically active molecules in rat milk between overfed and underfed pups is sufficient to elicit effects in vivo. It is likely that multiple growth factors act together to affect growth such as thyroxine potentiating the effect of EGF (119).

An alternative hypothesis is that the regulatory system for adipocyte differentiation and function is altered by infant under- and overnutrition when endocrine organs are developing in the rat pup (120). This hypothesis is based on observations that most rats overfed as pups are hyperinsulinaemic (60,64,70,121–124). Hyperinsulinaemia may be related to increases in adipocyte number. Committed fibroblasts from newborn rats require insulin for conversion to adipocytes (124). Concomitant with hyperinsulinaemia, hepatic fatty acid synthesis is also enhanced in adult rats overfed as pups (64,123,125). Adipocyte lipogenesis in adults is unaffected by infant nutrition (64,68). Oral intubation of glucose to overfed pups in a restricted litter enhanced hepatic lipogenesis earlier in adulthood than following overfeeding alone (125). It is likely that altered insulin metabolism induced by neonatal overfeeding combined with enhanced hepatic lipogenic activity contributes to the excessive body fat observed in rodents overfed as neonates.

During the preweaning period, administration of a high carbohydrate formula (122), reducing litter size (126) or feeding glucose supplements to pups in small litters (125) increased serum insulin, pancreatic insulin concentrations and pancreatic islet cell size and number of insulin-positive cells compared with pups reared in standard litters. The hyperinsulinaemia induced during the preweaning period was permanent and probably accounts for the continued hyperinsulinaemia observed during the post-weaning period and increased hepatic lipogenesis and adiposity in adult rats.

INFANT DIET AND ADULT ADIPOSITY EXPERIMENTS IN NON-HUMAN PRIMATES

Studies were initiated in baboons to test the hypothesis that infant overnutrition influences adiposity and adipocyte number (128–132). The baboon is more mature during the preweaning period than rodents, whose endocrine systems develop and mature near the end of the preweaning period (120). The formulas used to overfeed and underfeed are briefly summarized in Table 5. The degree of overfeeding with the concentrated formula is similar to that seen in human infants receiving over-concentrated feeds (37,39) or incorrectly reconstituted formulas (133). Compared with baboons receiving diluted or normal formula, those fed the concentrated formula grew faster

Table 5 Experimental diets used in baboon studies

	Formula[a]		
Nutrient (per litre)	*Diluted* (3:7)[b]	*Normal* (5:5)	*Concentrated* (7:3)
Energy (kcal)	405	675	945
Protein (g)	6.2	14.4	20.2
Fat (g)	15.7	36.4	51.0
Carbohydrate (g)	31.1	72.2	101.1
Sodium (mmol)	5	8	11
Water (ml)	938	898	857
Solute load (mOsm)	76	135	194

[a]Similac® concentrated liquid, Ross Laboratories, Columbus, Ohio, USA
[b]Parts formula: parts water

and were larger at weaning with greater absolute amounts of lean and fat mass, but they were not fatter based on percentage of total body weight (129,132,134). Body weight after weaning was similar for all three formula groups until 3–4 years, when the female overfed infants became heavier and fatter (130,131). Lean body mass and fat mass of underfed, normal and overfed infant baboons at 5 years of age are shown in Figure 3. The increased body fat in the female baboons resulted from larger adipocytes in every fat depot (130): adipocyte number was increased in only flank (subcutaneous) and mesenteric (visceral) fat depots. Increased adipocyte hyperplasia in the mesentery accounted for the significant increase in the mesenteric fat mass in male baboons overfed as infants (130,131). Although the extent of adipocyte hyperplasia in the mesenteric fat depot was not sufficient to influence total body fat in the male, the shift in fat distribution to the visceral region in overfed prepubertal baboons could be important to the development of abnormal metabolism (135). Adolescence is an important period for adult visceral fat development since this is when the number of adult visceral adipocytes is determined (136).

Potential mechanisms for effects of infant diet on adiposity in baboons

To identify the mechanism by which infant overfeeding increases adolescent adiposity, studies have focused on the preweaning period, when overnutrition occurs, and between 3 and 4 years of age, before the maximal effect of infant overfeeding is observed. Adipose tissue normally increases by hyperplastic and hypertrophic mechanisms during these two periods (136) (Fig. 4). How consumption of a concentrated formula results in adipocyte hypertrophy and, to a lesser extent, hyperplasia during adolescence in the baboon is not obvious. Neonatal exposure to exogenous thyroid and glucocorticoid hormones permanently changes adult endocrine function (137–143). Moreover, modulation of endogenous hormones by merely handling pups for 15 minutes per day was sufficient to influence glucocorticoid metabolism in adult rats (144). Similarly, fluctuations in hormones induced by the concentrated formula may influence preadipocyte differentiation, adipocyte function, or endocrine systems

***Figure 3** Total body lean body mass (a) and fat mass (b) in 5-year-old female and male baboons that were underfed (■), normally fed (▨) or overfed (▩) as infants. Body fat mass is equal to the triacylglycerol mass from 10 dissected fat depots plus triacylglycerol extracted from the remaining carcass excluding the adipocyte free tissues: blood, liver, kidneys, brain, eyes, spinal cord, lungs, spleen, tongue and hair. *Significantly ($p < 0.05$) different from normal and underfed groups. Data from Ref. 131*

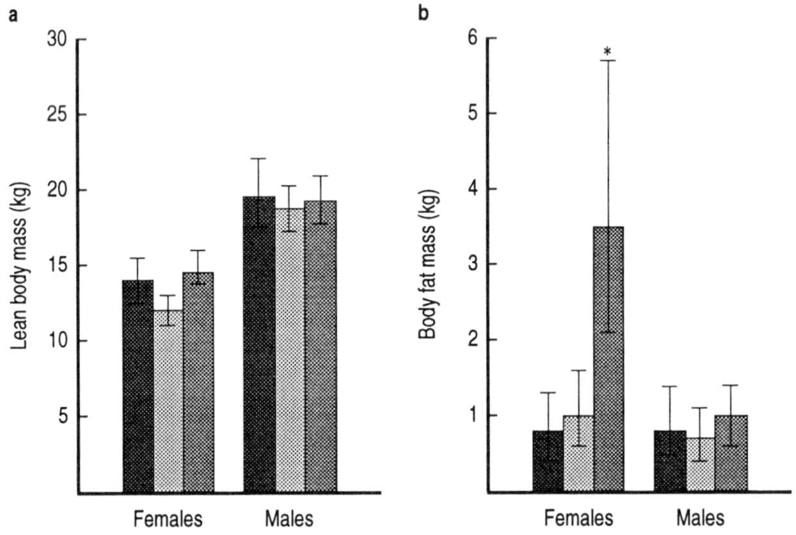

that regulate adipocyte function or some aspect of energy balance. Overfed infants have elevated thyroid hormones, fasting and postprandial insulin levels and depressed 24-h cortisol production and cortisol levels during the first 8 weeks of life (Fig. 5) (145). The most intense and long lasting response to overfeeding is the adaptation by infants to excrete less urine (75 *vs* 140 ml/kg per day, $p<0.05$) with higher osmolality (231 *vs* 135 mosmol/kg, $p<0.05$) in response to the increased solute load of the concentrated formula (146). This response is consistent with increased vasopressin synthesis and release. Neonatal exposure to vasopressin has long term effects on water metabolism in rats (147), although the relationship between vasopressin and adiposity is obscure.

Whether the metabolic processes affected by infant overnutrition and/or the associated hormone changes reside with regulation of adipocyte function or overall energy intake and expenditure is not clear. There is no evidence that regulation of energy balance is altered by infant overfeeding. There is no detectable differences in food intake from weaning until one year (Lewis and Jackson, unpublished observations). After weaning, body weights were similar between overfed and normally fed infants, suggesting similar energy balances. Infant overnutrition has no detectable

***Figure 4** Development of omental mean adipocyte volume (a) and adipocyte number (b) with age in the baboon. Data are expressed as mean ± 95% confidence intervals. Numbers of baboons: birth n = 3 male (M, ○) and 8 female (F, ●), at weaning n = 3M and 5F, 1 year n = 1M and 5F, 2 years n = 1M and 3F, 3 years n = 0M and 12F, 4 years n = 19M and 21F, 5 years n = 21M and 13F, 8 years n = 35M and 39F, and > 12 years n = 27M and 26F. All animals were either breast-fed or fed normal formula as infants. There were no differences between breast-fed and formula-fed animals. Data from Ref. 136*

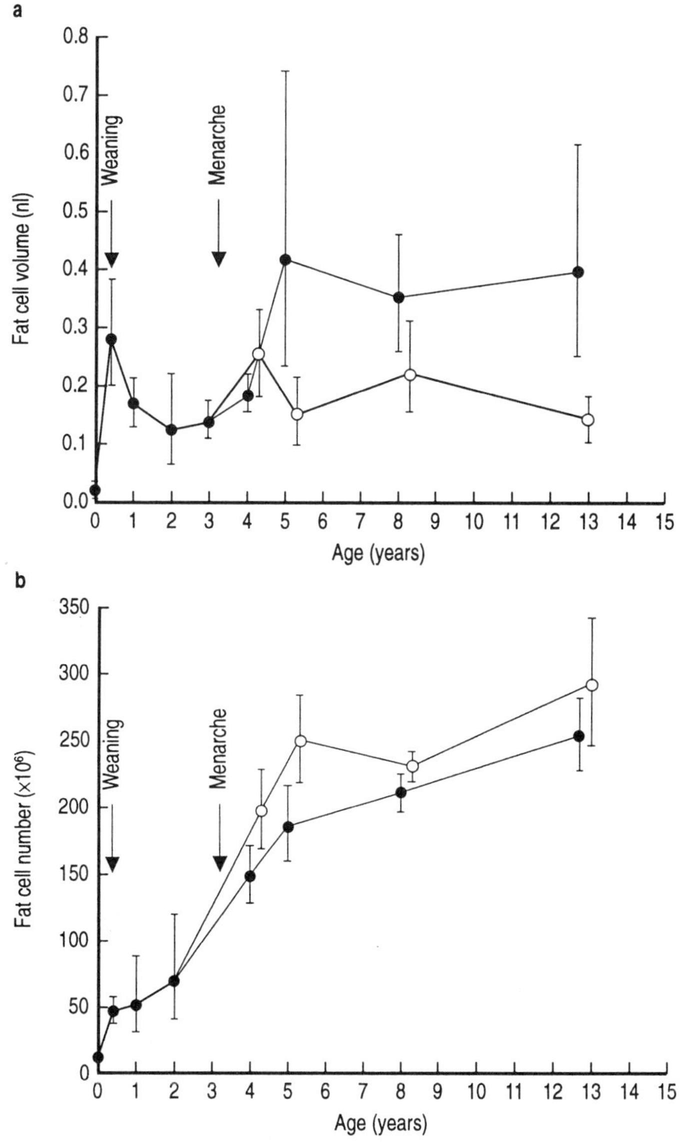

***Figure 5** Summary of hormone differences in infants fed concentrated formula (overfed) and those fed normal formula (normally fed). Plasma thyroid hormones were measured at fasting at 2 and 4 weeks and throughout the day over a 7 day period at 6–8 weeks of age. Plasma fasting insulin levels were measured at 2, 4 and 6–8 weeks of age. Postprandial insulin levels were measured after feeding at 2 and 4 weeks of age and throughout the day over a 7 day period at 6–8 weeks of age. Urine 17-hydroxycorticosteroids (OHCS) were determined in total urine excreted over two 24h periods and were normalized for urinary creatine content. Fasting plasma cortisol levels were determined at 2, 4 and 6–8 weeks of age. Daytime integrated cortisol levels were determined from plasma cortisol concentrations throughout the day over a 7 day period at 6–8 weeks of age. *$p < 0.05$. Data from Ref. 145*

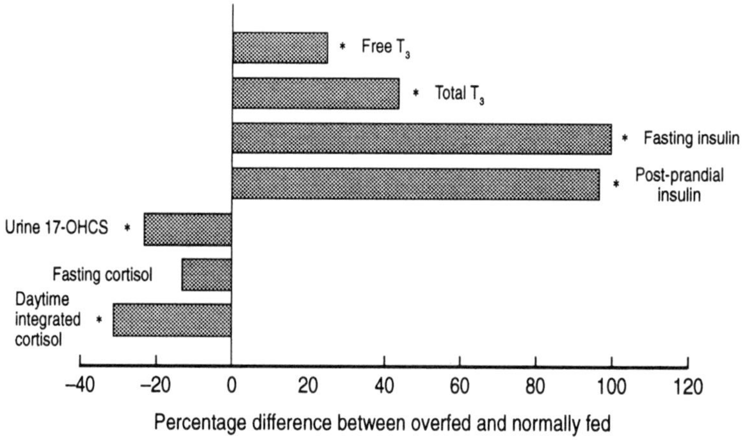

effect in adult rodents (62,64,65,78). Small differences in energy balance induced by infant overfeeding cannot be excluded because variability in energy expenditure and food intake between animals may be too high.

Altered adipocyte function inherent to the cell and regulation of adipocyte function by hormones are potential mechanisms underlying the effects of the infant diet for several reasons. First, most if not all adult adipocytes in baboons develop during two periods: the preweaning period when overfeeding occurs, and during adolescence when the effect of overfeeding occurs (Fig. 4). Hormone alterations induced by infant overfeeding could affect the development of adipocytes during these periods and influence adiposity. Second, uptake of dietary oleic acid into adipose tissue over a 24h period was greater in baboons overfed as infants than in baboons underfed (3.08 ± 1.05 g, $n=8$ *vs* 1.26 ± 0.29 g fatty acid, $n=8$). Uptake of dietary fatty acids was measured using a method described for use in humans (148). Net 24h esterification of dietary fatty acids into adipose tissue triacylglycerol was estimated using radioactive oleic acid, and total body fat was measured by dual energy X-ray absorptiometry. Incorporation of fatty acids into baboon adipose tissue was 2–3 mg, compared with 1–2 mg fatty acid uptake into subcutaneous adipose tissue in women (148). These

Table 6 Omental preadipocyte differentiation in pre-pubertal baboons that were normally fed or overfed as infants

Infant diet	*Percentage of cells differentiated*[a]	*GPDH activity* (mU/mg protein)[a]	*Triacyglycerol* (μg/mg protein)[a]
Normally fed (*n*=9)	35±2.0	573±33	561±43
Overfed (*n*=8)	50±3.0[b]	1395±100[b]	940±135[b]

[a]Mean±SEM
[b]Significantly different ($p<0.05$)

results suggest that utilization of dietary fat is more efficient in overfed baboons at 3 years of age compared with those normally fed as infants. Third, preadipocytes from 3-year-old baboons overfed as infants have enhanced differentiation potential in vitro (149) when cultured in a serum-free medium under optimal differentiation conditions (136) (Table 6). Preliminary data suggest that overfed infants enter puberty earlier (38 *vs* 46 months). Preadipocytes from baboons overfed as infants also differentiate to a greater extent in culture in the presence of progesterone compared with preadipocytes from baboons fed normally as infants (360±33 *vs* 210±57 mU glycerophosphate dehydrogenase activity/mg protein). It is not known whether the enhanced differentiation capability in preadipocytes from the baboons overfed as infants is due to differences within the preadipocyte or results from the endocrine environment of the preadipocytes in vivo prior to their isolation and culture. Preliminary evidence indicates that adipocytes are not different, but that the endocrine environment regulating preadipocyte differentiation and adipocyte function is affected by infant overnutrition (Lewis, unpublished observations).

SUMMARY

Animal studies consistently demonstrate long term effects of infant diets on adult adiposity. However, the magnitude of the effects of neonatal nutrition on adult adiposity in animals is modest, and well short of massive obesity. For example, in adult lean Zucker rats the amount of adipose mass expressed per body weight ranged from 6.6 to 9.9% for underfed, standard fed and overfed animals (60). The degree of adiposity in adult Osborn–Mendel rats from large and small litters was small, 10.9 and 13.1 g body fat/100 g body weight respectively (69). Mice reared in small litters were significantly fatter than mice from normal litters (36 *vs* 28% female; 28 *vs* 24% male) (64). Similarly, while some female baboons overfed as infants were considerably fatter, they were not obese. To relate these results to humans is difficult until the mechanisms underlying the effects of infant diet on animal adiposity are understood. Similarly, the demonstrated effects of various types of infant diet on adiposity in humans are modest, tempered by many studies in which no influence of infant diet on later adiposity was demonstrated. Identification of the causes for infant obesity and of the different types of infant obesities is necessary to determine the association between infant adiposity, overfeeding and underfeeding and later adiposity. Although it is likely that the potential effects of infant nutrition on later adiposity in humans are not large, in some individuals infant diet may contribute significantly to future weight problems.

ACKNOWLEDGEMENTS

Some of the work reported in this manuscript was supported in part by grants from the National Institute of Health (HD-23327) and the United States Department of Agriculture grant to Iowa State University Center for Designing Foods to Improve Nutrition, 94-34115-0269. Journal Paper number J-16631 of the Iowa Agriculture and Home Economics Experiment Station, Ames, Iowa, Project No. 3174, and supported by Hatch Act and State of Iowa Funds.

REFERENCES

1. Willett WC, Manson JE, Stampfer MJ, Colditz GA, Rosner B, Speizer FE, Hennekens CH 1995 Weight, weight change and coronary heart disease in women. Risk within the 'normal' weight range. J Am Med Assoc 273:461–465
2. Manson JE, Willett WC, Stampfer MJ, Colditz GA, Hunter DJ, Hankinson SE, Hennekens CH, Speizer FE 1995 Body weight and mortality among women. N Engl J Med 333:677–685
3. Iribarren C, Sharp DS, Burchfiel CM, Petrovich H 1995 Association of weight loss and weight fluctuation with mortality among Japanese American men. New Engl J Med 233:686–692
4. Williamson DF, Kahn HS, Remington PL, Anda RF 1990 The 10-year incidence of overweight and major weight gain in US adults. Arch Intern Med 150:665–672
5. Braddon FEM, Rodgers B, Wadsworth MEJ, Davies JMC 1986 Onset of obesity in a 36 year birth cohort. Br J Med 293:299–303
6. Dietz WH 1994 Critical periods in childhood for the development of obesity. Am J Clin Nutr 59:955–959
7. Hirsch J 1976 The adipose-cell hypothesis. N Engl J Med 295:389–390
8. Hirsch J, Batchelor B 1976 Adipose tissue cellularity in human obesity. Clin Endocrinol Metab 5:299–311
9. Hirsch J, Leibel RL 1991 Clinical Review 28: A biological basis for human obesity. J Clin Endocrinol Metab 73:1153–1157
10. Agras WS, Kraemer HC, Berkowitz RI, Hammer LD 1990 Influence of early feeding style on adiposity at 6 years of age. J Pediatr 116:805–809
11. Agras WC, Kraemer HC, Berkowitz RI, Korner AF, Hammer LD 1987 Does a vigorous feeding style influence early development of adiposity? J Pediatr 110:799–804
12. Poskitt EM 1977 Overfeeding and overweight in infancy and their relation to body size in early childhood. Second European Nutrition Conference, Munich 1976. Nutr Metab 21:54–55
13. Melbin T, Vuille JC 1976 Relationship of weight gain in infancy to subcutaneous fat and relative weight at 10.5 years of age. Br J Prev Soc Med 30:78–83
14. Vobecky JS, Vobecky J, Shapcott D, Demers PP 1983 Nutrient intake patterns and nutritional status with regard to relative weight in early infancy. Am J Clin Nutr 38:730–738
15. Poskitt EM, Cole TJ 1977 Do fat babies stay fat? Br Med J 1:7–9
16. Eid EE 1970 Follow-up study of physical growth of children who had excessive weight gain in first six months of life. Br Med J 2:74–76
17. Bindon JR 1984 The body build and composition of Samoan children: Relationships to infant feeding patterns and infant weight-for-length status. Am J Phys Anthropol 63:379–388
18. Shapiro LR, Crawford PB, Clark MJ, Pearson DL, Raz J, Huenemann RL 1984 Obesity prognosis: A longitudinal study of children from the age of 6 months to 9 years. Am J Publ Health 74: 968–972
19. Wolman PG 1984 Feeding practices in infancy and prevalence of obesity in preschool children. J Am Diet Assoc 84:436–438
20. Patterson RE, Typpo JT, Typpo MH, Krause GF 1986 Factors related to obesity in preschool children. J Am Diet Assoc 86:1376–1381

21. Baranowski T, Bryan GT, Rassin DK, Harrison JA, Henske J 1990 Ethnicity, infant feeding practices and childhood adiposity. J Dev Behav Pediatr. 11:234–239
22. Zive MM, McKay H, Frank-Spohrer GC, Broyles SL, Nelson JA, Nader PR 1992 Infant-feeding practices and adiposity in 4-y-old Anglo- and Mexican-Americans. Am J Clin Nutr 55:1104–1108
23. Dine MS, Gartside PS, Glueck CJ, Rheines L, Greene G, Khoury P 1979 Where do the heaviest children come from? A prospective study of white children from birth to 5 years of age. Pediatrics 63:1–7
24. Strbak V, Skultetyova M, Hromadova M, Randuskova A, Macho L 1991 Late effects of breast-feeding and early weaning: seven-year prospective study in children. Endocr Reg 25:53–57
25. Birkbeck JA, Buckfield PM, Silva PA 1985 Lack of long-term effect of the method of infant feeding on growth. Hum Nutr Clin Nutr 39C:39–44
26. Crisp AH, Douglas JW, Ross JM, Stonehill E 1970 Some developmental aspects of disorders of weight. J Psychom Res 14:313–320
27. Fomon SJ, Rogers RR, Ziegler EE, Nelson SE, Thomas LN 1984 Indices of fatness and serum cholesterol at age eight years in relation to feeding and growth during infancy. Pediatr Res 18: 1233–1238
28. Pollock JI 1994 Long-term associations with infant feeding in a clinically advantaged population of babies. Dev Med Child Neurol 36:429–440
29. Spahn U, Plenert W, Hesse V, Knoll G, Petrich E, Kauf E 1982 Overnutrition and obesity in childhood as a potential risk for chronic degenerative diseases in later life. Bibl Nutr Diet 31: 61–74
30. Dörner G, Hagen N, Witthuhn W 1976 Die frühpostnatale Überernährung als ätiopathogenetischer Faktor der Erwachsenenfettsucht. Acta Biol Med Ger 35:799–803
31. Law CM, Barker DJ, Osmond C, Fall CH, Simmonds SJ 1992 Early growth and abdominal fatness in adult life. J Epidemiol Commun Health 46:184–186
32. Mack RW, Johnston FE 1976 The relationship between growth in infancy and growth in adolescence: Report of a longitudinal study among urban black adolescents. Hum Biol 48: 693–711
33. Charney E, Goodman HC, McBride M, Lyon B, Pratt R 1976 Childhood antecedents of adult obesity: do chubby infants become obese adults? New Engl J Med 295:6–9
34. Kramer MS 1981 Do breast-feeding and delayed introduction of solid foods protect against subsequent obesity? J Pediatr 98:883–887
35. Ravelli GP, Stein ZA, Susser MW 1976 Obesity in young men after famine exposure in utero and early infancy. New Engl J Med 295:349–353
36. Dörner G 1973 Die mögliche Bedeutung der prä- und/oder perinatalen Ernährung für die Pathogenese der Obesität. Acta Biol Med Ger 30:19–22
37. Fomon SJ, Filer LJ, Thomas LN, Rogers RR, Proksch AM 1968 Relationship between formula concentration and rate of growth of normal infants. J Nutr 98:241–254
38. Fomon SJ, Filer LJ, Thomas LN, Anderson TA, Nelson SE 1975 Influence of formula concentration on caloric intake and growth of normal infants. Acta Paediatr Scand 64:172–178
39. Zoppi G, Mantovanelli F, Gobio Casali L, Astolfi R, Cecchettin M 1986 Effects of the composition and caloric value of infant formulas on intake and hormone levels. J Pediatr Gastroenterology Nutr 5:756–761
40. Forsum E, Sadurskis A 1986 Growth body composition and breast milk intake of Swedish infants during early life. Early Hum Dev 14:121–129
41. Dewey KG, Lönnerdal B 1983 Milk and nutrient intake of breast-fed infants from 1 to 6 months: Relation to growth and fatness. J Pediatr Gastroenterol Nutr 2:497–506
42. Fomon SJ, Thomas LN, Filer LJ, Anderson TA, Nelson SE 1976 Influence of fat and carbohydrate content of diet on food intake and growth of male infants. Acta Paediatr Scand 65:136–144
43. de Sweit MD, Fayer P, Cooper L 1977 Effect of feeding habit on weight in infancy. Lancet i: 892–894
44. Ferris AG, Laus MJ, Hosmer DW, Beal VA 1980 The effect of diet on weight gain during infancy. Am J Clin Nutr 33:2635–2642
45. Yeung DL, Pennell MD, Leung M, Hall J 1981 Infant fatness and feeding practices: A longitudinal assessment. J Am Diet Assoc 79:531–535
46. Shukla A, Forsyth HA, Anderson CM, Marwah SM 1972 Infantile overnutrition in the first year of life: A field study in Dudley, Worcestershire. Br Med J 4:507–515.

47. Dewey KG, Heinig MJ, Nommsen LA, Peerson JM, Lönnerdal B 1993 Breast-fed infants are leaner than formula-fed infants at 1 y of age: the DARLING study. Am J Clin Nutr 57:140–145
48. Rose HE, Mayer J 1968 Activity, caloric intake, fat storage, and the energy balance of infants. Pediatrics 41:18–29
49. Dubois S, Hill DE, Beaton GH 1979 An examination of factors believed to be associated with infantile obesity. Am J Clin Nutr 32:1997–2004
50. Sveger T, Lindberg T, Weibull B, Olsson UL 1975 Nutrition, overnutrition, and obesity in the first year of life in Malmö, Sweden. Acta Paediatr Scand 64:635–640
51. Huenemann RL 1974 I. Obesity in six-month-old children. Environmental factors associated with preschool obesity. J Am Diet Assoc 64:480–487
52. Taitz LS 1971 Infantile overnutrition among artificially fed infants in the Sheffield region. Br Med J 1:315–316
53. Heine W, Lapsien C 1982 Influence of early breast milk and formula feeding on body weight changes in children born in Rostock since 1945. Bibl Nutr Diet 31:17–18
54. Kramer MS, Barr RG, Leduc DG, Boisjoly C, Pless IB 1985 Infant determinants of childhood weight and adiposity. J Pediatr 107:104–107
55. Fomon SJ 1967 Body composition of the male reference infant during the first year of life. Pediatrics 40:863–870
56. Marcus C, Karpe B, Boine P, Sonnenfeld T, Arner P 1987 Changes in catecholamine-induced lipolysis in isolated human fat cells during the first year of life. J Clin Invest 79:1812–1818
57. Smith DW, Truog W, Rogers JE, Greitzer LJ, Skinner AL, McCann JJ, Harvey MA 1976 Shifting linear growth during infancy: Illustration of genetic factors in growth from fetal life through infancy. J Pediatr 89:225–230
58. Roberts SB, Savage J, Coward WA, Chew B, Lucas A 1988 Energy expenditure and intake in infants born to lean and overweight mothers. New Engl J Med 318:461–466
59. Roche AF, Siervogel RM, Chumlea WC, Reed RB, Eichorn D, McCammon RM 1982 Serial Changes in Subcutaneous Fat Thicknesses of Children and Adults. Basel Karger, New York
60. Johnson PR, Stern JS, Greenwood MRC, Zucker LM, Hirsch J 1973 Effect of early nutrition on adipose cellularity and pancreatic insulin release in the Zucker rat. J Nutr 103:738–743
61. Wurtman JJ, Miller SA 1976 Effect of litter size on weight gain in rats. J Nutr 106:697–701
62. Harris PM 1980 Changes in adipose tissue of the rat due to early undernutrition followed by rehabilitation I. Body composition and adipose tissue cellularity. Br J Nutr 43:15–26
63. Harris PM 1980 Changes in adipose tissue of the rat due to early undernutrition followed by rehabilitation II. Strain differences and adipose cellularity. Br J Nutr 43:27–31
64. Aubert R, Suquet J-P, Lemonnier D 1980 Long-term morphological and metabolic effects of early under- and over-nutrition in mice. J Nutr 110:649–661
65. Hausberger FX, Volz JE 1984 Feeding in infancy, adipose tissue cellularity and obesity. Physiol Behav 33:81–87
66. Obst BE, Schemmel RA, Czajka-Narins D, Merkel R 1981 Adipocyte size and number in dietary obesity resistant and susceptible rats. Am J Physiol 240:E47–E53
67. Widdowson EM, McCance RA 1962 Some effects of accelerating growth I. General somatic development. Proc R Soc B152:188–206
68. Knittle JL, Hirsch J 1968 Effect of early nutrition on the development of rat epididymal fat pads: Cellularity and metabolism. J Clin Invest 47:2091–2098
69. Faust IM, Johnson PR, Hirsch J 1980 Long-term effects of early nutritional experience on the development of obesity in the rat. J Nutr 110:2027–2034
70. Cryer A, Jones HM 1980 The development of white adipose tissue: Effect of litter size on the lipoprotein lipase activity of four adipose-tissue depots, serum immunoreactive insulin and tissue cellularity during the first year of life in male and female rats. Biochem J 186:805–815
71. Gaben-Cogneville A-M, Jahchan T, Swierczewski E 1981 Alteration by early underfeeding of cellular multiplication and differentiation in the inguinal fat pads of rats. J Nutr 111:2098–2105
72. Johnson PR, Zucker LM, Cruce AF, Hirsch J 1971 Cellularity of adipose depots in the genetically obese Zucker rat. J Lipid Res 12:706–714
73. Dugail I, Quignard-Boulange A, Dupuy F 1986 Role of adipocyte precursors in the onset of obesity induced by overfeeding in suckling rats. J Nutr 116:524–535
74. Bakke JL, Lawrence NL, Bennett J, Robinson S 1975 Late effects of neonatal undernutrition and overnutrition on pituitary-thyroidal and gonadal function. Biol Neonate 27:259–270

75. West DB, Diaz J, Woods SC 1982 Infant gastrostomy and chronic formula infusion as a technique to overfeed and accelerate weight gain of neonatal rats. J Nutr 112:1339–1343
76. Messer M, Thoman EB, Galofre A, Dallman T, Dallman PR 1969 Artificial feeding of infant rats by continuous gastric infusion. J Nutr 98:404–410
77. Auestad N, Korsak RA, Bergstrom JD, Edmond J 1989 Milk-substitutes comparable to rat's milk: their preparation, composition and impact on development and metabolism in the artificially reared rat. Br J Nutr 61:495–518
78. West DB, Diaz J, Roddy S, Woods SC 1987 Long-term effects on adiposity after preweaning nutritional manipulations in the gastrostomy-reared rat. J Nutr 117:1259–1264
79. Czajka-Narins DM, Hirsch J 1974 Supplementary feeding during the preweaning period: Effect on carcass composition and adipose tissue cellularity of the rat. Biol Neonate 25:176–185
80. Hiremagalur BK, Vadlamudi S, Johanning GL, Patel MS 1993 Long term effects of feeding high carbohydrate diet in preweaning period by gastrostomy: a new rat model for obesity. Int J Obesity 17:495–502
81. Roche AF 1981 The adipocyte-number hypothesis. Child Dev 52:31–43
82. Fiorotto ML, Burrin DG, Perez M, Reeds PJ 1991 Intake and use of milk nutrient by rat pups suckled in small, medium or large litters. Am J Physiol 260:R1104–R1113
83. Yagil R, Etzion Z, Berlyne GM 1976 Changes in rat milk quantity and quality due to variations in litter size and higher ambient temperature. Lab Anim Sci 26:33–37
84. Koldovsky O 1989 Critical review: Search for the role of milk borne biologically-active peptides for the suckling. J Nutr 119:1543–1551
85. Grota LJ 1973 Effects of litter size, age of young, and parity on foster mother behavior in *Rattus norvegicus*. Anim Behav 21:78–82
86. Deviterne D, Desor D, Krafft B 1990 Maternal behavior variations and adaptations, and pup development within litters of various sizes in Wistar rat. Dev Psychobiol 23:349–360
87. Barnes RH, Cunnold SR, Zimmermann RR, Simmons H, MacLeod RB, Krook L 1966 Influence of nutritional deprivations in early life on learning behavior of rats as measured by performance in a water maze. J Nutr 89:399–410
88. Winick M, Noble A 1966 Cellular response in rats during malnutrition at various ages. J Nutr 89:300–306
89. Pegg AE 1986 Recent advances in the biochemistry of polyamines in eukaryotes. Biochem J 234:249–267
90. Pairault J, Green H 1979 A study of the adipose conversion of suspended 3T3 cells by using glycerophosphate dehydrogenase as differentiation marker. Proc Natl Acad Sci USA 76: 5138–5142
91. Russell TR 1979 Differentiation of 3T3-L2 fibroblasts into adipose cells in bromodeoxyuridine-suppressed cultures. Proc Natl Acad Sci USA 76:4451–4454
92. Dijan P, Grimaldi P, Negrel R, Ailhaud G 1982 Adipose conversion of OB17 preadipocytes. Relationships between cell division and fat cell cluster formation. Exp Cell Res 142:237–281
93. Amrī EZ, Barbaras R, Doglio A, Dani C, Grimaldi P, Ailhaud G 1986 Role of spermidine in the expression of late markers of adipose conversion: effects of growth hormone. Biochem J 239: 363–370
94. Amrī EZ, Dani C, Doglio A, Etienne J, Grimaldi P, Ailhaud G 1986 Adipose cell differentiation: evidence for a two step process in the polyamine-dependent Ob 1754 clonal line. Biochem J 238:115–122
95. Bethel DR, Pegg AE 1981 Polyamines are needed for the differentiation of 3T3-L1 fibroblasts into adipose cells. Biochem Biophys Res Commun 102:272–278
96. Erwin BG, Bethel DR, Pegg AE 1984 Role of polyamines in differentiation of 3T3-L1 fibroblasts into adipocytes. Am J Physiol 246:C293–300
97. Schanberg SM, Evoniuk G, Kuhn CM 1984 Tactile and nutritional aspects of maternal care: Specific regulators of neuroendocrine function and cellular development. Proc Soc Exp Biol Med 175:135–146
98. Butler SR, Schanberg SM 1977 Effect of maternal deprivation on polyamine metabolism in preweanling rat brain and heart. Life Sci 21:877–884
99. Butler SR, Schanberg MR, Schanberg SM 1977 Maternal behavior as a regulator of polyamine biosynthesis in brain and heart of the developing rat pup. Science 199:445–446
100. Bell JM, Whitmore WL, Queen KL, Orband-Miller L, Slotkin TA 1987 Biochemical determinants

of growth sparing during neonatal nutritional deprivation or enhancement: ornithine decarboxylase, polyamines, and macromolecules in brain regions and heart. Pediatr Res 22: 599–604
101. Pollack PF, Koldovsky O, Nishioka K 1992 Polyamines in human and rat milk and in infant formulas. Am J Clin Nutr 56:371–375
102. Bardocz S, Brown DS, Grant G, Pusztai A 1990 Luminal and basolateral polyamine uptake by rat small intestine stimulated to grow by *Phaseolus vulgaris* lectin phytohaemagglutinin in vivo. Biochim Biophys Acta 1034:46–52
103. Osborne DL, Seidel ER 1990 Gastrointestinal luminal polyamines: cellular accumulation and enterohepatic circulation. Am J Physiol 258:G576–584
104. Seidel ER, Haddox MK, Johnson LR 1985 Ileal mucosal growth during intraluminal infusion of ethylamine or putrescine. Am J Physiol 249:G434–438
105. Dufour C, Dandrifosse G, Forget P, Vermesse F, Romain N, Lepoint P 1988 Spermine and spermidine induce intestinal maturation in the rat. Gastroenterology 95:112–116
106. Rasberg L, Nex E, Tollund L, Poulsen SS, Christensen SB, Christensen MS 1990 Epidermal growth factor reactivity in rat milk. Reg Pept 30:149–157
107. Donovan SM, Hintz RL, Wilson DM, Rosenfeld RG 1991 Insulin like growth factors I and II and their binding proteins in rat milk. Pediatr Res 29:50–55
108. Donovan SM, Odle J 1994 Growth factors in milk as mediators of infant development. Annu Rev Nutr 14:147–167
109. Carpenter G 1980 Epidermal growth factor is a major growth-promoting agent in human milk. Science 210:198–199
110. Koldovsky O, Philipps A, Rao RK, Schaudies P 1992 Possible role of milk-borne peptide growth factors for the breast-fed infant. Front Gastrointest Res 21:150–169
111. Berseth CL 1987 Enhancement of intestinal growth in neonatal rats by epidermal growth factor in milk. Am J Physiol 253:G662–665
112. Berseth CL, Go VLW 1988 Enhancement of neonatal somatic and hepatic growth by orally administered epidermal growth factor in rats. J Pediatr Gastroenterol Nutr 7:889–893
113. Baumrucker CR, Blum JW 1993 Secretion of insulin-like growth factors in milk and their effects on the neonate. Livestock Prod Sci 35:49–72
114. Baumrucker CR, Hadsell DL, Skaar TC, Campbell PG, Blum JW 1992 Insulin-like growth factor in mammary secretions: origins and implications in neonatal physiology, In: Picciasn MF, Lönnerdal B (eds) Mechanisms Regulating Lactation and Infant Nutrient Utilization. Wiley-Liss, New York, pp 285–308
115. Smith PJ, Wise LS, Berkowitz R, Wan C, Rubin CS 1988 Insulin-like growth factor-1 is an essential regulator of the differentiation of 3T3-L1 adipocytes. J Biol Chem 263:9402–9408
116. Serrero G, Mills D 1991 Physiological role of epidermal growth factor on adipose tissue development in vivo. Proc Natl Acad Sci USA 88:3912–3916
117. Bode HH, Vanjonack WJ, Crawford JD 1978 Mitigation of cretinism by breast-feeding. Pediatrics 62:13–16
118. Oberkotter LV 1988 Developmental changes in rat thyroid responsiveness to thyrotropin administered by the subcutaneous and peroral route. Proc Soc Exp Biol Med 187:360–365
119. Lakshmanan J, Perheentupa J, Hoath SB, Kim H, Gruters A, Odell C, Fisher DA 1985 Epidermal growth factor in mouse occular tissue: Effects of thyroxine and exogenous EGF. Pediatr Res 19:315–319
120. Hennig SJ 1981 Postnatal development: coordination of feeding, digestion, and metabolism. Am J Physiol 241:G199–21
121. Hahn P 1984 Effect of litter size on plasma cholesterol and insulin and some liver and adipose tissue enzymes in adult rodents. J Nutr 114:1231–1234
122. Lemmonnier D, Suquet J-P, Aubert R, Rosselin G 1973 Long-term effect of mouse neonate food intake on adult body composition, insulin, and glucose serum levels. Horm Metab Res 5:223–224
123. Vadlamudi S, Hiremagalur BK, Tao L, Kalhan SC, Kalaria RN, Kaung H-LC, Patel MS 1993 Long-term effects on pancreatic function of feeding a HC formula to rats during the preweaning period. Am J Physiol 265:E565–E571
124. Gaben-Cogneville A-M, Poussin B, Chamblier IU-C, Forgue-Lafitte M-E, Rosselin G 1988 Development of insulin and epidermal growth factors during the differentiation of rat adipocytes in primary culture. Biochim Biophys Acta 968:231–238

125. Duff DA, Snell K 1982 Effect of altered neonatal nutrition on the development of enzymes of lipid and carbohydrate metabolism in the rat. J Nutr 112:1057–1066
126. Asplund K 1972 Effects of postnatal feeding on the functional maturation of pancreatic islet β-cells on neonatal rats. Diabetologia 8:153–159
128. Lewis DS, Bertrand HA, Masoro EJ, McGill HC Jr, Carey KD, McMahan CA 1983 Preweaning nutrition and fat development in baboons. J Nutr 113:2253–2259
129. Lewis DS, Bertrand HA, Masoro EJ, McGill HC Jr, Carey KD, McMahan CA 1984 Effect of interaction of gender and energy intake on lean body mass and fat mass gain in infant baboons. J Nutr 114:2021–2026
130. Lewis DS, Bertrand HA, McMahan CA, McGill HC Jr, Carey KD, Masoro EJ 1986 Preweaning food intake influences the adiposity of young adult baboons. J Clin Invest 78:899–905
131. Lewis DS, Bertrand HA, McMahan CA, McGill HC Jr, Carey KD, Masoro EJ 1989 Influence of preweaning food intake on body composition of young adult baboons. Am J Physiol 257: R1128–R1135
132. Lewis DS, Coelho AM, Jackson EM 1991 Maternal weight and sire group, not caloric intake, influence adipocyte volume in infant female baboons. Pediatr Res 30:534–540
133. Jones RAK, Belsey EM 1978 Common mistakes in infant feeding: survey from a London Borough. Br Med J 2:112–114
134. Lewis DS, Rollwitz WL, Bertrand HA, Masoro EJ 1986 Use of NMR for measurement of total body water and estimation of body fat. J Appl Physiol 60:836–840
135. Krotkiewski M, Björntorp P, Sjöstrom L, Smith U 1983 Impact of obesity on metabolism in men and women. Importance of regional adipose tissue distribution. J Clin Invest 72:1150–1162
136. Lewis DS, Söderström P 1993 In vivo and in vitro development of visceral adipose tissue in a nonhuman primate (*Papio* species). Metabolism 42:1277–1283
137. Azizi F, Vagenakis AG, Bollinger J, Reichlin S, Braverman LE, Ingbar SH 1974 Persistent abnormalities in pituitary function following neonatal thyrotoxicosis in the rat. Endocrinology 94: 1681–1688
138. Bakke JL, Lawrence NL, Wilbur JF 1974 The late effects of neonatal hyperthyroidism upon the hypothalamic-pituitary-thyroid axis in the rat. Endocrinology 95:406–411
139. Dussault JH, Coulombe P, Walker P 1982 Effects of neonatal hyperthyroidism on the development of the hypothalamic-pituitary-thyroid axis in the rat. Endocrinology 110:1037–1042
140. Walker P, Courtin F 1985 Transient neonatal hyperthyroidism results in hypothyroidism in the adult rat. Endocrinology 116:2246–2250
141. Alm J, Lakshmanan J, Hoath S, Fisher DA 1988 Neonatal hyperthyroidism alters hepatic epidermal growth factor receptor ontogeny in mice. Pediatr Res 23:557–560
142. Miyabo S, Hisada T 1975 Sex difference in ontogenesis of circadian adrenocortical rhythm in cortisone-primed rats. Nature 256:590–592
143. Cost MG, Mann DR 1976 Neonatal corticoid administration: retardation of adrenal rhythmicity and desynchronization of puberty. Life Sci 19:1929–1935
144. Meaney MJ, Aitken DH, van Berkel C, Bhatnagar S, Sapolsky RM 1988 Effect of neonatal handling on age-related impairments associated with the hippocampus. Science 239:766–768
145. Lewis DS, Jackson EM, Mott GE 1992 Effect of energy intake on postprandial plasma hormones and triglyceride concentrations in infant female baboons (*Papio* species). J Clin Endocrinol Metab 74:920–926
146. Lewis DS, Jackson EM, Shade RE 1995 Infant adaptation to concentrated formula feeding. FASEB J 9:A731
147. Handelmann GE, Russel JT, Gainer H, Zerbe R, Bayorh M 1983 Vasopressin administration to neonatal rats reduces antidiuretic response in adult kidneys. Peptides 4:827–832
148. Marin P, Rebuffe-Scrive M, Björntorp P 1990 Uptake of trigylceride fatty acids in adipose tissue in vivo in man. Eur J Clin Invest 20:158–165
149. Lewis DS, Söderström P 1993 Enhanced differentiation potential in omental preadipocytes from pre-adolescent baboons overfed as infants. FASEB J 7:A388

Discussion Session 2

CHAIR: H.K.A. VISSER

CHAPTER 6

Elia One of the major considerations with any method for determining body composition is its accuracy. Given your attempt to demonstrate this accuracy by comparison with pig carcass analyses, I wonder if you can tell us a little bit more about the accuracy and how this might be affected by the performance of the TOBEC machine. For example there has been some concern about the uniformity of fields in the TOBEC machine and whether different body compartments or rather segments of the body may have equal contribution to the TOBEC results in terms of electrical conductivity. Does body shape make any difference and what about the concentration of electrolytes which presumably is responsible for the electrical conductivity?

De Bruin We found that in vivo TOBEC measurements are affected by ion concentration (non-isotonic hydration changes), geometry (deviations in body shape), temperature (fever, cooling) and tissue autolysis (measurements after death in pigs) (De Bruin et al. 1994 Phys Med Biol 39:1143–1156). Different electrolytes have a different contribution to the TOBEC value: for example, potassium has a greater contribution than sodium. We have to be careful with measurements of children with fever. Magnetic field plots have shown that TOBEC instruments may not be equivalent over the whole measuring range. They have to be calibrated and subjects should be placed within the homogeneous sections of the magnetic field (Fiorotto et al. 1995 Pediatr Res 37:94–100).

Hornstra I think that the simplified equations you have presented are very important for field studies. Are these equations specifically based on Caucasian babies or do they also apply to other babies and are there gender differences?

De Bruin We measured only Caucasian infants. We have to be cautious in applying these data to other races. There are no gender differences.

CHAPTER 7

Pencharz I just wanted to ask you about your fat data and hence lean body mass data. Outside the neonatal period there is a fair controversy as to how useful DEXA is in measuring the relative proportions of fat and lean body mass – it depends on what software and what kind of calibrations are built into the software. What kind of calibrations were built into the software of your machine in terms of measuring fat mass versus lean body mass?

J.G. Bindels, A.C. Goedhart and H.K.A. Visser (eds), Recent Developments in Infant Nutrition, 148–153.

Rigo In this machine we use the new infants' software with the infants' pad. This table pad helps to increase the linearity of the data and also reduces the radiation for the infants. In the software program we used a hydration factor of 83% for the lean body mass. In the piglet studies a change of 10% in hydration factor – as measured by chemical analysis – changed the calculation of fat mass by only 0.3%. I think that the DXA method is very robust for measuring fat mass, free fat mass and water content.

Cooke Are you measuring a total body scan or are you looking at the central skeleton?

Rigo We measure the total body scan.

Cooke We are working with preterm babies in their first year of life and what you can very clearly show is that the major differences in the patterns of bone mineralization relate to the skull and spine and less than 10% of the variation that you see relates, for example, to the upper limbs.

Fabris Recently it has been shown that summer-born newborns have lower bone mineral content than winter-born newborns. Have you see differences in bone mineral contents between summer-born and winter-born babies?

Rigo I did not analyse the data according to the time of birth, so it is impossible for me to give you the information. The data were only analysed according to gender.

Uauy What was the calcium intake and the phosphorus intake of your infants?

Rigo We are currently analysing the data according to diet and also according to treatment of the infant. We have some data from infants fed with human milk and human milk fortifier and with different calcium and phosphorus intake. I did not have time to present some of those data.

Elia You are considering centiles for bone minerals. I think it is important to remember that a number of workers using DXA have raised concerns about differences in bone mineral content and in calcium content of bones depending on the type of machine that is used for analysing the results. There are currently three machines commercially available, and when individual subjects, at least adult subjects, are compared by these different machines there can really be substantial differences of 10–20% in the results of measurements.

Rigo I agree completely that some differences arise according to the machine and the software that people use. I think it is difficult to compare the data from various studies. We were interested to see that bone mineral content from the piglet produced the same data as that from the preterm infant. It is better to express calcium data according to the specific equation conversion of one machine rather than express the results as bone mineral content. For each machine you have to find a conversion equation between BMC and calcium content.

Donzelli Do you have any data from intrauterine growth retarded babies, symmetrical or asymmetrical?

Rigo All the babies were healthy preterm infants with birth weight and length appropriate for gestational age. We have also some data on growth retardation. Most of the time in our unit we see symmetrically growth retarded patients. In those infants with symmetrical retardation, we obtained data according to body weight which are similar to the reference values.

CHAPTER 8

Eeckels When you calculate your SDS or Z-scores, what standards are you using?

Karlberg The standard scores are based on this material and we used the new Swedish reference values.

Eeckels Do you have any idea how your data will look when you use the NCHS (National Center for Health Statistics) curves which are used by WHO?

Karlberg People now do not recommend using NCHS values because they are based on mixed populations. For this age group they are probably valid but not for all age groups. It probably has no major impact on the results.

Wharton I suppose that body mass index is the index that has achieved most popularity, particularly in adults. One of its important aspects is its correlation with certain risk factors, blood pressure and so on, and we know they are associated with cardiovascular disease. Is there any need for indices like you have presented, unless we can show that they have some important clinical correlation with pathology? That is the only real use of an index. Also, what are your views on the body mass index? While accepting its empirical value in adults, it seems to make no physiological sense. Weight divided by length squared is weight divided by the area. I don't know what that means. The Ponderal index is weight divided by length cubed, so weight divided by volume. That has more physiological attraction to me because weight divided by volume is density. I wonder whether we should just accept empirical correlations. Finally coming back to the previous question. Is it worthwhile calculating these indices? Why don't we just use the weight for height standard deviation score?

Karlberg What I proposed here was an alternative to body mass index or some other index. Arguments against using these indices are the age dependency and the correlation with height. Body mass index is traditionally used in adult studies. It is very easy, people under-stand it, they know the numbers, but over the years many publications have shown that it is not good enough. If we are going to compare people of different heights we need to have some other index.

Desjeux You looked at the relationship between nutritional status and linear growth in the same period of time. Would you not expect a delay if you look for a causal relationship, say the nutritional status at one year of age and the outcome on height at 10 years of age?

Karlberg It is possible to study this using these data. I have concentrated here on the childhood years. I did not look at overnutrition in the infant period to see what happens much later on: that is another story.

Uauy Your presentation referred to your own data from Sweden. Are you willing to comment on the effect of specific nutrients such as protein and zinc on linear growth in childhood on the basis of data from say Hong Kong, Pakistan or other places?

Karlberg There are some studies from Guatemala and from other places which show that problems arise in children who do not have a balanced diet in their childhood years.

CHAPTER 9

Marini What is the protein content of baboon baby milk? Is it equal to human milk or different? Another question: you have a group with an atherogenic diet, but do you have a group without an atherogenic diet? How would they behave?

Lewis All the animals that we have been studying have been formula reared; they have not been on breast milk. Baboon breast milk is similar to human breast milk in protein composition.

Marini Is the amount of protein in the formula similar to the amount of the protein in the baboon milk?

Lewis Yes. Regarding your question about the atherogenicity of the diets, these experiments are very expensive and I could not convince a credit agency to double the costs of the experiment for a controlled study. However, studies are performed on baboons reared for 8 years on a low-fat diet (about 7% fat and virtually no cholesterol, compared to our atherogenic diet which is primarily lard, 40% of calories, with a lot of cholesterol) and there is very little difference in fatness. The baboon is really a lean animal. What we are talking about is a modest increase of fat. It is important, however, to realise that it has recently been shown that just being modestly overweight increases the risk for cardiovascular disease as well as mortality.

Koppe Did you control for physical activity?

Lewis No we did not control for it. We did measure total energy expenditure and one of the adaptations of those fat baboons is that they almost spend all the energy that they take in.

Büller Is there anything you can tell us from your experience about the genetic factor? Do you control for that in all your studies, fat mothers versus lean mothers?

Lewis A reasonable way for us to control the genetic factor in our studies is to use a sire group. We take sires who have produced a lot of babies over time who are not fat babies or very lean babies. That limits the variance. We do not use obese baboons for breeding. We have a wide range of fatness in the mothers. Baboons can range in body fat from as low as 2% to 35%. Most of them were below 10%. We found that body weight of the mother influenced how fat the baby becomes during the pre-weaning period, much more than the caloric intake during the pre-weaning period.

Carlson I have a question about the mechanism for overfeeding and underfeeding. Did you allow the baboons to self-regulate their volume?

Lewis Yes, during infancy, but within reason. They were always given enough formula at feeding so they would always leave something in the bottle. They were fed four times a day, so they got hungry during the night when they had nothing to eat.

Davies In one of your earlier studies, you emphasized how important it was to look to other environmental influences on growth. I am interested to know how the life-style of your baboon compares with the lifestyle of baboons in wild. The background to my question is whether you could really extrapolate your studies in baboons to studies in humans. How constricted were they? Were they emotionally disturbed? This affects hormones which may influence growth and adiposity.

Lewis The first thing is that the animals were put in incubators after they were born. They were then hand fed for about 2 weeks by a human caretaker and then they were given biweekly social peer interactions with other babies – animals could interact with each other, play and so on. They then went into an individual cage until they were weaned. Once they were weaned they went to a cage which allowed freedom of movement. They could play, they could run around and they were given access to an adult baboon. It is often critical to keep adult baboons with animals that have not been reared by an adult: we call them police adults. It keeps them from attacking each other. After one year of age they go to large outdoor cages where they have rocks, they can climb 20 feet high, there are trees, and there is plenty of movement for the animals. Whether they are socially abnormal, I cannot tell you. But animals do behave differently when reared with an artificial formula by aliens (basically that is what it must look like to them) to when they are left with their mother. Physiologically, they seem to be very similar and since all the animals were raised in the same way, there should not be that interference. You would have a problem if you compared breast-fed animals and formula-fed animals: then you do have to consider a possible difference in their behaviour.

Stephenson The effects of overfeeding only become apparent after sexual maturation. Does this also apply to the rodent studies that you reviewed initially?

Lewis Yes, this applies to rodents as well. In studies in rats the maximal effects in fatness occur well after 12 weeks of age when animals were manipulated in the first 3 weeks. These effects seem to become maximal after the animals have reached sexual maturity.

Zwiauer Protein intake is probably one of the important factors in the development of obesity. What was the protein intake in the overfed and the underfed baboons?

Lewis We diluted the concentrated formula so that the amount of protein per calorie or as a percentage of calories was the same. They just ate less protein, less fat, less carbohydrate, and so on. The overfed animals ate more protein, more carbohydrate, more fat, more vitamin A, more vitamin C, more everything. These animals maintained normal sodium plasma levels, and they did this by secreting a lot less urine of very high osmolality. Interestingly, the same animals urinate a lot more now that they are 3 years old, so they are receiving an osmotic load.

Berger You are feeding the baboons four times a day and only during the daytime. I imagine the logistics behind that is obvious, but what about the effect of this on fat metabolism?

Lewis That is a good question. We undertook a pilot study with normal formula. We gave some animals four feedings and others eight feedings and there was no difference in total intake. I have found in subsequent experiments that if you introduce a concentrated formula to a 4-week-old infant, the weight gain is different to that in an infant fed with formula from birth on. The animals in our study have been exposed to this formula from day 1 and there seems to be a different pattern of adaptation in an infant that has been on a normal formula and all of a sudden switches to a concentrated feed. There is a difference in weight gain but the composition of that weight gain is not clear.

Session 3
Long-chain polyunsaturated fatty acids (LCPUFA) in infant nutrition

10 Effects of $n-3$ fatty acid deficiency in nonhuman primates

S. REISBICK, M. NEURINGER and W.E. CONNOR

ABBREVIATIONS

AA	arachidonic acid, 20:4 $n-4$
ALA	α-linolenic acid, 18:3 $n-3$
DHA	docosahexaenoic acid, 22:6 $n-3$
EPA	eicosapentaenoic acid, 20:5 $n-3$
LA	linoleic acid, 18:2 $n-6$
LCPUFA	long-chain polyunsaturated fatty acids
PE	phosphaethanolamine

INTRODUCTION

Whether infant formulas should be supplemented with $n-3$ fatty acids and, if so, the optimal forms and amounts of fatty acids for both preterm and term infants have been the subjects of considerable attention and debate over the past few years. Studies in rhesus monkeys helped to establish the need for higher levels of α-linolenic acid (ALA; 18:3 $n-3$) in infant feeding. There is now strong and consistent evidence that a low intake of ALA is associated with delayed visual development, and the need for at least 0.7% of energy as ALA in infant formulas is now widely accepted. However, this consensus has not been universally translated into practice. As of 1994, formulas low in ALA, and specifically those containing corn oil as a source of polyunsaturates, were still widely available in many developing countries (Fig. 1; N. Zhu, W.E. Connor and S. Connor, unpublished data).

Much more controversial is the need for preformed sources of docosahexaenoic acid (DHA, 22:6 $n-3$), the major $n-3$ fatty acid in membranes of retinal photoreceptors and neural tissue. Addition of this fatty acid requires more caution, in part because the available sources have not been used in infant formulas before, because this fatty acid is highly susceptible to peroxidation, and because $n-3$ long-chain polyunsaturated fatty acids (LCPUFA) can suppress levels of arachidonic acid (AA, 20:4 $n-6$), the major $n-6$ fatty acid in biological membranes. Available evidence from human clinical trials now suggests that preformed DHA may be safe only when given with adequate AA. For preterm infants, a consensus is emerging that balanced

J.G. Bindels, A.C. Goedhart and H.K.A. Visser (eds), Recent Developments in Infant Nutrition, 157–172.

***Figure 1** Content of n−3 fatty acids in commercial infant formulas, 1994. Formulas sold in Mexico still contained corn oil as the source of polyunsaturates, while those sold in the USA generally contained oil blends which included soybean oil. Except for human milk, n−3 fatty acids were entirely in the form of α-linolenic acid*

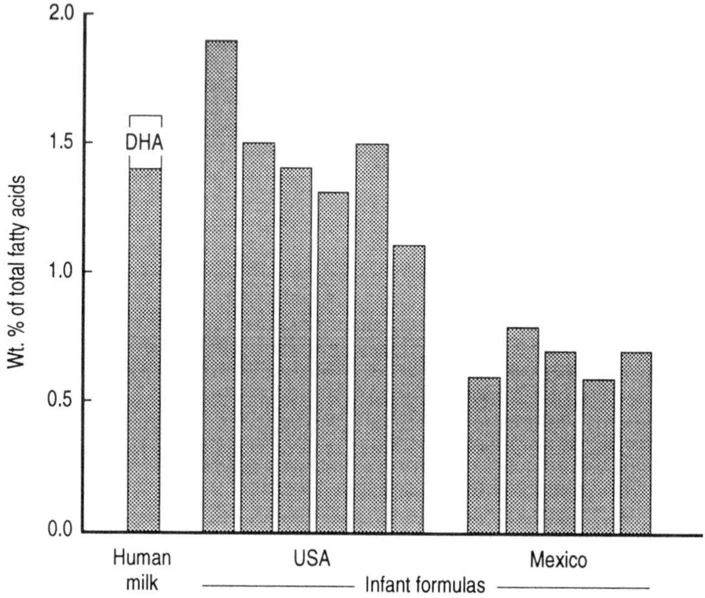

supplementation with DHA and AA is desirable. However, it is not yet clear whether these LCPUFA are effective in improving visual and behavioural development when added to formulas for term infants.

MATERIALS AND METHODS

Our previous and ongoing studies of rhesus monkeys have identified a number of changes in visual development and behaviour associated with low dietary levels of $n-3$ fatty acids (1–11). These studies have provided the opportunity to test for effects of deficiency on many aspects of visual, behavioural and cognitive function under controlled nutritional and environmental conditions and to correlate functional effects with brain and retinal lipid biochemistry. In these studies, female rhesus monkeys were fed diets differing in fatty acid composition for at least 2 months before breeding and throughout gestation, and their infants received analogous diets from birth. The control and $n-3$ fatty-acid-deficient diets (solid form for adults, liquid formulas for infants) differed only in the source of fat. Control diets supplied soybean oil, which contains abundant $n-6$ and $n-3$ fatty acids in the form of linoleic acid and α-linolenic acid (18:2 $n-6$, 54% by weight of total fatty acids; 18:3 $n-3$, 7.8%; $n-6/n-3$ ratio, 7). The deficient diets contained safflower oil, which is very low in α-linolenic acid but high in linoleic acid (18:2 $n-6$, 78%; 18:3 $n-3$, 0.3%; $n-6/n-3$ ratio, 260). Rhesus infants were raised in a primate nursery and received the experimental infant formulas 3–4 times per day. They were not deprived of food, were socialized with at least one age-mate for several hours per day and received extensive daily handling and interaction with human caretakers and experimenters. NIH, USDA and AAALAC guidelines for the care and use of animals were followed.

Rhesus infants are born at a later stage of brain and eye development than term human infants, and therefore the combination of maternal and postnatal nutritional manipulation in monkeys is necessary to model closely the situation of human infants, particularly those born prematurely. Our studies in monkeys have been shown to predict, both qualitatively and quantitatively, effects on visual and cognitive development in preterm human infants fed diets low in ALA and even in those fed formulas without preformed LCPUFA. This model may underestimate the severity of $n-3$ fatty acid deficiency in very-low-birth-weight preterm infants, however, because pregnant monkeys have substantial essential fatty acid stores and continue to supply some $n-3$ LCPUFA to the fetus in utero, even after $n-3$ deprivation for several months before conception.

In addition to the control group receiving a semipurified diet, we also used two standard comparison groups. Monkeys in the adult standard group were breast-fed by their mothers for at least 90 days and raised in a social group until at least 2 years of age (10). After weaning, they were fed Purina Monkey Chow®, which contains DHA from fish meal (18:2 $n-6$, 37% by weight of total fatty acids; 18:3 $n-3$, 2.3%; DHA, 0.3%; $n-6/n-3$ ratio, 12). Thus this group differed from the semipurified diet groups in both diet and social experience. Nursery standard infants were term births to mothers fed monkey chow, so that their levels of DHA were high at birth (11). These infants were raised in the same primate nursery as the control and deficient groups, fed a formula containing soybean oil (SMA, Wyeth) and supplemented from about 8 weeks with a primate chow (New World Primate Chow®) which also contained DHA (18:2 $n-6$, 31.9%; 18:3 $n-3$, 0.5%; DHA, 0.2%; $n-6/n-3$ ratio, 36). Their social experience was similar to that of the semipurified diet groups, but their nutritive input differed slightly in factors other than the source of fat. Thus, neither standard group was a pure control and each served as a comparison group in the same way that breast-fed infants serve as a 'gold standard' comparison in studies of human infants.

RESULTS

All studies showed no effects of $n-3$ fatty acid deficiency on general health. Body weights, coat condition and clinical serum measures did not differ between the semipurified diet groups or between these and the standard groups to 5 years of age, and there were no outward signs of nutritional deficiency. However, the fatty acid composition of the diets was reflected in erythrocytes and tissues. As previously reported (1,2), levels of $n-3$ fatty acids, particularly DHA, were very low in plasma erythrocytes and in retina and cerebral cortex of infants deficient in ALA.

***Figure 2** DHA (wt % of total fatty acids; mean ± SEM) in erythrocytes of breast-fed infant monkeys (■) and those fed soybean oil (○) or safflower oil (△) diets over the first 16 weeks*

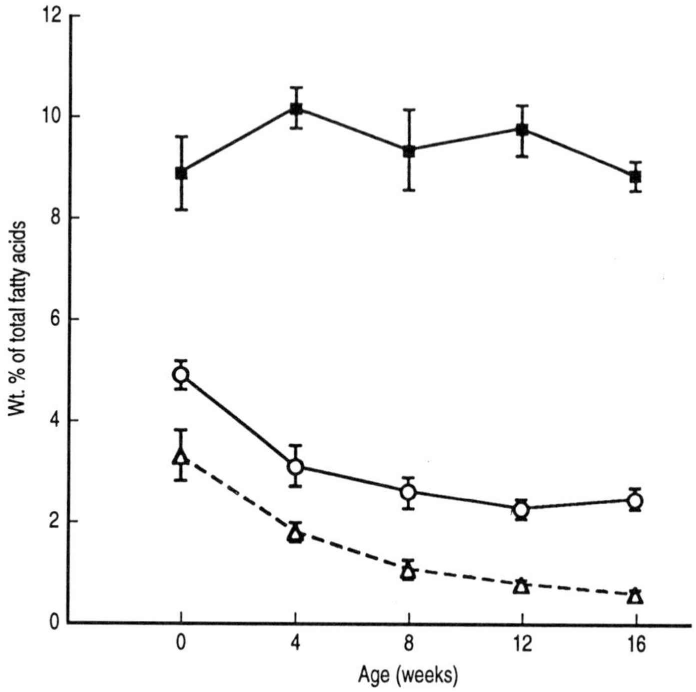

Preliminary biochemical data from a new normative developmental study of tissue fatty acid composition provide interesting comparison with the semipurified diet groups (Lin, Neuringer and Connor, unpublished data). As previously reported, erythrocyte DHA was higher in soybean-oil-fed controls than in deficient monkey infants and, as expected, breast-fed monkeys had much higher DHA levels than either of the groups fed only 18-carbon fatty acids (Fig. 2).

Preliminary data indicate that tissue levels of LCPUFA present a very different picture from that seen in erythrocytes. At birth, retinal levels of DHA were lower in infants of mothers fed soybean oil than in infants of mothers fed preformed LCPUFA (Fig. 3a), but levels in cerebral cortex were similar (Fig. 3b). At 2 years of age, DHA levels in both the retina and cortex of soybean-oil-fed infants were similar to those of breast-fed infants. Thus in neural tissues, a large difference in DHA was present between the soy control and deficient groups, but not between the soy control and LCPUFA-fed groups. This was also true for 22:5 $n-6$, which is a marker for specific $n-3$ fatty acid deficiency. Levels of 22:5 $n-6$ were higher only in the deficient group (Fig. 4a,b), while AA levels were highly conserved and did not differ among the three

***Figure 3** DHA (wt % of total fatty acids) in retina (a) and frontal cortex (b) in breast-fed/chow-fed monkeys (■) and those fed soybean oil (○) or safflower oil (△) diets*

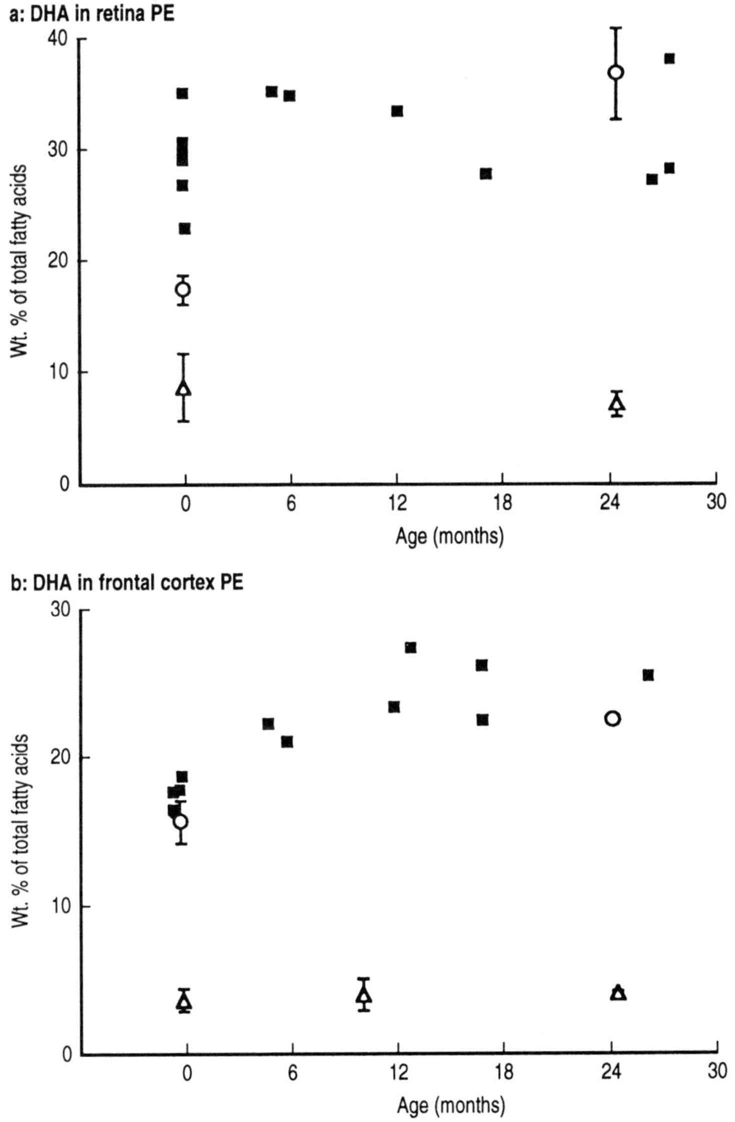

***Figure 4** 22:5 n−6 (wt % of total fatty acids) in retina (a) and frontal cortex (b) in breast-fed/chow-fed monkeys (■) and those fed soybean oil (○) or safflower oil (△) diets*

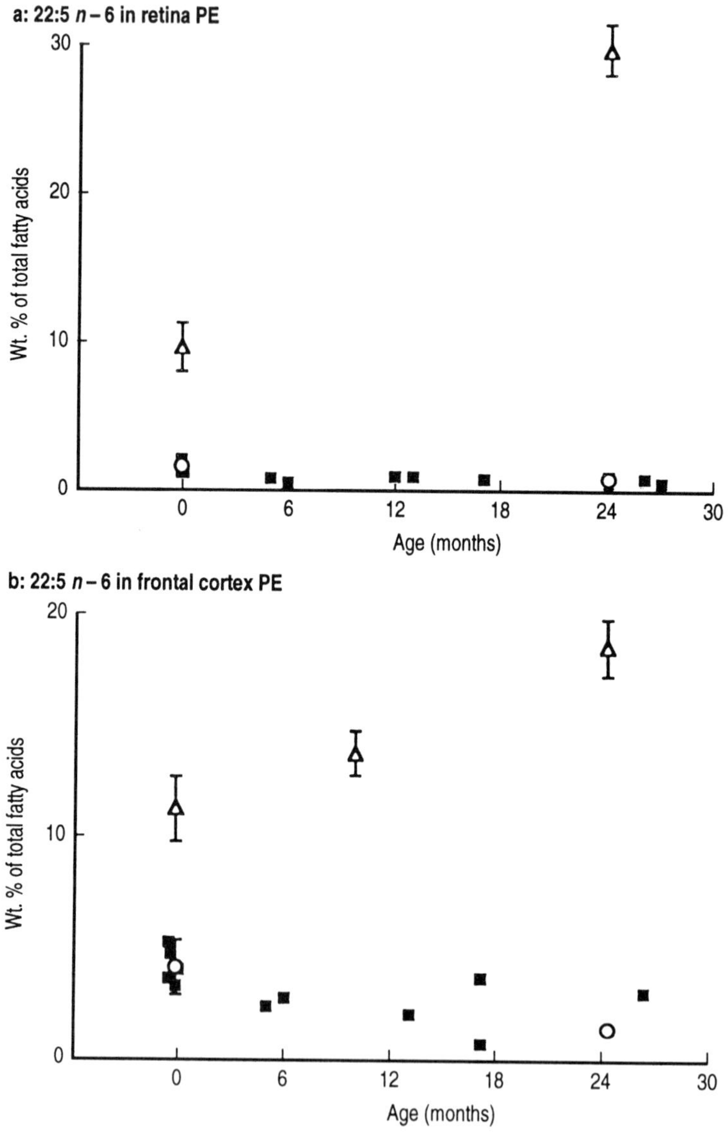

***Figure 5** AA (wt % of total fatty acids) in retina (a) and frontal cortex (b) in breast-fed/chow-fed monkeys (■) and those fed soybean oil (○) or safflower oil (△) diets*

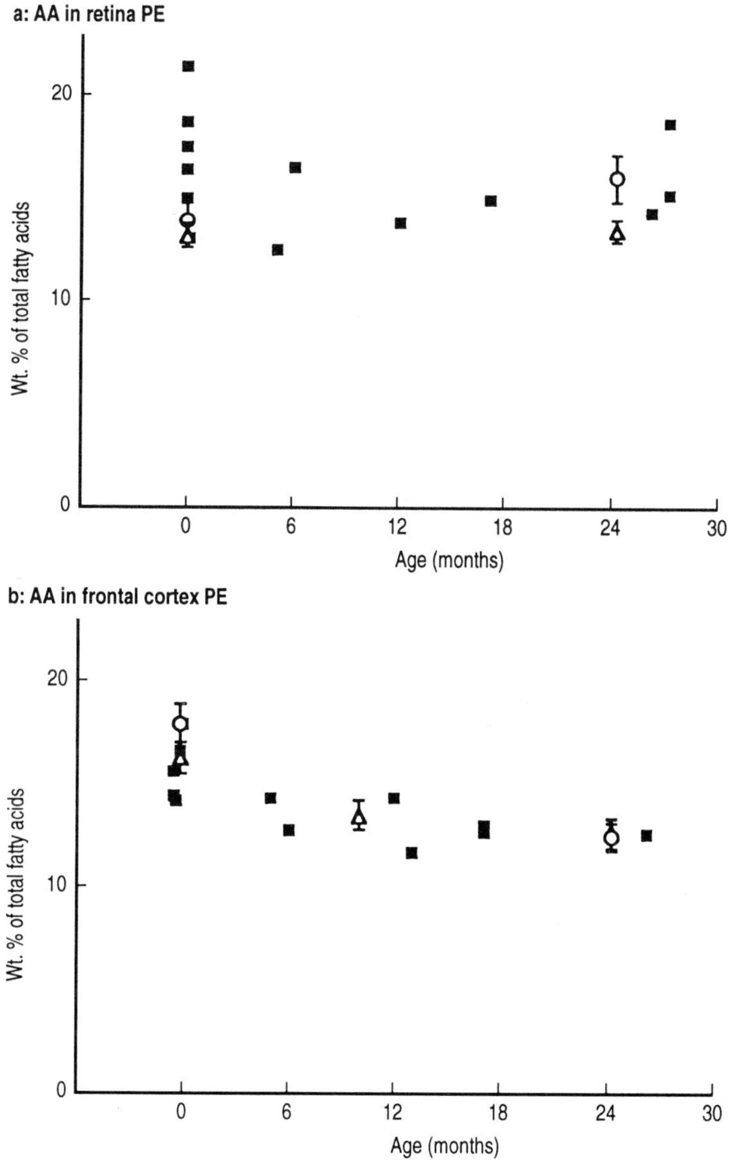

***Figure 6** Intake of water (mg/kg; mean ± SEM) over 24 h by monkeys fed soybean oil (control, n = 7) or safflower oil (deficient, n = 7) diets. Deficient intake differed from control intake measured hourly over 24 h [F(1,12) = 19.6, p < 0.01]. For comparison, values are also shown for standard adult monkeys (n = 4), and for monkeys in a different study fed two other soybean formulas with taurine (soy + T, n = 7) and without taurine (soy − T, n = 8)*

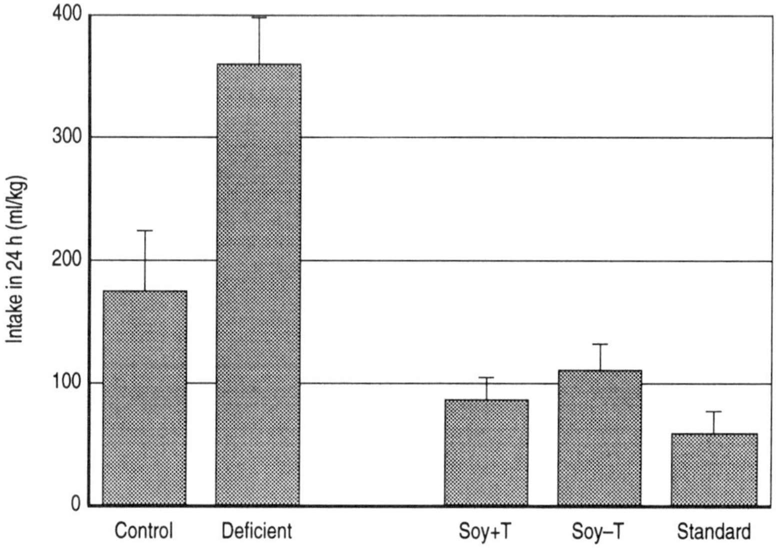

groups (Fig. 5a,b). Thus, in term monkeys, dietary supplementation with preformed LCPUFA has much less effect on tissues than on erythrocytes.

The effects of $n-3$ fatty acid deficiency on the visual system in monkeys have already been well characterized (2,5) and include delayed development of visual acuity and smaller amplitudes of retinal responses to light measured with the electroretinogram early in infancy. Both of these effects have been replicated in studies of preterm human infants fed formulas low in ALA and those without LCPUFA (see Chapters 12 and 15, this volume). Similar effects on acuity have also been found in some, but not all, studies of term infants fed formulas with 18-carbon essential fatty acids compared with those fed breast milk or LCPUFA-supplemented formulas (see Chapters 13 and 15, this volume). In monkeys, electroretinogram responses of deficient monkeys also show a slower time course after a flash and recover less quickly after an initial bright flash (2,5).

We have also identified a number of effects of $n-3$ fatty acid deficiency on behaviour. Juveniles and adults maintained long-term on deficient (safflower oil) diets ingested about twice as much water over 24 h as controls fed soybean oil (4,7). The intake of the deficient group was also higher than that of the standard comparison group and two other groups fed soybean oil formula as part of a separate study of dietary taurine

(Fig. 6). However, their serum chemistry was normal, and they concentrated urine normally. Deficient monkeys also ingested greater amounts of various saline solutions, indicating that the increased ingestion was not specific to water (6).

When observed in their home cages, deficient monkeys initiated more bouts of whole-body activity and of stereotyped behaviour than controls fed soybean oil, or adult standard monkeys (Fig. 7) (10). We hypothesized that the deficient monkeys might be generally more reactive to stimuli, and preliminary data support this hypothesis by demonstrating that they are more likely to approach and interact with stimuli than are controls fed soybean oil (Fig. 8). Following the findings of Delion and colleagues (12) that deficient rats had lower dopamine levels and dopamine receptor binding in frontal cortex, we speculate that these results are associated with similar changes in the prefrontal cortex of deficient monkeys. Prefrontal cortex dopamine circuits modulate reinforcement mechanisms in the nucleus accumbens (13), and stimulated or rewarded behaviour, including species-typical responsiveness, may be enhanced in deficient monkeys.

Although we have found several effects of $n-3$ fatty acid deficiency on behaviour in monkeys, clear effects of $n-3$ fatty acid deficiency on learning and cognitive function have been more elusive. Before an effect on learning ability can be validly claimed, comparisons need to be made with performance in a series of control tasks in order to distinguish effects on learning per se from effects on performance variables such as motivation, reactivity, sensory threshold or motor competence. We would also like to emphasize the idea that effects on these performance variables may be just as important as effects on learning itself, as is true following protein malnutrition (14). For example, it is now accepted that protein-malnourished children have disabilities in attention and social interaction which affect their ability to apply their intelligence. Dietary variables affect performance in both monkeys and humans. Experiments which differentiate performance from learning variables will best further our understanding of the effects of dietary $n-3$ fatty acids.

Several learning tasks in monkeys showed large individual differences, but no effects of $n-3$ fatty acid deficiency. Diet did not affect learning of visual discriminations in an operant chamber by adults or learning to uncover hidden objects by infants. Adults in deficient and soy control groups did not differ in learning to vary their response patterns on two levers or to wait for a fixed amount of time before responding for a food reward. There were no diet effects for either infants or older monkeys in learning to retrieve treats from behind a transparent barrier (object retrieval) or in a measure of short-term or working memory (the AnotB task). Thus it is clear that long-term severe deficiency of all $n-3$ fatty acids did not lead to a general deficit in learning or memory, even though it altered species-typical behaviours in the same individuals. However, we have found a consistent effect of $n-3$ fatty acid deficiency on visual processing and/or attention, which promises to provide important insight into the effects of $n-3$ fatty acids on cognitive development (9,11).

One measure of visual attention is average look duration (the average duration of individual fixations). When assessed in habituation tests, average look duration decreases with age and shorter looks have been associated with faster processing (15). Look duration in infancy correlates moderately and negatively with childhood measures

Figure 7 Bouts of activity (bouts/40 min; mean ± SEM) in 4-year-old standard ($n = 4$), soybean oil control ($n = 5$) and deficient ($n = 4$) monkeys and in 2-year-old control ($n = 2$) and deficient ($n = 3$) monkeys. (a) Whole-body activity. Both control and deficient 4 year olds differed from the standard 4 year olds (Tukey tests, $p < 0.05$). Combining both ages, controls had fewer bouts than deficient monkeys [$F(1,10) = 5.917$, $p = 0.035$]. (b) Stereotyped behaviour. Both control and deficient 4 year olds differed from the standard 4 year olds (Tukey tests, $p < 0.05$). Combining both ages, controls had fewer bouts than deficient monkeys [$F(1,10) = 6.63$, $p = 0.028$] and, combining both diets, 2 year olds initiated fewer bouts than 4 year olds [$F(1,10) = 12.50$, $p = 0.005$]. Within the deficient group, 4 year olds initiated more bouts than 2 year olds (Tukey tests, $p < 0.05$), while bouts did not increase significantly with age in control monkeys

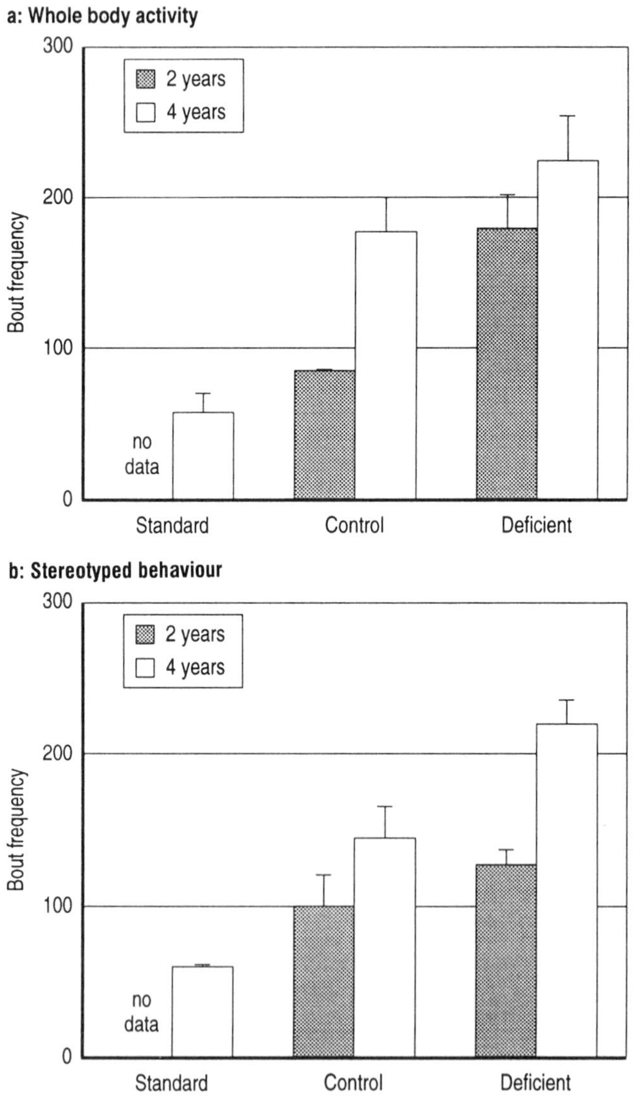

***Figure 8** Scores (mean ± SEM) reflecting interactions or attempts to interact with stimuli presented to the monkeys in their home cages in standard adult monkeys (n = 7) and in those fed soybean oil (control, n = 8) or safflower oil (deficient, n = 8) diets. Deficient monkeys differed from both standard and control groups (Tukey tests, $p < 0.05$)*

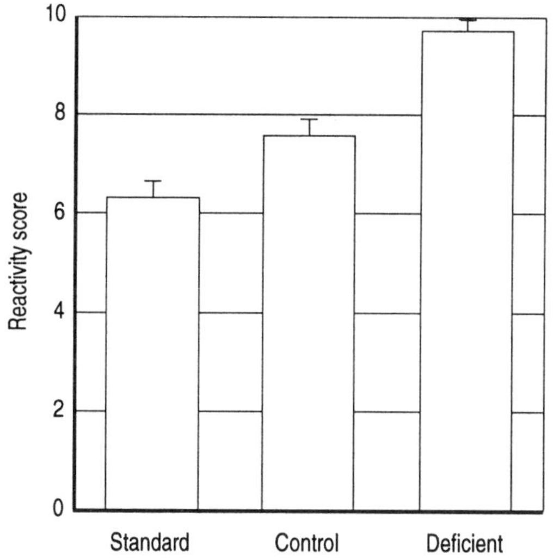

of IQ and with some aspects of language development in the same children (16–19). Another way to measure look duration is with paired comparison tests, described below (20). We chose the second method because it also provides a measure of visual recognition memory in infants; a version of this test, the Fagan Test of Infant Intelligence, has been used for this purpose in human infants. Subjects for this study were nine deficient infants and eight standard nursery-raised infants, as described above. For the paired comparison tests, monkey infants were shown two discrete sets of stimuli, one of patterns and one of photographs of primate faces. The pattern pairs were all hand-drawn colour representations of objects or animals. Each set of stimuli contained four or five subsets (each with six pattern pairs and six face pairs) so that each monkey was presented with a new subset at each age.

The paired comparison tests involved two experimenters. One set up the stimuli and held the infant, while the other observed the infant and recorded its looks. The tests included two phases. During the familiarization phase, two identical stimuli were presented until the infant accumulated a fixed amount of fixation time (in this case 30 s). With these stimuli, 30 s is a long exposure and is followed by significant visual recognition memory even in very young infants. During the test phase, the familiarized stimulus was presented together with a novel stimulus. The observer, blind to the location of the stimuli, recorded looks to each stimulus. In this study, each pair of stimuli was presented for two 10-s tests, each beginning with the first fixation to one

***Figure 9** Average look durations (mean ± SEM) to familiar and novel stimuli during paired comparison tests in standard (▩) nursery infants (n = 8) compared to those fed a safflower oil (deficient, □, n = 9) diet. Deficient infants had longer looks to both novel and familiar stimuli (F(1,15) = 8.95, p = 0.009 and F(1,15) = 9.51, p = 0.008, respectively)*

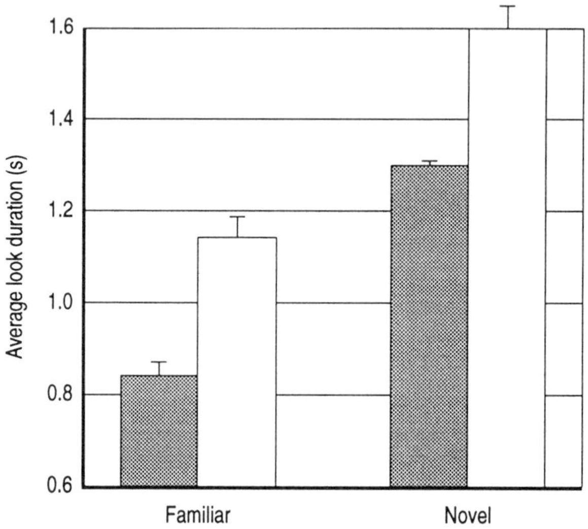

of the stimuli. The position of the novel stimulus was determined semi-randomly for the first test and reversed on the second test. Tests were given both immediately after familiarization and after a 24-h delay. Attention to novel stimuli during the test phase was assessed by a percentage novel score (% novel = time fixating the novel stimulus divided by total time fixating either stimulus). Visual recognition memory was defined as percentage novel significantly greater than chance (50%). Average look duration was calculated by dividing the total looking time by the number of looks for each stimulus pair for each monkey. Attention to novelty and average look duration were evaluated by mixed ANOVA with diet and sex as between-group variables and several within-group variables (age, stimulus set and delay before test).

The diet groups did not differ in attention to novel stimuli or visual recognition memory for either set of stimuli or for either delay, so that there was no evidence for an effect of diet on visual memory. However, there was consistently an effect on average look duration. Deficient infants had longer looks than nursery standard infants both to novel and to familiar stimuli (Fig. 9). Recent preliminary data from our laboratory comparing deficient monkey infants with controls fed soybean oil again found that deficient infants had longer look durations, in this case during familiarization, when familiarization was only 10 s (mean ± SEM: control, 0.928 ± 0.03; deficient, 1.079 ± 0.026 s). Short looks may be dependent on the integrity of the prefrontal cortex, which is necessary for some kinds of attentional shifts (21).

This effect, like the effects of $n-3$ fatty acid deficiency on visual acuity and the electroretinogram, has now been confirmed in preterm human infants. Carlson et al. (22–24) found that look duration, measured in the Fagan Test of Infant Intelligence, was shortened by supplementation with marine oil (Chapter 15, this volume). Another potential confirmation of this effect was presented at this symposium. Dr Forsyth reported that LCPUFA supplementation influenced look duration in those infants which had their longest look late in the session (late peakers) in a habituation test (Chapter 16, this volume). In infants, look duration has been interpreted as a measure of speed of processing and encoding visual information (15,16). However, longer look durations might also indicate greater engagement with stimuli (21), an effect which could be consistent with effects on prefrontal dopamine.

DISCUSSION

Recommendations for LCPUFA supplementation are based on the premise that preformed dietary LCPUFA are necessary to support optimal levels of their incorporation into the developing retina and brain. The placenta provides a rich supply of LCPUFA to the fetus, and preterm infants are deprived of this source. In monkeys, levels of DHA in tissues at birth were affected by the mother's diet during pregnancy. Levels of DHA in both retina and cerebral cortex were substantially reduced in infants of mothers fed a diet low in ALA compared with infants of mothers fed diets containing either soybean oil or a chow containing LCPUFA (1, Fig. 3). DHA levels in the retina, but not cerebral cortex, at birth were also lower in infants of mothers fed soybean oil than in infants of mothers fed the chow diet. Thus, maternal levels of LCPUFA affected levels in infants at birth, and tissue levels at birth may interact with the postnatal dietary supply of LCPUFA to determine the rate of tissue accumulation.

After birth, breast-feeding provides enough DHA and AA to account for their rapid accumulation in the infant brain (25). In monkey infants fed a diet low in ALA for 2 years, levels of DHA in retina and cortex did not increase above those present at birth (1), while in those fed soybean oil, tissue DHA levels increased postnatally, and by 2 years of age did not differ from those of breast-fed infants weaned onto the chow diet. However, we do not know whether DHA accretion proceeded more slowly in those fed soybean oil than in the breast-fed group. In human infants, levels of LCPUFA in erythrocytes are used as an index of the levels in neural tissue. Our data from monkeys suggest that diet-induced differences in erythrocyte DHA can greatly overestimate differences in retinal and cerebral cortex DHA. Nevertheless two postmortem studies of human infants have shown lower DHA levels in the cerebral cortex of term formula-fed infants than in those who were breast fed (26,27). Thus, infants fed diets lacking LCPUFA may show a postnatal delay in DHA accretion which is correlated with their slower visual development.

Functional data from monkeys deficient in $n-3$ fatty acids have provided strong evidence for the importance of these nutrients in human infant development. We have found consistent effects of ALA deficiency on retinal function and development of visual acuity, and we have also recently found alterations in visual attention. All of these effects have now been replicated in preterm human infants fed formulas low in

ALA or lacking $n-3$ LCPUFA. Slower development of both the electroretinogram and visual acuity has been found in corn-oil-fed preterm infants (28), supporting the importance of higher levels of α-linolenic acid (a minimum of ~0.7% energy) in infant formulas. However, as noted in the introduction, the more controversial current issue is whether infant formulas should contain preformed LCPUFA. Results in rhesus monkeys predict not only the effects of feeding corn oil formulas to human preterm infants, but also the effects of feeding formulas without LCPUFA. Thus, slower visual acuity development and longer look durations have been found in preterm infants fed soybean oil formulas compared with those who were either breast fed or fed marine-oil-supplemented formulas (23,29–32), and similar effects on visual acuity have been found in some, but not all, studies of term infants (33–38).

Although this evidence supports the addition of DHA to formulas for preterm infants, the type and balance of LCPUFA must be carefully considered. Detrimental effects have been produced by supplementing with marine oil, particularly that containing relatively high levels of eicosapentaenoic acid (EPA, 20:5 $n-3$). In a trial in preterm human infants, marine oil (0.3 wt % EPA, 0.2% DHA) supplementation until 9 months past term was associated with lowered levels of erythrocyte AA, slower growth and lower Bayley Psychomotor Development scores at 12 months (30,39). These negative outcomes were not found with a marine oil supplement lower in EPA (0.06% EPA, 0.2% DHA) and supplied until only 2 months past term in a nutrient-enriched formula (40). A recent trial in term infants evaluated formulas supplemented with either marine oil (0.07% EPA, 0.2% DHA) or an egg phospholipid source of DHA and AA (0.12% and 0.4%, respectively) to 12 months of age. Compared with control, egg-phospholipid-supplemented or breast-fed infants, the marine oil group had significantly lower scores for both production and comprehension of language as measured by the MacArthur Communicative Development Inventory at 14 months of age (41). These results indicate that if DHA is to be added to either preterm or term formulas, the simultaneous addition of AA is advisable.

The duration of supplementation and the kind of formula supplemented may also be important considerations. In both of the trials which found negative effects of supplementation, addition of LCPUFA was continued for at least 9 months past term. In the second study of Carlson and colleagues (40), preterm infants were supplemented only until 2 months past term, and the infants received a nutrient-enriched preterm formula until 2 months past term. In this case there was a significant improvement in the Bayley Mental Development Index (40; see also Chapter 15, this volume). Thus, duration of administration and/or enrichment of formulas may be important factors for both the safety and efficacy of LCPUFA supplementation. Additional studies are needed to test these possibilities.

ACKNOWLEDGEMENTS

This work was funded by NIH grants DK-29930, DK-40566 and RR-00163 and a grant from the Retinitis Pigmentosa Foundation – Fighting Blindness. We are grateful for the contributions of D. Lin, G. Anderson and N. Zhu at OHSU and R. Hasnain, K. Saunders, D. Daigle, G. Houghton, A. Jones and M. Graham at ORPRC.

REFERENCES

1. Neuringer M, Connor WE, Van Petten C, Barstad L 1984 Dietary omega-3 fatty acid deficiency and visual loss in infant rhesus monkeys. J Clin Invest 73:272–276
2. Neuringer M, Connor WE, Lin DS, Barstad L, Luck S 1986 Biochemical and functional effects of prenatal and postnatal ω3 fatty acid deficiency on retina and brain in rhesus monkeys. Proc Natl Acad Sci USA 83:4021–4025
3. Neuringer M, Anderson GJ, Connor WE 1988 The essentiality of $n-3$ fatty acids for the development and function of the retina and brain. Annu Rev Nutr 8:517–541
4. Reisbick S, Neuringer M, Hasnain R, Connor WE 1990 Polydipsia in rhesus monkeys deficient in omega-3 fatty acids. Physiol Behav 47:315–323
5. Neuringer M, Connor WE, Lin DS, Anderson GJ, Barstad L 1991 Dietary omega-3 fatty acids: Effects on retinal lipid composition and function in primates. In: Anderson RE, Hollyfield JG, LaVail MM (eds) Retinal Degenerations. CRC Press, New York, pp 117–129
6. Reisbick S, Neuringer M, Connor WE, Iliff-Sizemore S 1991 Increased intake of water and NaCl solutions in omega-3 fatty acid deficient monkeys. Physiol Behav 49:1139–1146
7. Reisbick S, Neuringer M, Connor WE, Barstad L 1992 Postnatal deficiency of omega-3 fatty acids in monkeys: Fluid intake and urine concentration. Physiol Behav 51:473–479
8. Neuringer M, Reisbick S, Janowski J 1994 The role of $n-3$ fatty acids in visual and cognitive development: Current evidence and methods of assessment. J Pediatr 125:S39–S47
9. Reisbick S, Neuringer M, Gohl E 1994 Increased look duration in paired comparisons by rhesus monkey infants with $n-3$ fatty acid deficiency. Soc Neurosci Abstr 20:1696
10. Reisbick S, Neuringer M, Hasnain R, Connor WE 1994 Home cage behavior of rhesus monkeys with long-term deficiency of omega-3 fatty acids. Physiol Behav 55:231–239
11. Reisbick S, Neuringer M, Gohl E, Wald R, Anderson GJ 1996 Visual attention in infant monkeys: Effects of dietary fatty acids and age. Dev Psychol (in press)
12. DeLion SS, Chalon S, Herault J, Guilloteau D, Besnard J-C, Durand G 1994 Chronic dietary α-linolenic acid deficiency alters dopamine and serotoninergic neurotransmission in rats. J Nutr 124:2466–2476
13. Doherty MD, Gratton A 1994 Medial prefrontal cortical D1 receptors modulate stress-induced dopamine release in nucleus accumbens. Soc Neurosci Abstr 20:821
14. Tonkiss J, Shukitt-Hale B, Formica RN, Rocco FJ, Galler JR 1990 Prenatal protein malnutrition alters response to reward in adult rats. Physiol Behav 48:675–680
15. Colombo J, Mitchell DW, Coldren JT, Freeman LJ 1991 Individual differences in infant visual attention: Are short lookers faster processors or feature processors? Child Dev 62:1247–1257
16. Jacobson SW, Jacobson JL, O'Neill JM, Padgett RJ, Frankowski JJ, Bihun JT 1992 Visual expectation and dimensions of infant information processing. Child Dev 63:711–724
17. Jacobson SW, Jacobson JL, Sokol RJ, Martier SS, Ager JW 1993 Prenatal alcohol exposure and infant information processing ability. Child Dev 64:1706–1721
18. McCall RB, Carriger MS 1993 A meta-analysis of infant habituation and recognition memory performance as predictors of later IQ. Child Dev 64:57–79
19. Thompson LE, Fagan JF, Fulker DW 1991 Longitudinal predictions of specific cognitive abilities from infant novelty preference. Child Dev 62:530–538
20. Rose SA, Feldman JF, McCarton CM, Wolfson J 1988 Information processing in seven-month-old infants as a function of risk status. Child Dev 59:589–603
21. Johnson MH, Posner MI, Rothbart MK 1991 Components of visual orienting in early infancy: Contingency learning, anticipatory looking, and disengaging. J Cogn Neurosci 3:335–344
22. Carlson SE, Werkman SH 1995 Preterm infants fed formula with compared to without docosahexaenoic acid (DHA) have shorter look duration ten months after DHA is discontinued. Pediatr Res 37:14A
23. Carlson SE, Werkman SH 1996 A randomized trial of visual attention of preterm infants fed docosahexaenoic acid until two months. Lipids 31:85–90
24. Werkman SH, Carlson SE 1996 A randomized trial of visual attention of preterm infants fed docosahexaenoic acid until nine months. Lipids 31: 91–97
25. Clandinin MT, Chappel JE, Heim T 1981 Do low weight infants require nutrition with chain elongation–desaturation products of essential fatty acids? Prog Lipid Res 20:901–904
26. Farquharson J, Cockburn F, Patrick WA, Jamieson EC, Logan RW 1992 Infant cerebral cortex

phospholipid fatty-acid composition and diet. Lancet 340:810–813
27. Makrides M, Neumann MA, Byard RW, Simmer K, Gibson RA 1994 Fatty acid composition of brain, retina, and erythrocytes in breast and formula fed infants. Am J Clin Nutr 60:189–194
28. Birch DG, Birch EE, Hoffman DR, Uauy RD 1992 Retinal development in very-low-birth-weight infants fed diets differing in omega-3 fatty acids. Invest Ophthalmol Vis Sci 33: 2365–2376
29. Birch EE, Birch DG, Hoffman DR, Uauy RD 1992 Dietary essential fatty acid supply and visual acuity development. Invest Ophthalmol Vis Sci 33:3242–3253
30. Carlson SE 1993 Lipid requirements of very-low-birth-weight infants for optimal growth and development. In: Dobbing J (ed) Lipids, Learning and the Brain: Fats in Infant Formulas. Ross Laboratories, Columbus, pp 188–207
31. Carlson SE, Werkman SH, Rhodes PG, Tolley EA 1993 Visual-acuity development in healthy preterm infants: effect of marine-oil supplementation. Am J Clin Nutr 58:35–42
32. Carlson SE, Werkman SH, Tolley EA 1996 The effect of long chain $n-3$ fatty acid supplementation on visual acuity of preterm infants with and without bronchopulmonary dysplasia. Am J Clin Nutr 63:687–697
33. Birch EE, Birch DG, Hoffman DR, Hale L, Everett M, Uauy R 1993 Breast-feeding and optimal visual development. J Pediatr Ophthalmol Strabismus 30:33–38
34. Makrides M, Simmer K, Goggin M, Gibson RA 1993 Erythrocyte docosahexaenoic acid correlates with the visual response of healthy, term infants. Pediatr Res 33:425–427
35. Makrides M, Neumann M, Simmer K, Pater J, Gibson R 1995 Are long-chain polyunsaturated fatty acids essential nutrients in infancy? Lancet 345:1463–1468
36. Jorgensen MH, Jonsbo F, Holmer G, Hernell O, Michaelsen KF 1994 Breast-fed (BF) infants have a better visual acuity than formula fed (FF) infants at the age of 2 and 4 mo. FASEB J 8:460
37. Innis SM, Nelson CM, Rioux MF, King DJ 1994 Development of visual acuity in relation to plasma and erythrocyte $\omega-6$ and $\omega-3$ fatty acids in healthy term gestation infants. Am J Clin Nutr 60:347–352
38. Hartmann EE, Neuringer M 1995 Longitudinal behavioral measures of visual acuity in full-term human infants fed different dietary fatty acids. Invest Ophthalmol Vis Sci 36:S869
39. Carlson SE, Werkman SH, Peeples JM, Cooke RJ, Tolley EA 1993 Arachidonic acid status correlates with first year growth in preterm infants. Proc Natl Acad Sci USA 90:1073–1077
40. Carlson SE, Werkman SH, Peeples JM, Wilson WM III 1994 Growth and development of premature infants in relation to $\omega3$ and $\omega6$ fatty acid status. World Rev Nutr Diet 75:63–69
41. Janowski JS, Scott DT, Wheeler RE, Auestad N 1995 Fatty acids affect early language development. Pediatr Res 37:310A

11 Essential fatty acid metabolism in neonates

V.P. CARNIELLI and P.J.J. SAUER

ABBREVIATIONS

AA	arachidonic acid, 20:4 $n-6$
ALA	α-linolenic acid, 18:3 $n-3$
APE	atom percent excess
DHA	docosahexaenoic acid, 22:6 $n-3$
EFA	essential fatty acid
EPA	eicosapentaenoic acid, 20:5 $n-3$
FA	fatty acid
GC-IRMS	gas chromatography-isotope ratio mass spectrometry
LA	linoleic acid, 18:2 $n-6$
LCPUFA	long-chain polyunsaturated fatty acid
PUFA	polyunsaturated fatty acid

BACKGROUND TO POLYUNSATURATED FATTY ACID METABOLISM

The concept of essential fatty acids (EFA) was recognized at the end of the 1920s when George and Mildred Burr (1,2) described growth retardation and skin lesions in rats fed a fat-free diet, which could be prevented by administration of some fat fractions. It was not until the 1950s, however, that Hansen et al. demonstrated the role of linoleic acid (LA; 18:2 $n-6$) in human infants (3,4).

Studies in animals and humans conducted mainly in the 1950s and 1960s showed that the fatty acid composition of tissues is greatly influenced by the dietary intake of fatty acids (FA), and that an increase in LA in the diet results in an elevation of arachidonic acid (AA, 20:4 $n-6$) and a simultaneous reduction of a 20 carbon atom trienoic fatty acid (20:3 $n-9$). The ratio between 20:3 $n-9$ and 20:4 $n-6$, called the triene/tetraene ratio, was used as an indicator of EFA deficiency (5). During this period LA was shown to be the precursor of AA and α-linolenic acid (ALA, 18:3 $n-3$) the precursor of eicosapentaenoic acid (EPA, 20:5 $n-3$) and docosahexaenoic acid (DHA, 22:6 $n-3$) (5), and it was demonstrated that no conversion is possible between the $n-6$ and $n-3$ families of the FA. At the end of this period, however, the term essential fatty acids was reserved for those compounds which seemed to be important for growth and for the integrity of the skin: LA and AA (6).

In the 1980s there was a renewed interest in the EFA, mainly because of the observations that diets rich in $n-3$ fatty acids were associated with a low incidence of

J.G. Bindels, A.C. Goedhart and H.K.A. Visser (eds), Recent Developments in Infant Nutrition, 173–181.

coronary heart disease in the Greenland Eskimos (7,8). The distinct biological effects of the $n-3$ fatty acids on several physiological processes such as platelet function, inflammation, lowering of plasma lipids etc. became well accepted (9,10). During the 1980s and early 1990s a large body of evidence was collected on the key role played by the $n-3$ FA during the perinatal period. The importance of $n-3$ fatty acids for fetal growth, for vascular reactivity of the mother and the fetus and for initiation of labour and delivery was clearly recognized and, perhaps even more importantly, the pivotal role of $n-3$ FA for the development of neonatal brain and visual function (11–14) became accepted. The 1990s have shown that the synthetic pathway for the long-chain polyunsaturated fatty acids (LCPUFA) if profoundly different from that accepted in the 1980s. Tools are available with which the metabolism of the EFA can be investigated safely in humans: these include essential fatty acids labelled with stable isotopes, and isotope ratio mass spectrometers, which are highly sensitive detection systems for isotopic enrichment. In the following sections we will present data from our ongoing studies particularly on these two areas outlined above.

INTERCONVERSION BETWEEN 20 AND 22 CARBON n–3 AND n–6 FATTY ACIDS VIA THE δ4-DESATURASE INDEPENDENT PATHWAY

Until recently it was generally accepted that three position-specific desaturases were used for metabolizing dietary LA and ALA to 22:5 $n-6$ and DHA (22:6 $n-3$), respectively. When LA and ALA are incubated with liver microsomes and appropriate co-factors they are desaturated at position 6. Similarly, both 20:3 $n-6$ and 20:4 $n-3$ are desaturated at position 5 to yield AA (20:4 $n-6$) and EPA (20:5 $n-3$) respectively. Considerable evidence has accumulated in the literature supporting the existence of specific δ6- and δ5-desaturases (15). It was therefore logical to assume that a third enzyme (i.e. a δ4-desaturase) might be responsible for converting 22:4 $n-6$ to 22:5 $n-6$ and 22:5 $n-3$ to 22:6 $n-3$. There was no direct support however for this commonly accepted hypothesis. Voss and Sprecher (16) have shown that when [1-^{14}C]22:5 $n-3$ is incubated with rat liver microsomes it is not possible to detect the product of the putative δ4-desaturase. When malonyl-CoA is included in the incubations, however, there is a sequential chain elongation of 22:5 $n-3$ to 24:5 $n-3$, followed by its desaturation at position 6 to yield 24:6 $n-3$. It has been known for some time that 22-carbon polyunsaturated fatty acids are substrates for partial β-oxidation (retroconversion), with subsequent esterification of chain shortened products into membrane lipids. It has never been established, however, whether the retro-conversion process is localized exclusively in mitochondria or peroxisomes. A revised pathway is now used to illustrate the synthesis of the long-chain derivatives of the EFA (Fig. 1). According to this modified pathway 22:5 $n-3$ and 22:4 $n-6$ are further chain elongated into 24:5 $n-3$ and 24:4 $n-6$ and these are then desaturated at position 6, possibly by the same δ6-desaturase which converts LA into 18:3 $n-6$ and ALA into 18:4 $n-3$. The most likely metabolic fate of 24:6 $n-3$ and 24:5 $n-6$ once formed, according to this pathway, would be retroconversion to DHA (22:6 $n-3$) and 22:5 $n-6$ respectively, since they are relatively poor substrates for membrane lipid biosynthesis.

***Figure 1** Metabolic pathways of polyunsaturated fatty acids. n−6 and n−3 fatty acids compete for the same microsomal enzyme system for desaturation and chain elongation. Recent evidence suggests that the conversion of 22:5 n−3 to 22:6 n−3 and 22:4 n−6 to 22:5 n−6 is an indirect conversion including chain elongation to 24 carbon atom fatty acids, δ6-desaturation and peroxisomal chain shortening*

	n−6		*n*−3	
Linoleic acid	18:2 *n*−6		18:3 *n*−3	α-Linolenic acid
	−2H	δ6	−2H	
	18:3 *n*−6		18:4 *n*−3	
	+2C	elongation	+2C	
	20:3 *n*−6		20:4 *n*−3	
	−2H	δ5	−2H	
Arachidonic acid	20:4 *n*−6		20:5 *n*−3	Eicosapentaeonic acid
	+2C	elongation	+2C	
	22:4 *n*−6		22:5 *n*−3	
	+2C	elongation	+2C	
	24:4 *n*−6		24:5 *n*−3	
	−2H	δ6	−2H	
	24:5 *n*−6		24:6 *n*−3	
	−2C	shortening	−2C	
	22:5 *n*−6		22:6 *n*−3	Docosahexaenoic acid

LONG-CHAIN POLYUNSATURATED FATTY ACIDS IN FETAL TISSUES

During the last trimester of pregnancy the human fetal brain experiences a rapid growth spurt, increasing in weight by four to five times (17). The most abundant fatty acids in brain cellular membranes are AA, adrenic acid (C22:4 *n*−6) and DHA (18). DHA is particularly abundant in the brain cortex phosphatidylethanolamine and in retinal photoreceptors (19). Animal studies indicate that retinal function and learning ability are impaired if DHA fails to accumulate during growth (20). Since the brain does not use the more commonly found C18 fatty acids but only incorporates the C20 and C22 fatty acids, the elongation and desaturation mechanisms of the parent fatty acids may become the limiting reactions in supplying sufficient amounts of LCPUFA for the normal growth and development of the central nervous system. The current view is that the human fetus obtains most of its LCPUFA by placental transfer. Very little is known about the endogenous synthesis of LCPUFA from LA and ALA in the human fetus, however. Concentrations of AA and DHA are higher in the umbilical blood than in the maternal circulation, while the concentration of LA is higher on the maternal side (21). Little is known about the mechanism of LCPUFA transfer through the human placenta. Selective incorporation of particular FA into phospholipids seems to be a more likely mechanism than fatty acid elongation in the placenta itself (22). A similar selective mechanism of incorporation of LCPUFA into phosphoglycerides could also explain the availability of AA and DHA to the brain of the fetus and of the newborn infant (23).

We are currently studying changes in EFA in plasma lipids and tissues of newborns

Figure 2 Molar percentages of 24:6 n−3 and 22:6 n−3 in human retina total lipids plotted against gestational age

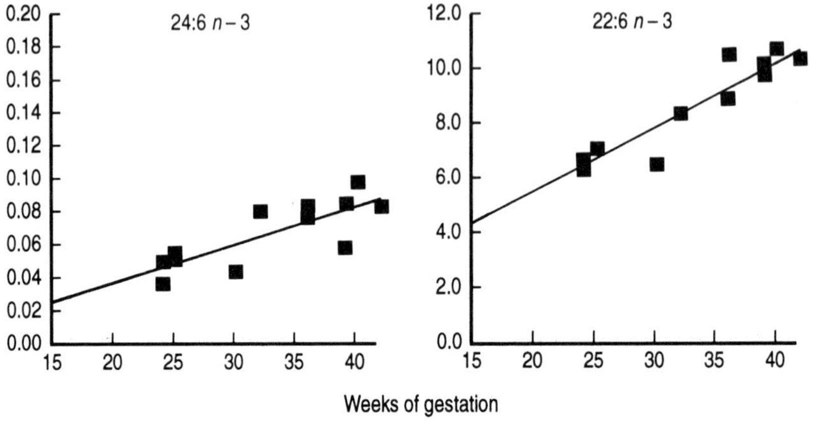

of gestational ages between 20 and 42 weeks who died during the first 72 h of life. The aim of this study is to ascertain whether there is a relation between the EFA status of different tissues and to determine whether 24-carbon atom polyunsaturated fatty acids (C24-PUFA) are found in human tissues during fetal life. Preliminary data from a small number of subjects indicate that the C24-PUFA are present, although in small amounts, in almost all fetal tissues. Figure 2 depicts the mol% levels of DHA (right panel) and its precursor C24:6 $n-3$ (left panel) in the total lipids of the retina, plotted against the duration of gestation. Both DHA and C24:6 $n-3$ increase as gestation progresses. This finding, if confirmed in a larger number of infants, could indicate either that there is an efficient placental transport for the C24-PUFA or that LCPUFA synthesis is active in fetal tissues. Because the C24-PUFA were almost undetectable in cord plasma phospholipids of a similar group of infants (data not shown), the latter seems a more likely hypothesis than the former.

LONG-CHAIN POLYUNSATURATED FATTY ACIDS IN INFANTS WITH ZELLWEGER SYNDROME

Recently, a marked deficiency of DHA has been observed in brain, retina, liver and kidney from patients with defective peroxisomal biogenesis (24). It is noteworthy that a decrease in brain DHA of the magnitude found in these patients could be responsible for many of their symptoms. Since DHA synthesis in rat liver has been shown to occur by retroconversion of the very-long-chain fatty acid 24:6 $n-3$ and since it is not known whether this retroconversion (partial β-oxidation) takes place in either mitochondria or peroxisomes, we studied the FA composition of several tissues obtained from patients with Zellweger syndrome, who died shortly after birth. Zellweger

Figure 3 Molar percentages of 24:6 n−3 in the brain and liver total lipids from newborn infants suffering from Zellweger syndrome and age-matched controls

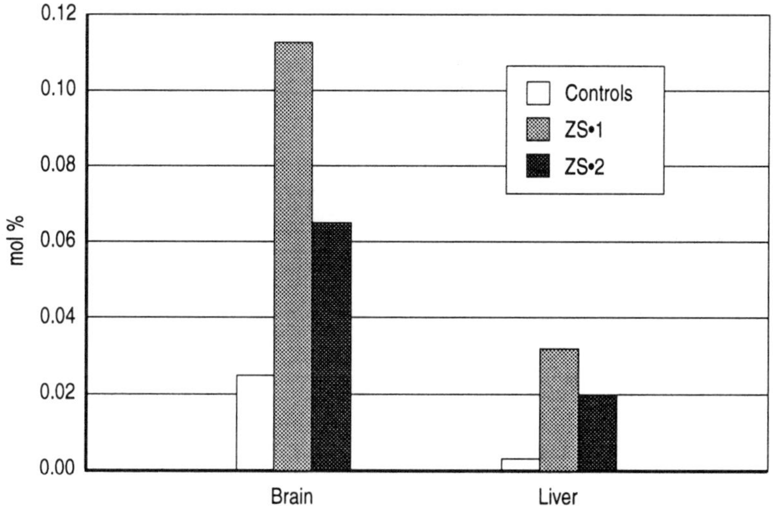

syndrome is an inborn error of metabolism characterized by the absence of the peroxisomes. Figure 3 illustrates the molar percentages of 24:6 $n-3$ in brains of 2 patients suffering from Zellweger syndrome and of control infants. Both patients died within the first 3 months of life, and their brains contained much higher levels of 24:6 $n-3$ than were found in the controls. Our current interpretation of these data is either that active LCPUFA synthesis from shorter chain EFA occurs in the fetus before birth, or that C24-PUFA are taken up by the fetus from the maternal circulation and incorporated into the brain, where they cannot be retroconverted to DHA or to 22:5 $n-6$. We favour the first hypothesis since the C24-PUFA exhibit a very low tendency for incorporation into plasma lipid classes. Ongoing studies using stable isotope technology and a high sensitivity tracer detection using gas chromatography-isotope ratio mass spectrometry show that the conversion of ALA into all the $n-3$ intermediates, including the C24-PUFA, is present in Zellweger patients, but also that DHA synthesis is severely depressed (data not shown).

MEASUREMENTS OF THE METABOLIC CONVERSION OF LA AND ALA IN PRETERM INFANTS WITH STABLE ISOTOPE LABELLED FATTY ACIDS

After birth, preterm infants who are fed formulas which contain ample amounts of LA and ALA, but unlike breast milk are devoid of LCPUFA, exhibit rapidly declining plasma and red blood cell LCPUFA concentrations (25). Term breast-milk-fed infants have higher plasma, red blood cell and brain LCPUFA concentrations than bottle-fed infants (11,26–29). However, small preterm infants fed exclusively with expressed preterm human milk exhibit declining levels of AA and DHA from birth up to 28 days of life (30). A decreased DHA status in infancy is associated with poorer retinal development and visual function (29,31–33). Infants fed with DHA-supplemented formulas have better neurodevelopmental quotients at 4 months (34) and have higher cognitive scores at 12 months (35) than those fed standard formula (see also Chapters 14 and 15, this volume).

The conversion of LA and ALA to LCPUFA in mammals remains largely unexplored (36,37) and data from human adults are few (38–41). The consistent observations of declining LCPUFA levels after birth pose the question as to whether the newborn infant is capable of synthesizing AA and DHA. We investigated in vivo whether the preterm infant can synthesize LCPUFA by administering ^{13}C-labelled LA and ALA and by measuring the ^{13}C enrichment for their metabolic products in plasma phospholipids by gas chromatography-isotope ratio mass spectrometry (GC-IRMS) (42). We present here the preliminary results obtained from three infants fed a standard preterm formula devoid of LCPUFA. The isotope study consisted of the simultaneous and continuous oral administration of [U-^{13}C]LA and [U-^{13}C]ALA (Martek, Columbia, MD, USA) for 48 h. The isotopes were mixed with the formulas and were added in amounts such that the isotope represented approximately 2.5% of the dietary LA and ALA. All infants received the formulas via continuous intragastric infusion, as is customary in our neonatal intensive care unit. Blood was obtained at baseline and at 12, 24, 48, 72, 96 and 168 h after the start of isotope administration. Plasma lipid classes and their fatty acids were analysed as described previously (43). ^{13}C enrichment of the plasma phospholipid fatty acids was measured by gas chromatography-isotope ratio mass spectrometry (42). The GC-IRMS system used was an Isochrom III (VG Isotech, Middlewich, Cheshire, UK). Results obtained from the instrument are based on the ^{13}C enrichment of the CO_2 of the combusted sample and are expressed in atom percentage excess (APE) compared with a reference CO_2 gas of known mass 45/44 ratio. Results obtained from the patients are given in APE that expresses an increase in enrichment above baseline following the isotope infusion. A calibration curve was constructed by adding different amounts of enriched LA and ALA to their naturally enriched counterparts.

Figure 4 shows the ^{13}C enrichment of LA (top left panel), AA (top right panel), ALA (bottom left panel) and DHA (bottom right panel). Both tracers were rapidly incorporated in plasma phospholipids, although neither LA nor ALA reached a plateau at the 48 h time point, corresponding to the end of the isotope administration. The APE of all $n-6$ and $n-3$ fatty acids (not shown) increased markedly in all patients after the administration of the tracers and this clearly shows that all infants were capable of actively synthesizing the long-chain polyunsaturated fatty acids from their dietary precursors. The enrichment curves of AA and DHA peaked between 72 and 96 h, with a marked delay in comparison with the time curves of their precursors. These preliminary results show for the first time that small preterm infants are capable of converting both LA and ALA into LCPUFA. A second important technical achievement

Figure 4 ***^{13}C enrichment of LA (18:2 n−6), AA (20:4 n−6), ALA (18:3 n−3) and DHA (22:6 n−3) of plasma phospholipids of three preterm infants after administration of ^{13}C-LA and ^{13}C-ALA. (from left to right and top to bottom)***

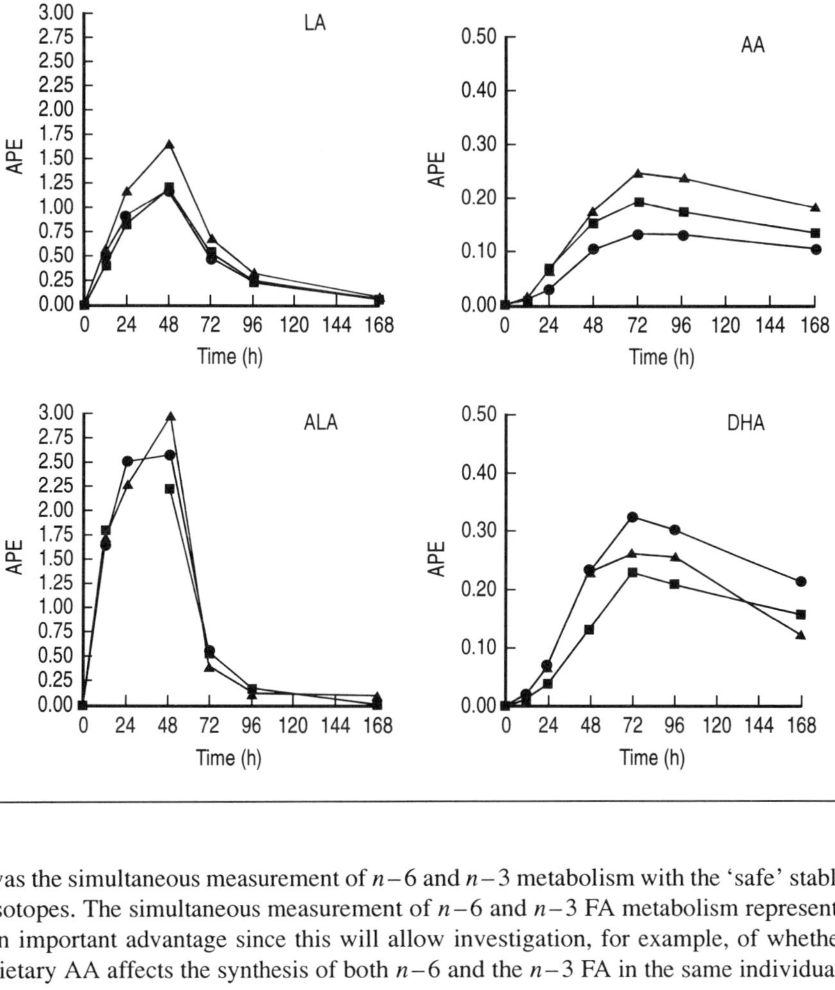

was the simultaneous measurement of $n-6$ and $n-3$ metabolism with the 'safe' stable isotopes. The simultaneous measurement of $n-6$ and $n-3$ FA metabolism represents an important advantage since this will allow investigation, for example, of whether dietary AA affects the synthesis of both $n-6$ and the $n-3$ FA in the same individual.

CONCLUSIONS

Our knowledge of EFA metabolism has increased dramatically. It is foreseeable that much more information will be gained in the coming years, since we have now the tools to investigate EFA metabolism in infants in vivo. Data will be gathered on regulation of FA synthesis in free-living subjects, on the effect of diet on the endogenous synthesis of LCPUFA, and perhaps on maternal–fetal EFA metabolism. Hormonal as well as pharmacological influences will also be studied. Clinical and functional studies will definitely play a key role and much will also be learned by studying patients with peroxisomal dysfunction.

REFERENCES

1. Burr GO, Burr MM 1929 A new deficiency disease produced by the rigid exclusion of fat from the diet. J Biol Chem 82:345–367
2. Burr GO, Burr MM 1930 On the nature and role of the fatty acids essential in nutrition. J Biol Chem 86:587–621
3. Hansen AE, Haggard ME, Boelsche AN, Adam DJD, Wiese HF 1958 Essential fatty acids in infant nutrition. III. Clinical manifestation of linoleic acid deficiency. J Nutr 66:565–576
4. Hansen AE, Wiese HF, Boelsche AN, Haggard ME, Adam DJD, Davis H 1963 Role of linoleic acid in infant nutrition: clinical and chemical study of 428 infants fed on milk mixtures varying in kind and amount of fat. Pediatrics 31:171–192
5. Widmer C, Holman RT 1950 Polyethenoid fatty acid metabolism. II. Deposition of polyunsaturated fatty acids in fat-deficient rats upon single fatty acid supplementation. Arch Biochem Biophys 25:1–12
6. Holman RT 1970 Essential fatty acid deficiency. Prog Chem Fats Lipids 9:275–348
7. Dyerberg J, Bang HO, Stoffersen E, Moncadu S, Vane JR 1978 Eicosapentaenoic acid and prevention of thrombosis and atherosclerosis. Lancet ii:117–119
8. Dyerberg J, Bjerregaard P 1987 Mortality from ischemic heart disease and cerebrovascular disease in Greenland. In: Lands WEM (ed) Polyunsaturated Fatty Acids and Eicosanoids. American Oil Chemists' Society, Champaign, IL, pp 2–8
9. Nelson GJ 1991 Health Effects of Dietary Fatty Acids. American Oil Chemists' Society, Champaign, IL
10. Hoak JC, Glinsmann WH, Judd JT 1992 Workshop on dietary fatty acids and thrombosis. Am J Clin Nutr 56:783S–826S
11. Innis SM 1991 Essential fatty acids in growth and development. Prog Lipid Res 30:39–103
12. Simopoulos AP 1991 Omega-3 fatty acids in health and disease and in growth and development. Am J Clin Nutr 54:438–463
13. Willis AL 1984 Essential fatty acids, prostaglandins and related eicosanoids. In: Olson RE (ed) Present Knowledge in Nutrition. Nutrition Foundation, Washington, DC, pp 90–113
14. Uauy R, Hoffman DR 1991 Essential fatty acid requirements for normal eye and brain development. Semin Perinatol 15:449–455
15. Sprecher H 1991 Enzyme activities affecting tissue lipid fatty acid composition. In: Simopoulos A, Martin RE, Barlow SM (eds) Health Effects of ω-3 Polyunsaturated Fatty Acids in Seafood, Karger, New York, pp 166–176
16. Voss A, Reinhart M, Sankarappa S, Sprecher H 1991 The metabolism of 7,10,13,16,19-docosapentaenoic acid to 4,7,10,13,16,19-docosahexaenoic acid in rat liver is independent of a 4-desaturase. J Biol Chem 266:19995–20000
17. Dorovini-Zis K, Dolman CL 1977 Gestational development of brain. Arch Pathol Lab Med 101: 192–195
18. Martinez M, Ballabriga A, Gil-Gibernau JJ 1988 Lipids of the developing human retina. I. Total fatty acids, plasmalogens and fatty acid composition of ethanolamine and choline phosphoglycerides. J Neurosci Res 20:484–490
19. Bourre JM, Dumont O, Piciotti M, Clement M, Chaudiere J, Bonneil M, Nalbone G, Lafont H, Pascal G, Durand G 1991 Essentiality of ω3 fatty acids for brain structure and function. World Rev Nutr Diet 66:103–117
20. Neuringer M, Anderson GJ, Connor WE 1988 The essentiality of $n-3$ fatty acids for the development and function of the retina and brain. Ann Rev Nutr 8:517–541
21. Olegard R, Svennerholm L 1970 Fatty acid composition of plasma and red cell phosphoglycerides in full term infants and their mothers. Acta Paediatr Scand 59:637–647
22. Kuhn D, Crawford M 1986 Placental essential fatty acid transport and prostaglandin synthesis. Prog Lipid Res 23:345–353

23. Leyton J, Drury PJ, Crawford MA 1987 In vivo incorporation of labeled fatty acids in rat liver lipids after oral administration. Lipids 22:553–558
24. Martinez M 1990 Severe deficiency of docosahexaenoic acid in peroxisomal disorders: a defect of delta 4 desaturation? Neurology 40:1292–1298
25. Carlson SE, Rhodes PG, Ferguson MG 1986 Docosahexaenoic acid status of preterm infants at birth and following feeding with human milk or formula. Am J Clin Nutr 44:798–804
26. Clark KJ, Makrides M, Neumann MA, Gibson RA 1992 Determination of the optimal ratio of linoleic to alpha linolenic acid in infant formulas. J Pediatr 120:S151–S158
27. Putnam JC, Carlson SE, de Voe P, Barness LA 1982 The effect of variations in dietary fatty acids on the fatty acid composition of erythrocyte phosphatidylcholine and phosphatidyletanolamine in human infants. Am J Clin Nutr 36:106–114
28. Farquharson J, Cockburn F, Patrick WA, Jamieson EC, Logan RW 1992 Infant cerebral cortex phospholipid fatty-acid composition and diet. Lancet 340:810–813
29. Makrides M, Neumann MA, Byard RW, Simmer K, Gibson RA 1994 Fatty acid composition of brain, retina, and erythrocytes in breast- and formula-fed infants. Am J. Clin Nutr 60:189–194
30. Carnielli VP, Pederzini F, Vittorangeli R, Luijendijk IHT, Pedrotti D, Sauer PJJ 1996 Plasma and red blood cell fatty acid of very-low-birth-weight infants fed exclusively with expressed preterm human milk. Pediatr Res 39:671–679
31. Uauy RD, Birch DG, Birch EE, Tyson JE, Hoffman DR 1990 Effect of dietary omega-3 fatty acids on retinal function of very-low-birth-weight neonates. Pediatr Res 28:485–492
32. Birch DG, Birch EE, Hoffman DR, Uauy RD 1992 Retinal development in very-low-birth-weight infants fed diets differing in omega-3 fatty acids. Invest Ophthalmol Vis Sci 32: 2365–2376
33. Makrides M, Neumann M, Simmer K, Pater J, Gibson RA 1995 Are long-chain polyunsaturated fatty acids essential nutrients in infancy? Lancet 345:1463–1468
34. Agostoni C, Bellù R, Trojan S, Riva E, Giovannini M 1995 Neurodevelopmental quotient of healthy term infants at 4 months and feeding practice: the role of long-chain polyunsaturated fatty acids. Pediatr Res 38:262–266
35. Carlson SE, Werkman SH, Peeples JM, Wilson MW 1994 Growth and development of premature infants in relation to $\omega 3$ and $\omega 6$ fatty acid status. World Rev Nutr Diet 75:63–69
36. Pawlosky R, Salem N Jr 1992 Metabolism of essential fatty acids in mammals. In: Sinclair A, Gibson R (eds) Essential Fatty Acids and Eicosanoids. American Oil Chemists' Society, Champaign, IL pp 26–30
37. Sheaff RC, Su HM, Keswick LA, Brenna JT 1995 Conversion of α-linolenate to docosahexaenoate is not depressed by high dietary levels of linoleate in young rats: Tracer evidence using high precision mass spectrometry. J Lipid Res 36:998–1008
38. Emken EA, Rohwedder WK, Adlof RO, Rakoff H, Gulley RM 1987 Metabolism in humans of cis-12,trans-15-octadecadienoic acid relative to palmitic, stearic, oleic and linoleic acids. Lipids 22:495–504
39. Emken EA, Adlof RO, Rakoff H, Rohwedder WK 1989 Metabolism of deuterium labelled linoleic, linolenic, oleic, stearic and palmitic acid in human subjects. In: Baillie TA, Jones JR (eds) Synthesis and Applications of Isotopically Labeled Compounds, Proceedings of the Third International Symposium. Elsevier, Amsterdam, pp 713–716
40. Emken EA, Adlof RO, Gulley RM 1994 Dietary linoleic acid influences desaturation and acylation of deuterium-labeled linoleic and linolenic acids in young adult males. Biochim Biophys Acta 1213:277–288
41. El Boustani S, Causse J, Descomps B, Monnier L, Mendy F, Crastes de Paulet A 1989 Direct in vivo characterization of delta 5 desaturase activity in humans by deuterium labeling: effect of insulin. Metabolism 38:315–321
42. Carnielli VP, Sulkers EJ, Moretti C, Wattimena JLD, Van Goudoever JB, Degenhart HJ, Zacchello F, Sauer PJJ 1994 Conversion of octanoic acid into long-chain saturated fatty acids in premature infants fed a formula containing medium-chain triglycerides. Metabolism 43:1287–1292
43. Carnielli VP, Luijendijk IHT, Van Beek RHT, Boerma GJM, Degenhart HJ, Sauer PJJ 1995 Effect of dietary triacylglycerol fatty acid positional distribution on plasma lipid classes and their fatty acid composition in preterm infants. Am J Clin Nutr 62:776–781

12 LCPUFA requirements for preterm infants: neurophysiological studies

R. UAUY, P. PEIRANO, P. MENA, D. HOFFMAN, D. BIRCH and E. BIRCH

ABBREVIATIONS

AA	arachidonic acid, 20:4 $n-6$
ALA	α-linolenic acid, 18:3 $n-3$
ANS	autonomic nervous system
AS	active sleep
CNS	central nervous system
DHA	docosahexaenoic acid, 22:6 $n-3$
DPA	docosapentaenoic acid, 22:5 $n-6$
EEG	electroencephalogram
EFA	essential fatty acid
EPA	eicosapentaenoic acid, 20:5 $n-3$
ERG	electroretinogram
HF	variability of the high frequency
HM	human milk
HRV	heart rate variability
IS	indeterminate sleep
LA	linoleic acid, 18:2 $n-6$
LCPUFA	long-chain polyunsaturated fatty acid
LF	variability of the low frequency
MAR	minimal angle of resolution
PNS	parasympathetic nervous system
PUFA	polyunsaturated fatty acid
QS	quiet sleep
RBC	red blood cell
REM	rapid eye movements
SNS	sympathetic nervous system
SWS	slow-wave sleep
TPN	total parenteral nutrition
VEP	visual evoked potential

J.G. Bindels, A.C. Goedhart and H.K.A. Visser (eds), Recent Developments in Infant Nutrition, 182–201.

INTRODUCTION

Although the need for specific components as a prerequisite for the proper growth and development of animals and possibly humans was introduced in the 1930s (1), essential fatty acids (EFA) were considered of marginal importance until the 1960s when signs of clinical deficiency became apparent in infants fed skim-milk-based formula and in those receiving lipid-free parenteral nutrition (2–5). During the past decade, it has become generally accepted that $n-6$ as well as $n-3$ fatty acids play a key role in perinatal nutrition, especially for the developing CNS. Clinical manifestations of EFA deficiency disappear after the administration of diets that provide 2% or more of the calories as linoleic acid (LA, 18:2 $n-6$) (2–7). More subtle clinical symptoms appear in $n-3$ EFA deficiency: these include skin changes unresponsive to LA supplementation, abnormal visual function and peripheral neuropathy (8). Nervous system manifestations are probably caused by an insufficient supply of docosahexaenoic acid (DHA, 22:6 $n-3$), the metabolic derivative of α-linolenic acid (ALA, 18:3 $n-3$).

Dietary EFA deficiency has become a significant issue for the nutrition of preterm infants since they do not receive an intrauterine supply of DHA and arachidonic acid (AA, 20:4 $n-6$) during the third trimester. Even term infants are at risk of DHA deficiency, since most formulas are devoid of this critical EFA. Recent investigations in term infants who died suddenly with no clear explanation have shown a dietary dependent fatty acid composition of the brain cortex. Infants who are breast fed have higher levels of DHA and lower levels of AA and docosapentaenoic acid (DPA, 22:5 $n-6$) in their cortex phospholipids relative to infants receiving cow's-milk-based formulas (9,10). DHA content increases with advancing age in breast-fed infants and is proportional to the duration of human milk feeding, while AA content remains stable. Furthermore the dietary ratio of $n-6$ to $n-3$ fatty acids appears to have significant effects, since the lowest DHA and highest DPA levels in brain cortex phospholipids were observed in preterm infants fed formulas with a high LA to ALA ratio (9).

BIOLOGICAL BASIS FOR THE EFFECTS OF DIETARY ESSENTIAL FATTY ACIDS ON NEURAL DEVELOPMENT

The functional implications of diet-induced changes in structural lipids have been extensively studied (11–13). Membrane physiologists have suggested that the fatty acid composition of structural membrane lipids can affect membrane function by modifying overall membrane fluidity, by affecting membrane thickness, by changing lipid phase properties, by changing the membrane microenvironment or by interacting with membrane proteins (14–16). The changes in neural membranes of greatest potential significance to the human infant are those related to changes in physical properties and those that affect membrane excitability (11–13).

Deficiency of $n-3$ fatty acids can change membrane physical properties including activity of membrane-bound enzymes and receptors, antigenic recognition and signal transduction (14–16). $n-3$ fatty acid deficiency also affects rotational mobility, as measured by fluorescent diphenylhexatriene probes, to a lesser degree than lateral mobility, as measured by pyrene dimer formation within the lipid bilayer (11,14–17).

Recently, we found that excited dimer (eximer) pyrene formation in human retinal cells, which occurs when two pyrene monomers collide, is increased by DHA supplementation of the culture media. Changes of membrane composition in human retinoblastoma and neuroblastoma cell lines also affect membrane transport mechanisms: a higher DHA content increases the affinity and transport rates for choline and taurine (17,18). Preliminary studies suggest that the dietary $n-3$ fatty acid supply may affect nucleotide cyclase activity and ribosylation of guanine nucleotide binding proteins (19), while protein kinase C phosphorylation activity is modified by the species and degree of unsaturation of the phospholipid moieties in the lipid bilayer. There is progressive activation with unsaturation, but subsequent increases eventually result in reduced activity (20). These significant membrane-related phenomena may explain the changes in electrophysiological responses induced by $n-3$ fatty acid supply.

Most dietary $n-3$ fatty-acid-induced membrane changes are not reflected by an overall change in membrane fluidity but rather result in selective changes in membrane microenvironment affecting specific domains. The replacement of DHA by DPA usually results in the same overall lipid unsaturation level, and average fluidity thus remains unchanged. Furthermore, the main changes in the physical state of lipid bilayers induced by changes in the fatty acid composition occur after the first and second double bonds are introduced; namely when a saturated fatty acid such as stearic acid (18:0) is replaced by oleic acid (18:1 $n-9$) or by linoleic acid (16,17). Others have suggested that DHA supply modifies the phospholipid molecular species present in neural tissues, thus possibly affecting overall function (21). More recently it has been demonstrated that acyl chain packing free volume, which is determined in part by degree and position of unsaturated bonds, is linearly correlated with formation of metarhodopsin II, the form of photoactivated rhodopsin which is responsible for G protein activation (22). This indicates that the phototransduction cascade is potentially modulated by lipid membrane composition and offers a plausible explanation for the observed effects on photoreceptor responses.

Diet-induced changes in structural lipids affect the functional characteristics of excitable membranes in several animal species and in human neural cell lines (11–13, 23,24). Electrocardiographic abnormalities such as a notching in the QRS complex, indicating impaired electrical conduction, occur in linoleic acid and α-linolenic acid deficiency before clinical signs appear (25). Administration of LA or ALA corrects these abnormalities. More recently it has been shown that dietary fat influences the susceptibility to cardiac arrhythmia, and its incidence and severity (26), while studies with myocardial preparations have indicated a reduced vulnerability to catecholamine-induced arrhythmia in animals fed either $n-6$ or $n-3$ PUFA-enriched diets (27). Feeding oil derived from bluefin tuna rather than sunflower oil and saturated fat markedly reduced induced arrhythmias in two animal species and in isolated papillary muscle (26). Changes in cardiac electrophysiologic responses to β-mimetics, reduced excitability of cardiac myocytes and reduced susceptibility to arrhythmias have also been noted (28,29). Myocytes form minimal amounts of cyclo-oxygenase products and no lipo-oxygenase products, thus the changes in excitability and conduction are probably related to structural lipid composition and reflect changes in the function of ion channels (24). Supplementation with $n-3$ fatty acids ameliorates the fluidifying effect of ethanol on neural membranes. While LA and ALA deficiency potentiates the action of volatile anaesthetics in rats, supplementation specifically with LA reverses this effect (30).

The role of membrane lipid composition in determining the electrical properties of cultured neuronal cells exposed to exogenous fatty acids has been investigated (21,23). Both $n-3$ and $n-6$ fatty acids induced slower rates of rise, and to a lesser extent a lower amplitude, of Na^+ action potentials. The opposite effects were observed when saturated or *trans*-monoenoic fatty acids were added. These effects are probably mediated by a change in the number of active Na^+ channels, which may be mediated by alterations in membrane composition or altered fatty acid availability to the cells (23). Eicosanoid and docosanoid production is another mechanism which may explain the effect of $n-3$ fatty acid supplementation on neural function. Phospholipases liberate AA, EPA, and DHA from membrane lipids and make them available to cyclo-oxygenase or lipo-oxygenase action for production of eicosanoids and docosanoids (31). Oxidative metabolism of lipids may also affect brain function. Changes in sleep organization such as increased slow-wave sleep have been associated with increased lipolysis in fasting geese and emperor penguins (32–35). In these species, lipids account for most of the energy consumed and are the main oxidative fuel. In rats, slow wave sleep is increased during acute experimental epinephrine infusion, a phenomenon which appears to be mediated by activation of lipolytic activity. In contrast, slow-wave sleep decreases when lipogenesis is enhanced during acute insulin infusion (32). Chronic intracerebroventricular infusion of insulin also caused selective increases of slow-wave sleep in rats (33). Furthermore, if epinephrine is infused during the normal lipogenic dark period and insulin during the lipolytic light period, counter-acting the usual circadian rhythm, a complete inversion of the normal circadian distribution of sleep occurs (32).

ELECTROPHYSIOLOGICAL METHODS AND THE EFFECT OF ESSENTIAL FATTY ACIDS ON RETINAL AND BRAIN DEVELOPMENT

Sleep–wake cycle organization

Since intrauterine life is characterized by the virtual absence of wake periods (36) and postnatally infants spend most of their time sleeping, the study of sleep–wake cycles is an important component in the evaluation of central nervous system (CNS) development. Specific developmental patterns in infants have been described and long term effects of changes in sleep–wake cycles have been documented. The sleep–wake cycle has been well characterized as a function of age during the first months of life (37–40). It is defined by a combination of physiological variables, the electro-encephalogram (EEG), electrooculogram, electromyogram, the cardiorespiratory pattern, and motor activity. Several striking and precise changes, with specific timing for each, occur during this period, reflecting CNS maturation and adaptation to the extrauterine environment. Typically, the full-term newborn has significantly less indeterminate sleep (IS), with well-differentiated active (AS) and quiet sleep (QS), than do preterm infants, whose indeterminate sleep may represent up to 30% of total

***Figure 1** Distribution of sleep states over total sleep time according to age. Quiet sleep or non-REM sleep (QS), indeterminate sleep or transitional sleep (IS) and active sleep or REM sleep (AS). Age ranges in weeks gestational age (w) or months postnatal age (m). Bars indicate from top to bottom: % of QS, IS and AS. Data derived from Refs. 78 and 79*

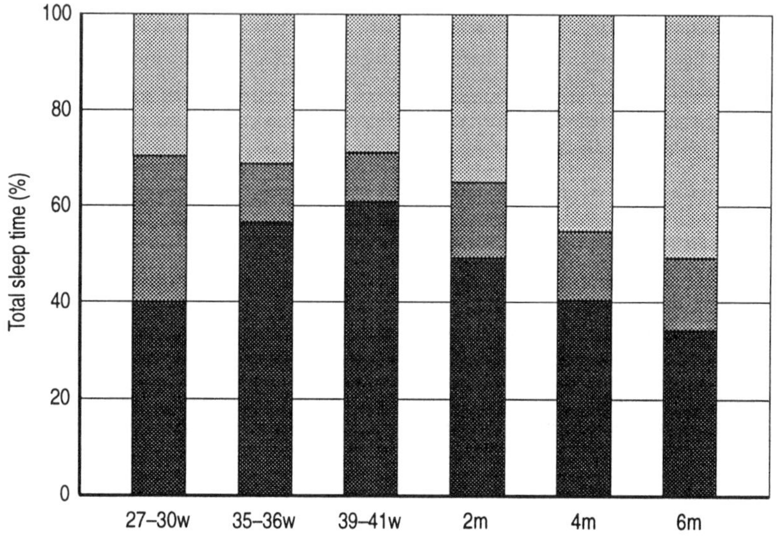

sleep time. The architecture of the sleep–wake cycle in infants differs markedly from that of older age groups. Day–night differences in sleep–wake cycles, with an increase in the ability to remain asleep during the night and remain awake during the day for progressively longer periods, begin to arise, reflecting the emergence of distinctive adult sleeping traits. These trends are relevant developmental indices of brain activity and CNS functional integrity (Fig. 1). A full description of developmental patterns for these changes is constrained by the scant data available between 6 and 24 months of age, and by the limited duration of recording periods conducted in most studies (usually less than 12 h of total recording).

Unlike sleep in the healthy adult, which is characterized by a recurrence of well-defined sleep stages, sleep in the immature animal is characterized by a gradual coalescence of a number of physiological variables. Preterm infants, and to some extent term infants, exhibit an appreciable lack of concordance between characteristic EEG patterns, presence or absence of rapid eye movements (REM), tonic chin electromyographic and motor activities, and cardiorespiratory patterns. These characteristics have led to a yet controversial debate about definition of sleep states during early life. It is well known, however, that the extent of indeterminate sleep (i.e. lack of concordance between variables and longer duration of sleep state transitions) shows a clear decline with normal progression of central nervous system maturation (40). The prevailing view is that the persistence of a poorly consolidated sleep–wake cycle may be more than a simple sign of immaturity. Several authors have reported altered sleep–wake cycle development in infants with brain pathology, those born to drug-addicted mothers or those subjected to severe maternal malnutrition or early postnatal malnutrition (41–46). In this context, the differentiation of REM and non-REM sleep during early development may not be sufficient to assess infant maturity, and particularly to identify infants at high risk for poor neurological development. We support the nomenclature that differentiates sleep episodes in active (also called paradoxical), quiet and indeterminate states and suggest that the corresponding duration of each be quantified. This approach includes the separation of the four stages (I–IV) of quiet sleep beyond 3–5 months of age. Stages III and IV of QS are also referred to as slow-wave sleep (SWS) and are related to the release of hypothalamic hormones which mediate the growth response. Even in the absence of differences in the organization of overall sleep states, subtle neurovegetative differences may be apparent in different infant groups. For instance, healthy small for gestational age (SGA) infants are similar to appropriate for gestational age (AGA) infants as far as general sleep state organization is concerned, whereas the control of neurovegetative variables (qualitative aspects) within sleep states differs (47,48). SGA infants show a higher incidence of respiratory pauses, more periodic breathing, less developed patterns of cardiac rhythm modulation, faster heart rate and reduced spontaneous heart rate variability.

Fagioli et al. (49) attempted to characterize the effects of long term dietary essential fatty acid deficiency on sleep organization in children 5–12 years of age maintained on fat-free total parenteral nutrition (TPN) for 2–6 months. These children had consistently less slow-wave sleep than children who received TPN plus essential lipids (Fig. 2a). The main difference was a lower cumulative amount of slow-wave sleep time (stages III and IV of QS) from the second hour of night sleep onwards (Fig. 2b). It is during these stages that growth hormone secretion occurs. These results suggest that early lipid nutrition may modulate sleep regulation and may have implications for whole body energy expenditure during sleep since slow-wave sleep is associated with a lower metabolic rate than other stages of sleep (50). Interestingly, no significant differences were found in either total sleep time or intra-sleep waking between infant groups. These few studies provide evidence of impaired CNS function when there is no or low intake of LA and ALA. However, no data in the literature specifically relate sleep development to dietary $n-3$ fatty acid supply.

Differences in the early functional organization of the CNS between breast-fed and bottle-fed infants have also been reported. At 4 months of age breast-fed infants spend a greater proportion of nocturnal sleep time in QS and a lower proportion of time in AS when compared with formula-fed infants (51). Furthermore, heart rates were slower in all sleep–wake states in breast-fed infants (52). Although total energy expenditure during AS and QS did not differ between feeding groups, the different organization of sleep states throughout the night might account for differences in energy expenditure noted between the groups (51).

In addition to sleep EEG recordings, we have also measured spontaneous activity using piezoelectric transducers which permit continuous recording during sleep and wakefulness. Healthy infants demonstrate a progressive decrease in activity with

***Figure 2a** Distribution of sleep states over total sleep time in children given essential fatty acids (EFA) in their total parenteral nutrition (TPN) or given a lipid-free infusion. Quiet sleep has been divided into stage 1, 2 and 3 and 4 combined as slow wave sleep (SWS). AS and IS are the same as in Figure 1. *Indicates statistically significant difference between lipid groups $p < 0.05$. Data derived from Ref. 49*

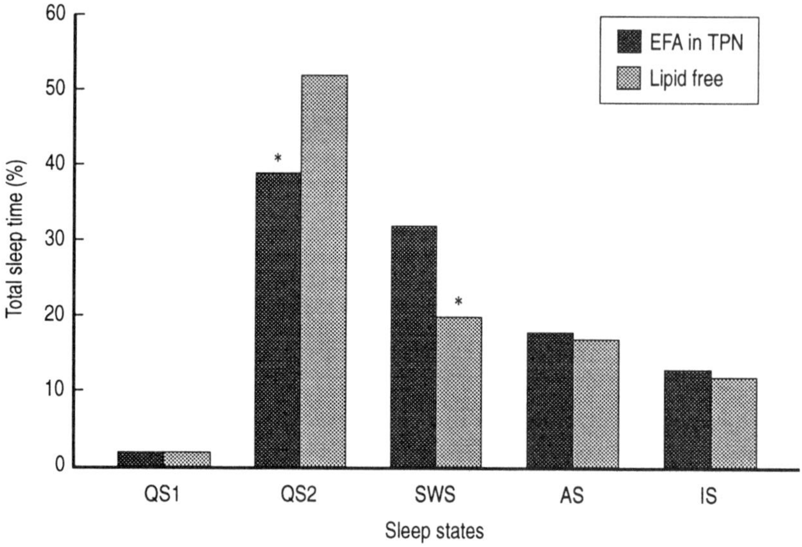

***Figure 2b** Slow wave sleep (SWS) accumulation over a 10-hour night sleep period in children given essential fatty acids (EFA) in their total parenteral nutrition (TPN) or given a lipid-free infusion. *Indicates statistically significant difference between lipid groups $p < 0.05$, **$p < 0.01$. Data derived from Ref. 49*

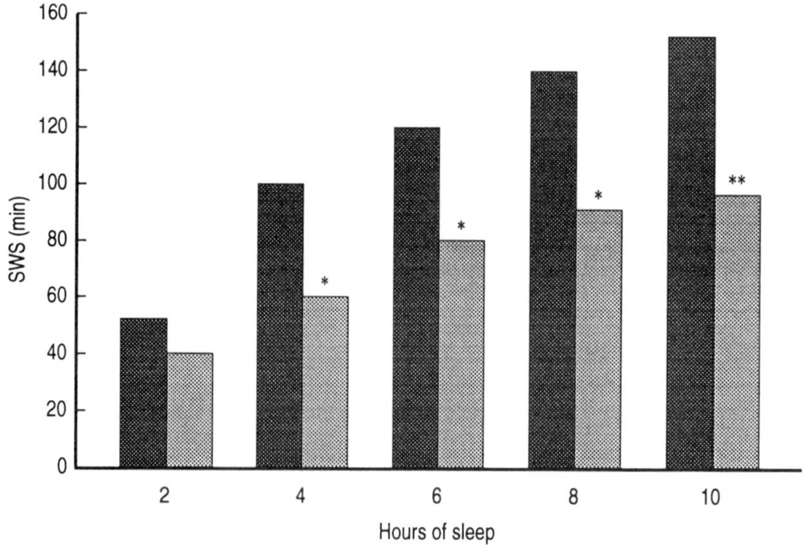

advancing age in both active and quiet sleep. We have found (unpublished observations) that time spent moving is no different between diet groups but the duration of motor pauses is longer in breast-fed infants. Differences in activity pattern during sleep could account for the observed effect of similar growth despite lower energy intake in breast-fed infants.

Heart rate variability patterns as indicators of brain maturation

The autonomic nervous system (ANS) regulates homeostatic function including heart rate. It is composed of two permanent competing subsystems, the parasympathetic nervous system (PNS) and the sympathetic nervous system (SNS). In general, the PNS promotes functions associated with anabolism and is concerned with the restoration and conservation of bodily energy and the resting cycle of vital organs, while the SNS promotes increased metabolic output and is responsible for much of the stress response. Thus, the ANS plays an important role in the individual's capacity to organize physiological resources and to respond appropriately to changing demands (53).

The spontaneous variability patterns of heart rate (HRV) during development have been closely linked to the permanent interaction of sympathetic and parasympathetic activity (Fig. 3a). In healthy neonates the variability of the high frequency (HF) component (vagal tone) is higher in QS than in AS, while the low frequency (LF) component is higher in AS (54). During development there is a progressive increase in HRV during both AS and QS, with specific trends for each type of HRV (55). In general, the sympathetic tone dominates during AS while there is a predominance of the parasympathetic vagal tone in QS. These patterns are not only related to sympathetic and parasympathetic controls, but are also indicators of normal CNS maturation in infants. HRV shows a typical maturational pattern in the human infant. In order to assess HRV the ECG signal is digitized on-line at a sample rate of 250 Hz and processed off-line by a signal-to-noise ratio algorithm that detects the peak of the R wave for each heart beat and measures sequential RR intervals in ms. In short, a method of derivates comparison with an adaptive threshold is used to construct a continuous signal of RR intervals that contains the time between each R wave and the one that precedes it. The raw RR signal is processed to correct for artefacts recognized by specific criteria (RR intervals below or above threshold values and large percentage difference from the previous RR value). Artefacts are removed and replaced by corresponding values interpolated from the trend in RR values. Epochs containing $< 10\%$ artefacts before correction are processed (56). After this procedure, the RR interval signal, considered as a digital signal sampled at a rate of one sample per beat, is filtered using a Fast Fourier Transform to separate spontaneously occurring variations in three non-overlapping frequency bands (Fig. 3b): a high-frequency band HRV (HF-HRV), with variation periods ranging from 3 to 8 heart beats, a mid-frequency HRV (MF-HRV), with variation periods from 10 to 25 heart beats; and a low-frequency HRV (LF-HRV) that varies with periods ranging from 30 to 100 heart beats (57,58). Over the first months of life the amplitude or extent of all HRV patterns increases, but there is a specific developmental trend for each kind of variability. These differences indicate physiological temporal asynchrony between the parasympathetic and

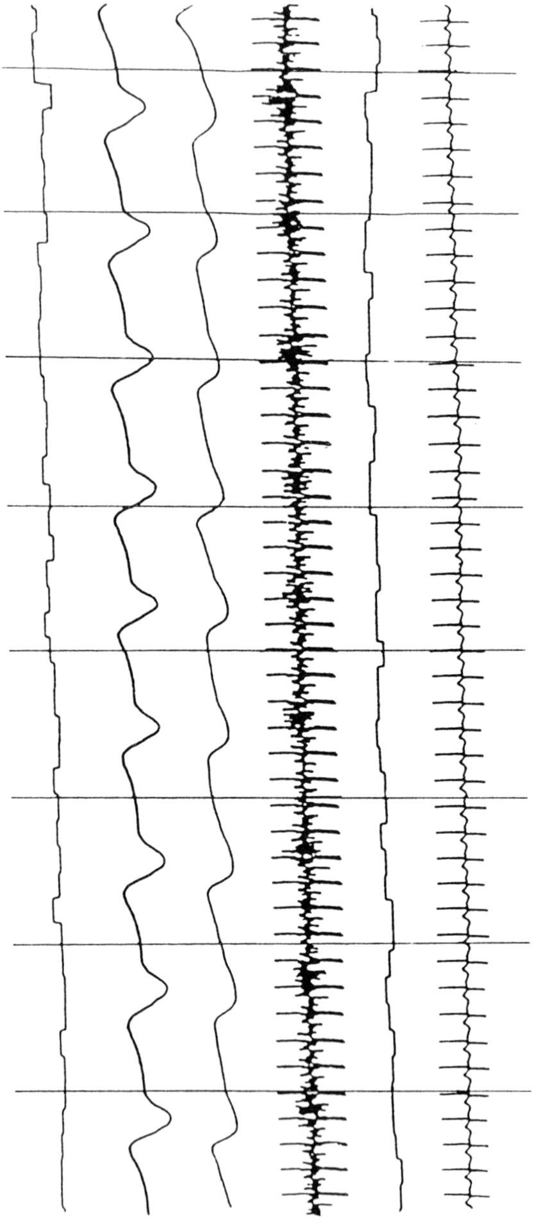

***Figure 3a** Polygraphic recording of an infant during quiet sleep. From top to bottom: top channel electrocardiographic recording, instantaneous heart rate, diaphragmatic electromyographic tracing, abdominal respiratory movements, nasal air flow, oxygen saturation. Time frame is 3 seconds between vertical lines*

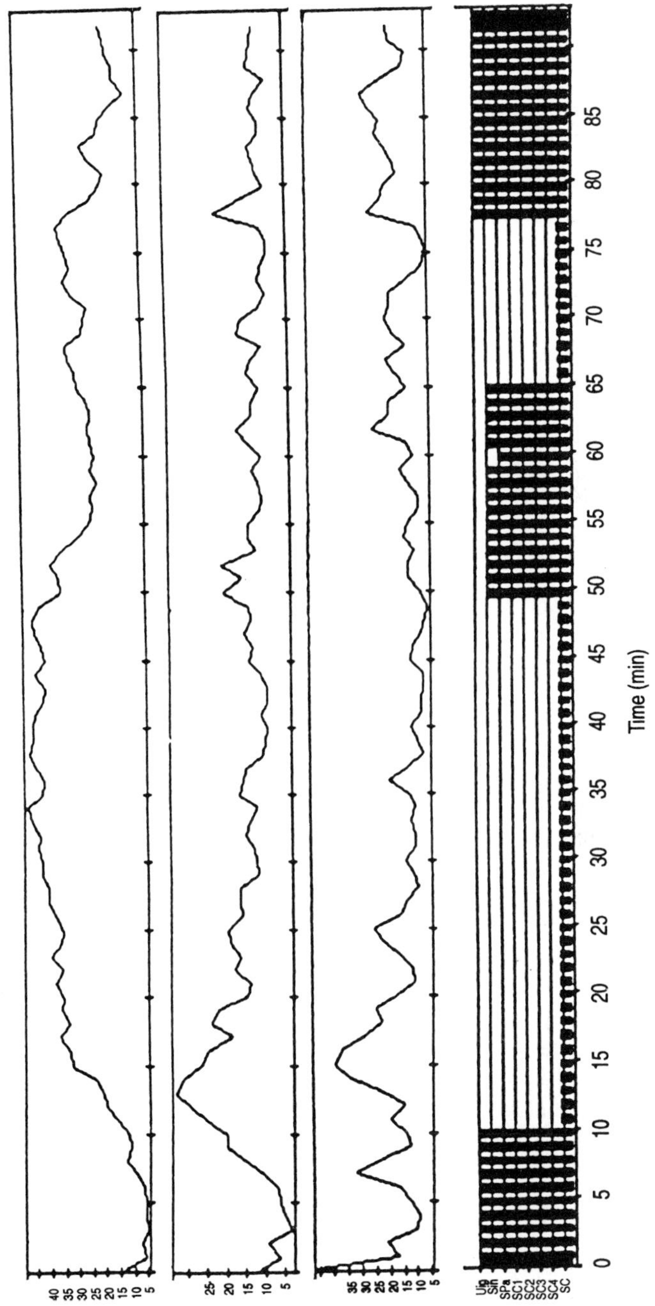

***Figure 3b** Heart rate variability (HRV) in 3 different non-overlapping frequency bands: from top to bottom high-, mid- and low-frequency HRV. Vertical axis indicates amplitude in milliseconds. The organization of sleep–wake states over time is shown at the bottom; sequential changes in sleep state indicate initial wake state, then quiet sleep, active sleep, quiet sleep and final wake period. Horizontal axis denotes time in minutes, marks in horizontal axis indicate 5 minute intervals (for details see text)*

***Figure 4** High frequency heart rate variability during quiet sleep and active sleep, amplitude in milliseconds, according to weeks conceptional age (wCA). Modified from Ref. 59*

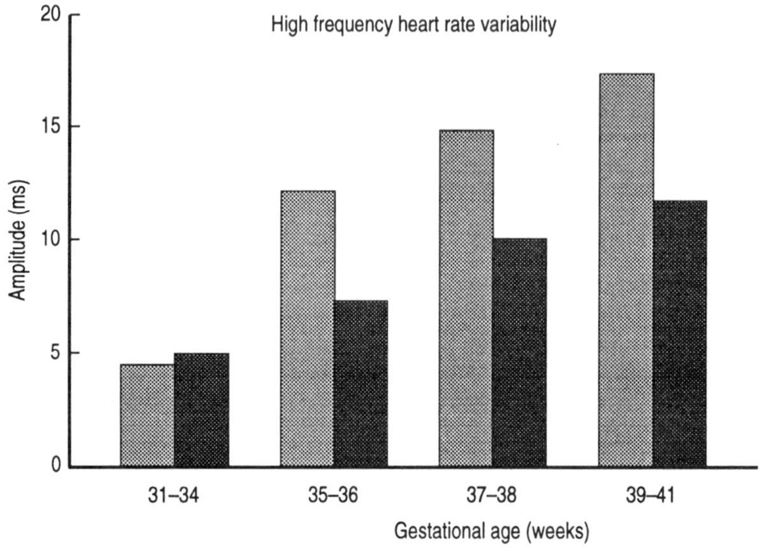

sympathetic subsystems in autonomic nervous system maturation (54): the parasympathetic system matures later than the sympathetic system. Heart rate and HRV can be modified by both central nervous system alterations and metabolic changes. Deviations from normal maturational patterns in variability have been observed in high risk infants (e.g. protein energy malnutrition, intrauterine growth retardation and those at high risk for sudden infant death syndrome). Thus, maturational patterns of heart rate and HRV provide useful noninvasive tools for investigating central nervous system functioning and integrity during early human development (Fig. 4) (59).

Decreased vagal tone has been associated with poor developmental outcome and with negative reactivity to mildly stressful conditions; it has been interpreted to indicate vulnerability to stress. Increased vagal tone, however, parallels an increase in self-regulatory and exploratory behaviours and is associated with better visual recognition memory. Several investigators have suggested that these physiological and behavioural processes are dependent on the functional integrity of the autonomic nervous system. Accordingly, infants with reduced development of vagal tone would be at higher risk for altered behaviour and poorer developmental outcome (60). Assessment of the ANS may therefore become a useful tool for evaluating stressful conditions, especially in young preverbal infants.

Our initial findings in preterm infants (unpublished observations) support published reports of a higher HRV during AS and QS in breast-fed infants, indicating a more mature pattern of CNS development. The effect is observed at all frequency bands but is more marked in the high frequency component, suggesting a higher parasympathetic vagal tone in the breast-fed group. Infants fed a mixture of formula and human milk show results closer to those exclusively breast fed. Within the randomized formula groups there are differences in HRV which are no longer apparent during wake periods.

In other studies, term breast-fed neonates, even at 2 days postpartum, are physiologically more organized than bottle-fed neonates (61,62). They exhibit a slower heart rate in both AS and QS and also present a higher incidence of low-frequency HRV, greater heart rate variability and elevated vagal tone, indicating a more mature HRV pattern (61). In a separate study of older infants, breast-fed infants had lower heart rates during both active and quiet sleep and more organized heart rate rhythm patterns (62).

Electrophysiological measures of retinal development

The full-field electroretinogram (ERG) represents the summed activity of the retina in response to a light flash. A typical response to white light contains an early, corneal-negative component (a-wave), a slower, corneal-positive component (b-wave) and small sub-components on the ascending slope of the b-wave (oscillatory potentials). The a-wave is known to be generated within the photoreceptor layer and directly reflects photoreceptor activity. The b-wave is generated at the level of the inner nuclear and reflects bipolar cell activity. Oscillatory potentials are generated at the ganglion cell level, and reflect ganglion cell–amacrine cell feedback interactions (63,64). The advent of fast, low-cost computers has led to rapid improvements in clinical electroretinography. It is now possible to test neonates within the special care nursery. Infants under 6 months of age can be routinely tested without sedation by lowering a Ganzfeld dome over the isolette. Full-field stimulators use a Grass photostimulator or equivalent as the light source. The advantage of this xenon gas discharge source is that each flash lasts only 10 ms, and flashes can be repeated to produce a flicker response. Brief flashes are preferable to longer flashes because they do not cause light-adaptation of the retina and do not produce a second ERG at flash offset.

Full-field ERGs are typically measured with a contact lens electrode. The most reliable and widely used lens is the Burian-Allen bipolar contact lens electrode, which is available in several sizes to fit subjects from very-low-birth-weight (VLBW) infants to adults. Although each lens is expensive, it can be reused indefinitely with occasional restoration by the manufacturer. In light of present concern about the transmission of the HIV virus through tears, we sterilize lenses immediately after each recording session. Contact lens electrodes are well tolerated by infants through the age of 6 months. An alternative to the Burian-Allen lens is a disposal monopolar lens. Pupils should be maximally dilated for ERG recording to minimize variations in retinal illuminance. To minimize discomfort, the infant should be dark-adapted for at least 30 min prior to inserting the lens. Two drops of topical anaesthetic (proparacaine hydrochloride) are applied before inserting the lens. A silver chloride disc electrode filled with electrode cream serves as a ground on the forehead. The contact lens electrode is inserted in the eye in dim red illumination provided by a head-mounted flashlight. Responses can be obtained from preterm infants at 30–36 weeks postconception

(65). The responses, however, are extremely small and cannot be readily quantified without computer averaging. A response must be approximately 10 mV in amplitude to be detectable without computer averaging, while responses as small as 2 mV can be reliably measured with computer averaging and the resulting increased signal–noise ratio.

To facilitate the comparable recording of ERGs throughout the world, a standard for clinical electroretinography has been developed (66). The five responses that comprise the standard protocol are:

(1) Rod response. The first response that should be recorded following a least 30 min of dark adaptation is a rod response. The preferred procedure for obtaining this response is to attenuate the maximum white flash with a 2.5 log unit neutral density filter. Such a stimulus is below cone threshold and produces a 'pure' rod ERG. The waveform of the rod ERG is simple. Frequently there is no a-wave, even in normal subjects, but when it is present, it is generally <5 mV and has an implicit time of approximately 40 ms. The b-wave is generally slow and smooth, without subcomponents or oscillations, and peaks at approximately 80 ms.

(2) Maximal response. This response is also elicited following at least 30 min of dark adaptation. In the standard protocol, it is obtained immediately after the rod response by removing the 2.5 log unit filter. The response is complex in shape and contains both rod and cone components in the normal eye. The ascending limb contains subcomponents which can be selectively filtered to produce oscillatory potentials.

(3) Oscillatory potentials. Oscillatory potentials are obtained from the dark-adapted eye immediately after recording the maximal response and with the same stimulus conditions. The only difference is the high pass filter, set to 75–100 Hz. The oscillatory potential generally contains at least two prominent peaks termed 01 and 02. The summed amplitude of the components is often used as an index of inner retinal activity (67).

(4) Single-flash cone response. 'Pure' cone responses can be obtained by suppressing rod activity with a white background light. The luminance of the background should be 34 cd/m^2 and the patient should be light-adapted for 10 min before recording. This period of light adaptation is necessary because the cone signal grows during the first few minutes of light adaptation. The stimulus used to record the light-adapted cone response is the same as that used for the maximal response. The single-flash cone response in normal subjects generally contains a prominent a-wave with an implicit time of approximately 15 ms, a b-wave with an implicit time of approximately 30 ms and at least two subsequent oscillations after the b-wave.

(5) Light-adapted flicker response. Light-adapted flicker responses should also be recorded in the presence of a 34 cd/m^2 background and after at least 10 min of light adaptation. The response is elicited by the maximum intensity white stimulus flickering at 30 Hz. Generally, the first few responses are larger than subsequent responses, so recording should not begin until steady-state conditions have been reached. b-Wave implicit time is measured from stimulus onset to b-wave peak. Amplitude is measured from response minimum to response maximum.

Standard responses from a preterm infant at 36 weeks postconception are at least an order of magnitude smaller than those from an adult. Cone responses are relatively more developed at this age than rod responses. Oscillatory potentials are not present at this age or at birth (65). By 4 months of age (57 weeks postconception) there has been considerable growth in ERG amplitude: rod responses in particular show rapid development between birth and 4 months. Oscillatory potentials are present at 4 months but lag considerably behind the b-wave in amplitude. The findings are consistent with anatomical findings that cone development precedes rod development, and that the distal retinal development precedes proximal retinal development. The rod response in the standard protocol is only one of a family of responses that can be obtained by varying the retinal illuminance of the stimulus. While the standard rod response is sufficient for many purposes, considerably more information about the condition of the retina can be gained by recording the serial response to graded illuminances. Rod b-wave implicit time decreases with increasing retinal illuminance, while amplitude increases and plateaus at high retinal illuminances.

Results from our studies of retinal function in VLBW infants have been described elsewhere and will only be summarized here (69–71) (Figs 5 and 6). Retinal function responses demonstrated significantly higher threshold values from rod photoreceptors in the $n-3$ deficient group (corn-oil-fed) at the 36-week follow-up relative to groups receiving $n-3$ FA. This indicates that the sensitivity and maturity of the rod photoreceptors in the $n-3$ FA-deficient group were significantly reduced (69,70). A group of infants of 35 weeks' gestational age underwent visual maturity testing 3–5 days after birth and served as a normal standard for comparison. This group of infants receiving transplacentally acquired EFA was nearly identical in all rod ERG functional indices to the breast-fed or marine-oil-fed preterm infants of equivalent post-conceptional age. Infants fed soy oil (containing ALA but no DHA) had higher threshold values than those fed soy/marine oil. Cone photoreceptor function was not significantly affected by diet, although the trends were similar to the results obtained from rods. At the 57 week follow-up, a- and b-wave ERG indices did not differ between diet groups but the corn-oil-fed infants consistently had longer implicit times in light adapted oscillatory potentials. The soy-oil-fed infants had some response peaks that differed significantly from the HM-fed infants while those fed soy and marine oil or breast milk showed similar oscillatory potential responses (70).

Visual evoked potential acuity

Cortical electrophysiological responses to temporal changes in patterns can provide an estimate of visual acuity. Typically, gold cup electrodes are used to provide one or more channels at the occipital pole, as well as a ground and a reference. The infant views patterns on a video monitor, and evoked responses are led from the electrodes through a pre-amplifier, which also acts as a bandpass filter. The signal is led from the pre-amplifier to a device capable of averaging, filtering, and analysing visual evoked potentials (VEP). The main purpose of signal averaging and filtering is to improve the signal-to-noise ratio and, thus, the detectability of small amplitude signals (72,73).

The patterns employed are high contrast black-and-white checkerboards or gratings, which undergo contrast reversal or pattern onset from a homogeneous background field. The reversal or onset can be scheduled to occur infrequently (< 1 reversal/s) or very frequently (> 10 reversals/s). When the pattern changes occur infrequently, a transient VEP is obtained. Peak-to-peak amplitude and latency of the primary positive peak are usually measured for the transient response (72–74). When the pattern changes occur frequently, a steady-state VEP is obtained. Response amplitude and phase are usually measured for the steady-state response. The shape of the VEP waveform changes with age, infants 2 months of age or younger have simple, nearly sinusoidal, steady-state responses while older infants show multiple peaks. For acuity testing, a range of checkerboards or gratings from coarse to fine is used in each test. Acuity is estimated by linear regression of VEP amplitude on pattern element size; extrapolation of the best fit line to zero microvolts determines the acuity estimate. Several rapid methods of assessing the VEP acuity have been developed specifically for use with infants and in clinical populations (72–75).

Results of VEP acuity testing in infants must be evaluated by comparison with acuities of age-matched normal infants rather than in terms of categories designed for use in adults. Several alternative strategies are currently used for determining whether the acuity test result for an individual infant indicates normal vision or visual impairment. The most obvious, and most common, method for determining abnormality is comparison of the infant's acuity with the range of acuities obtained from an age-matched normal population; this method is sensitive to acuity reduced by a factor of 2–3. Since normal infants have poor acuity and large individual differences during the first 2 months of life, this method is insensitive to mild or moderate acuity deficits at these early ages. It is important to adjust for the gestational age of the infant at birth and, given the effect of early diet on visual development, a breast-fed infant should possibly be used as the only valid true reference control.

Our results from preterm infant studies published elsewhere demonstrated that after 5–6 months of controlled feeding, infants receiving soy/marine oil or breast milk had lower values of log MAR (minimal angle of resolution), that is better acuity, relative to infants fed formulas devoid of DHA (Figs. 5 and 6) (76,77). A significant negative correlation was found between RBC DHA levels and log MAR values; that is, when erythrocyte DHA levels were high, log MAR values were corresponding low (i.e. better visual acuities). The breast-fed and soy/marine-oil-fed infants had the highest DHA/DPA ratio in RBC and also had the best visual acuity. The group that received soy oil as a source of $n-3$ FA had poorer VEP acuity at 57 weeks postconception relative to the soy/marine-oil-fed group, indicating that despite ample ALA, visual function was less mature than in the DHA supplemented infants. A group of healthy full-term breast-fed infants matched by conceptional age were used as controls. Visual acuities of the breast-fed and soy/marine-oil-fed preterm infants were virtually identical to term, breast-fed controls, while the corn and soy oil preterm groups had poorer visual acuities at 57 weeks postconception (76).

CONCLUSION

Experimental animal and human studies indicate that the high DHA content of neural tissues such as the retina is functionally important, and that replacement of this fatty

***Figure 5** DHA in total red blood cell (RBC) lipids and plasma phospholipids (PL) of preterm infants on study diets at 36 weeks postconception. Inset: Rod ERG thresholds (intensity units = log scotopic troland seconds) and VEP visual acuity (Snellen equivalents) in study infants at 36 weeks. Arrows indicate significant differences between diet groups using Newman–Keuls multiple comparison analysis ($p < 0.05$)*

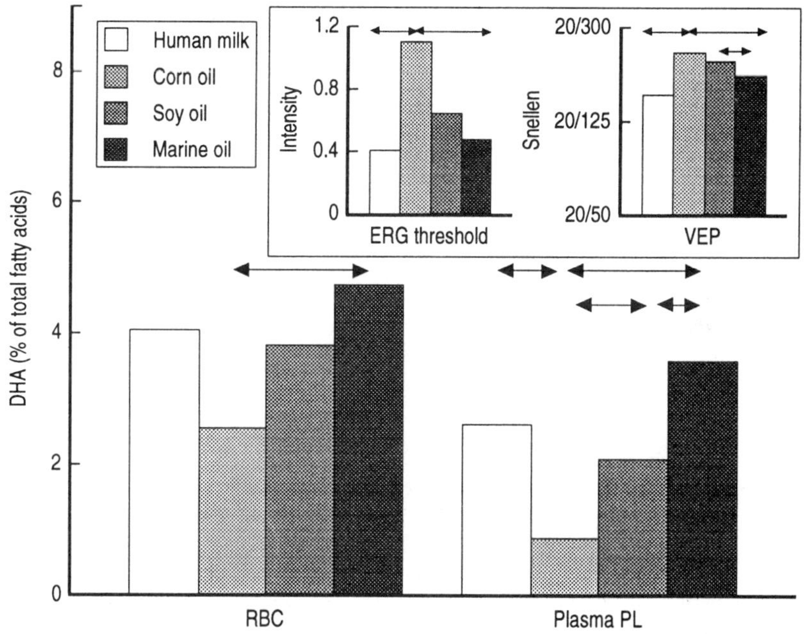

acid by $n-6$ or $n-9$ fatty acids has significant effects on brain and retinal function. Preterm infants require long-chain $n-3$ fatty acids for normal retinal functional development. Rod a- and b-wave responses are affected by $n-3$ deficiency while cones appear less vulnerable. Threshold illuminances are significantly higher in $n-3$-deficient infants. α-Linolenic acid is only partially effective in reversing these changes, while DHA supplementation induces responses similar to those obtained in human-milk-fed infants or those from infants of equivalent gestation who remained in utero.

Visual acuity measured by visual evoked potentials in preterm infants fed $n-3$ LCPUFA is similar to that of infants fed human milk, while those receiving soy oil have poorer visual acuity relative to breast-fed preterm infants and full term breast-fed controls of equivalent gestational age. Functional development of the retina and visual cortex correlates well with biochemical markers of DHA status. DHA should be considered essential for normal retinal and brain development in preterm infants.

***Figure 6** DHA in total RBC lipids and plasma phospholipids (PL) of preterm infants on study diets at 57 weeks postconception.* Inset: *Rod ERG thresholds (intensity units = log scotopic troland seconds) and VEP visual acuity (Snellen equivalents) in study infants at 57 weeks. Arrows indicate significant differences between diet groups using Newman–Keuls multiple comparison analysis ($p < 0.05$)*

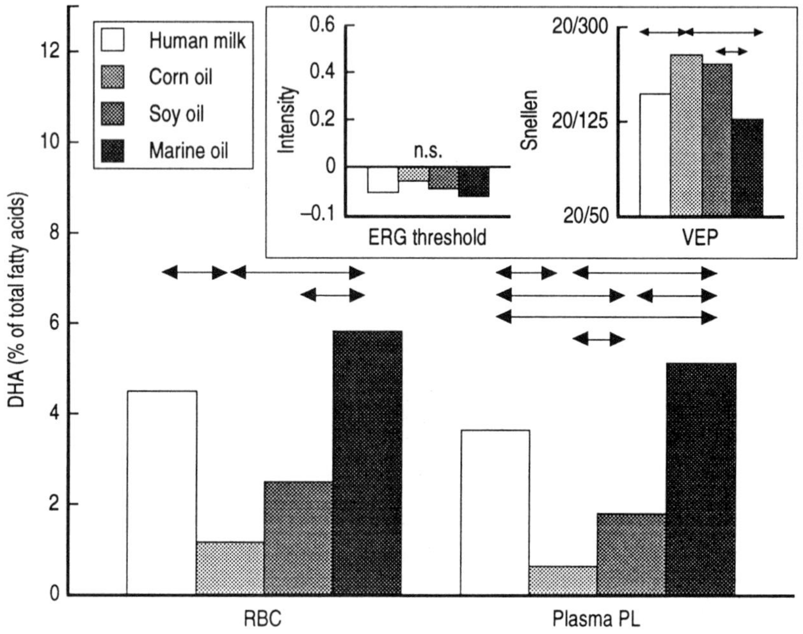

ACKNOWLEDGEMENTS

Work by the authors of this publication is supported in part by Fondecyt CHILE Grant 1930820, and by NIH Grants HD 22380 and EY 09076.

REFERENCES

1. Burr GO, Burr MM 1929 A new deficiency disease produced by rigid exclusion of fat from the diet. J Biol Chem 82:345–367
2. Hansen AE, Wiese HF, Boelsche AN, Haggard ME, Adam DJD, Davis H 1963 Role of linoleic acid in infant nutrition: clinical and chemical study of 428 infants fed on milk mixtures varying in kind and amount of fat. Pediatrics 31:171–192
3. Caldwell MD, Johnsson HT, Othersen HB 1972 Essential fatty acid deficiency in an infant receiving prolonged parenteral alimentation. J Pediatr 81:894–898
4. White HB, Turner MD, Turner AC, Miller RC 1973 Blood lipid alterations in infants receiving intravenous fat-free alimentation. J Pediatr 83:305–313
5. Paulsrud JR, Pensler L, Whitten CF, Stewart S, Holman KT 1972 Essential fatty acid deficiency in infants induced by fat-free intravenous feeding. Am J Clin Nutr 25:897–904

6. Friedman Z 1980 Essential fatty acids revisited. Am J Dis Child 134:397–408
7. Friedman Z, Shochat SJ, Maisels MJ, Marks KH, Lamberth EL 1976 Correction of essential fatty acid deficiency in newborn infants by cutaneous application of sunflower seed oil. Pediatrics 58: 650–654
8. Holman RT, Johnson SB, Hatch TF 1982 A case of human linolenic acid deficiency involving neurological abnormalities. Am J Clin Nutr 35:617–623
9. Farquharson J, Cockburn F, Patrick WA, Jamieson EC, Logan RW 1992 Infant cerebral cortex phospholipid fatty-acid composition and diet. Lancet 340:810–813
10. Makrides M, Neumann MA, Byard RW, Simmer K, Gibson RA 1994 Fatty acid composition of brain, retina, and erythrocytes in breast- and formula-fed infants. Am J Clin Nutr 60:189–194
11. Bourre JM, Francois M, Youyou A, Dumont O, Piciotti M, Pascal G 1989 The effects of dietary alfa-linolenic acid on the composition of nerve membranes, enzymatic activity, amplitude of electrophysiological parameters, resistance to poisons and performance of learning tasks in rats. J Nutr 119:1880–1892
12. Wheeler TG, Benolken RM, Anderson RE 1975 Visual membranes: specificity of fatty acid precursors for the electrical response to illumination. Science 188:1312–1314
13. Neuringer M, Connor WE, van Petten C, van Barstad L 1984 Dietary omega-3 fatty acid deficiency and visual loss in infant rhesus monkeys. J Clin Invest 73:272–276
14. Stubbs CD, Smith AD 1984 The modification of mammalian membrane polyunsaturated fatty acid composition in relation to fluidity and function. Biochim Biophys Acta 779:89–137
15. Lee AG, East JM, Froud RJ 1986 Are essential fatty acids essential for membrane function? Prog Lipid Res 25:41–46
16. Salem N Jr, Shingu T, Kim HY, Hullin F, Bougnoux P, Karanian JW 1988 Specialization in membrane structure and metabolism with respect to polyunsaturated lipids. In: Karnovsky ML, Leaf A, Bollis LC (eds) Biological Membranes: Aberrations in Membrane Structure and Function. Alan R. Liss Inc., New York, pp 319–333
17. Treen M, Uauy RD, Jameson D, Thomas V, Hoffman DR 1992 Effect of docosahexaenoic acid on membrane fluidity and function in intact cultured Y-79 retinoblastoma cells. Arch Biochem Biophys 294:564–570
18. Yorek MA, Bohnker RR, Dudley DT, Spector AA 1984 Comparative utilization of N-3 polyunsaturated fatty acids by cultured human Y-79 retinoblastoma cells. Biochim Biophys Acta 795: 277–285
19. Ahmad SN, Alam SQ, Alam BA 1990 Influence of dietary omega-3 fatty acids on transmembrane signalling in rat submandibular salivary gland. Cell Sign 2:29–41
20. Slater SJ, Kelly MK, Taddeo FJ, Ho C, Rubin E, Stubbs CD 1994 The modulation of protein kinase C activity by membrane lipid bilayer structure. J Biol Chem 269:4866–4871
21. Lin DS, Connor WE, Anderson GJ, Neuringer M 1990 The effects of dietary $n-3$ fatty acids on phospholipid molecular species of monkey brain. J Neurochem 55:1200–1207
22. Mitchell DC, Litman BJ 1995 A role for phospholipid polyunsaturation in modulating membrane protein function. Proceedings of the 2nd International Congress of the ISSFAL (International Society for the Study of Fatty Acids and Lipids). NIH, Bethesda, MD, June 7–16
23. Love JA, Saurn WR, McGee R 1985 The effects of exposure to exogenous fatty acids and membrane fatty acid modification on the electrical properties of NG108-15 cells. Cell Mol Neurobiol 5:333–352
24. Holh CM, Rosen P 1987 The role of arachidonic acid in rat heart cell metabolism. Biochim Biophys Acta 921:356–363
25. Caster WO, Ahn P 1963 Electrocardiographic notching in rats deficient in EFA. Science 139: 1213
26. Charnock JS 1991 Anti-arrhythmic effects of fish oil. In: Simopoulos AP (ed) Health Effects of $\omega3$ Polyunsaturated Fatty Acids in Seafoods. Karger, Basel, pp 278–291
27. Abeywardena MY, McLennan PL, Charnock JS 1987 Diet and cardiac arrhythmia: involvement of eicosanoids. In: Lands WEM (ed) Proceedings of the AOCS Short Course of Polyunsaturated Fatty Acids and Eicosanoids. American Oil Chemists' Society, Champaign, IL, pp 62–64
28. Hallaq H, Smith TW, Leaf A 1992 Modulation of dihydropyridine-sensitive calcium channels in hearts cells by fish oil fatty acids. Pro Natl Acad Sci USA 89:1760–1764
29. Kang JX, Xiao YF, Leaf A 1995 Free, long-chain polyunsaturated fatty acids reduce membrane electrical excitability in neonatal rat cardiac myocytes. Proc Natl Acad Sci USA 92:3997–4001

30. Evers AS, Elliott WJ, Lefkowith JB, Needleman P 1986 Manipulation of rat brain fatty acid composition alters volatile anesthetic potency. J Clin Invest 77:1028–1033
31. Bazan NG 1989 The metabolism of omega-3 polyunsaturated fatty acids in the eye: The possible role of docosahexaenoic acid and docosanoids in retinal physiology and ocular pathology. Prog Clin Biol Res 312:95–112
32. Danguir J, Nicolaidis S 1980 Circadian sleep and feeding patterns in the rat: possible dependence on lipogenesis and lipolysis. Am J Physiol 238:E223–E230
33. Danguir J, Nicolaidis S 1984 Chronic intracerebroventricular infusion of insulin causes selective increases of slow wave sleep in rats. Brain Res 306:97–106
34. Dewasmes G, Cohen-Adad F, Koubi H, Le Maho Y 1984 Sleep changes in long-term fasting geese in relation to lipid and protein metabolism. Am J Physiol 247:R663–R671
35. Dewasmes G, Buchet C, Geloen A, Le Maho Y 1989 Sleep changes in emperor penguins during fasting. Am J Physiol 256:R476–R480
36. Rigatto H 1992 Maturation of breathing. Clin Perinatol 19:739–756
37. Coons S, Guilleminault C 1982 The development of sleep-wake patterns and non-rapid eye movement sleep stages during the first six months of life in normal infants. Pediatrics 69:793–798
38. Fagioli I, Salzarolo P 1982 Sleep states development in the first year of life assessed through 24 hour recording. Early Hum Dev 6:215–228
39. Anders TF, Keener M 1985 Developmental course of nighttime sleep-wake patterns in full term and premature infants during the first year of life. Sleep 8:173–192
40. Curzi-Dascalova L, Peirano P, Morel-Kahn F 1988 Development of sleep states in normal premature and full term newborns. Dev Psychobiol 21:431–444
41. Monod N, Guidasci S 1976 Sleep and brain malformation in the neonatal period. Neuropaediatrie 7:229–249
42. Schulman CA 1969 Alterations of the sleep cycle in heroin-addicted and 'suspect' newborns. Neuropaediatrie 1:89–100
43. Bhatia VP, Katiyar GP, Agarwal KN, Das TK, Dey PK 1980 Sleep cycle studies in babies of undernourished mothers. Arch Dis Child 55:134–138
44. Salzarulo P, Fagioli I, Salomon F, Ricour C 1982 Developmental trend of quiet sleep is altered by early human malnutrition and recovered by nutritional rehabilitation. Early Hum Dev 7:257–264
45. Peirano P, Fagioli I, Singh BB 1989 Effect of early human malnutrition on waking and sleep organization. Early Hum Dev 20:67–76
46. Peirano P, Fagioli I, Singh BB 1990 Quiet sleep and slow wave sleep in malnourished infants. Brain Dysfunct 3:80–83
47. Curzi-Dascalova I, Peirano P 1989 Sleep organisation in small-for-gestational age human neonates. Brain Dysfunct 2:45–54
48. Spassov L, Curzi-Dascalova L, Clairambault J, Kauffmann F, Eiselt M, Médigue C, Peirano P 1994 Heart rate and heart rate variability during sleep in small for gestational age newborns. Pediatr Res 35:500–505
49. Fagioli I, Baroncini P, Ricour C, Salzarulo P 1989 Decrease of slow-wave sleep in children with prolonged absence of essential lipids intake. Sleep 12:495–499
50. Stothers JK, Warner RM 1977 Oxygen consumption and sleep state in the newborn. J Physiol (Lond) 269:57–58
51. Butte NF, Jensen CL, Moon JK, Glaze DG, Frost JD 1992 Sleep organization and energy expenditure of breast-fed and formula-fed infants. Pediatr Res 32:514–519
52. Butte NF, Smith EO, Garza C 1991 Heart rate of breast-fed and formula-fed infants. J Pediatr Gastroenterol Nutr 13:391–396
53. Porges SW 1992 Vagal tone: a physiologic marker of stress vulnerability. Pediatrics 90:498–504
54. Curzi-Dascalova L, Clairambault J, Kauffmann C 1991 Cardiorespiratory variability and development of sleep state organization. In: Gaultier C, Escourrou P, Curzi-Dascalova L (eds) Sleep and Cardiorespiratory Control. John Libbey, London, pp 155–163
55. Schechtman VL, Harper RM, Kluge KA 1989 Development of heart rate variation over the first 6 months of life in normal infants. Pediatr Res 26:343–346
56. Kauffmann F, Cauchemez B 1991 Extraction of cardiorespiratory parameters. In: Gaultier C, Escourrou P, Curzi-Dascalova L (eds) Sleep and Cardiorespiratory Control. John Libbey, London, pp 105–112
57. Eiselt M, Curzi-Dascalova L, Clairambault J, Kauffmann F, Médigue C, Peirano P 1993 Heart-

rate variability in low-risk prematurely born infants reaching normal term: A comparison with full-term neonates. Early Hum Dev 32:183–195
58. Spassov L, Curzi-Dascalova L, Clairambault J, Kauffmann F, Biselt M, Médigue C, Peirano P 1994 Heart rate and heart rate variability during sleep in small-for-gestational age newborns. Pediatr Res 35:500–505
59. Curzi-Dascalova L, Spassov L, Eiselt M, Peirano P, Kauffmann F, Clairambault J, Médigue C 1994 Development of cardiorespiratory control and sleep in newborns. In: Cosmi AV, Renzo GC (eds) Current Progress in Perinatal Medicine. Parthenon, London, pp 303–308
60. Porges SW 1995 Cardiac vagal tone: A physiological index of stress. Neurosci Biobehav Rev 19:225–233
61. DiPetro JA, Larson KS, Porges SW 1987 Behavioral and heart rate pattern differences between breast-fed and bottle-fed neonates. Dev Psychol 23:467–474
62. Zeskind PS, Marshall TR, Goff DM 1992 Rhythmic organization of heart rate in breast-fed and bottle-fed newborn infants. Early Dev Parent 1:79–87
63. Granit R 1993 The components of the retinal action potential in mammals and their relation to the discharge of the optic nerve. J Physiol 77:207–239
64. Brown KT 1968 The electroretinogram; its components and their origins. Vision Res 8:633–637
65. Birch EE, Birch DG, Petrig B, Uauy R 1990 Retinal and cortical function of infants at 36 and 57 weeks post-conception. Clin Vision Sci 5:363–373
66. Marmor MF, Arden GB, Nilsson SEG, Zrenner E 1989 Standard for clinical electroretinography. Arch Ophthalmol 107:816–881
67. Ogden TE 1973 The oscillatory waves of the primate electroretinogram. Vision Res 13:1059–1074
68. Naka KI, Rushton WAH 1966 S-potentials from luminosity units in the retina of fish (Cyprinidae). J Physiol (Lond) 185:587–599
69. Uauy RD, Birch DG, Birch EE, Tyson JE, Hoffman DR 1990 Effect of dietary omega-3 fatty acids on retinal function of very-low-birth-weight neonates. Pediatr Res 28:485–492
70. Birch DG, Birch EE, Hoffman DR, Uauy R 1992 Retinal development in very-low-birth-weight infants fed diets differing in omega-3 fatty acids. Invest Ophthalmol Vis Sci 33:2365–2376
71. Hoffman DR, Uauy R 1992 Essentiality of dietary omega-3 fatty acids for premature infants: plasma and red blood cell fatty acid composition. Lipids 27:886–895
72. Birch EE 1989 Visual acuity testing in infants and young children. Ophthalmol Clin North Am 2:369–389
73. Uauy R, Birch E, Birch D, Peirano P 1992 Visual and brain function measurements in studies of $n-3$ fatty acid requirements of infants. J Pediatr 120:S168–S180
74. Sokol S, Hansen VC, Moskowitz A, Greenfield P, Towle VL 1983 Evoked potential and preferential looking estimates of visual acuity in pediatric patients. Ophthalmology 9:552–556
75. Norcia AM, Tyler CW 1985 Spatial frequency sweep VEP: Visual acuity during the first year of life. Vision Res 25:1399–1408
76. Birch EE, Birch DG, Hoffman DR, Uauy R 1992 Dietary essential fatty acid supply and visual acuity development. Invest Ophthalmol Vis Sci 33:3242–3253
77. Uauy R, Hoffman DR, Birch EE, Birch DG, Jameson DM, Tyson J 1994 Safety and efficacy of omega-3 fatty acids in the nutrition of very low birth weight infants: soy oil and marine oil supplementation of formula. J Pediatr 124:612–620
78. Curzi-Dascalova L 1995 Développement du sommeil et des fonctions sous contrôle du système nerveux autonome chez les nouveau-nés prématurés et à terme. Arch Pédiatr 2:255–262
79. Hoppenbrouwers T, Hodgman J, Arakawa K, Geidel SA, Sternman MB 1988 Sleep and waking states in infancy: normative studies. Sleep 11:387–401

13 Defining the LCPUFA requirement of term infants

M. MAKRIDES and R.A. GIBSON

ABBREVIATIONS

AA	arachidonic acid, 20:4 $n-6$
ALA	α-linolenic acid, 18:3 $n-3$
DHA	docosahexaenoic acid, 22:6 $n-3$
EPA	eicosapentaenoic acid, 20:5 $n-3$
GLA	γ-linolenic acid, 18:3 $n-6$
LA	linoleic acid, 18:2 $n-6$
LCPUFA	long-chain polyunsaturated fatty acid
VEP	visual evoked potential

INTRODUCTION

Much has been written in both the scientific literature and the lay press about the need for long-chain polyunsaturated fatty acids (LCPUFA) in the diets of term infants. In part this has been spurred by the finding that preterm infants fed diets containing LCPUFA have better neural development than preterm infants fed unsupplemented diets. Several studies in term infants have produced mixed results. The controversy can be divided into two separate but related topics: first, is there a difference in outcome between breast-fed and formula-fed term infants, and if so, can this be attributed to LCPUFA content or are there other confounding factors? Second, does the addition of LCPUFA to infant formula actually result in a measurable improvement in developmental outcomes? Establishing a need for LCPUFA in the diet of term infants requires adequate proof of causation. This paper examines the evidence relating to this issue with particular reference to the adequacy of the data in establishing this proof.

The fats of human milk provide a complex mixture of fatty acids including the essential 18 carbon polyunsaturates and many long-chain (20 and 22 carbon) polyunsaturates that have biological activity. In contrast, most infant formulas chiefly contain vegetable oils that provide only 18 carbon essential fatty acids, α-linolenic acid (ALA, 18:3 $n-3$) and linoleic acid (LA, 18:2 $n-6$). Despite this, infant formulas supply the growing infant with a basic level of nutrition that results in growth rates which are similar to or higher than those observed in breast-fed infants (1). The use of neurophysiological end points has raised the possibility of limitations in the fat supply of formula-fed infants. Of particular interest are the results of research on the roles of two LCPUFA, docosahexaenoic acid (DHA, 22:6 $n-3$) and arachidonic acid

J.G. Bindels, A.C. Goedhart and H.K.A. Visser (eds), Recent Developments in Infant Nutrition, 202–211.

***Table 1** Non-randomized studies comparing the visual outcome of breast- and formula-fed infants*

Reference	*Result*
5	HM VEP acuity>formula fed at 5 months
6	HM VEP and FCPRL acuities>formula fed at 4 months
8	HM PRL acuity = formula fed at 3 months
7	HM PRL acuity>formula fed at 2 and 4 months

HM, human milk; VEP, visual evoked potential; FCPRL, forced choice preferential looking; PRL, preferential looking

(AA, 20:4 $n-6$). Biochemical studies assessing the relationship between brain fatty acids and diet in human infancy have demonstrated that breast-fed infants (receiving DHA and AA) have a greater proportion of DHA in their brain cortex than those fed formula (2,3). Furthermore, the accumulation of DHA in brain cortex until 40 weeks postnatal age is dependent on the length of breast-feeding, indicating a partial dependence on performed DHA in the diet (2). In contrast, there are no differences in the level of brain AA between breast-fed and formula-fed infants, suggesting that either AA is aggressively conserved in brain tissue or the amount of AA produced from the precursor (LA) in formula is sufficient to meet the AA requirements of the rapidly growing brain.

Clinical studies that have attempted to find an association between dietary DHA and various aspects of brain development have largely focused on the maturation of the visual pathway, since photoreceptor outer segment membranes in the retina contain uniquely high concentrations of DHA, which is important for normal photochemical activity of the visual pigment rhodopsin (4). The processes of myelination, synaptogenesis and the development of several enzyme systems all occur during the postnatal brain growth spurt and are reflected in the changing functional responses of the visual system. For example, visual acuity improves rapidly during the first year of life. Indeed, the most successful animal and human models for demonstrating an effect of the physiological importance of DHA have been based on different aspects of the visual system, probably because aspects of visual function are relatively easy to measure objectively.

BREAST VERSUS FORMULA

Over the last 5 years, four non-randomized studies have compared physiological parameters, such as visual acuity, between breast-fed and formula-fed term infants and have related the results to infant fatty acid profiles (Table 1). Three studies (5–7) reported better visual function indices and higher DHA levels in breast-fed than in formula-fed infants, while the other demonstrated differences in erythrocyte DHA levels but no difference in visual acuity between breast-fed and formula-fed infants (8). There are, of course, many nutritional differences between breast milk and infant formula other than the supply of LCPUFA. Various constituents of breast milk, such as hormones and growth factors, may affect neural function. To determine whether

Table 2 Dietary treatments of randomized clinical trials of fatty acid supplementation of term infants

	Supplemental PUFA (wt %)				Background PUFA (wt %)	
Study	*DHA*	*AA*	*Other*	*Form*	*LA*	*ALA*
Makrides et al. (9)						
Breast[a]	0.21	0.40	n−3 0.23 n−6 0.61	TG	13.9	0.94
Formula 1[a]	ND	ND	ND	–	16.8	1.58
Formula 2[a]	0.36	ND	EPA 0.58 GLA 0.27	TG	17.4	1.52
Agostoni et al. (12)						
Breast	NM	NM	NM	TG	NM	NM
Formula 1	ND	ND	ND	–	11.1	0.70
Formula 2	0.30	0.44	GLA 0.30	egg PL TG	10.8	0.73
Carlson et al. (10)						
Breast≤3months	NM	NM	NM	TG	NM	NM
Formula 1	ND	ND	ND	–	21.9	2.2
Formula 2	0.1	0.43	ND	egg PL	21.8	2.0
Auestad et al. (11)						
Breast≤3months	NR	NR	NR	–	NR	NR
Formula 1	ND	ND	ND	–	21.9	2.2
Formula 2	0.12	0.43	ND	egg PL	21.7	1.9
Formula 3	0.23	ND	EPA 0.07	TG	20.7	1.9

TG, triglyceride; PL, phospholipid; ND, not detected (<0.05% total fatty acids); NM, not measured; NR, not reported

dietary LCPUFA contribute to the visual and neural development of infants, randomized studies comparing infant diets with and without DHA have been undertaken to establish whether there is a causative link between dietary DHA and neural development.

RANDOMIZED CONTROLLED TRIALS

Four placebo-controlled, randomized studies of formula feeding with and without LCPUFA have been published (9–12) and many more are in progress. These trials have proved difficult to interpret because different research groups have used diverse dietary sources of PUFA (Table 2), the definition of breast-feeding reference groups has often been inadequate (Table 2) and developmental assessments have differed (Table 3). Not surprisingly, therefore, some studies have demonstrated a beneficial effect of LCPUFA supplementation (9,10,12), while another reported no effect (11).

Experimental design

Most studies have enrolled and followed 20–30 infants per treatment group. Although these numbers are sufficient to detect small changes in the fatty acid profiles of

Table 3 Neurodevelopmental outcomes of randomized clinical trials of fatty acid supplementation of term infants

Reference	*Diet*	*Test*	*Age*	*Results*
9	HM	transient VEP	16 weeks	HM & FO, EPO>std formula
	std formula		30 weeks	HM & FO, EPO>std formula
	+FO, EPO			
12	HM	DQ	4 months	HM & egg PL, EPO>std formula
	std formula			
	+egg PL, EPO			
10	HM≤3 months	acuity cards	2 months	HM & egg PL>std formula
	std formula		4,6,9,12 months	no difference with diet
	+egg PL			
11	HM≤3 months	sweep VEP	2,4,6,9,12 months	no difference with diet
	std formula	acuity cards	2,4,6,9,12 months	no difference with diet
	+egg PL	Bayley scales	12 months	no difference with diet
	+FO	McArthur CDI	14 months	negative correlation between RBC DHA & CDI

HM, human milk; std, standard; PL, phospholipid; EPO, evening primrose oil; FO, fish oil; VEP, visual evoked potential; DQ, developmental quotient; CDI, communicative developmental inventory; RBC, red blood cell plasma and erythrocyte membranes, they may be inadequate to determine true differences in developmental scores and growth. Although an important consideration in prospective, long term clinical trials has been maintaining a balance between statistical power and the cost associated with the upkeep of compliance and consistent follow-up of infant–mother pairs, it is also important to avoid the risk of type II errors by ensuring adequate numbers.

Attempts to improve the power of studies inevitably result in the use of multiple study centres to recruit and assess subjects. However, multicentre trials involve many people in the recruitment and assessment of subjects and this naturally increases the variability associated with each aspect of the study. Randomization from a single centre, strict adherence to a single study protocol and careful monitoring of inter-centre variability of assessment methods are vital so that the increase in subject numbers is not achieved at the expense of increased variability in assessment methods.

Dietary treatments

The alternative strategies used to improve the PUFA blends of infant formulas have largely been governed by the availability of different oils and LCPUFA supplements to the various research groups. For example, at the commencement of our study the only feasible DHA source was a fish oil containing high levels of eicosapentaenoic acid (EPA, 20:5 $n-3$; Table 2). Because of the emerging work with preterm infants which suggested that feeding high-EPA fish oils was associated with poorer growth than that achieved with standard formula feeding (13), we also supplemented with evening primrose oil, which contains γ-linolenic acid (GLA, 18:3 $n-6$), a metabolic precursor of AA, that bypasses the supposed rate-limiting δ-6 desaturase enzyme. More recent studies have tried to avoid the addition of EPA by using fish oils high in DHA, such as tuna oil (11), or by supplementing with fractions of egg phospholipids (10–12). Egg phospholipids have provided the opportunity to mimic the levels of DHA and AA found in the breast milks of Western women (10–12), although they provide these fatty acids in a form different to that in breast milk and contain extra phosphatidylcholine and cholesterol (14). Further investigations are now necessary to determine whether the addition of these compounds is of any consequence to the infant.

The base formulas used in various studies also vary (Table 2). Although most groups have used fat blends based on vegetable oil, there are notable differences in absolute levels of LA and ALA, the ratio of LA to ALA and proportions of saturates and monounsaturates. All of these parameters have been associated with effects on LCPUFA incorporation (15).

Interestingly, only one of the randomized clinical trials involving term infants reports the fatty acid composition of reference breast-milk-fed infants (9,16). Other studies have relied on breast milk fatty acid composition data reported in the scientific literature (10–12). Although these are representative, the fatty acid intake of particular infants cannot be related directly to the study results. A further complicating factor with some breast-fed reference groups (10,11) is mixed feeding, as some trial protocols permit a large proportion of energy to be derived from formula.

Assessments

Visual function

Visual acuity can be measured both electrophysiologically and behaviourally, and the acuity values obtained from such tests may be measuring different phenomena. For example, the visual evoked potential (VEP) is the visually evoked response of the electroencephalogram recorded at the occipital lobe and is evoked by a reversing pattern: the different methods of recording VEP largely depend on the nature of the pattern stimulator. Different patterns and different reversal rates elicit disparate brain responses (see also Chapter 12, this volume). Our results that demonstrated differences in VEP acuity related to $n-3$ LCPUFA status have extrapolated acuity thresholds from checkerboard patterns reversing at a slow temporal rate (9). Auestad et al. (11), who showed no effect of diet on VEP acuity development, assessed VEP acuity using a rapid sweep technique with bar gratings. The response component of the VEP that was measured also varied between the two studies, and the contrasting results may be explained by the differences in the two methodologies, which may be measuring different neural mechanisms with different developmental time-courses (17).

Behavioural measures of acuity can be rapid and easy to use but are also subject to procedural differences. Carlson et al. (10), using a rapid acuity card technique (Teller), showed a beneficial effect of DHA on visual acuity at 2 months postnatal age, but not at later ages. This result was in direct contrast to that of Auestad et al. (11), who used similar dietary treatments and apparently identical assessments but found no difference with diet. These data highlight the complexity of this field of endeavour and raise the possibility that other influencing or confounding factors may be masking the true result.

Developmental indices

Results of preliminary studies of the effects of DHA supply on global assessments of development and have been inconsistent. For example, while Agostoni and co-workers reported a direct relationship between DHA status and performance on the Brunet-Lézine neurodevelopmental scale at 4 months (12, Chapter 14, this volume), Janowsky and co-workers demonstrated a negative correlation between DHA status and vocabulary production and comprehension at 14 months (18) and no effect of dietary DHA on the Bayley Scales of Mental and Psychomotor Development (18). However, neither trial made any attempt to control for variables, such as social class or maternal intelligence, which are known to influence a baby's performance. Studies with larger numbers of infants and more rigorous controls are necessary to elucidate the effect of dietary LCPUFA on global indices of development.

Growth

There have been no reports of reduced growth associated with the dietary supply of DHA in term infants, even following administration of fish oil rich in EPA (9).

However, adequate numbers may once again be a limiting factor. The growth of infants varies widely between boys and girls and between breast-fed and formula-fed babies. Since growth remains one of the most sensitive indicators of nutritional status, and is therefore an important assessment outcome, future studies must enrol adequate numbers of boys and girls in each of their feeding groups.

Confounding influences

Developmental outcomes are known to be multifactorial and much can be learnt from studies that have examined the inter-relationships between breast-feeding and formula feeding and a child's subsequent development. Such studies have consistently shown a beneficial effect of breast-feeding on cognitive abilities when compared with non-breast-fed peers, but this positive effect was largely diminished when confounding influences were considered (19–22). For example, children who were breast fed were from more advantaged environments and had more intelligent mothers than those who were bottle fed (19–22). Indeed, one study reported that the developmental advantage of breast-fed over formula-fed children could be fully accounted for by maternal general mental ability, maternal education, socio-economic level and childhood developmental experiences (23).

Although differences in social class, education and parenting styles between mothers who choose to breast feed or formula feed would be reduced by comparing the outcomes of formula-fed infants randomized to receive either a LCPUFA-supplemented or an unsupplemented formula, there still remain noteworthy confounding environmental and biological variables, including gender (24), race, gestational age, birth weight, incidence of infection or illness, parity, bilirubin levels in the first week of life (25), alcohol use in pregnancy, smoking, maternal nutrition, strenuous exercise during pregnancy, maternal weight, age and marital status. To control for any number of these variables will require large numbers of subjects and/or stringent inclusion and exclusion criteria.

Are LCPUFA essential for term infants?

The hypothesis that dietary LCPUFA are essential for optimal development of term infants has largely stemmed from two observations. First, breast-fed infants perform better than formula-fed infants on tests that assess neurodevelopmental outcomes, and breast milk contains LCPUFA while formula milks do not. It is, therefore, necessary to demonstrate that any beneficial effects of breast milk on infant development are attributable purely to the presence of LCPUFA in breast milk and their absence from formula. Proving this hypothesis has been difficult, and perhaps the most noteworthy impediment has been our inability to define adequately our dietary extremes – the breast-fed and the standard-formula-fed infants. Over the last decade, DHA levels in breast milk have declined with the decreased popularity of seafoods, meats and eggs (16), while formula manufacturers have also made attempts to improve their formula fat blends by including greater amounts of ALA, and hence possibly enhancing metabolism to DHA. However, despite the moving together of these two extreme reference groups, the reports of positive associations between dietary DHA or infant DHA status with improved visual or developmental outcomes (5–7,9,10,12) outweigh those demonstrating no effect (8,11).

Bradford-Hill (26) proposed that an association between two variables should fulfil certain criteria before the effect can be considered causal. For example, the hypothesis that dietary DHA is associated with developmental outcome needs to be plausible, the effect must be consistent, specific and independent of confounding factors. Our hypothesis is certainly plausible: DHA is avidly incorporated and retained in brain cerebral phospholipids. Breast-fed infants (receiving DHA) have higher brain DHA levels than formula-fed infants (receiving no DHA) and animal studies have directly related brain cortex phospholipid DHA to performance on visual and learning tasks (27–30). However, studies in term infants that have attempted to improve the DHA supply of infant formula and hence infant development have not yielded consistent results. Some studies demonstrate a benefit (9,10,12) while another shows no effect (11). Of the studies reporting a positive result with DHA supplementation, one indicates that the effect is transient (10) while another suggest that there is no catch-up with age (9). Supplemental regimens have not specifically added DHA and have included other LCPUFA (Table 2), although only DHA consistently correlated with outcome (9,10,31). Finally, no report adequately considers the enormous number of variables that are capable of impinging on complex functions such as cognitive processes. Thus, based on these criteria, the evidence to support the use of LCPUFA in formulas for term babies is far from conclusive.

In summary, the most consistent, and perhaps most disturbing, finding has been the lower level of cerebral DHA in the brains of formula-fed infants relative to those fed breast milk (2,3). The formula-fed infants in these studies were generally fed formulas with adequate ALA levels and, although controversial, it is possible that the lower DHA status of formula-fed infants is reflected in aspects of suboptimal neural development. The four randomized studies of LCPUFA supplementation of formula-fed term infants have failed to demonstrate consistently positive findings in favour of LCPUFA. However, no study has specifically assessed the impact of dietary DHA in the absence of other LCPUFA, consistently used the same assessment methodologies, or made adequate provision for valid confounding factors. Despite these flaws, economic imperatives may force many manufacturers to include LCPUFA in term formulas in the near future. LCPUFA, as was the case with taurine (32), will be added to many formulas as a measure of prudence to provide improved nourishment with the same margin of safety as that found in human milk. We may thus be left with the task of ensuring that the various fat blends that are used to mimic the effects of breast milk at least do no harm.

ACKNOWLEDGEMENTS

Aspects of this work were supported by the NH&MRC Public Health Research and Development Committee, and the SIDS Association of South Australia.

REFERENCES

1. Dewey KG, Heinig MJ, Nommsen LA, Peerson JM, Lönnerdal B 1992 Growth of breast-fed and formula-fed infants from 0 to 18 months: The DARLING study. Pediatrics 89:1035–1041
2. Makrides M, Neumann MA, Byard RW, Simmer K, Gibson RA 1994 Fatty acid composition of brain, retina, and erythrocytes in breast- and formula-fed infants. Am J Clin Nutr 60:189–194
3. Farquharson J, Cockburn F, Patrick WA, Jamieson EC, Logan RW 1992 Infant cerebral cortex phospholipid fatty-acid composition and diet. Lancet 340:810–813
4. Wiedmann TS, Pates RD, Beach JM, Salmon A, Brown MF 1988 Lipid-protein interactions mediate the photochemical function of rhodopsin. Biochemistry 27:6469–6474
5. Makrides M, Simmer K, Goggin M, Gibson RA 1993 Erythrocyte docosahexaenoic acid correlates with the visual response of healthy, term infants. Pediatr Res 33:425–427
6. Birch EE, Birch DG, Hoffman DR, Hale L, Everett M, Uauy R 1993 Breast-feeding and optimal visual development. J Pediatr Ophthalmol Strabismus 30:33–38
7. Jørgensen MH, Hernell O, Lund P, Holmer G, Fleischer, Michaelsen KF 1996 Visual acuity and erythrocyte docosahexaenoic acid status in breast-fed and formula-fed infants during the first 4 months of life. Lipids 31:99–105
8. Innis SM, Nelson CM, Rioux MF, King DJ 1994 Development of visual acuity in relation to plasma and erythrocyte ω-6 and ω-3 fatty acids in healthy term gestation infants. Am J Clin Nutr 60:347–352
9. Makrides M, Neumann M, Simmer K, Pater J, Gibson R 1995 Are long-chain polyunsaturated fatty acids essential nutrients in infancy? Lancet 345:1463–1468
10. Carlson SE, Ford AJ, Werkman SH, Peeples JM, Koo WWK 1996 Visual acuity and fatty acid status of term infants fed human milk and formulas with and without docosahexaenoate and arachidonate from egg yolk lecithin. Pediatr Res (in press)
11. Auestad N, Montalto MB, Wheeler RE, Fitzgerald KR, Hall RT, Neuringer M, Connor WE, Hartmann EE, Taylor JA 1995 Visual acuity, RBC fatty acids and growth in term infants fed formulas with and without long chain polyunsaturated fatty acids (LCP). Pediatr Res 37:302A
12. Agostoni C, Trojan S, Bellù R, Riva E, Giovannini M 1995 Neurodevelopmental quotient of healthy term infants at 4 months and feeding practice: The role of long-chain polyunsaturated fatty acids. Pediatr Res 38:262–266
13. Carlson SE, Cooke RJ, Werkman SH, Tolley EA 1992 First year growth of preterm infants fed standard compared to marine oil $n-3$ supplemented formula. Lipids 27:901–907
14. Agostoni C, Riva E, Bellù R, Trojan S, Luotti D, Giovannini M 1994 Effects of diet on the lipid and fatty acid status of full-term infants at 4 months. J Am Coll Nutr 13:658–664
15. Innis SM 1991 Essential fatty acids in growth and development. Prog Lipid Res 30:39–103
16. Makrides M, Simmer K, Neumann M, Gibson R 1995 Changes in the polyunsaturated fatty acids of breast milk from mothers of full-term infants over 30 weeks of lactation. Am J Clin Nutr 61:1231–1233
17. Orel-Bixler DA, Norcia AM 1987 Differential growth of acuity for steady-state pattern reversal and transient pattern onset-offset VEP's. Clin Vis Sci 2:1–9
18. Janowsky JS, Scott DT, Wheeler RE, Auestad N 1995 Fatty acids affect early language development. Pediatr Res 37:310A
19. Taylor B, Wadsworth J 1984 Breast feeding and child development at five years. Dev Med Child Neurol 26:73–80
20. Rogan WJ, Gladen BC 1993 Breast-feeding and cognitive development. Early Hum Dev 31: 181–193
21. Morrow Tlucak M, Haude RH, Ernhart CB 1988 Breastfeeding and cognitive development in the first 2 years of life. Soc Sci Med 26:635–639
22. Bauer G, Ewald LS, Hoffman Y, Dubanoski R 1991 Breastfeeding and cognitive development of three-year-old children. Psychol Rep 68:1218
23. Silva PA, Buckfield P, Spears GF 1978 Some maternal and child developmental characteristics associated with breast feeding: a report from the Dunedin Multidisciplinary Child Development Study. Aust Paediatr J 14:265–268
24. Lucas A, Morley R, Cole TJ, Lister G, Leeson-Payne C 1992 Breast milk and subsequent intelligence quotient in children born preterm. Lancet 339:261–264
25. Chen Y-J, Kang W-M 1995 Effects of bilirubin on visual evoked potentials in term infants. Eur J Pediatr 154:662–666

26. Bradford-Hill A 1965 The environment and disease: association or causation? Proc R Soc Med 58:295–300
27. Neuringer M, Connor WE, Van Petten C, Barstad L 1984 Dietary omega-3 fatty acid deficiency and visual loss in infant rhesus monkeys. J Clin Invest 73:272–276
28. Neuringer M, Connor WE, Lin DS, Barstad L, Luck S 1986 Biochemical and functional effects of prenatal and postnatal $\omega 3$ fatty acid deficiency on retina and brain in rhesus monkeys. Proc Natl Acad Sci USA 83:4021–4025
29. Lamptey MS, Walker BL 1976 A possible essential role for dietary linolenic acid in the development of the young rat. J Nutr 106:86–93
30. Bourre J-M, Francois M, Youyou A, Dumont O, Piciotti M, Pascal G, Durand G 1989 The effects of dietary α-linolenic acid on the composition of nerve membranes, enzymatic activity, amplitude of electrophysiological parameters, resistance to poisons and performance of learning tasks in rats. J Nutr 119:1880–1892
31. Agostoni C, Riva E, Trojan S, Bellù R, Giovannini M 1995 Docosahexaenoic acid status and developmental quotient of healthy term infants. Lancet 346:638
32. Gaull GE 1989 Taurine in pediatric nutrition: review and update. Pediatrics 83:433–442

14 LCPUFA status and developmental quotient in term infants fed different dietary sources of lipids in the first months of life

C. AGOSTONI, S. TROJAN, R. BELLÙ, E. RIVA, D. LUOTTI and M. GIOVANNINI

ABBREVIATIONS

AA	arachidonic acid, 20:4 $n-6$
DHA	docosahexaenoic acid, 22:6 $n-3$
DQ	developmental quotient
FA	fatty acid
HM	human milk
LCPUFA	long-chain polyunsaturated fatty acid

INTRODUCTION

Dietary long-chain polyunsaturated fatty acids (LCPUFA) such as arachidonic acid (AA) and docosahexaenoic acid (DHA) have been suggested among the possible causes of higher neurodevelopmental scores obtained by children fed human milk as infants (1–4). Experimental data show that dietary supplementation with or deprivation of linoleic acid and α-linolenic acid, the precursors of LCPUFA, during the early stages of growth affect both the LCPUFA composition of the nervous tissue and sensorimotor development (5–8). Decreased levels of DHA, the major $n-3$ LCPUFA derived from α-linolenic acid, have been associated with altered learning patterns in animals, while dietary supplementation with preformed DHA increases DHA levels in nervous tissues and improves learning skills (9,10). Recent studies also found a significant relationship between DHA concentrations in erythrocytes and performance of full-term infants in visual acuity tests (11,12)

The milk fat of European women has an AA content of 0.2–1.2%, and a DHA content of 0.1–0.6% of total fatty acids (13). In contrast to human milk, the standard vegetable-oil-based infant formulas currently available lack LCPUFA, the fatty acids most characteristic of animal fats. Consequently, levels of erythrocyte-bound DHA are particularly low among formula-fed infants (14). A blend of lipids containing a comprehensive selection of parent $n-3$ and $n-6$ polyunsaturated FA and their

J.G. Bindels, A.C. Goedhart and H.K.A. Visser (eds), Recent Developments in Infant Nutrition, 212–217.

Table 1 Composition of breast milk and formulas used in this study

Fatty acids (g/100 g fat)	*F1*[a]	*F2*	*European HM*[b]
Saturated	51.6	48.2	39.0–51.3
18:1 *n*–9	28.7	38.8	34.2–44.9[c]
18:2 *n*–6	10.8	11.1	6.9–16.4
18:3 *n*–6	0.30	–	0.1–0.9
18:3 *n*–3	0.73	0.70	0.7–1.3
20:4 *n*–6	0.44	–	0.2–1.2
20:5 *n*–3	0.05	–	0.0–0.6
22:6 *n*–3	0.30	–	0.1–0.6

F1 = LCPUFA supplemented formula, F2 = standard formula
[a]As the raw materials used in the lipid blend are from natural sources, FA content may vary
[b]Adapted from Ref. 1
[c]Given as total monounsaturated FA

derivatives could be used to raise LCPUFA concentrations among these formula-fed infants. Accordingly, we compared in a randomized, placebo-controlled prospective study the performance on the Brunet–Lézine psychomotor development scale of breast-fed infants and those fed formulas with and without LCPUFA, and related the results to infant fatty acid profiles.

SUBJECTS AND METHODS

The original study design included 90 healthy, full-term infants. Sixty infants whose mothers decided not to breast feed were randomly assigned to receive either a formula enriched with LCPUFA (Aptamil with Milupan, F1) or a standard infant formula (Aptamil, F2; Table 1), and were examined in a double-blind, prospective study with a parallel group of 30 breast-fed infants (HM group) (15). The neurodevelopmental response was tested in all subjects at 4 months of age by means of the Brunet–Lézine psychomotor development test (16). This test, commonly used in Western Europe, derives from Gesell's approach and explores four developmental areas: posture and gross motor function, adaptation and fine motor function, social reactions and language. In 59 infants whose parents had given consent for blood sampling, the fatty acid profile of circulating and erythrocyte lipids and fatty acid concentrations were measured by gas-chromatography (17) and expressed as weight%. Four infants (all sampled for the fatty acid status) were too agitated at developmental quotient (DQ) testing to be properly evaluated. Thus, 86 infants underwent DQ testing (27 fed LCPUFA formula, 29 standard infant formula, and 30 human milk) and assessment of fatty acid status/DQ relationship was undertaken in 55 (21 fed LCPUFA formula, 19 standard infant formula, and 15 human milk).

After the 4-month assessment all infants received the first solid foods and within the fifth month of life a standard follow-on formula replaced both starting formulas. Lactating mothers were encouraged to complement their milk progressively with

***Table 2** Levels of the major long-chain polyunsaturated fatty acids (% wt/wt) in plasma and erythrocyte phospholipids*[a]

	F1 (*n*=21)	*F2* (*n*=23)	*HM* (*n*=15)	*F*	*p*
LCPUFA in plasma phospholipids					
20:4 *n*−6	9.0±1.4[a] (8.4−9.6)	6.2±1.3[b] (5.6−6.8)	10.8±2.1[c] (9.7−11.9)	41.6	<0.001
22:6 *n*−3	2.7±0.6[a] (2.4−3.0)	0.9±0.3[b] (0.8−1.0)	2.8±0.8[a] (2.4−3.2)	73.6	<0.001
LCPUFA in erythrocyte phospholipids					
20:4 *n*−6	16.3±1.2[a] (15.8−16.8)	15.1±1.8[b] (14.3−15.9)	14.6±2.3[b] (13.4−15.8)	4.61	0.01
22:6 *n*−3	4.1±0.6[a] (3.8−4.4)	1.8±0.4[b] (1.6−2.0)	4.1±1.1[a] (3.5−4.7)	75.1	<0.001

[a]Values are mean±SD (95% confidence interval)
Different superscripts indicate significantly different values ($p<0.05$)
F1 = LCPUFA supplemented formula, F2 = standard formula

solid foods. In this group the follow-on formula was introduced when the mother's milk became insufficient.

At 12 months of age a second neurodevelopmental evaluation was performed by means of the Brunet–Lézine scale in 81 subjects (26 LCPUFA formula, 29 standard formula, 26 human milk). Of the 9 subjects who dropped out, two left the study at parental request and seven did not present due to common infectious disorders of infancy.

Differences between DQ scores and FA levels were tested by analysis of variance (ANOVA). Categorical variables for socioeconomic indicators were compared using the χ^2 test. Comparison between group means was tested by the Newman–Keuls method. Linear regression was used to investigate the relationship between single fatty acids and the developmental scores. All statistical analyses were performed on the SPSS/PC+ statistical package (SPSS/PC+3.1; SPSS Inc., Chicago, IL).

RESULTS

There were no differences between the three dietary groups in terms of weight at birth, parity, mother's age, father's age, gestational age, postconceptional age at DQ, real age at DQ and socioeconomic indicators (parents' education and occupation).

At 4 months of age both the breast-fed and the enriched-formula-fed groups displayed higher *n*−6 and *n*−3 LCPUFA levels in plasma phospholipids than the standard-formula-fed infants (Table 2). DHA levels in erythrocyte phospholipids among the infants fed the standard formula were <50% of those found in the other two groups. Erythrocyte AA levels were higher in the infants fed the enriched formula than in the two other groups (Table 2). At 4 months of age the enriched-formula-fed

Table 3 Results of the neurodevelopmental assessments in the first year of life[a]

	F1	n	F2	n	HM	n	ANOVA (by milk) F	ANOVA (by milk) p
DQ at 4 months	105.3±9.4[a] (101.8–108.8)	27	96.5±10.9[b] (92.6–100.4)	29	102.2±11.5[a] (98.1–106.3)	30	4.93	0.009
DQ at 12 months	101.5±9.2 (97.8–105.2)	26	101.2±8.0 (98.1–104.3)	29	105.4±7.7 (102.2–108.5)	26	2.05	0.13

[a]Mean ± SD (95% confidence interval)
Different superscripts indicate significantly different values ($p<0.05$)
F1 = LCPUFA enriched formula, F2 = standard formula

infants scored significantly higher on the Brunet–Lézine scale than infants who received the standard formula (Table 3). Breast-fed infants also performed better than those fed the standard formula.

A significant relationship was found between DHA levels in erythrocyte phospholipids and the DQ score ($r=0.32$, $p=0.01$) (18). This correlation persisted when considering the randomized groups of bottle-fed infants ($r=0.47$, $p<0.01$). No other fatty acid of erythrocyte phospholipids correlated positively with DQ scores. Retesting 81 subjects at 12 months showed no significant difference between the DQ scores of the three groups, although a trend towards higher scores in the previously breast-fed group was evident (Table 3).

DISCUSSION

Our study supports the hypothesis that the supply of LCPUFA through an enriched infant formula helps artificially fed infants to maintain plasma and erythrocyte LCPUFA levels close to those of breast-fed infants at 4 months of age, when it can be assumed that the composition of circulating and erythrocyte lipids depends largely on the type of feeding and not on the status at birth.

Our findings also suggest that dietary LCPUFA can benefit the rate of neurodevelopment of term infants, at least by influencing the speed of functional maturation. The reported observations parallel the findings reported by Makrides et al (11,12) on the relationship between LCPUFA status and the visual response assessed by the evoked potential technique.

Docosahexaenoic acid is probably the most important of the LCPUFA for its functional role in nerve cells. When the diet of term infants does not include DHA from breast milk the LCPUFA pattern of the central nervous system is affected, as has been found in infants suffering cot death (19,20). The biochemical evidence of a long synthetic pathway (elongation plus retroconversion by peroxisomal processes) (21) for DHA furthers the examination of this molecule as a 'semi-essential' nutrient.

The finding that LCPUFA circulating in plasma phospholipids are related to DQ suggests that the functional response could depend on the quality of both the membrane and the circulating phospholipid pools, which are in continuous metabolic exchange (22,23). The observation that psychomotor development at 4 months of age is directly influenced by the dietary intake of both AA and DHA does not ensure a predictive value of these values for childhood trends, and even less for development in later childhood, because psychometric tests show a poor relationship with the intelligence quotient of older children (24). Moreover, it has been shown that the intelligence quotient mostly depends on heritability (70% of the variation in the intelligence quotient score is explained by genetic variability) (25), and consequently the influence of environment on the quality of development is minor. Nevertheless, while genetics can hardly be modified, the environment (including nutritional habits) can be strongly influenced.

The diet groups we studied are currently being followed to determine long-term implications, to better define the dietary essential fatty acid requirements of formula-fed term infants.

ACKNOWLEDGEMENT

The contribution of Maria Grazia Bruzzese, PhD, in the laboratory findings is gratefully acknowledged.

REFERENCES

1. Lucas A, Morley R, Cole TJ, Lister J, Leeson-Payne C 1992 Breast milk and subsequent intelligence quotient in children born preterm. Lancet 339:261–264
2. Rogan WJ, Gladen BC 1993 Breast-feeding and cognitive development. Early Hum Dev 31: 181–193
3. Pollock JI 1994 Long-term associations with infant feeding in a clinically advantaged population of babies. Dev Med Child Neurol 36:429–440
4. Riva E, Agostoni C, Biasucci G, Trojan S, Luotti D, Fiori L, Giovannini M 1996 Early breast-feeding is linked to higher intelligence quotient scores in dietary treated phenylketonuric children. Acta Paediatr 85:56–58
5. Yamamoto N, Saitoh M, Moriuchi A, Nomura M, Okuyama H 1987 Effect of dietary alpha-linolenate/linoleate balance on brain lipid composition and learning ability of rats. J Lipid Res 28:144–151
6. Coscina DV, Yehuda S, Dixon LM, Kish SJ, Leprohon-Greenwood C 1986 Learning is improved by a soybean oil diet in rats. Life Sci 38:1789–1794
7. Yehuda S 1989 Behavioral effects of dietary fats. In: Chandra RK (ed) Health Effects of Fish and Fish Oils. ARTS Biomedical Publishers, St John's, Newfoundland, pp 327–335
8. Enslen M, Milon H, Malnoe A 1991 Effect of low intake of $n-3$ fatty acids during development on brain phospholipids, fatty acid composition and exploratory behavior in rats. Lipids 26: 203–208
9. Fujimoto K, Yao K, Miyazawa T, Hirono H, Nishikawa M, Kimura S, Maruyama K, Nonaka M 1989 The effect of dietary docosahexaenoate on the learning ability of rats. In: Chandra RK (ed) Health Effects of Fish and Fish Oils. ARTS Biomedical Publishers, St John's, Newfoundland, pp 275–284
10. Wainwright PE, Huang YS, Bulman-Fleming B, Mills DE, Redden P, McCutcheon D 1991 The role of $n-3$ fatty acids in brain and behavioral development: a cross-fostering study in the mouse. Lipids 26:37–45
11. Makrides M, Simmer K, Goggin M, Gibson RA 1993 Erythrocyte docosahexaenoic acid correlates with the visual response of healthy, term infants. Pediatr Res 33:425–427
12. Makrides M, Neumann M, Simmer K, Pater J, Gibson R 1995 Are long-chain polyunsaturated fatty acids essential nutrients in infancy? Lancet 345:1463–1468
13. Koletzko B, Thiel I, Abiodun PO 1992 The fatty acid composition of human milk in Europe and Africa. J Pediatr 120:S62–S70

14. Ponder DL, Innis SM, Benson JD, Siegman JS 1992 Docosahexaenoic acid status of term infants fed breast milk or infant formula containing soy oil or corn oil. Pediatr Res 32:683–688
15. Agostoni C, Trojan S, Bellù R, Riva E, Giovannini M 1995 Neurodevelopmental quotient of healthy term infants at 4 months and feeding practice: the role of long-chain polyunsaturated fatty acids. Pediatr Res 38:262–266
16. Brunet O, Lézine I 1966 Le dévelopement psychologique de la première enfance. 2nd ed (Italian translation 1967). Presses Universitaires de France, Paris
17. Agostoni C, Riva E, Bellù R, Trojan S, Luotti D, Giovannini M 1994 Effects of diet on the lipid and fatty acid status of full-term infants at 4 months. J Am Coll Nutr 13:658–664
18. Agostoni C, Riva E, Trojan S, Bellù R, Giovannini M 1995 Docosahexaenoic acid status and developmental quotient of healthy term infants. Lancet 346:638
19. Makrides M, Neumann MA, Byard RW, Simmer K, Gibson RA 1994 Fatty acid composition of brain, retina and erythrocytes in breast- and formula-fed infants. Am J Clin Nutr 60:189–194
20. Farquharson J, Cockburn F, Patrick WA, Jamieson EC, Logan RW 1992 Infant cerebral cortex fatty acid composition and diet. Lancet 340:810–813
21. Voss A, Reinhart M, Sankarappa S, Sprecher H 1991 The metabolism of 7,10,13,16,19-docosapentaenoic acid to 4,7,10,13,16,19-docosahexaenoic acid in rat liver is independent of a 4-desaturase. J Biol Chem 266:19995–20000
22. Clandinin MT, Jumpsen J, Suh M 1994 Relationship between fatty acid accretion, membrane composition and biologic function. J Pediatr 125:S25–S32
23. Innis SM 1992 Plasma and red blood cell fatty acid values as indexes of essential fatty acids in the developing organs of infants fed with milk or formulas. J Pediatr 120:S78–S86
24. Amiel Tison C, Grenier A 1980 Evaluation Neurologique du Nouveau-né et du Nourrisson. Masson, Paris
25. Bouchard TJ, Lykken DT, McGue M, Segal NL, Tellegen A 1990 Sources of human psychological differences: the Minnesota study of twins reared apart. Science 250:223–228

15 LCPUFA and functional development of preterm and term infants

S. CARLSON

ABBREVIATIONS

BPD	bronchopulmonary dysplasia
DHA	docosahexaenoic acid
EPA	eicosapentaenoic acid
LCPUFA	long-chain polyunsaturated fatty acids
MDI	mental developmental index
PDI	psychomotor developmental index
PMA	postmenstrual age

INTRODUCTION

Monkey infants with $n-3$ fatty acid deficiency have lower docosahexaenoic acid (DHA) levels and differ functionally in several ways from monkeys fed $n-3$ fatty acids. For example, they have poorer retinal physiology, poorer visual acuity, and longer look duration (see Table 1), which has been associated with slower information processing. Infants fed α-linolenic acid also have poorer DHA status than infants fed DHA from human milk. On the basis of this observation, we hypothesized that infants born before the normal intrauterine accumulation of brain DHA has been completed could have a conditional need for $n-3$ long-chain polyunsaturated fatty acids ($n-3$ LCPUFA) that was not met by dietary α-linolenic acid. Outcomes used in the monkey studies of $n-3$ deficiency have been adapted for studying the essential nature of $n-3$ LCPUFA in preterm and term babies. In two randomized clinical trials DHA status was improved by adding $n-3$ LCPUFA to diets containing α-linolenic acid. In each case, the control group received a diet with generous amounts of α-linolenic acid but no $n-3$ LCPUFA. The effects of $n-3$ LCPUFA supplementation on visual acuity, look duration, visual recognition memory, and mental and psychomotor development were determined.

METHODS

Preterm infants

Clinical trials included preterm infants of very low birth weight, but who were appropriate for gestational age and did not have medical problems that could adversely

J.G. Bindels, A.C. Goedhart and H.K.A. Visser (eds), Recent Developments in Infant Nutrition, 218–224.

Table 1 Positive effects of feeding diets with α-linolenic acid compared to α-linolenic acid-deficient diets: rhesus monkeys

Retinal physiology	Higher amplitudes of cone and rod A-waves at 3–4 months but not at later ages (1)
	Cone- and rod-dominated B-wave implicit times significantly shorter in older monkeys (1)
	Recovery of dark-adapted A- and B-wave amplitudes significantly better at all ages. The difference between groups became more pronounced with age (2)
Visual acuity	Higher at 1, 2 and 3 months of age (the only ages tested) (3)
Look duration	Shorter look duration (4; see also Chapter 10, this volume)

affect growth and/or development. The infants were randomized to receive either a marine-oil-supplemented formula (containing 0.2 g DHA/100 g fatty acids) or a standard preterm formula until 2 months (this and all subsequent ages relative to expected term). Growth, plasma and erythrocyte phospholipid LCPUFA levels, visual acuity, look duration and visual recognition memory (novelty preference) and the Bayley Mental and Psychomotor development were studied at specific ages until 12 months corrected age. Visual acuity was measured by the Teller Acuity Card procedure, look duration and novelty preference were determined with an experimental version of the Fagan Infantest, and LCPUFA levels were determined by capillary gas liquid chromatography.

The first and second trials differed in several ways, including the level of eicosapentaenoic acid (EPA; 20:5$n-3$) in the experimental formula, the duration of LCPUFA supplementation, the nutrient composition of the formula given after discharge from the hospital, and the age of the infants at randomization. In the first trial, relative to the second, infants received DHA from marine oil with five times as much EPA (0.3 vs. 0.06% of total fatty acids), were fed DHA until 9 rather than 2 months corrected age, were discharged on a term formula instead of a nutrient-enriched preterm formula, and were older when randomized to their diet group (3.5 weeks vs 3 days). In addition, infants in the first trial were a very select group of the healthiest infants in our nursery. Only three of 67 had bronchopulmonary dysplasia (BPD) whereas 40% of the preterm infants in the second trial had this disease, the incidence typical for 750–1350 g infants in our nursery.

Term infants

We also determined the effect of LCPUFA supplementation on visual acuity, look duration and Bayley Mental and Psychomotor developmental indices in healthy term infants randomized to receive either an egg-yolk-supplemented formula (containing 0.1% DHA and 0.43% arachidonic acid of total fatty acids) or a term formula that was the same except that it did not contain the supplement. Visual acuity (Teller Acuity Cards) and look duration were determined at select ages during the first year of life.

RESULTS – PRETERM INFANTS

Visual acuity

In the first trial, visual acuity was better in infants fed the marine-oil-supplemented formula than in the control group at 2 and 4 months of age. After 4 months of age (at 6, 9 and 12 months), visual acuity in control and supplemented infants was similar. Similar results were found in the second trial except that the supplemented group had better visual acuity only at 2 months (when supplementation stopped) and only among supplemented infants who did not have BPD. Regardless of diet, infants with BPD had poorer visual acuity than infants without BPD at 2 and 4 months of age. Just as in the first study, the supplemented and control groups had similar visual acuity for most of infancy. There are at least two explanations for this finding. The control infants might have caught up in visual function and become the same as the supplemented infants. Alternatively, supplemented infants might have continued to have higher visual function different from grating acuity.

Novelty preference and look duration

Table 2 shows the effects of DHA supplementation on visual attention of preterm babies in our first study. We found no effect of LCPUFA on novelty preference, but marine-oil-supplemented infants did have an increased number of looks compared with controls at 6.5, 9 and 12 months of age. We could not interpret this finding until Reisbick and co-workers showed that longer look duration was associated with poorer DHA status in rhesus monkeys (4, see also Chapter 10, this volume). When we calculated look duration by dividing the total time spent looking by the number of looks, we found a significantly longer look duration in preterm babies fed the control formula (and that we speculated had a conditional need for $n-3$ LCPUFA) than in those receiving the marine-oil-supplemented formula containing $n-3$ LCPUFA. This effect was seen at 6.5, 9 and 12 months corrected age (5).

In our second trial, we also measured look duration and novelty preference at 12 months of age (7). Unfortunately, only half of the planned sample received the version of the Fagan Infantest that was used in the first study. The other half received a newer version of the Fagan Infantest, which included looks of less than 1 s duration. This test measures many more looks in the same looking time, affecting both the apparent number and duration of looks compared with the first test. Despite the loss in power from being able to use only half of the intended study sample, we found a significantly shorter look duration in the DHA-supplemented group than in controls at 12 months. Not all outcomes that were significantly different in the first study were different, again because of a loss of power, but the numbers were similar to those from the first study (7). Shorter look duration has been associated with more rapid information processing. Because preterm infants fed formulas with $n-3$ LCPUFA

Table 2 The effect of n−3 fatty acid supplementation on visual attention of preterm infants during paired comparisons of a novel and a familiar picture†

		Post-conceptional age (weeks)			
	Diet	*69*	*79*	*92*	*ANOVA*
Novelty preference (%)	Control	60.4	62.2	64.6	
	n−3 supplement	59.8	62.0	60.5*	N.S.
Total looks (no.)	Control	41.0	40.3	42.4	
	n−3 supplement	48.2**	47.0**	47.7**	$p<0.02$
Looks to novel (no.)	Control	22.2	22.1	23.6	
	n−3 supplement	26.0**	25.5**	26.0*	$p<0.03$
Looks to familiar (no.)	Control	18.8	18.2	18.8	
	n−3 supplement	22.1*	21.4*	21.7*	$p<0.03$
Total time (s)	Control	72.0	51.1	48.9	
	n−3 supplement	72.6	50.9	51.2	N.S.
Time to novel (s)	Control	45.3	32.9	32.6	
	n−3 supplement	45.9	32.6	31.9	N.S.
Time to familiar (s)	Control	26.6	18.2	16.3	
	n−3 supplement	26.6	18.3	19.3*	N.S.
Time/look (s)	Control	1.75	1.30	1.18	
	n−3 supplement	1.62	1.13*	1.11	$p<0.03$
Time/novel look (s)	Control	2.03	1.51	1.43	
	n−3 supplement	1.88	1.33	1.27	$p<0.03$
Time/familiar look (s)	Control	1.42	1.04	0.85	
	n−3 supplement	1.31	0.91	0.91	N.S.

*$p<0.05$ compared with control infants at the same age; **$p<0.01$ compared with control infants at the same age

†Least squares means from repeated-measures ANOVA corrected for subject effect

Adapted from Ref. 5

had shorter look duration than those fed α-linolenic acid, these data could be interpreted to mean that they had more rapid information processing. Shorter look duration was seen as long as 10 months after DHA supplementation was discontinued.

Bayley MDI and PDI

In the first trial there were no differences in Bayley MDI and PDI between the supplemented and control infants. In the second trial, however, infants fed the marine-oil-supplemented formula had significantly higher Bayley MDI scores at 12 months of age than those fed the control formula. In the second trial compared with the first, all babies received a nutrient-enriched formula for approximately 3 months after hospital discharge (until 2 months adjusted age). One possible explanation for the higher Bayley MDI among supplemented infants was that DHA supplementation had an effect on neural development that was dependent upon improving nutritional status of preterm infants after discharge.

A summary of the functional outcomes of DHA supplementation in preterm infants is given in Table 3.

***Table 3** Positive effects of adding docosahexaenoic acid (DHA) to formula containing α-linolenic acid (experimental) compared with α-linolenic acid alone (control): randomized trials in preterm infants*

Retinal physiology	Enhanced at 36 weeks PMA (8)
	Unchanged at 57 weeks PMA (9)
Visual acuity	Higher at 48 and/or 57 weeks PMA (9–11)
	Unchanged at 69, 79 and 92 weeks PMA (6,11)
Look duration	Shorter look duration at 69, 79 and 92 weeks PMA (the only ages studied) (5)
	Shorter look duration at 92 weeks PMA (the only age studied) in a second trial (7)
Bayley MDI	Higher 12-month scores in infants discharged from the hospital and fed a nutrient-enriched (preterm) formula until 48 weeks PMA (both control and experimental groups had normal biochemical indices of vitamin A, zinc and protein status by 48 weeks PMA) (12)
	No effect on 12-month Bayley MDI in infants discharged from the hospital on term formula (both control and experimental groups had evidence of marginal vitamin A, zinc and protein status at 48 weeks PMA) (12)

RESULTS – TERM INFANTS

Visual acuity

We also found an effect of DHA supplementation on visual acuity in term infants. At 2 months past term, visual acuity was better in infants receiving the LCPUFA-supplemented formula than in the control group. Visual acuity was also better in a group of infants fed human milk for at least the first 3 months of life: in fact the supplemented and human-milk-fed groups had similar visual acuity at 2 months. We were somewhat surprised to find higher visual acuity in the supplemented and breast-fed infants for two reasons. First, the formula contained only 0.1% DHA (it was originally supposed to have provided 0.15–0.2%), and second, we considered term infants less likely to need DHA because of their opportunity to accumulate DHA in utero.

These same formulas were used in a larger randomized trial, but no effect of the supplemented formula on visual acuity was found at any age (13). One of the major differences between our study and the latter was that we determined visual acuity at 2, 4, 6, 9 and 12 months past term (=40 weeks postmenstrual age (PMA)) whereas they measured visual acuity at 2, 4, 6, 9 and 12 months postnatal age. Because grating acuity increases exponentially in the first half of infancy, it may have been important to control for true age rather than postnatal age. The normal range of true age at term birth ranges from 37 to 43 weeks PMA.

Table 4 Positive effects of adding docosahexaenoic acid (DHA) to formula containing α-linolenic acid (experimental) compared with α-linolenic acid alone (control): randomized trials in term infants

Retinal physiology	No effects noted (14)
Visual acuity	Higher visual evoked potential acuity at 4 and 7 months (15)
	Higher Teller Acuity Card visual acuity at 2 months, but not at 4, 6, 9 and 12 months (16)
	No effect at 1, 2, 4, 6, 9 and 12 months using Teller Acuity Cards or pattern sweep-visual-evoked-potential acuity (age not stated) (13)
Look duration	Data not yet available
Bayley MDI	No effect on 12 months Bayley (Carlson et al., unpublished)

Look duration

Unfortunately, we had the same problem with the determination of look duration in term infants as in the second half of the preterm study. Using the new Fagan test, many looks of very short duration are recorded that are not considered in the older version of the test. We are still hopeful that by subtracting these looks, we can determine whether look duration in these term infants was affected by LCPUFA. The recoding of looking data for each of these 60 infants has not yet been completed.

Bayley MDI and PDI

We observed no effect of LCPUFA supplementation on 12-month Bayley MDI and PDI scores in term infants. Either there is no effect of $n-3$ LCPUFA on these indices in infants born at term or the level of DHA in the formula was too low to have an effect. Table 4 summarizes functional outcomes observed in other randomized trials of LCPUFA supplementation in term infants.

SUMMARY

Preterm infants fed DHA-supplemented formula had better visual acuity at 2 and 4 months past expected term, and a shorter look duration in the second half of infancy. In addition, preterm infants supplemented with DHA until 2 months and provided with a nutrient-enriched preterm formula for several months after discharge from the hospital had higher Bayley MDI scores at 12 months past expected term than control infants fed the same nutrient intake without $n-3$ LCPUFA. As in preterm infants, term infants fed LCPUFA-supplemented formula had higher visual acuity at 2 months than those receiving control formula, but LCPUFA supplementation did not improve the 12-month Bayley MDI and PDI scores of term infants. We do not know whether LCPUFA supplementation decreased look duration in term infants.

REFERENCES

1. Neuringer M, Connor WE, Lin DS, Anderson GJ, Barstad L 1991 Dietary omega-3 fatty acids: effects on retinal lipid composition and function in primates. In: Anderson RE, Hollyfield JG, LaVail MM (eds) Retinal Degenerations. CRC Press, New York, pp 117–129
2. Neuringer M, Connor WE, Lin DS, Barstad L, Luck S 1986 Biochemical and functional effects of prenatal and postnatal omega 3 fatty acid deficiency on retina and brain in rhesus monkeys. Proc Natl Acad Sci USA 83:4021–4025
3. Neuringer M, Connor WE, Petten van C, Barstad L 1984 Dietary omega-3 fatty acid deficiency and visual loss in infant rhesus monkeys. J Clin Invest 73:272–276
4. Reisbick S, Neuringer M, Gohl E 1994 Increased look duration in paired comparisons by rhesus monkey infants with $n-3$ fatty acid deficiency. Soc Neurosci Abstr 20:1696
5. Werkman SH, Carlson SE 1996 A randomized trial of visual attention of preterm infants fed docosahexaenoic acid until nine months. Lipids 31:91–97
6. Carlson SE, Werkman SH, Tolley EA 1996 The effect of long chain $n-3$ fatty acid supplementation on visual acuity and growth of preterm infants with and without bronchopulmonary dysplasia. Am J Clin Nutr 63 (in press)
7. Carlson SE, Werkman SH 1996 A randomized trial of visual attention of preterm infants fed docosahexaenoic acid until two months. Lipids 31:85–90
8. Uauy RD, Birch DG, Birch EE, Tyson JE, Hoffman DR 1990 Effect of dietary omega-3 fatty acids on retinal function of very-low-birth-weight neonates. Pediatr Res 28:485–492
9. Birch EE, Birch DG, Hoffman DR, Uauy R 1992 Dietary essential fatty acid supply and visual acuity development. Invest Ophthalmol Vis Sci 33:3242–3253
10. Carlson SE, Cooke R, Werkman S, Peeples J 1989 Docosahexaenoate (DHA) and eicosapentaenoate (EPA) supplementation of preterm (PT) infants: effects on phospholipid DHA and visual acuity. FASEB J 3:A1056
11. Carlson SE, Werkman SH, Rhodes PG, Tolley EA 1993 Visual-acuity development in healthy preterm infants: effect of marine-oil supplementation. Am J Clin Nutr 58:35–42
12. Carlson SE, Werkman SH, Peeples JM, Wilson WM III 1994 Growth and development of premature infants in relation to $\omega 3$ and $\omega 6$ fatty acid status.World Rev Nutr Diet 75:63–69
13. Auestad N, Montalto MB, Wheeler RE, Fitzgerald KR, Hall RT, Neuringer M, Connor WE, Hartmann EE, Taylor JA 1995 Visual acuity, RBC fatty acids and growth in term infants fed formulas with and without long chain polyunsaturated fatty acids. Pediatr Res 37:302A
14. Neuringer M, Fitzgerald KM, Weleber RG, Murphey WH, Giambrone SA, Cibis GW, Arends J 1995 Electroretinograms in four-month-old fullterm human infants fed diets differing in long-chain $n-3$ and $n-6$ fatty acids. Invest Ophthalmol Vis Sci 36:S48
15. Makrides M, Neumann M, Simmer K, Pater J, Gibson R 1995 Are long-chain polyunsaturated fatty acids essential nutrients in infancy? Lancet 453:1463–1468
16. Carlson SE, Ford AJ, Werkman SH, Peeples JM, Koo WWK 1996 Visual acuity and fatty acid status of term infants fed human milk and formulas with and without docosahexaenoate and arachidonate from egg yolk lecithin. Pediatr Res 39:1–7

16 Do LCPUFA influence infant cognitive behaviour?

J.S. FORSYTH and P. WILLATTS

ABBREVIATIONS

AA	arachidonic acid, 20:4 $n-6$
DHA	docosahexaenoic acid, 22:6 $n-3$
EPA	eicosapentaenoic acid, 20:5 $n-3$
IQ	intelligence quotient
LCPUFA	long-chain polyunsaturated fatty acids
MAC	mid-arm circumference
OFC	occipital–frontal circumference

INTRODUCTION

Several observational studies have indicated that breast-fed infants may be advantaged in their subsequent development and intelligence quotient (IQ) compared with formula-fed infants (1–3). This advantage persisted after adjustments were made for social, demographic and educational status, factors which are known to be related to developmental scores. Various constituents of human milk, including hormones, growth promoting factors and specific nutrients such as fatty acids, have been postulated as potential factors which may positively influence neural development. It has also been suggested that the advantage to the infant is not due to the breast milk *per se* but related to the mother's decision to provide milk. A mother sufficiently concerned about her child's welfare to choose breast-feeding may have a greater desire to stimulate and facilitate her baby's developmental progress (4). This issue was addressed by Lucas et al. who showed that preterm infants who received their mother's milk had a substantial advantage in subsequent IQ at age 7.5–8 years compared to those who did not receive mother's milk (5). This difference persisted after adjustment for a wide range of factors which might have confounded this comparison. The study showed a significant dose–response relationship between the proportion of mother's milk consumed and later development, and as the infants received the milk by nasogastric tube the observed advantage could not have been due to interaction between mother and child during the process of suckling. Correspondence following this publication raised the question of whether adjustment for potentially confounding variables such as maternal education attainment and social class adequately eliminated the parental factors (6).

J.G. Bindels, A.C. Goedhart and H.K.A. Visser (eds), Recent Developments in Infant Nutrition, 225–234.

LONG-CHAIN POLYUNSATURATED FATTY ACIDS

Although all of the studies hitherto supporting the view that breast milk confers intellectual advantage to the infant have limitations, there is sufficient evidence to encourage investigation of those factors in milk which may initiate this benefit. Currently, the most favoured explanation involves long-chain polyunsaturated fatty acids (LCPUFA). Sixty percent of the structural material of the brain is lipid, and an important functional component of this lipid is LCPUFA (7,8). These are derived from the essential fatty acids, linoleic acid and α-linolenic acid through elongase and desaturase enzyme systems. It was assumed that the term and preterm infant would be able to synthesize the longer chain fatty acids but there is increasing evidence that, during the first few months of life, preterm infants and young term infants may not be able sufficiently to elongate or desaturate linoleic and linolenic acids to their functional derivatives arachidonic acid (AA) and docosahexaenoic acid (DHA) respectively (9). As a result, there may be a deficiency of LCPUFA at a critical time of brain growth and development.

Carlson et al. demonstrated that preterm infants reared on mother's milk had near normal levels of DHA, whereas those fed on standard milk formula or standard formula to which α-linolenic acid was added continued to show low DHA levels (10). Only when preterm infants received formula supplemented with fish oil containing DHA did their DHA level reach that found in infants fed mother's milk (11). Similar differences have also been demonstrated in term infants, supporting the view that formulas containing only linoleic acid and α-linolenic acid may not meet the full essential fatty acid requirements of healthy term infants (9). The studies on blood lipids are relevant because it has been shown that the fatty acid composition of the erythrocyte membrane reflects that of neural tissue. In support of this, recent reports on the brain lipid composition of infants fed human milk or formula milk have shown a higher concentration of DHA in breast-fed infants (12)

LCPUFA AND COGNITIVE FUNCTION

Evidence that specific fatty acids play a major role in neurodevelopment and cognitive function was first presented over 20 years ago. In a series of studies on learning behaviour in rats, Lamptey and Walker demonstrated that essential fatty acid deficiency, particularly of α-linolenic acid, impaired learning abilities (13–15); from these and other studies the hypothesis of a close correlation between brain lipid composition and learning behaviour has evolved. Following these early studies artificial formulas were supplemented with linoleic and α-linolenic acid on the assumption that longer chain polyunsaturated fatty acids would be synthesised sufficiently in vivo. Recent clinical studies of visual and cortical function cast doubt upon this assumption (16–18). Formula-fed preterm infants have a measurable reduction in neuronal function of the visual pathway compared with breast-fed preterm infants, as measured by electroretinogram (16), visual evoked potentials and behavioural measures of visual acuity (17,18). The physiological parameters correlated with DHA status of the infants and were corrected by the addition of fish oil, which contains DHA, to the infant formula.

Makrides et al. (19) reported a correlation between visual evoked potential acuity of healthy term infants and the level of erythrocyte DHA, suggesting that term infants may be equally susceptible to the altered visual development demonstrated in preterm infants. The longer term effects of these changes are as yet unknown. Studies of the effect of LCPUFA on cortical function are few. Carlson et al. investigated the effect of supplementation and eicosapentaenoic acid (EPA) and DHA on growth and development of very-low-birth-weight infants (750–1350 g) and found that, using the Fagan Infantest, the percentage of time the control infants spent looking at a novel stimulus at 79 weeks of age was highly correlated with plasma arachidonic acid level (20). More recently, Agostoni et al. (21) reported an advantage in Brunet–Lézine psychomotor development scores at 4 months of age in term infants who were fed a formula supplemented with DHA and AA, compared with infants who were fed an identical formula containing no supplement (Chapter 14, this volume).

ASSESSMENT OF COGNITIVE FUNCTION

In assessing cortical function in infants it is vital that the most appropriate tests are applied. Standard tests of infant development such as the Bayley scales (22), when applied to normal populations during the first 2 years of life, show virtually no correlation with intelligence when the same children are tested several years later (23). In recent years, developmental psychologists have turned to other measures of infant ability and developed tests which more closely predict later intelligence. Infant habituation, which is described as a decrease in the attention a baby pays to a stimulus over repeated presentations, has become one of the basic tools for studying perception, memory and learning in infancy (24,25). Measures of looking time during habituation reflect the infant's efficiency at processing and learning information and show a significant correlation with childhood IQ scores. Normal, term infants who habituate with shorter looking times achieve higher childhood IQ scores than infants who habituate with longer looking times (26–29). Infants also demonstrate distinct patterns of habituation organization which remain stable across several months and have been interpreted as representing differences in efficiency at acquiring stimulus information (30–33). Efficient habituation is characterized by a pattern of the longest look occurring early in the sequence of looks, a rapid decline in attention, and a shorter overall looking time. Less efficient habituation is shown by a pattern of the longest look occurring later in the sequence, a slower decline in attention, and a longer overall looking time. Habituation offers a method of assessing efficiency of infant learning and attention, and was therefore chosen as the assessment of infant cognitive behaviour in a study which we have recently completed.

Infant habituation procedure

In our study, infant habituation was assessed using the Infant Control Procedure described by Horowitz et al. (34). This was performed in a room with a low level of background lighting and when the infant was judged to be in a quiet, alert state. The infant sat on the parent's lap in front of a rear projection screen (90 cm high × 70 cm wide) 46 cm from the infant's eyes. Two white screens (123 cm high × 106 cm wide) positioned on either side of the infant concealed two observers who monitored the infant's fixations through small holes in the screens. Presentation of stimuli and recording of fixations was controlled by a computer connected to a random access carousel projector.

The infant's attention was first directed to the centre of the screen with a flashing light. When the infant fixated, the light went off and the habituation stimulus appeared. This was an abstract geometric pattern (random coloured bars) and the same habituation stimulus was used for all infants. The habituation stimulus was 13 cm square and subtended a visual angle of 16°. Infant fixation on the stimulus was recorded by the observer using a hand held button push connected to the computer. Individual 'trials' ended when the infant fixated away from the stimulus for 2 s or longer. The stimulus was presented constantly and remained in view throughout the habituation phase. The baseline duration of fixation was calculated as the average duration of the first two trials, and a floating point procedure was adopted in which the baseline was recalculated if the average of any two consecutive trials increased beyond the baseline. Habituation criterion was reached when the average duration of fixation on any two consecutive trials fell to 50% or less of the baseline. A minimum of four trials was required to achieve habituation criterion and the maximum permitted was 10 trials. Infants who failed to reach criterion within the maximum number of trials were dropped from the study. When habituation criterion was achieved, the stimulus was removed and two novelty preference tests were administered. These tests provided evidence that infants had processed information about the habituation stimulus and could recognize it by showing a preference for fixating the novel stimulus. On the first novelty preference test, infants were shown a pair of stimuli with the habituation stimulus on the left and the novel stimulus (four black diamonds arranged in a square) on the right. After 10 s of fixation had accumulated on either stimulus, the pair was presented in reversed positions until another 10 s fixation had accumulated. This procedure was repeated for the second novelty preference test with two further 10 s presentations of the same stimuli. Five standard measures of habituation performance were obtained. Total fixation time was defined as the sum of the duration of fixation on all trials, up to and including the final two criterion trials. Mean fixation duration was defined as the total time divided by the total number of trials. First fixation duration was defined as the total fixation of duration on the first trial. Peak fixation duration was defined as the longest total duration of fixation on any trial. Number of trials was the total number of trials to reach criterion. Two measures of habituation organization were also obtained. Ordinal position of the peak fixation (31) was calculated as the number of the trial with the longest fixation divided by the total number of trials. Infants were also classified as having their peak fixation either on trial 1 or on a later trial (30). Finally, the infant's response to novelty following habituation was indexed by a novelty preference score. Novelty preference was a proportion score and was defined as the time spent fixating the novel stimulus divided by the total time fixating both the familiar and the novel stimuli. Novelty preference scores were calculated for each of the two novelty preference tests, averaged across the two 10 s periods. Estimates of the reliability of experimenter scoring were obtained by having a second independent observer score a random sample of 29% of all habituation sessions. Interobserver reliability was high on Pearson's *r* calculated for the measures of total fixation time ($r=0.99$), peak fixation duration ($r=0.99$), and ordinal position of the peak fixation ($r=0.98$). The proportion of agreements divided by the number of agreements plus disagreements on whether infants had their peak fixation on trial 1 or on a later trial was 95%.

LCPUFA and infant habituation measures

Forty-eight term infants were randomized to a formula supplemented with LCPUFA or a formula which did not contain LCPUFA but was otherwise identical. A breast-fed group was included for comparison. Infant habituation was assessed at 3 months of age (34). Although there was a trend for total fixation time, mean fixation duration and peak fixation duration to be shorter for the LCPUFA group compared to the no-LCPUFA group, the differences were not significant. Similarly, the differences between the groups in first fixation duration, number of trials to habituation criterion, ordinal position of the peak fixation and the proportion of infants with an early peak fixation were not significant. The breast-fed group had the shortest total fixation time and lowest ordinal position of the peak fixation. The breast-fed group also had means for total fixation time, mean fixation duration and peak fixation duration that were closer to the corresponding means for the LCPUFA group than the no-LCPUFA group. Although these differences were not statistically significant, and this may be related to the small sample size (48 formula-fed infants, 37 breast-fed infants), there is an indication that the no-LCPUFA group was the least efficient at processing information and the reference breast-fed group was the most efficient.

Early and peak fixation

Although there were no overall differences in habituation duration measures between the LCPUFA and no-LCPUFA groups, the position of the peak fixation varied between infants. Further analyses of the groups showed that half the infants in each of the formula groups had an early peak fixation while the remainder had a late peak fixation. Infants with an early peak fixation had significantly higher scores on all anthropometric measures taken at birth, and on crown–heel length at 3 months (Table 1). There were no differences between infants with an early or late peak fixation on habituation measures of total fixation time, mean fixation duration or peak fixation duration, but infants with an early peak fixation had a significantly longer first fixation time ($p<0.001$) and habituated in fewer trials ($p<0.001$). Comparison of the means for infants who had an early peak fixation showed that the LCPUFA and no-LCPUFA groups did not differ significantly on any of these measures. In contrast, comparisons of the means for infants who had a late peak fixation showed that the no-LCPUFA group had a significantly longer total fixation time ($p<0.01$), mean fixation duration ($p<0.01$), and peak fixation duration ($p<0.05$; Table 2). Because the position of the peak fixation was associated with differences on several anthropometric measures, the analyses were repeated with the inclusion of mid upper-arm circumference at birth

Table 1 Anthropometric measurements at birth of infants with early and late peak fixations (mean (SEM))

	Early peak fixation	*Late peak fixation*
Crown–heel length (cm)	50.4 (0.4)	49.0 (0.4)*
Occipital-frontal circumference (cm)	34.7 (0.3)	34.0 (0.2)*
Subscapular skinfold (mm)	4.7 (0.1)	4.1 (0.2)*
Triceps skinfold (cm)	4.9 (0.2)	4.2 (0.2)*
Mid upper arm circumference (cm)	11.3 (0.2)	10.1 (0.2)***
Weight (g)	3479 (68)	3159 (75)**
MAC/OFC ratio	0.33 (0.03)	0.30 (0.03)**

Student's *t*: *$p<0.05$; **$p<0.01$; ***$p<0.001$

as a co-variate. This was the only variable which a multiple regression analysis showed to be significantly related to position of the peak fixation ($p<0.0005$). After adjusting for this co-variate, no changes in the results were found with any of the habituation measures.

Speed of information processing or development of attention control?

The finding that the late peak fixation group who received LCPUFA supplementation had shorter duration scores than infants with late peak fixation who were fed unsupplemented formula is of particular interest because there is a recognized inverse relationship between fixation duration scores and childhood intelligence (26–29). One possible explanation for the shorter durations in the LCPUFA-supplemented group is that accumulation of LCPUFA in cell membranes of the central nervous system may increase the speed at which infants are able to process information. Measures of fixation duration have been interpreted as reflecting speed of information processing (25,28), and infants who show characteristics of faster processing (i.e. shorter looking times) have been shown to have faster reaction times (35). An alternative explanation is that LCPUFA influence the development of attention control. Young infants show considerable difficulty in releasing attention and disengaging from one stimulus to orient to another; there is substantial improvement in disengagement after 3 months of age (36–38). Development of several cortical structures, including parietal cortex, prefrontal cortex, and frontal eye fields, is thought to be involved in this change (36,39). Improved maturation of these structures in infants who received LCPUFA supplement may have enhanced their capacity to release attention after sufficient information had been processed, which in turn would reduce their looking time. Although it is not possible to distinguish between the speed of processing or release of attention hypotheses in our present study, the use of appropriate assessments in future research should help to identify the specific effects of LCPUFA on infant attention.

Efficiency of processing may also depend on an infant's ability to inhibit attention to irrelevant stimuli. Infants are thought to process stimulus information primarily on their peak fixation (28,33,40,41) and it has been suggested that infants who produce

***Table 2* Habituation scores for infants with an early or late peak fixation in the LCPUFA and no-LCPUFA groups**

	Early peak fixation		*Late peak fixation*	
	LCPUFA (n=12)	*No-LCPUFA* (n=12)	*LCPUFA* (n=12)	*No-LCPUFA* (n=12)
Total fixation time (s)[a]	89.5 (50.7, 244.3)	83.4 (34.0, 146.4)	65.6 (46.5, 109.5)	158.5 (89.2, 355.6)**
Mean fixation duration (s)[a]	22.4 (8.9, 61.1)	20.8 (8.5, 35.5)	13.0 (7.6, 21.1)	30.5 (21.2, 51.7)**
Peak fixation duration (s)[a]	60.0 (23.6, 134.5)	57.8 (27.1, 93.5)	34.1 (23.7, 49.4)	102.8 (36.9, 177.6)*
First fixation duration (s)[a]	60.0 (23.6, 134.5)	57.8 (27.1, 93.5)	8.1 (3.8, 26.2)	20.6 (8.6, 36.6)
Mean (SEM) number of trials	4.5 (0.3)	4.1 (0.1)	6.0 (0.5)	5.6 (0.6)

[a]Median (quartiles)
Late peak fixation, LCPUFA vs No-LCPUFA: *$p<0.05$; **$p<0.01$

an early peak fixation are immediately able to sustain attention on the stimulus because they are better able to inhibit attention to other stimuli (31,33,41). In contrast, infants who produce a late peak fixation take longer to sustain attention on the stimulus, possibly because they have poorer inhibitory control and are more easily distracted by other stimuli. Thus, infants who produce a late peak fixation may be slower to begin processing information because they initially engage in 'blank looking' on trials prior to the trial of peak fixation (31).

Late peak fixation and impaired intrauterine growth

There was no corresponding effect of LCPUFA supplementation on habituation performance in infants who had better attention control and produced an early peak fixation. This may be explained by these infants already being efficient at processing information and therefore not benefiting from additional LCPUFA supplementation. These infants were found to weigh more, have a large head circumference, and have a higher mid-arm circumference to head circumference ratio at birth than infants who subsequently showed poorer organization of habituation and produced a late peak fixation. This new finding suggests that factors affecting intrauterine nutrition and growth may also play an important role in the early development of attention and cognitive function. A positive correlation between head circumference and fetal growth with circulating concentrations of LCPUFA has been demonstrated previously (42). Infants who are better nourished in utero may therefore be sufficiently endowed with LCPUFA at birth to sustain adequate development during the first months of life despite the absence of LCPUFA in their feed.

CONCLUSIONS

Our preliminary data suggest that measures of infant habituation are influenced by LCPUFA supplementation. Total fixation times tended to be shorter in infants receiving LCPUFA, although the differences were not statistically significant. However, in those infants with poorer attention control, as defined by a late peak fixation, the fixation times were significantly shorter in infants receiving LCPUFA. Those infants with late peak fixations had significantly lower scores for all growth parameters measured at birth. These findings suggest that term infants with evidence of intrauterine growth retardation may develop more efficient information processing if supplemented with LCPUFA. Further studies are required to determine if LCPUFA supplementation enhances intellectual abilities in later childhood.

ACKNOWLEDGEMENT

The study was approved by the Tayside Committee on Research Ethics. The investigators wish to acknowledge the support they received from Milupa Ltd, UK, and thank Alan Slater and Liz Brown for provision of computer software and advice on the habituation assessment, and Lyn Galbraith for assistance with reliability coding.

REFERENCES

1. Hoefer A, Hardy MC 1929 Later development of breast fed and artificially fed infants. JAMA 92:615–619
2. Taylor B 1977 Breast versus bottle feeding. NZ Med J 85:235–238
3. Rodgers B 1987 Feeding in infancy and later ability and attainment: a longitudinal study. Dev Med Child Neurol 20:421–426
4. Pollock JI 1989 Mother's choice to provide breast milk and developmental outcome. Arch Dis Child 64:763–764
5. Lucas A, Morley R, Cole TJ, Lister G, Leeson-Payne C 1992 Breast milk and subsequent intelligence quotient in children born preterm. Lancet 339:261–264
6. Wright P, Deary IJ 1992 Breast feeding and intelligence. Lancet 339:612
7. Crawford MA, Sinclair AJ 1971 Nutritional influences in the evolution of mammalian brain. In: Elliot K, Knight J (eds) Lipids, Malnutrition and the Developing Brain. Ciba Foundation symposium. Elsevier, Amsterdam, pp 267–292
8. Martinez M 1992 Tissue levels of polyunsaturated fatty acids during early human development. J Pediatr 120:S129–S138
9. Clark KJ, Makrides M, Neumann MA, Gibson RA 1992 Determination of the optimal ratio of linoleic and alpha-linolenic acid in infant formulas. J Pediatr 120:S151–S158
10. Carlson SE, Rhodes PG, Ferguson MG 1986 Docosahexaenoic acid status of preterm infants at birth and following feeding with human milk or formula. Am J Clin Nutr 44:798–804
11. Carlson SE, Rhodes PG, Rao VS, Goldgar DE 1987 Effect of fish oil supplementation on the n-3 fatty acid content of red blood cell membranes in preterm infants. Pediatr Res 21:507–510
12. Farquharson J, Jamieson EC, Abbasi KA, Patrick WJA, Logan RW, Cockburn F 1995 Effect of diet on the fatty acid composition of the major phospholipids of infant cerebral cortex. Arch Dis Child 72:198–203
13. Lamptey MS, Walker BL 1976 A possible essential role for dietary linolenic acid in the development of the young rat. J Nutr 106:86–93
14. Lamptey MS, Walker BL 1978 Physical and neurological development of the progeny of female rats fed an essential fatty acid-deficient diet during pregnancy or lactation. J Nutr 108:351–357
15. Lamptey MS, Walker BL 1978 Learning behaviour and brain lipid composition in rats subjected to essential fatty acid deficiency during gestation, lactation, and growth. J Nutr 108:358–367
16. Uauy RD, Birch DG, Birch EE, Tyson JE, Hoffman DR 1990 Effect of dietary omega-3 fatty acids on retinal function of very-low-birth-weight neonates. Pediatr Res 28:485–492
17. Birch DG, Birch EE, Hoffman DR, Uauy RD 1992 Retinal development in very-low-birth-weight infants fed diets differing in omega-3 fatty acids. Invest Ophthalmol Vis Sci 33: 2365–2376
18. Carlson SE, Werkman SH, Rhodes PG, Tolley EA 1993 Visual-acuity development in healthy preterm infants: effects of marine-oil supplementation. Am J Clin Nutr 58:35–42
19. Makrides M, Simmer K, Goggin M, Gibson RA 1993 Erythrocyte docosahexaenoic acid correlates with the visual response of healthy, term infants. Pediatr Res 33:425–427
20. Carlson SE, Werman SH, Peeples JM, Cooke RJ, Wilson WM 1992 Plasma phospholipid arachidonic acid and growth and development of preterm infants. In: Koletzko B, Okken A, Rey J, Salle B (eds) Recent Advances in Infant Feeding. Georg Thieme Verlag, Stuttgart, pp 22–27
21. Agostoni C, Trojan S, Bellu R, Riva E, Giovannini M 1995 Neurodevelopmental quotient of healthy term infants at 4 months and feeding practice: The role of long-chain polyunsaturated fatty acids. Pediatr Res 38:262–266
22. Bayley N 1969 The Bayley scales of infant development: Birth to two years. The Psychological Corporation, New York
23. McCall RB 1979 The development of intellectual functioning in infancy and the prediction of later IQ. In: Osofsky JD (ed) Handbook of Infant Development. Wiley, Chichester, pp 707–741
24. Bornstein MH 1985 Habituation of attention as a measure of visual information processing in human infants: Summary, systematisation, and synthesis. In: Gottleib G, Krasnegor NA (eds) Measurement of Audition and Vision in the First Year of Postnatal Life: A Methodological Overview. Ablex, Norwood, NJ, pp 253–300
25. Colombo J, Mitchell DW 1990 Individual differences in early visual attention: Fixation time and information processing. In: Colombo J, Fagen JW (eds) Individual Differences in Infancy. Lawrence Erlbaum Associates, Hillsdale, NJ, pp 193–228

26. Bornstein MH, Sigman MD 1986 Continuity in mental development from infancy. Child Dev 57:251–274
27. Rose D, Slater A, Perry H 1986 Prediction of childhood intelligence from habituation in early infancy. Intelligence 10:251–263
28. Colombo J 1993 Infant Cognition: Predicting Later Intellectual Functioning. Sage, London
29. McCall RB, Carriger MS 1993 A meta-analysis of infant habituation and recognition memory performance as predictors of later IQ. Child Dev 64:57–79
30. Bornstein MH, Benasich AA 1986 Infant habituation: Assessments of individual differences and short term reliability at five months. Child Dev 57:87–99
31. Lecuyer R 1989 Habituation and attention, novelty and cognition: where is the continuity? Hum Dev 32:148–157
32. Mayes L, Kessen W 1989 Maturational changes in measures of habituation. Infant Beh Dev 12:437–450
33. McCall RB 1994 What process mediates predictions of childhood IQ from infant habituation and recognition memory? Speculations on the role of inhibition and the rate of information processing. Intelligence 18:107–125
34. Horowitz FD, Paden LY, Bhana K, Self PA 1972 An infant control procedure for the study of infant visual fixations. Dev Psychol 7:90
35. Jacobson SW, Jacobson JL, O'Neill JM, Padgett RJ, Frankowski JJ, Bihun JT 1991 Visual expectation and dimensions of infant information processing. Child Dev 63:711–724
36. Johnson MH, Posner MI, Rothbart MK 1991 Components of visual orienting in early infancy: Contingency learning, anticipatory looking, and disengaging. J Cogn Neurosci 3:335–343
37. Hood BM, Atkinson J 1993 Disengaging visual attention in the infant and adult. Infant Behav Dev 16:405–422
38. Hood BM 1995 Shifts of visual attention in the human infant: A neuroscientific approach. Adv Infancy Res 9:163–216
39. Johnson MH 1995 The development of visual attention: A cognitive neuroscience perspective. In: Gazzaniga MS (ed) The Cognitive Neurosciences. MIT Press, Cambridge, MA, pp 735–747
40. Colombo J, Mitchell DW, O'Brien M, Horowitz FD 1987 The stability of visual habituation during the first year of life. Child Dev 57:474–488
41. McCall RB 1979 Individual differences in the pattern of habituation at 5 and 10 months of age. Dev Psychol 15:559–569
42. Leaf AA, Leighfield MJ, Costeloe KL, Crawford MA 1992 Long chain polyunsaturated fatty acids and fetal growth. Early Hum Dev 30:183–191

17 Supply and metabolism of LCPUFA in term infants

B. KOLETZKO, T. DECSI, H. DEMMELMAIR and T. SAUERWALD

ABBREVIATIONS

AA	arachidonic acid, 20:4 $n-6$
ALA	α-linolenic acid, 18:3 $n-3$
DGLA	dihomo γ-linolenic acid, 20:3 $n-6$
DHA	docosahexaenoic acid, 22:6 $n-3$
GC-C-IRMS	gas chromatography combustion isotope ratio mass spectrometry
LA	linoleic acid, 18:2 $n-6$
LCPUFA	long-chain polyunsaturated fatty acids
PDB	PeeDeeBelemnite carbonate
RBC	red blood cell

INTRODUCTION

Long-chain polyunsaturated fatty acids with 20 and 22 carbon atoms (LCPUFA) are essential for early human growth and development (1). The available evidence strongly suggests that a dietary supply of LCPUFA is desirable for preterm infants (2). The capacity for endogenous synthesis of these long-chain metabolites seems limited in low birth weight infants, and the feeding of preterm infants with formulas devoid of LCPUFA results in rapid LCPUFA depletion of plasma and tissue lipids (3,4), which has been associated with reduced visual acuity during the first postnatal months and with altered Bayley mental developmental index scores at one year of age (see Chapter 15 by Dr Carlson, this volume). The degree to which a dietary source of preformed LCPUFA is also essential for term infants is an area of active investigation. As in preterm infants, levels of docosahexaenoic acid (DHA; 22:6 $n-3$) in plasma and red blood cell (RBC) phospholipids of term, formula-fed infants are lower than those of breast-fed infants. Controversy exists over whether the absence of AA from the diet affects plasma and erythrocyte arachidonic acid (AA; 20:4 $n-6$) levels in term infants. Some studies report that AA levels in plasma and RBC are very stable, while others report decreased levels in infants not receiving dietary AA (reviewed in Ref. 5).

We studied the effects of feeding human milk or a standard term formula on the fatty acid composition of plasma lipids in healthy infants during the first 2 months of life (5). Formula-fed infants had significantly lower median plasma phospholipid values of AA at 2 and 4 weeks and of DHA at 4 and 8 weeks of age than breast-fed infants. Thus, term infants fed formula without LCPUFA were unable to match the

J.G. Bindels, A.C. Goedhart and H.K.A. Visser (eds), Recent Developments in Infant Nutrition, 235–241.

$n-6$ and $n-3$ LCPUFA status of breast-fed infants until at least 2 months after birth. This suggests that in term infants endogenous LCPUFA synthesis may be insufficient to meet the needs for tissue growth.

It has recently become possible to measure the endogenous synthesis of LCPUFA by using stable isotope techniques. For such studies, the essential fatty acid precursors linoleic acid (LA; 18:2 $n-6$) and α-linolenic acid (ALA; 18:3 $n-3$) with an increased content of the stable, non-radioactive carbon isotope ^{13}C are administered and the ^{13}C-enrichment of their long-chain derivatives in plasma is measured by mass spectrometric methods. The application of stable isotope techniques for such metabolic studies in vivo is safe for the participating subjects (6). To study the conversion of the parent fatty acids LA and ALA to their longer chain metabolites, either fatty acids with an artificially elevated ^{13}C content or fatty acids which are naturally enriched in ^{13}C can be used. Several plants from arid climates classified as C_4 plants, such as corn and sugar cane, differ in their metabolic pathways for carbon dioxide fixation and thus have a higher natural ^{13}C content than C_3 plants. In our first study of endogenous LCPUFA synthesis in newborn infants, we used such naturally ^{13}C-enriched LA from corn oil (8). In our second, ongoing study we used uniformly ^{13}C-enriched tracer materials from a biotechnological source to measure conversion of the precursor essential fatty acids. Data from these two studies are summarized in this paper.

We also present results of a study in term infants investigating the effects of supplementation with LCPUFA from egg yolk phospholipids on growth, fatty acid composition of plasma lipid classes, and plasma retinol and α-tocopherol levels in term infants (7). Following these biochemical evaluations, we began an evaluation of possible functional effects in the Munich intervention trial, in which the effect of LCPUFA supplementation on visual recognition memory of term infants is being tested.

METHODS

Stable isotope studies

In our first stable isotope study, we utilized the natural variation of ^{13}C content in four term infants with newly diagnosed phenylketonuria (PKU) at a mean age of 18 days. Three infants had been fully breast fed, and one infant had been formula fed. Following the diagnosis of PKU, all infants were changed to a phenylalanine-free diet with corn oil as the only fat source (8). Compared with oils derived from C_3 plants, corn oil LA is naturally enriched in ^{13}C. The infants consumed 600–800 ml/day of the phenylalanine-free diet (3.5 g fat/100 ml) providing 12–16 g of linoleic acid per day. The ^{13}C content of LA and AA were determined in 0.25–0.5 ml plasma before and 3–4 days after the diet change with gas chromatography combustion isotope ratio mass spectrometry (GC-C-IRMS). ^{13}C contents of substances were expressed as $\delta^{13}C$ values relative to the international standard PDB (*PeeDeeBelemnite* carbonate). These $\delta^{13}C$ values are calculated by dividing the difference of the $^{13}C/^{12}C$ ratio of the sample and the $^{13}C/^{12}C$ ratio of the PDB standard by the ratio of the standard multiplied by 1000. In order to determine $\delta^{13}C$ values relative to PDB for the fatty acid methyl esters, samples of reference CO_2 with a known $\delta^{13}C$ value were introduced at appropriate time points during chromatography. A simplified isotopic balance calculation was used to assess the contributions of dietary LA to serum LA and of endogenously synthesised AA to serum AA, respectively (see Ref. 8).

In a further, ongoing stable isotope study, we used uniformly ^{13}C-labelled fatty acid tracers with high ^{13}C-enrichment (Martek Biosciences Corp., Columbia, MD, USA) to determine endogenous AA and DHA synthesis in term infants. Uniformly ^{13}C-enriched LA and ALA were administered to term infants at various ages, and enrichment of serum long-chain metabolites was measured by GC-C-IRMS.

LCPUFA supplementation study: biochemical effects

In this trial, healthy infants born at term were randomly assigned to receive either a standard term formula ($n=10$) or the same formula enriched with egg lipids and evening primrose oil ($n=12$) to achieve the LCPUFA levels in their diets typically found in human milk (DHA 0.3% and AA 0.5% of total fatty acids). Anthropometric measurements were carried out and fatty acid compositions of plasma lipids as well as plasma retinol and α-tocopherol levels were assessed at 5 days and at 1, 2, 3, and 4 months of age.

LCPUFA supplementation study: functional effects

In the Munich intervention trial we are looking at the effect of LCPUFA supplementation on functional outcome measures in both term and preterm infants. In the term infant study, among other tests the Fagan Infantest is used to assess visual recognition memory (novelty preference) at 9 months of age of healthy term infants fed one of human milk, a standard term formula, or the same formula supplemented with $n-6$ and $n-3$ LCPUFA. The Fagan test is based on the infant's ability to recognize visual stimuli and, consequently, the infant's differential fixation to novel over previously shown pictures. Recognition is tested by presenting the infant with one picture to study for a short period of time, after which the studied picture is paired with a new picture, which constitutes a novelty problem. The mean proportion of total looking time paid to the novel picture over pairings constitutes the infant's novelty preference score.

RESULTS

Stable isotope studies

Baseline δ^{13}C values were -31.5 ± 1.1% (mean$\pm$SD) for serum LA and -30.1 ± 1.2% for AA. The ^{13}C-enriched phenylalanine-free diet induced changes of δ^{13}C values over baseline on days 1 and 4 of 8.9 ± 1.0% and 12.7 ± 0.7% for LA, and of 0.5 ± 0.7% and 2.7 ± 0.7% for AA, respectively (Figure 1). Using the simplified isotopic balance calculation, it was estimated that the contribution of dietary LA to serum LA was about 61% on day 1 and 84% on day 4. The contribution of endogenously synthesized AA to serum AA was calculated to be about 23% at day 4 (Table 1). This would mean

***Figure 1** Time course of the $\delta^{13}C$ values of linoleic acid (LA; C18:2n−6, solid line), arachidonic acid (AA; C20:4n−6, dashed line), and dihomo-gamma-linolenic acid (DGLA; C20:3n−6, dotted line) methyl esters in serum lipids after change of the dietary fat to corn oil at day 0. $\delta^{13}C$ values of the methyl esters at day 0: LA, −31.5 ± 0.6‰; AA, −30.1 ± 0.6‰; DGLA, −31.4 ± 0.9‰*

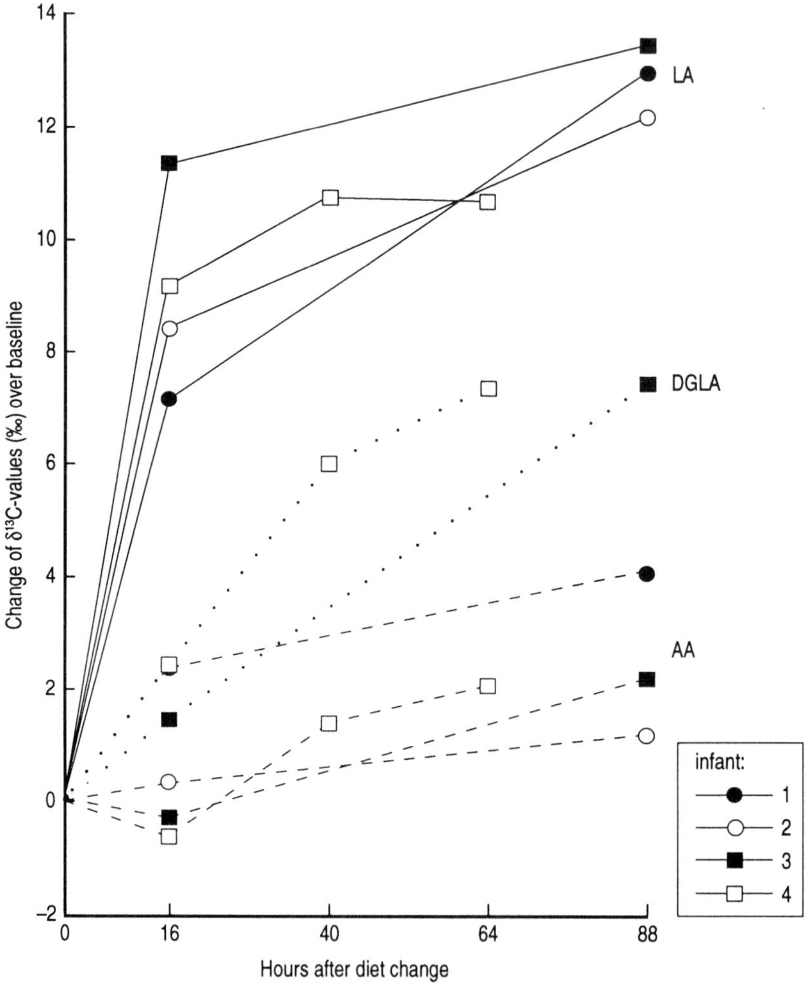

that the daily renewal of serum AA from endogenous synthesis is about 6% of the total serum AA pool.

The limited data available so far from the ongoing stable isotope study in term infants, in which we use uniformly labelled LA and ALA, suggest that the ability of

Table 1 Estimated contribution of dietary LA (%) to total plasma LA on days 1 and 4 and the contribution of AA synthesized from plasma LA to total plasma AA on days 1 and 4, based on carbon isotope balances

	Proportion (%) of serum LA contributed by dietary LA		*Proportion (%) of serum AA contributed by endogenous LA conversion*	
	Day 1	*Day 4*	*Day 1*	*Day 4*
Infant 1	46	84	24	36
Infant 2	52	77	0	17
Infant 3	79	95	0	21
Infant 4[a]	66	79	0	19
Average	60.8	83.8	6.0	23.3
SEM	7.4	4.0	6.0	4.3

[a]Infant 4 was investigated on day 3 instead of day 4

the term infant to convert the parent fatty acids to their long-chain derivatives does not increase during the first year of life. Relative rates of AA synthesis did not differ between a 2-week-old neonate and an infant 12 months of age, whereas the relative rate of DHA synthesis appeared to be almost four-fold higher at 2 weeks than at 12 months.

LCPUFA supplementation study: biochemical effects

Gains in weight, length and head circumference did not differ between the LCPUFA-supplemented and the control groups throughout the study period. Between 1 and 4 months of age relative plasma phospholipid levels of AA and DHA were significantly higher in the LCPUFA-supplemented infants than in controls (Table 2). Plasma retinol and α-tocopherol concentrations did not differ between the two groups during the study period.

LCPUFA supplementation study: functional effects

At present, about 100 term infants are enrolled in this study. Interim data on the Fagan test scores at 9 months of age are available from 61 infants. Novelty preference scores of the infants fed the LCPUFA-enriched formula were similar to those of the breast-fed infants, whereas the test scores of the infants fed the control formula were significantly lower (data not shown).

CONCLUSIONS

Although term neonates are able to actively convert LA and ALA to their long-chain derivatives, feeding term infants formulas without dietary LCPUFA results in a marked depletion of both AA and DHA in plasma lipids. Feeding term infants a formula enriched with AA and DHA from egg phospholipids completely normalized

***Table 2** Fatty acid content (mg/l) and composition (%wt/wt) of plasma phospholipids in infants receiving conventional formula (F, n = 10) or a formula enriched with long-chain polyunsaturated fatty acids (LCP-F, n = 12) (median (interquartile range)).*

		Age		
		5 days	*1 month*	*4 months*
Sum (mg/l)	F	1106 (264)	1239 (199)	1197 (247)
	LCP-F	1020 (239)	1295 (168)	1351 (121)
Saturated fatty acids				
Total	F	47.38 (1.90)	45.18 (1.14)**	45.85 (0.64)
	LCP-F	46.46 (0.74)	46.36 (1.02)	45.14 (1.19)
Monounsaturated fatty acids				
Total	F	15.53 (0.79)	14.88 (0.67)**	14.39 (2.08)
	LCP-F	15.76 (1.04)^{c}	13.85 (1.04)^{c}	13.66 (0.28)
n−6 polyunsaturated fatty acids				
C18:2*n*−6	F	14.28 (2.78)^{c}	25.22 (2.17)^{c**}	25.78 (3.91)
	LC-F	13.67 (1.87)^{d}	22.39 (0.97)^{d}	24.43 (1.63)
C18:3*n*−6	F	ND	0.13 (0.03)	0.09 (0.08)
	LCP-F	0.08 (0.05)	0.12 (0.02)	0.12 (0.02)
C20:2*n*−6	F	0.22 (0.05)^{c}	0.42 (0.05)^{c**}	0.33 (0.02)
	LCP-F	0.26 (0.05)	0.28 (0.04)	0.28 (0.03)
C20:3*n*−6	F	2.88 (0.61)	3.06 (0.46)	2.92 (0.74)
	LCP-F	2.91 (0.31)	2.85 (0.43)	2.74 (0.11)
C20:4*n*−6	F	14.23 (1.84)^{c}	6.97 (1.27)^{c***}	6.61 (1.01)**
	LCP-F	15.05 (2.55)^{c}	9.67 (0.82)^{ac}	8.68 (0.55)^{a}
C22:4*n*−6	F	0.56 (0.08)	0.45 (0.09)	0.48 (0.11)
	LCP-F	0.50 (0.04)^{d}	0.37 (0.02)^{d}	0.41 (0.04)
n−6 LCP	F	17.66 (1.55)^{c}	11.23 (1.78)^{c**}	10.95 (1.47)**
	LCP-F	18.83 (2.00)^{d}	13.52 (0.57)^{cd}	12.47 (0.52)^{c}
n−6 PUFA	F	33.25 (1.41)^{c}	37.01 (1.27)^{c*}	36.83 (2.33)
	LCP-F	33.55 (1.05)^{c}	35.91 (0.61)^{c}	37.22 (1.58)
n−3 polyunsaturated fatty acids				
C18:3*n*−3	F	ND	0.02 (0.05)	0.06 (0.08)
	LCP-F	ND	0.06 (0.02)	0.10 (0.03)
C20:5*n*−3	F	0.15 (0.07)	0.12 (0.02)**	0.12 (0.02)**
	LCP-F	0.16 (0.03)	0.16 (0.02)	0.15 (0.03)
C22:5*n*−3	F	0.22 (0.05)^{c}	0.30 (0.05)^{c}	0.44 (0.08)
	LCP-F	0.19 (0.05)^{c}	0.28 (0.06)^{ac}	0.47 (0.10)
C22:6*n*−3	F	3.04 (0.42)^{a}	1.57 (0.31)^{ab***}	0.89 (0.35)**
	LCP-F	3.33 (0.82)	2.91 (0.47)	2.90 (0.40)
n−3 LCP	F	3.45 (0.47)^{c}	1.98 (0.29)^{c***}	1.55 (0.43)
	LCP-F	3.54 (0.85)	3.41 (0.49)	3.53 (0.53)
N-3 PUFA	F	3.46 (0.49)^{c}	2.02 (0.41)^{c***}	1.61 (0.52)**
	LCP-F	3.56 (0.79)	3.50 (0.48)	3.76 (0.51)

Effects of diet (*$p<0.05$, **$p<0.01$, ***$p<0.001$) and age (identical superscripts indicate differences $^{a,b}p<0.05$, $^{c}p<0.01$, $^{d}p<0.001$).

ND = Not detectable

plasma LCPUFA levels, while no appreciable adverse effects occurred and growth and lipid soluble antioxidant status were not altered. Preliminary results show that healthy, term infants fed this LCPUFA-supplemented formula during the first months of life have improved novelty preference scores at 9 months of age compared with those fed control formula. Thus, LCPUFA supplementation of infant formula has a potential benefit for the infant. In view of these results, further detailed clinical studies are desirable to assess the optimum levels and duration of dietary LCPUFA supply in infancy and their possible longer term functional effects (9).

ACKNOWLEDGEMENTS

These studies have been supported in part by Bundesministerium für Bildung und Forschung, Bonn, Germany (07 ERG 08/1), Deutsche Forschungsgemeinschaft, Bonn, Germany (Ko 912/4-3) and Milupa GmbH, Friedrichsdorf, Germany.

REFERENCES

1. Decsi T, Koletzko B 1994 Polyunsaturated fatty acids in infant nutrition. Acta Paediatr Suppl 395: 31–37
2. Aggett P, Haschke F, Heine W, Hernell O, Koletzko B, Launiala K, Rey J, Rubino A, Schöch G, Senterre J, Tormo R 1991 Comment on the content and composition of lipids in infant formulas. Acta Paediatr Scand 80:887–896
3. Koletzko B, Schmidt E, Bremer HJ, Haug M, Harzer G 1989 Effects of dietary long-chain polyunsaturated fatty acids on the essential fatty acid status of premature infants. Eur J Pediatr 148: 669–675
4. Koletzko B, Edenhofer S, Lipowsky G, Reinhardt D 1995 Effects of a low birthweight infant formula containing docosahexaenoic and arachidonic acids at human milk levels. J Pediatr Gastroenterol Nutr 21:200–208
5. Decsi T, Thiel I, Koletzko B 1995 Essential fatty acids in full term infants fed breast milk or formula. Arch Dis Child 72:F23–F28
6. Sauerwald T, Demmelmair H, Koletzko B 1996 Zur Unbedenklichkeit von stabilen Isotopen in der pädiatrischen Forschung und Diagnostik. Monatsschr Kinderheilkd (in press)
7. Decsi T, Koletzko B 1995 Growth, fatty acid composition of plasma lipid classes, and plasma retinol and α-tocopherol concentrations in full-term infants fed formula enriched with ω-6 and ω-3 long-chain polyunsaturated fatty acids. Acta Paediatr 84:725–732
8. Demmelmair H, von Schenck U, Behrendt E, Sauerwald T, Koletzko B 1995 Estimation of arachidonic acid synthesis in full term neonates using natural variation of ^{13}C content. J Pediatr Gastroenterol Nutr 21:31–36
9. Koletzko B, Decsi T, Demmelmair H 1996 Arachidonic acid supply and metabolism in human infants born at full term. Lipids 31:79–83

Discussion Session 3

CHAIR: SIR DAVID HULL

CHAPTER 10

Elia One of the few positive effects of $n-3$ fatty acid supplementation that you observed was a decrease in look duration. What is the significance of this decrease in look duration? Is it a good thing or a bad thing? One could argue that as you look longer, you look at more details and interpret things better from the environment. On the other hand, you can argue that a longer look duration means that you need more time to process the information.

Reisbick So far we think it is a good thing. Measures of look duration have been shown to reflect the infant's efficiency at processing information: short lookers need less time to make a novelty preference judgement. Furthermore, look duration in infancy appears to correlate negatively with childhood IQ scores and with some aspects of language development.

Hull Were the earlier results that you presented indicating that the animals reared under the experimental situation showed changes in behaviour similar to those found with long-chain polyunsaturated fatty acid deficiency? Could this mean that your markers of deficiency are also expressed in part if the animals are reared in an unusual situation?

Reisbick Yes, both we and other people who have raised animals in nurseries have found that they are more stereotyped than animals that are raised in groups.

Hull Could the happiness of the baby monkeys therefore be as important as their intake of LCPUFA in the test that you were using?

Reisbick We do not know.

CHAPTER 11

Elia The fatty acid composition of membranes is considered to be important functionally. In the results that you presented about the fatty acid compositions of various tissues, did you attempt to distinguish between membrane lipids and non-membrane lipids? There could be major differences, particularly for tissues that store lipid. Second, you showed that the preterm infant is able to convert linoleic acid and α-linolenic acid to their long-chain derivatives. How does this synthesizing capacity compare with term infants?

Carnielli It is extremely difficult to differentiate between lipids incorporated in membranes and lipids that are free in the cytosol or derived from membrane

J.G. Bindels, A.C. Goedhart and H.K.A. Visser (eds), Recent Developments in Infant Nutrition, 242–247.

degradation around death. I cannot give you any in vivo data on the differences in fatty acid composition between membrane and non-membrane lipids. I question whether other investigators have been able to really differentiate between the two, other than assessing the fatty acid composition of phospholipids and assuming that phospholipids are membranes. With respect to the difference in endogenous synthesis of LCPUFA between term infants and preterm infants, I just presented the data available from preterm infants, and these data are encouraging in that the biosynthetic capacity of these infants is considerable. I expect that in term infants this capacity would be equal or even better.

Koletzko What is the adequate level of LCPUFA in formula for preterm infants? In the formula you used in your study, you chose to aim for levels as found in preterm human milk at day 7, if I remember correctly: that is not different from term milk. Studies from Finland and from our group show that during the first months of life preterm and term milk are equal in LCPUFA composition. If you look at your plasma data, levels of arachidonic acid are clearly lower in the infants fed the LCPUFA-enriched formula than in those fed human milk at day 7 and they are higher at day 28. Can you draw any conclusions on the bio-availability of the two different LCPUFA sources from that?

Carnielli The plasma LCPUFA concentrations we found in preterm infants receiving LCPUFA-supplemented formula were still about 50% lower than fetal LCPUFA levels. Both the percentages and absolute levels of LCPUFA were significantly higher in utero. In view of this, I would not worry too much about a 10–20% difference in LCPUFA content between formulas. Second, I would not really focus on the relatively low plasma levels of LCPUFA at day 7, because all our infants received parenteral nutrition during the first postnatal week, which may have affected plasma LCPUFA levels at day 7. It is therefore more reliable to look at plasma LCPUFA levels at day 28, when complete enteral intake was achieved and the infants were stabilized.

Koletzko You showed that the relative LCPUFA levels in plasma triglycerides decreased postnatally. It is known, however, that plasma triglyceride levels increase after birth. Did the absolute amounts of LCPUFA in plasma triglycerides also fall?

Carnielli Yes, the absolute amounts of LCPUFA in plasma triglycerides also decreased. We did not expect this, because we all know that plasma triglyceride and phospholipid levels are low in cord blood and rise postnatally. Once we corrected for this factor, however, the absolute DHA concentrations in plasma triglycerides were higher in cord blood than at day 28.

Koletzko Did your fat balance study provide any data on the bioavailability of the LCPUFA?

Carnielli We have fat balance data comparing different types of formulation and different sources of LCPUFA. It appears that in spite of fears about fat malabsorption in the preterm infant, a fair amount of fat is absorbed. Of course there were differences between the different sources, but they were not so dramatic that this could change the picture completely.

CHAPTER 15

Cooke How can you explain the reversal of findings between your first and your second study if you look at the Bayley scores of the control and treatment groups?

Carlson It is not a reversal. We performed two randomized trials and in the second trial 40% of the children had bronchopulmonary dysplasia, which does influence Bayley scores. Despite that, the Bayley scores of the control group of the second study were basically identical to those of both the control and the DHA-supplemented group of our first study. The only group with significantly higher Bayley scores was the DHA-supplemented group of the second trial. I speculate that the increase in Bayley MDI seen in the second trial was due to the simultaneous improvement of DHA status and nutritional status. In this second trial infants were discharged on a nutrient-enriched preterm formula, and these infants had normal plasma levels of vitamin A, while in the first trial many infants were vitamin A deficient. Furthermore, the infants receiving the preterm formula after discharge had normal plasma zinc concentrations, whereas it has been shown that preterm infants fed term formula after hospital discharge have low plasma zinc levels.

Cooke I agree that there are many other things that might effect the nutritional status of babies. Until we can see the larger picture, which I do not think we can yet, we will not be able to pin down the fine details. One can argue that prematurely born babies are nutritionally at risk, and that we should be adding something to obviate this risk. I do not accept that. I think we ought to have data that show that something is both effective and safe. The absence of information does not mean that something is safe.

CHAPTER 17

Carlson I was surprised to see that your control group had low Fagan test scores, whereas the new Fagan test, if anything, gives high scores. Did you use the new or the old version of the Fagan test? If you used the new Fagan test, these test scores are really abnormally low, unless these were developmentally delayed infants.

Koletzko We followed your advice to wait for the new version of the Fagan test, and we studied healthy term infants, who were randomized to the two different formulas and tested blindly.

GENERAL DISCUSSION

Hull Could I ask a general question to the whole group? Many of you have talked about premature babies. There is clearly a distinction between babies born under 28 weeks, those born between 28 and 32 weeks, and those born between 33 and 36 weeks of gestation. It may be that there is argument to supplement LCPUFA for one group and not for the other. What is the mean gestational age of the babies that you call premature and are you confident that your conclusions extend right the way down to the lowest border of viability?

Uauy We have only studied relatively healthy babies with a birth weight between 1000 and 1500 g and a mean gestational age of 30.5 weeks. One reason why we selected preterm infants is that these infants are more vulnerable to LCPUFA deficiency, as they did not receive the third trimester intrauterine supply of LCPUFA. I would expect that infants with a birth weight below 1000 g would be progressively more vulnerable to the effects of LCPUFA deficiency.

Hull Would these infants be even more vulnerable to deficiency if they were very sick and/or if they were given intravenous nutrition and not receiving oral feed for the first 2–3 weeks?

Uauy In our studies we only included infants who received at least 70 kcal/kg body weight of enteral nutrition by day 10. We also excluded infants with congenital infection or malformation, retinopathy of prematurity, grade 3 or 4 intraventricular haemorrhage, or respirator treatment for more that 7 days. We thus studied a select, relatively healthy group of preterm infants. About 40% of the preterm babies with birth weights 1000–1500 g born in Dallas at the time met our inclusion criteria.

Carlson In our first trial we studied a select healthy group of appropriate for gestational age infants with no bronchopulmonary dysplasia (BPD) or intraventricular haemorrhage. In our second trial we did allow children with BPD to enter into the study. These infants constituted about 40% of the children we studied, and that is the incidence that is typical in our nursery. The average gestational age of the infants we studied was around 29 weeks.

Sauer I think we have to make a distinction between term, preterm and perhaps very preterm infants. The available evidence suggests that preterm infants need dietary LCPUFA. Are the levels of AA and DHA we are giving to these infants sufficient? Looking at Carnielli's data one might argue that we have to give more. I doubt whether a dietary supply of LCPUFA is essential for the term infant for two reasons. First, there is a wide variation in brain LCPUFA levels, and there is even an overlap between the levels of breast-fed and formula-fed infants. Therefore, I am not impressed by the data on the LCPUFA content of cortex phospholipids. Second, when DHA really is an important long-chain polyunsaturated fatty acid, you should see nice correlation between DHA level and developmental quotient, Fagan test score or whatever. When Agostoni plotted erythrocyte phospholipid DHA against developmental quotient at 4 months of age he found a significant correlation between the two. However, there was a wide scatter, and with such a scatter you can always find some kind of correlation. I really doubt whether this correlation has any significant meaning. To summarize, LCPUFA supplementation is justified for very preterm infants but we are probably not giving sufficiently high levels, and I am not convinced that term infants require dietary LCPUFA.

Hull Can someone tell whether babies will benefit from dietary LCPUFA by just looking at them?

Forsyth No, the mothers did not tell us that their babies had poor attention control, but from the habituation assessment it became obvious that the somewhat growth retarded babies benefited from LCPUFA supplementation. I agree with Professor Sauer and Professor Koletzko that we should monitor these babies and assess them at a later stage, particularly around the age of 3–4 years, to see if there are any longer term effects of LCPUFA supplementation.

Carnielli I agree with the first statement of Professor Sauer that dietary LCPUFA are essential for preterm infants. We have evidence to suggest that these infants may become deficient without dietary LCPUFA, but I would not underestimate the endogenous synthesis of LCPUFA in these infants. It would be interesting if we could assess whether there are interindividual differences in biosynthetic capacity. Furthermore, we are not yet able to assess how much of the dietary DHA is taken up by the brain. The tremendous differences in brain DHA content at birth among infants may be related to differences in the maternal intake of LCPUFA, in placental synthesis and/or transport, and in biosynthetic capacity.

Reisbick Recently, an interesting paper was published in the American Journal of Clinical Nutrition reporting that a proportion of hyperactive children have low $n-3$ essential fatty acid plasma levels. The investigators hypothesize that even in adolescence and adulthood there may be rather large interindividual differences in the ability to synthesize LCPUFA, and that a fraction of hyperactive children may be unable sufficiently to produce the long-chain derivatives, which may explain their hyperactivity.

Battaglia As someone who is very interested in brain development my conclusion, having heard all these presentations, is that if you have to work this hard to find differences even in preterm infants, I am not that impressed. A clear understanding of the importance of dietary LCPUFA for neonates will depend on future research. I would further like to emphasize that the more variables you measure, the more differences you will find in hyperactive children or in any other group. You have to rigorously control for this issue of multiple comparisons. The third point I would like to make, addresses the two people that used stable isotopes and tried to do substrate/product analyses. You cannot do that by looking at the enrichment of the product alone: this will result in large errors in interpretation.

Koletzko I would like to hear the reasoning why you cannot do this. To come back to your previous point and that of Professor Sauer, would we really expect that a complicated outcome variable as neurological function would be closely linearly related to one nutrient? Would you believe a study with a correlation of 0.9 of one nutrient to neurological outcome? I would not and I think that we can only expect nutrients to have a fine-tuning effect on neurological outcome variables which are mostly determined by other variables such as genetic factors. Furthermore, I am not convinced that a variation of a biochemical parameter excludes a biological meaning. If you extrapolate that reasoning then you would conclude that proteins and amino acids are probably not important biologically, because there is such variation in plasma levels of amino acids. You would also conclude that you would not need to supplement vitamin D or vitamin K because most non-supplemented infants will not have marked deficiency symptoms. I am not convinced that variation in biological parameters excludes biological effects, and even if only a fraction of the infants would benefit from LCPUFA supplementation it still justifies looking at these parameters very closely.

Hull I think we would all accept that. I would like to ask all experts whether they think that dietary LCPUFA are essential for preterm and term infants. You could answer with yes, no, or neutral.

Reisbick Neutral.

Agostoni For DHA in preterm infants, I would say 'ycs'. We need more data on eicosanoid production from arachidonic acid and on the balance of eicosanoids in preterm babies. For the term infant, I would say 'neutral'.

Carlson Since I have performed two of the randomized trials in preterm infants and we found many significant effects, I would be crazy if I would say 'no' to supplementation for the preterm infant. I believe that preterm babies do need dietary LCPUFA. We have observed one negative effect on growth, and we now know how to prevent that. As far as term infants are concerned, I am still neutral, but I am starting to be convinced. Certainly I have not seen any negative effects of LCPUFA supplementation that I believe.

Forsyth I would say 'yes' for preterm infants, but we do not yet know whether we are providing these infants with the right quantities of LCPUFA. For term infants I think we need to hold off at the moment. We need further studies and we also need to look at how we are assessing these babies to determine the effects of LCPUFA.

Makrides For preterm infants, the answer is probably 'yes'. For term infants I would say 'neutral'. We certainly need more studies to account for confounding factors and we should rationalize the studies with assessment, methodologies and numbers.

Koletzko We have indications for beneficial effects of LCPUFA supplementation both in preterm and in term infants. We have much more data in the preterm infant, however, so we need more studies in term babies.

Uauy I say 'yes' for the preterm infant and I think we need more studies in term babies before we can make official recommendations.

Carnielli I would say 'yes' for preterm infants, and for term infants more studies need to be undertaken, which should also look at the maternal diet during pregnancy and at the LCPUFA status of the infants at birth.

Session 4
Selected nutrients in infancy: essentiality and deficiency

18 Is dietary cholesterol beneficial to the infant?

D.L. HACHEY, W.G. POND and W.W. WONG

ABBREVIATIONS

BA	bile acid
CEH	cholesterol ester hydrolase
FSR	fractional synthesis rate
LDL-R	low density lipoprotein receptor
MFGM	milk fat globule membrane

INTRODUCTION

Cholesterol is essential to life: it is required by all eukaryotes and it is produced in sufficient amounts to meet the functional needs of the animals. Rarely do animals depend solely on exogenous sources for such a critical component of life (1). Cholesterol fulfils two basic roles in physiology: it is a structural element in the lipid layer of cell membranes, and is also a substrate for the production of steroid hormones, vitamin D and bile acids (BA). Cholesterol is a key component of both the plasma membrane and the outer mitochondrial membrane of all cells. The brain and central nervous system are especially rich in cholesterol, which forms the myelin sheaths that insulate nerve fibres. In the human, brain myelination accelerates rapidly during the third trimester and continues at a high level during the first 2 years of life. The teleological argument is often invoked that because human milk is high in cholesterol, it must be beneficial to the infant. This notion has proven remarkably resistant to experimental validation. The high content of milk cholesterol may represent nothing more than an adaptation of maternal physiology necessary to package and secrete fat as stable globules (2).

The issue of whether dietary cholesterol in human milk offers any special benefits to infants has been a topic of speculation for several decades. The problem is essentially difficult to study because the effects of cholesterol cannot easily be separated from those of essential fatty acids when breast-fed infants are used as the control group. This review will focus on three aspects of dietary cholesterol: the role of early dietary cholesterol on infant growth and development, especially neurological development; the cardioprotective role of early cholesterol feeding, i.e. the Reiser hypothesis; and the converse hypothesis that early dietary cholesterol initiates atherosclerosis and leads to premature coronary artery disease.

J.G. Bindels, A.C. Goedhart and H.K.A. Visser (eds), Recent Developments in Infant Nutrition, 251–259.

CHOLESTEROL INTAKE OF BREAST-FED INFANTS

In order to appreciate the dimensions of cholesterol homeostasis in infants, some discussion of cholesterol intake and 'requirement' is helpful. Human milk contains about 250–300 μmol/l (10–12 mg/dl) of cholesterol. About 80% of this is found in the outer milk fat globule membrane (MFGM), which is formed predominantly from the plasma membrane of epithelial cells in the breast. The central core of the MFGM contains 15–20% of total cholesterol as esters (3). Although the range of cholesterol intake is broad, a 6 kg breast-fed infant consumes about 260 μmol/day (100 mg/day). A small additional intake of desmosterol, about 10% of total sterols, is probably converted rapidly to cholesterol by the liver. Thus, cholesterol intake is about 43 μmol/kg/day (16.7 mg/kg/day) in a breast-fed infant, which is far in excess of intakes recommended for adults (11 μmol/kg/day, 4.3 mg/kg/day) by the National Cholesterol Education Program (4). Absorption of dietary cholesterol by adults is about 45–50% (5), which is also likely to hold true for infants, but to the best of our knowledge no such measurements have been reported. These crude calculations suggest that human milk provides about 130 μmol/day (50 mg/day) of absorbed cholesterol.

Estimates of the cholesterol requirement of infants for tissue growth, steroidogenesis and bile acid production are difficult to find in the literature. The growth rate of a 4-month-old infant is about 20 g/day and the average cholesterol content of extraneural tissues is about 5 μmol/g (2 mg/g) wet weight, resulting in 100 μmol/day (40 mg/day) needed for growth. A conservative estimate of desquamation losses is about 25 μmol/day (10 mg/day) and steroidogenesis requires 12.5 μmol/day (5 mg/day). According to measurements made by Watkins et al., the bile acid production rate of preterm infants at 2.40 kg is about 13 μmol/kg/day (5.1 mg/kg/day) (6). However, this value may underestimate BA production in older infants, so a value of 20 μmol/kg/day, between that of adults and preterm infants, may give a more reliable estimate of BA production. Thus, a 4-month-old infant needs 120 μmol/day (48 mg/day) for BA synthesis. Total cholesterol required for growth and metabolism by a 4-month-old infant is about 267 μmol/day (103 mg/day), excluding that required for the brain and nervous system. In situ steroidogenesis by the CNS is the dominant source of cholesterol in these tissues, so they have been excluded from this discussion of 'requirements'. This rather crude analysis illustrates that human milk supplies less than half of the cholesterol required for growth and metabolic function. Commercial infant formulas based on casein or mixed casein/whey protein also supply cholesterol to some degree. However, the intakes are less than 20% of that supplied by human milk, leading to some speculation that it should be supplied in the formula. In order to justify addition of cholesterol to infant formula, convincing proof is required that its absence results in metabolic, functional or growth deficits.

THE EFFECT OF CHOLESTEROL SUPPLEMENTATION ON GROWTH AND DEVELOPMENT

The human infant may need cholesterol for growth and development, and to improve cognitive, visual and neurological functions (7). These functions of cholesterol have not achieved as much emphasis as the issue of $n-3$ fatty acid metabolism and chain elongation and desaturation. The brain contains relatively high levels of low density lipoprotein receptors (LDL-R) and cholesterol ester hydrolase (CEH) activity (8,9), both of which are necessary to deliver cholesterol to tissues. Isotopic tracer studies consistently show that most cholesterol in the brain is derived from synthesis in situ, with little or none derived from circulating lipoproteins (10). This anomaly seems inconsistent with the presence of LDL receptors and CEH in the brain. Long-term feeding studies of pregnant Rhesus monkeys with ^{14}C- or ^{3}H-labelled cholesterol have shown that dietary cholesterol crosses the placenta and contributes to fetal tissue cholesterol (42.6%) and to the fetal brain (4.6%) (11). In a study of one 13-week human fetus, [^{14}C]cholesterol was incorporated into the fetal brain to a specific activity of 21.5% of the maternal serum (12). This study was performed at a stage of fetal development before most brain growth had taken place, and it may not be representative of events during the third trimester. For obvious reasons, few studies have been conducted in human fetuses, so our knowledge in this area is limited.

Only a few cholesterol supplementation studies of infant formulas have been reported (6,13–19). Jarvenpaa et al. (13) in Finland studied the effects of cholesterol and taurine supplementation on growth and metabolism over a 16-week period in four groups of low birth weight (LBW) infants. Seventy-one LBW infants were fed human milk, a 60:40 whey:casein formula, or the same formula supplemented with taurine or with taurine plus cholesterol. They found no consistent difference among the feeding groups with respect to weight gain, length, head circumference or haematological status. As expected, the formula-fed groups all had higher blood urea nitrogen and total serum protein levels than the human-milk-fed group. In a second report, these authors found that human-milk-fed infants had a higher serum cholesterol than each of the formula groups (14). Curiously, the cholesterol plus taurine-supplemented formula groups consistently had the lowest serum cholesterol level ($p<0.05$), even though the formula contained as much cholesterol as did human milk. In a final report from this study, Watkins et al. (6) found that human-milk-fed infants had a greater bile acid pool size ($p<0.001$) than any of the formula groups, but there was no difference in BA production rate. They speculate that some unique factor in human milk affects hepatic and/or intestinal sterol metabolism.

Van Biervliet et al. (15) studied the effect of human milk, infant formula and the same formula supplemented with γ-linolenic acid or cholesterol. No differences in growth and development were noted among any of the groups. The human-milk-fed group had the highest serum cholesterol level ($p<0.05$) and infants receiving cholesterol-supplemented formula had cholesterol levels identical to those of the other formula-fed groups. The only other finding of note was an elevated level of HDL-2b cholesterol in the infants fed human milk or γ-linolenic acid-supplemented formula ($p<0.05$).

Wong et al. (16) and Cruz et al. (17) studied the kinetics of cholesterol synthesis in 4- to 5-month-old breast-fed and formula-fed infants. The methodology and experimental design were similar in both studies, as were the conclusions on the metabolic effects of dietary cholesterol. No differences in growth and development were found between groups in either study. High intakes of cholesterol by the infants fed human

***Figure 1** Incorporation of deuterium from body water (top panel) into erythrocyte membrane cholesterol of soy protein formula (○), soy protein with cholesterol (◇), cow milk formula (△), and human milk (□). Adapted from Ref. 17*

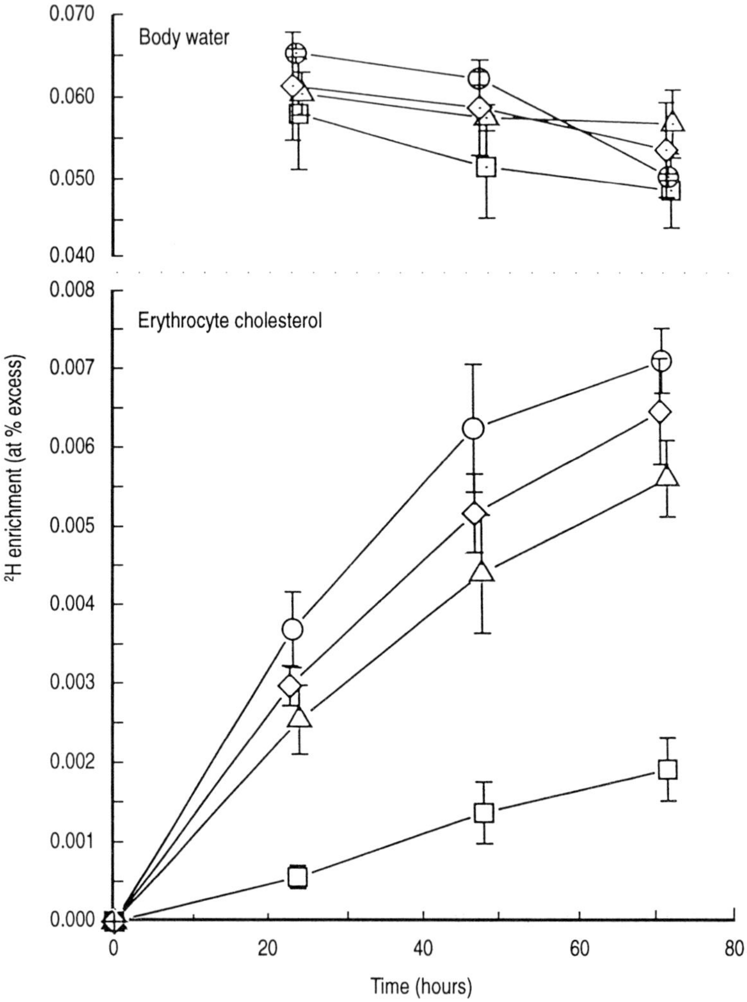

milk suppressed cholesterol synthesis (Fig. 1) compared with the formula-fed groups ($p<0.001$). Serum cholesterol and LDL-cholesterol levels were higher among breast-fed infants than any of the formula-fed groups, and the serum cholesterol level of all formula groups was similar (Table 1). There was a significant positive correlation ($r^2=0.59$, $p<0.001$) between cholesterol intake and serum cholesterol, and there was

Table 1 Effect of human milk, cow milk formula, soy protein formula, and cholesterol modified soy milk formula on serum cholesterol and fractional synthesis rates in infants

Diet	*Cholesterol* (mg/dl)	*HDL-cholesterol* (mg/dl)	*LDL-cholesterol* (mg/dl)	*FSR* (%/day)
Human milk (13)	172±7.7*	51±3.5	98±8.1*	2.97±0.49*
Cow's milk (10)	129±9.3	47±4.2	56±9.7	9.43±1.12
Soy protein (10)	120±10.0	55±4.6	38±10.4	12.0±1.36
Soy+cholesterol (6)	131±10.8	49±5.0	58±11.2	10.6±1.05

Adapted from Ref. 17
*$p<0.05$ compared with formula groups

a negative correlation ($r^2=0.44$, $p<0.002$) between intake and cholesterol fractional synthesis rate (FSR) (16). Similarly, Cruz et al. (17) found a negative correlation ($r^2=0.45$, $p<0.003$) between LDL-cholesterol level and the FSR. When the soy protein formula was modified with additional cholesterol to make it comparable to the casein formula, serum cholesterol levels rose and FSR fell, making the two formula groups indistinguishable. Both studies demonstrate that the human infant is fully able to down-regulate cholesterol synthesis when challenged with high intakes of dietary cholesterol and can thus control the amount of cholesterol available for cell function. Neither study was able to determine absolute synthesis rates because the magnitude of the central cholesterol pool was not determined. The fractional synthesis rate of cholesterol probably reflects metabolism in the central rapidly miscible pool (18). Whether these metabolic adaptations persist beyond the neonatal period is the major question posed by proponents of the Reiser hypothesis.

Most of the infant feeding studies have focused on differences in biochemical measures or on growth. Lucas et al. (19,20) have taken a different approach and concentrated on functional outcomes. They report that at 18 months of age, no differences were apparent between preterm infants fed human milk or formula on indices of psychomotor and mental development (19). The trials are still in progress, and the preliminary results were encouraging (20) but they have not attributed the differences in function to a specific component in milk. Many investigators feel that any functional differences are more likely to be due to higher intakes of 22:6 $n-3$ and 20:4 $n-6$ fatty acids by breast-fed infants, rather than cholesterol.

THE REISER HYPOTHESIS

The conjecture that cholesterol in milk may endow the nursing infant with an enhanced ability to metabolize dietary cholesterol later in life has become known as the Reiser hypothesis (21,22). Rats, pigs and baboons fed diets high in cholesterol had lower serum cholesterol levels later in life than did animals fed a low cholesterol diet (22). The studies show a decided species difference: those in rats were inconclusive, studies in pigs generally supported the hypothesis and those in baboons were negative. In the 1980s, Mott and co-workers began a series of studies in infant baboons to test this hypothesis in a species with patterns of lipid metabolism similar to that of the human (23–27). After weaning, serum cholesterol levels of both groups were identical until 52 weeks of age (27). Thereafter, serum cholesterol and HDL_2 cholesterol were lower in the breast-fed group ($p<0.05$). Breast-fed baboons had a 25% lower bile acid synthesis rate ($p<0.02$) than formula-fed animals, and a 44% higher hepatic LDL-R mRNA level ($p<0.003$). No difference in BA pool size was found. The results are somewhat discordant with our current concept of hepatic sterol metabolism. Generally, higher bile acid synthesis rates require more cholesterol, and thus upregulate LDL-R synthesis. The overall results of these baboon studies point to differences in sterol metabolism being mediated through catabolic pathways rather than through synthesis.

Despite the inconsistencies in the animal studies, available data from humans have uniformly failed to support the Reiser hypothesis. The earliest studies in this area were those of Glueck et al. (28) and Friedman and Goldberg (29), neither of which showed any long term benefits. A study by Hodgson et al. (30) found that infants fed a low cholesterol commercial formula had lower serum cholesterol levels at age 7–12 than those fed either human milk or cow's milk ($p<0.05$). Perhaps the definitive study in this area is that by Fomon et al. (31), who studied 469 children until 8 years of age who had been fed human milk or a low cholesterol formula as infants. Although a transient rise in serum cholesterol was observed during the neonatal period in both males and females fed human milk, the difference in cholesterol levels was not significant in any group at 8 years of age. This conclusion has generally been supported by later studies (32).

The difficulty of designing and implementing controlled studies to address this issue properly over the span of a human life is a formidable barrier. The theoretical basis for the Reiser hypothesis was formulated in an era when our understanding of the regulation of sterol and lipoprotein metabolism was primitive. The hypothesis fails to explain adequately how the effects of a diet stimulus in early infancy on the genes that govern lipid and lipoprotein metabolism will persist into adulthood when the intracellular concentration of regulatory factors has returned to basal levels.

PAEDIATRIC DETERMINANTS OF ATHEROSCLEROSIS RISK

The converse of the Reiser hypothesis is that high intakes of dietary cholesterol early in life may form pre-atherosclerotic lesions and fatty streaks in the arterial wall and thereby induce premature atherosclerosis in genetically predisposed individuals. Few studies in the literature have addressed this issue directly because of the difficulty of correlating long term outcome with early infant nutrition. Two large multicentre trials are currently in progress in the USA and the UK (33,34). In the USA, the Pathobiological Determinants of Atherosclerosis in Youth (PDAY) study will investigate genetic and environmental factors (i.e. diet and smoking) that influence atherogenesis in young adults who die of exogenous causes (33). The programme is in the early stages and a few quantitative conclusions are available with respect to infant feeding and atherosclerosis. The PDAY project is mentioned here so that interested readers can follow its progress as more results become available.

Using a different strategy, and as also described in Chapter 1, Fall et al. (34) studied a male population ($n = 5718$) born in Hertfordshire, England, between 1911 and 1930 for whom the local health service kept accurate records of birth weight, abdominal circumference, mode of infant feeding and weight during the first year of life. They correlated these factors with mortality from atherosclerosis and with serum cholesterol and lipoprotein levels at 59–70 years of age. Breast-fed male infants who had been weaned by 1 year of age had lower ($p < 0.05$) serum cholesterol and LDL-cholesterol than bottle-fed infants or breast-fed infants who had not been weaned by 1 year. In addition, the breast-fed infants weaned by one year of age had a lower age-adjusted mortality ratio than the other two groups. A similar analysis of the female Hertfordshire population showed similar trends, but low rates of infant growth did not predict levels of risk factors in women (35). Another study of males in Sheffield, England, obtained results similar to those of the Hertfordshire study (36). The conclusions from these studies are mixed, which may possibly be due to a gender effect in the response to infant feeding. A clear conclusion from this study, however, is that rates of intrauterine growth and neonatal growth may affect serum cholesterol homeostasis later in life. Fall and his colleagues have been cautious and unwilling to attribute these effects to any specific agent in human milk. Until convincing data become available to support either viewpoint, dietary cholesterol must be regarded as neither beneficial nor harmful for the majority of infants, and any cardioprotective benefits require further study.

CONCLUSIONS

The evidence to date is that dietary cholesterol in human milk offers no long-term benefit for the overwhelming majority of infants. None of the studies have found any benefits of dietary cholesterol on growth and development, and the purported benefits to cardiovascular health have been difficult to elicit from epidemiological data. Although studies are still under way, the early evidence is that cholesterol in human milk is associated with no increased risk of cardiovascular disease. However, dietary cholesterol may offer special benefits to individuals with very low serum cholesterol and to patients with Smith–Lemli–Opitz syndrome (37,38).

ACKNOWLEDGEMENTS

Preparation of this manuscript has been funded in part with funds from the United States Department of Agriculture (USDA), Agricultural Research Service, under Cooperative Agreement No. 58-6250-1-003. The contents of this manuscript do not necessarily reflect the views or policies of the USDA, nor does mention of trade names, commercial products, or organizations imply endorsement by the US Government.

REFERENCES

1. Bloch K 1985 Cholesterol: evolution of structure and function. In: Vance DE, Vance JE (eds) Biochemistry of Lipids and Membranes. Benjamin/Cummings Publishing Co., Menlo Park, pp 1–24
2. Jensen RG 1989 The Lipids of Human Milk. CRC Press, Boca Raton, FL, pp 187–192

3. Bitman J, Wood DL, Mehta NR, Hamosh P, Hamosh M 1986 Comparison of the cholesteryl ester composition of human milk from preterm and term mothers. J Pediatr Gastroenterol Nutr 5: 780–786
4. Anonymous 1991 National Cholesterol Education Program. Report of the expert panel on population strategies for blood cholesterol reduction. Circulation 83:2154–2232
5. Zilversmit DB, Hughes LB 1974 Validation of a dual-isotope plasma ratio method for measurement of cholesterol absorption in rats. J Lipid Res 15:465–473
6. Watkins JB, Järvenpää A-L, Szczepanik-Van Leeuwen PA, Klein PD, Rassin DK, Gaull G, Raïhä NCR 1983 Feeding the low birth weight infant. V. Effects of taurine, cholesterol and human milk on bile acid kinetics. Gastroenterology 85:793–800
7. Fomon SJ 1974 Infant Nutrition, 2nd edn. WB Saunders, Philadelphia, PA, pp 172–174
8. Dehouck B, Dehouck MP, Fruchart JC, Cecchelli R 1994 Upregulation of the low density lipoprotein receptor at the blood–brain barrier: intercommunications between brain capillary endothelial cells and astrocytes. J Cell Biol 126:465–473
9. Pitkanen AS, Halonen TO, Kilpelainen HO, Riekkinen PJ 1986 Cholesterol esterase activity in cerebrospinal fluid of multiple sclerosis patients. J Neurol Sci 74:45–53
10. Edmond J, Korsak RA, Morrow JM, Totok-Both G, Catlin DH 1991 Dietary cholesterol and the origin of cholesterol in the brain of developing rats. J Nutr 121:1323–1330
11. Pitkin RM, Connor WE, Lin DS 1972 Cholesterol metabolism and placental transfer in the pregnant rhesus monkey. J Clin Invest 51:2584–2592
12. Lin DS, Pitkin RM, Connor WE 1977 Placental transfer of cholesterol into the human fetus. Am J Obstet Gynecol 128:735–739
13. Järvenpää A-L, Raïhä NCR, Rassin DK, Gaull GE 1983 Feeding the low-birth-weight infant. I. Taurine and cholesterol supplementation of formula does not affect growth and metabolism. Pediatrics 71:171–178
14. Rassin DK, Gaull GE, Järvenpää A-L, Raïhä NCR 1983 Feeding the low-birth-weight infant. II. Effect of taurine and cholesterol supplementation on amino acids and cholesterol. Pediatrics 71: 179–186
15. Van Biervliet J-P, Vinaimont N, Vercaemst R, Rosseneu M 1992 Serum cholesterol, cholesteryl ester, and high-density lipoprotein development in newborn infants: response to formulas supplemented with cholesterol and γ-linolenic acid. J Pediatr 120:S101–108
16. Wong WW, Hachey DL, Insull W, Opekun AR, Klein PD 1993 Effect of dietary cholesterol on cholesterol synthesis in breast-fed and formula-fed infants. J Lipid Res 34:1403–1411
17. Cruz MLA, Wong WW, Mimouni F, Hachey DL, Setchell KDR, Klein PD, Tsang RC 1994 Effects of infant nutrition on cholesterol synthesis rates. Pediatr Res 35:135–140
18. Goodman DS, Smith FR, Seplowitz AH, Ramakrishnan R, Dell RB 1980 Prediction of the parameters of whole body cholesterol metabolism in humans. J Lipid Res 21:699–713
19. Lucas A, Morley R, Cole TJ, Gore SM 1994 A randomised multicentre study of human milk versus formula and later development in preterm infants. Arch Dis Child 70:F141–146
20. Morley R, Lucas A 1994 Influence of early diet on outcome in preterm infants. Acta Paediatr Suppl 405:123–126
21. Reiser R, Sidelman Z 1972 Control of serum cholesterol homeostasis by cholesterol in the milk of suckling rat. J Nutr 102:1009–1016
22. Reiser R, O'Brien BC, Henderson GR, Moore RW 1979 Studies on a possible function for cholesterol in milk. Nutr Rep Int 19:835–849
23. Mott GE, McMahan CA, Kelley JL, Farley CM, McGill HC 1982 Influence of infant and juvenile diets on serum cholesterol, lipoprotein cholesterol, and apolipoprotein concentrations in juvenile baboons (*Papio* sp.). Atherosclerosis 45:191–202
24. Mott GE, Jackson EM, McMahan CA, Farley CM, McGill HC 1985 Cholesterol metabolism in juvenile baboons. Influence of infant and juvenile diets. Arteriosclerosis 5:347–354
25. Jackson EM, Lewis DS, McMahan CA, Mott GE 1993 Preweaning diet affects bile lipid composition and bile acid kinetics in infant baboons. J Nutr 123:1471–1479
26. Lewis DS, McMahan CA, Mott GE 1993 Breast feeding and formula feeding affect differently plasma thyroid hormone concentration in infant baboons. Biol Neonate 63:327–335
27. Mott GE, Jackson EM, DeLallo L, Lewis DS, McMahan CA 1995 Differences in cholesterol metabolism in juvenile baboons are programmed by breast- versus formula-feeding. J Lipid Res 36:299–307

28. Glueck CJ, Tsang R, Balistreri W, Fallat R 1972 Plasma and dietary cholesterol in infancy: effects of early low or moderate dietary cholesterol intake on subsequent response to increased dietary cholesterol. Metabolism 21:1181–1192
29. Friedman G, Goldberg SJ 1975 Concurrent and subsequent serum cholesterol of breast- and formula-fed infants. Am J Clin Nutr 28:42–45
30. Hodgson PA, Elletson RD, Elveback LR, Harris RH, Nelson RH, Weidman WH 1976 Comparison of serum cholesterol in children fed high, moderate or low cholesterol milk diets during neonatal period. Metabolism 25:739–746
31. Fomon SJ, Rogers RR, Ziegler EE, Nelson SE, Thomas LN 1984 Indices of fatness and serum cholesterol at age eight years in relation to feeding and growth during early infancy. Pediatr Res 18:1233–1238
32. Wagner V, von Stockhausen HB 1988 The effect of feeding human milk and adapted milk formulae on serum lipid and lipoprotein levels in young infants. Eur J Pediatrics 147:292–295
33. Wissler RW 1994 New insights into the pathogenesis of atherosclerosis as revealed by PDAY. Pathobiological Determinants of Atherosclerosis in Youth. Atherosclerosis 108(Suppl.):S3–20
34. Fall CHD, Barker DJP, Osmond C, Winter PD, Clark PMS, Hales CN 1992 Relation of infant feeding to adult serum cholesterol concentration and death from ischaemic heart disease. Br Med J 304:801–805
35. Fall CHD, Osmond C, Barker DJP, Clark PMS, Hales CN, Stirling Y, Meade TW 1995 Fetal and infant growth and cardiovascular risk factors in women. Br Med J 310:428–432
36. Barker DJP, Martyn CN, Osmond C, Hales CN, Fall CHD 1993 Growth in utero and serum cholesterol concentration in adult life. Br Med J 307:1524–1527
37. Nwokoro NA, Hyde B, Mulvihill JJ 1994 Smith–Lemli–Opitz syndrome: biochemical before clinical diagnosis; early dietary management. Am J Med Genet 50:375–376
38. Irons M, Elias ER, Tint GS, Salen G, Frieden R, Buie TM, Ampola M 1994 Abnormal cholesterol metabolism in the Smith–Lemli–Opitz syndrome: report of clinical and biochemical findings in four patients and treatment in one patient. Am J Med Genet 50:347–352

19 Nutrient deficiency in the breast-fed infant

B. WHARTON

INTRODUCTION

Nutritional pathology in breast-fed infants may be related to all or to particular nutrients, and it may be a problem for individual mothers and their babies, or potentially a problem for the whole population. A possible classification of low nutrient intake in breast-fed children is shown in Table 1. This review is concerned mainly with the 'all mothers, particular nutrients' section but the other two are considered briefly. It is not a systematic review, but attempts to indicate points of practical management of the mother–children dyad, and to raise some questions for the infant food industry. I have also aimed to be more speculative, even contentious, than in a systematic review.

INSUFFICIENT VOLUME

General reviews concerning the physiological factors affecting breast milk volume were presented at a symposium on 'maternal–offspring interactions' held in Southampton in 1995 (1–3). Dr Motil reviews the subject in Chapter 3. In daily practice, an infant is judged to be breast-fed satisfactorily if it is satisfied and if growth is satisfactory. However, both methods of assessment have difficulties.

Satiety

Whether a baby is 'satisfied' is a complex linkage of signs and their interpretation between a mother and her baby. Some mothers interpret frequent feeding, crying shortly after a feed or waking up within an hour of a feed as evidence that the baby is not satisfied, but this may not be so. If a high frequency of feeding is accepted as a normal variant but crying without adequate cause persists, most mothers will seek advice. If these behaviours are abolished when the baby receives infant formula ad lib, a reasonable interpretation is that some aspect of the breast-feeding was not giving satisfaction, and this may have been volume. However, if the mother has perceived one method of feeding to be unsatisfactory there is an inherent likelihood that the new method will be more acceptable, i.e. a change to bottle-feeding may have an important placebo effect on the mother's emotions and worries, as well as having a direct nutritional effect. On the other hand, it is well established that the occasional breast-fed infant who is growing poorly by any standard will nevertheless be quite satisfied and contented (4). When infant formula ad libitum is introduced, such a child not only

J.G. Bindels, A.C. Goedhart and H.K.A. Visser (eds), Recent Developments in Infant Nutrition, 260–271.

Table 1 Nutrient limitations in breast milk

	Individual mothers	*All mothers*
Particular nutrients	Maternal deficiency Prolonged lactation Mammary abnormality	Vitamin K Vitamin D Iron
All nutrients	Insufficient volume	

shows catch up growth but becomes more active and alert. Whether deficiency of a specific nutrient could affect satiety is not clear. The role of tryptophan in satiety and sleep has received increasing attention. Some bottle-fed babies have lower plasma tryptophan concentrations than breast-fed babies (5,6) and some speculation has arisen over whether this may affect behaviour. Dr Pencharz discusses amino acids in more detail in Chapter 21.

Growth

Comparison with a growth chart may suggest failure to thrive, but a problem arises over which reference standard should be used. Many charts are based on a largely bottle-fed population. A number of cohort studies have shown that exclusively breast-fed children are lighter than bottle-fed children, but similar in length and head circumference in later infancy (7–11). Intriguing as this is for auxologists and nutritionists, it causes little difficulty of interpretation when monitoring the progress of an individual child. If length or head circumference as well as weight fall below the reference then possible causes, including insufficient volume of feed, must be considered. If the head circumference and length continue to increase at a satisfactory velocity and the baby is satisfied, then a weight gain below that seen in bottle-fed babies is acceptable. However, defects in length and head circumference take much longer to appear than a low weight velocity, which may therefore signal problems earlier. Apart from plotting measurements on distance growth charts, an additional rule of thumb is to regard a weight gain of less than 15 g per day in the early months of life as unusual and requiring an explanation.

PARTICULAR NUTRIENTS IN INDIVIDUAL MOTHERS

Table 2 shows the concentrations of micronutrients in a British representative sample of one breast full of milk 6–10 weeks post partum, together with some figures for preterm milk (12). Three deficiency situations may be defined: that where a low concentration of the nutrient in the breast milk reflects a general deficiency in the mother, or it appears as lactation progresses, or it is a specific mammary phenomenon not associated with a general maternal deficiency. The book edited by Berger (13) summarizes much of the literature published prior to 1988 and contains an amount of previously unpublished data.

Table 2 Vitamin and mineral content of human breast milk

	Expressed			*Preterm milk* (per 100 ml)	
	per 100 ml	per 100 kcal	Drip (per 100 ml)	6–15 days	16–196 days
Vitamin					
Retinol (μg)	60	86			
Vitamin D (μg)	0.01	0.014			
α-Tocopherol (μg)	0.35	0.5			
Vitamin K (μg)	1.5	2.1			
Thiamine (μg)	16	23		5	9
Riboflavin (μg)	31	44		36	27
Nicotinic acid (μg)	230	329		114	210
Vitamin B_6 (μg)	6	9		3	6
Vitamin B_{12} (μg)	0.01	0.014		0.05	0.02
Folic acid (μg)	5.2	7.4		2.1	3.1
Pantothenic acid (μg)	260	371		164	230
Biotin (μg)	0.8	1.1		0.2	0.5
Vitamin C (mg)	3.8	5.4			
Mineral					
Sodium (mg)	15	22	13	51	29
Potassium (mg)	60	86	63	69	60
Chlorine (mg)	43	61		88	59
Calcium (mg)	35	50	28	25	22
Magnesium (mg)	2.8	4	2.9	3.1	2.5
Phosphorus (mg)	15	21		14	14
Iron (μg)	76	109			
Copper (μg)	39	56		78	63
Zinc (μg)	295	421		475	392
Manganese (μg)	2.0	2.8			
Selenium (μg)	2–3	2.8–4.3			
Iodine (μg)	7	10			
Fluorine (μg)	5–30	7–43			
Chromium (ng)	340–430				
Molybdenum (μg)	0.5–25				

Abstracted from Wharton (12); full references given in that book

Maternal deficiency

Infantile beri-beri occurring in the breast-fed child of a mother eating polished rice in South East Asia is the classical example of this condition. Both cardiac and neurological (aphonic forms) have been described. This syndrome is now uncommon, although thiamine deficiency may complicate kwashiorkor when it presents in older children. The concentration of thiamine in breast milk of Swiss and Finnish mothers remains more or less constant during the first 6 months of lactation (14). As vegetarianism becomes more common there has been increased interest in the breast milk composition of vegetarian women. A vegetarian diet includes more vegetable oils,

Table 3 Micronutrient deficiencies and growth faltering

Nutrient	*Zinc*[22]	*Pyridoxine*[23]
Country	France	Finland
Subjects	Immigrants from North Africa, breast-fed for 4–9 months	Exclusively breast-fed for 6 months
Number studied	58	44
Intervention	Placebo controlled trial of oral zinc 5 mg daily given to infants	All mothers received pyridoxine 1 mg daily after delivery
Biochemistry		7 had low vitamin B_6 status by 6 months age
Anthropometry	Treatment group had significantly greater increases in weight and length during 3 months of supplementation	These 7 had a slower length velocity and a deeper fall in length for age from 6–9 months

which have a higher level of mono and polyunsaturated fatty acids, and this is reflected in the fatty acid composition of the breast milk (15). Only in vegans (i.e. mothers not consuming any animal products whatsoever) has there been any concern regarding low concentrations of nutrients in breast milk, namely vitamin B_{12} and taurine. In practice, vegans take vitamin B_{12} supplements and presumably are able to manufacture taurine from methionine via the transulphuration pathway, so that deficiencies do not occur in their babies. A detailed bibliography is given in the briefing paper of the British Nutrition Foundation and other reviews are available (16,17). Vegetarian mothers from the Indian subcontinent who have emigrated to Britain do not commonly choose to breast-feed, but when they do, they appear to have no more, and probably fewer, problems than European omnivore mothers. The health of the sucklings of these immigrant mothers was good; for their weanlings it was not so good (18–21).

Changes during lactation

The concentrations of many nutrients change during lactation. For most the concentration falls slightly and biochemical evidence of 'insufficiency' may be found in mothers and their babies as lactation progresses. The concentration of nutrients in breast milk and the biochemical tests in the baby may be 'corrected' with maternal supplementation; but there is little evidence of clinical problems in association with such abnormalities (13,14). Two exceptions to this general rule are zinc and pyridoxine deficiencies which are associated with and probably cause growth faltering (22–24, see Table 3).

Possible deficiencies associated with exclusive breast-feeding for longer periods than usual in a population seem to be associated with both extremes of social privilege.

Table 4 Hereditary trace element deficiency

Iron	Iron malabsorption?
Zinc	Acrodermatitis enteropathica
Copper	Menke's syndrome ('kinky hair disease')
	Copper malabsorption?
Iodine	Goitrous cretinism
Molybdenum	Sulphite oxidase deficiency

Abstracted from Wharton (26); full references given in that review

Zinc deficiency occurs both in the underprivileged North African immigrants to France (22) and, perhaps, in an affluent group in Copenhagen (24). These 'deficiencies' therefore probably reflect a true effect of long lactation rather than represent the nutritional exhaustion of the mother. In practical terms, should we recommend mothers who have breast fed for more than (say) 4 months to take supplements of micronutrients? This would be impractical for most of the world's population, and for the majority of mothers in the developed world it would be of little value although the above studies may cause some thought (22,24). General advice might be that it is probably not necessary if the baby is well and growing normally, but since supplements do help a few mothers and their babies they should be taken if the mother so wishes. Calcium intake should be adequate, particularly after the end of lactation, in case a recovery of bone mass was necessary (25).

Possible specific mammary abnormality

The occasional mother produces breast milk which contains very small amounts of zinc, even in the early weeks of the lactation. One child believed to have acrodermatis enteropathica in fact had normal intestinal transport of zinc, and the condition was due to the very low concentration of zinc in the mother's breast milk (26). This does seem to be a specific mammary effect since the mothers have no other evidence of zinc deficiency themselves, and administration of zinc supplements causes no increase in zinc concentration in the breast milk. Similar mechanisms as yet undiscovered almost certainly affect other micronutrients. Table 4 summarizes the hereditary abnormalities of trace element metabolism (26). Genetic differences causing pathology in one tissue often result in parallel but asymptomatic changes in other tissues (e.g. cystinuria results in symptomatic renal stones, but also causes asymptomatic intestinal malabsorption of certain amino acids). Alternatively, related but non-identical gene defects may result in disorders of widely differing clinical severity, such as Gilbert's disease (benign) or Crigler–Najjar disease (often disastrous) as a cause of unconjugated hyperbilirubinaemia. It is possible that some of the disorders shown in Table 4 have milder variants affecting transport processes in the mammary gland.

LOW NUTRIENTS IN ALL MOTHERS

The breast milk from all mothers contains very low concentrations of vitamin D and iron, and the concentration of vitamin K is well below that found in cow's milk. Com-

Table 5 Nutrient concentration (μg per 100 ml)

	Breast milk	*Typical infant formula*
Vitamin K	1.5	5.0
Vitamin D	0.01	1.0
Iron	76	700

pared with breast milk, all modern infant formulas contain more vitamin K and most of them contain substantially more vitamin D and iron (see Table 5).

Vitamin K

Haemorrhagic disease of the newborn, whether the early classical variety or later intracranial haemorrhage, has been seen with increasing frequency in more recent years. It is due to inadequate vitamin K-dependent coagulation factors (e.g. prothrombin) and occurs almost exclusively in breast-fed infants. Babies receiving an infant formula have a shorter prothrombin time, similar to that seen in babies who have received intramuscular vitamin K (2 mg) at birth (27). Since the majority of breast-fed babies, including those who do not receive perinatal vitamin K, do not develop the disorder, it presumably occurs only when the amount in breast milk is even lower than usual, or if the infant has unusually high requirements for some reason. Logical as this may seem, however, there is no evidence for or against either possibility. A third well-known possibility is that the endogenous supply of vitamin K from intestinal bacteria may be reduced. Figure 1 summarizes some of our own studies of the relationships between faecal flora and diet in the early weeks of life. In our unit, breast-fed babies had a faecal flora predominantly composed of bifidobacteria and lactobacilli, whereas babies fed formulas of many different varieties had an adult type flora with significant populations of *Bacteroides*, streptococci and coliforms (28–31). One study of this topic (32) was interesting in that the faecal flora of both breast-fed and bottle-fed babies born in a London hospital were similar in showing the 'adult' pattern, but babies born in a Nigerian hospital showed the same breast/bottle difference seen in our studies. It is tempting to suggest that the antiseptic environment of modern perinatal care (e.g. vaginal antiseptics during labour, nipple sprays used by some lactating mothers) alters the colonization of the neonatal intestine and hence the vitamin K synthesizing ability. Dr Hudault discusses intestinal colonization in further detail in Chapter 22. Intriguing as this possibility is, however, there is no evidence to support or refute it. The vitamin K-generating capabilities of a particular organism and the actual amounts absorbed have received little attention.

Some caution regarding the administration of unphysiological amounts of vitamin K is necessary. Very large doses (10 mg intramuscular given on many days) results in kernicterus because it displaces bilirubin from the plasma albumin. Some concern was expressed in Britain that more modest doses (1 mg intramuscular once only) was associated with malignancy in later childhood. However, these very large doses are

***Figure 1** Faecal flora (mean proportion of total bacterial counts) at 14 days of age in breast-fed babies and those receiving various routine and experimental infant formulas. Bacteria: B/L, bifidobacteria and lactobacilli; Ec, Escherichia coli; Se, Staphylococcus epidermidis; Bac, Bacteroides; Sf, Streptococcus faecalis; Cl, Clostridia. Diets: whey predominant infant formula, casein predominant formula, + F, added iron, 0.7–0.9 mg/100 ml. + L, added bovine lactoferrin 280 mg/100 ml. Based on data in Balmer and Wharton (28–31)*

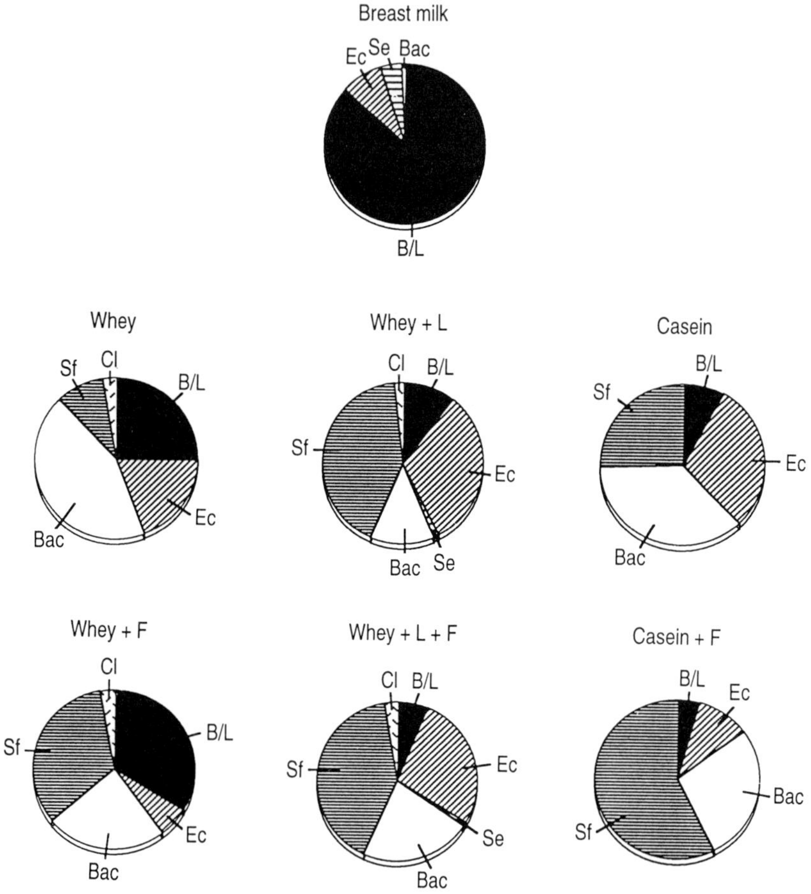

no longer given and the malignancy association has not been confirmed in other studies. McNinch and Tripp (33) have reviewed the risk benefit ratio of vitamin K. The death of a previously healthy, thriving, breast-fed baby from catastrophic intracranial haemorrhage is a chastening experience. If a baby is to be bottle fed, parenteral vitamin K supplementation is unnecessary. If the mother plans to breast feed then the balance of evidence regarding the risk benefit ratio, together with the logistic uncertainty of giving repeated oral doses of vitamin K, favours a single parenteral dose of 1 mg.

Vitamin D

Breast milk contains very small amounts of vitamin D. The suggestion that significant amounts are present as water soluble vitamin D sulphate has been refuted and breast milk has no healing effect in rachitic rats (34,35). While rickets and hypocalcaemic convulsions may occur in breast-fed children, they are uncommon. Presumably the reason is that just as with iron, babies have stores in the liver at birth and, unlike iron, there is a continuing supply of vitamin D from the sun. It is reasonable, therefore, to suggest that prevention should concentrate on ensuring an adequate vitamin D status in the mother, and appropriate exposure to sunlight after birth. Sunlight may be a problem in northern countries but rickets is also well described in sunny climates where fear of burning or heat means that children have only limited exposure. Supplementation programmes exist in countries as climatically diverse as Scotland and Saudi Arabia. Ensuring a good vitamin D status in young women is a matter of dietary supplements and sunlight. In Moslem countries clothing customs and rules of modesty may limit exposure to sunlight, particularly if the family lives in modern apartments rather than in a traditional house with an inner private courtyard. Various studies have shown that vitamin D supplements in pregnancy reduce the risk of hypocalcaemic convulsions in the neonatal period (36–38). Presumably this also reduces the risk of rickets many months later, but longitudinal studies to support that presumption are not available.

Although it may be ideal to concentrate on the health of young women and exposure of infants to sun as strategies to prevent infant and toddler rickets, there may nevertheless be a need for a safety net, and this may be achieved by fortifying infant formula with vitamin D. Despite the lack of firm evidence, however, there are arguments against exposing young infants to very much higher doses of a nutrient than is found in breast milk. It has been suggested that man probably evolved in the tropics and had little use for dietary vitamin D, since this was provided by sunlight. There is an argument that the persistence of intestinal lactase into adult life was a useful mutation for man as he migrated from the tropics to the temperate areas and that this enabled him to absorb calcium more effectively. Is it permissible to argue that this same migration would also have favoured a mutation which led to the presence of vitamin D in breast milk, unless the vitamin has some still unidentified adverse effects in early infancy. Vitamin D has molecular similarities to cholesterol, bile acids and the steroid hormones. Is it too speculative to wonder whether one might affect the metabolism and handling of another?

Greater attention to the vitamin D status of young women and pregnant women might remove the need for vitamin D supplementation in early infancy. In later infancy, however, while rickets is now unusual, low plasma concentrations of vitamin D are common (21). There is, therefore, a public health argument for adding extra vitamin D to the weaning diet and a follow-on milk is a suitable vehicle.

Iron

There is little iron in milk, but this causes no problems in early infancy because total body iron does not increase at that time, remaining around 250 mg despite a weight gain of 3.5 kg in the first 4 months of life (39). Iron stored from the physiological haemolysis which occurs postnatally is stored and used as the baby grows and the total mass of circulating haemoglobin increases. Furthermore there is some evidence of an advantage in avoiding iron in early life. Many pathogens require iron for multiplication, and an iron-deficient environment means that growth of pathogens is less prolific. The very limited amounts of dietary iron in breast milk contribute to this low-iron environment, and its availability to the microorganisms is limited further by the presence of proteins with a high affinity for iron, such as lactoferrin in the gut, transferrin in plasmas and ferritin within cells. There is also circumstantial evidence that large doses of iron (usually parenteral) result in more severe or more frequent episodes of meningitis, septicaemia and enteritis (40–42).

Some studies have shown a slightly different faecal flora in babies receiving iron-fortified formulas. Figure 1 shows that the addition of iron to a formula (whether whey, casein or bovine lactoferrin were present) encouraged the growth of *Streptococcus faecalis*. Studies in Holland have also shown that the addition of iron to a formula moves the faecal flora further away from that of a breast-fed baby (43,44). This does not necessarily imply any clinical problem but does show an intestinal effect of the unabsorbed iron.

The breast milk model, changes in body composition, the absence of clinical deficiency and the possibility of adverse effects all suggest that iron fortification of formulas used in early infancy is unnecessary. Iron-fortified formulas should be seen as a convenient, effective vehicle for adding iron to the diet later in infancy, and perhaps as a way of excluding other possible deleterious effects of pasteurized cows' milk, such as gastrointestinal blood loss and high concentrations of calcium and phosphorus which could interfere with iron absorption. Dr Dallman discusses iron deficiency in further detail later in Chapter 20.

In view of concerns about the prevalence and effects on psychomotor development of iron deficiency in the second year of life (45,46) this argument for a consideration of unfortified formulas may seem surprising. It has been presented before, however (47), and the logic is self evident – give the extra dietary iron when demands of growth lead to an increase in total body iron, and not before then.

NUTRIENT LIMITATIONS IN LACTATION

If the volume of breast milk is limited then all nutrients will be deficient, but it is not clear which will be the first limiting nutrient nor are the signs of insufficient volume (i.e. absence of satiety and insufficient growth) easy to interpret. Individual mothers may secrete milk which leads to a nutrient deficiency in her baby if she herself is deficient, if the lactation is prolonged or occasionally because there is a specific mammary defect in the secretion of the nutrient.

A conundrum is the low concentrations of vitamin K, vitamin D and iron in breast milk despite the occurrence of clinical disease due to deficiencies of these nutrients. Arguments are presented that in early infancy there is little need for dietary sources of iron and vitamin D, and their low concentration in breast milk therefore gives little cause for concern. There may be arguments for starting infant formulas to have similarly low concentrations. The fortification of formulas used in later infancy should then be seen as an effective way of improving the weaning diet, not of improving breast milk. These arguments support the concept of a starting infant formula, a breast milk substitute, using the composition of breast milk as the gold standard for use in the first 4–6 months of life. This would then be followed by a milk designed to be one part of the weaning diet, a weaning milk, a follow-on milk aiming to prevent the known nutrient deficiencies and excesses of that age.

Vitamin K remains an enigma. We have no option but to give supplements to breast-fed babies and to fortify infant formulas. Nevertheless the need for such a policy may indicate some aspect of modern perinatal practice which is unsatisfactory.

REFERENCES

1. Hartmann P, Sherriff J, Kent J 1995 Maternal nutrition and the regulation of milk synthesis. Proc Nutr Soc 54:379–389
2. Prentice AM, Prentice A 1995 Evolutionary and environmental influences on human lactation. Proc Nutr Soc 54:391–400
3. Wilde CJ, Prentice A, Peaker M 1995 Breast feeding: matching supply with demand in public lactation. Proc Nutr Soc 54:401–406
4. Evans TJ, Davies DP 1977 Failure to thrive at the breast: an old problem revisited. Arch Dis Child 52:974–976
5. Wharton BA 1992 Food for the brain. Proc R Coll Phys Edin 22:336–346
6. Scott PH, Sandham S, Balmer SE, Wharton BA 1990 Diet related reference values for plasma amino acids in newborns measured by reversed-phase HPLC. Clin Chem 36:1922–1927
7. Salmanperä L, Perheentupa J, Siimes MA 1985 Exclusively breast-fed healthy infants grow slower than reference infants. Pediatr Res 19:307–312
8. Whitehead RG, Paul AA, Cole TJ 1989 Diet and the growth of healthy infants. J Hum Nutr Diet 2:73–84
9. Dewey KG, Heinig MJ, Nommsen LA, Peerson JM, Lönnerdal B 1992 Growth of breast-fed and formula-fed infants from 0 to 18 months: the DARLING study. Pediatrics 89:1035–1041
10. Dewey KG, Heinig MJ, Nommsen LA, Peerson JM, Lönnerdal B 1993 Breast-fed infants are leaner than formula-fed infants at 1 yr of age: the DARLING study. Am J Clin Nutr 57:140–145
11. Hitchcock NE, Gracey M, Gilmour AI 1985 The growth of breast fed and artificially fed infants from birth to twelve months. Acta Paediatr Scand 74:240–245
12. Wharton BA 1983 (ed) Nutrition and Feeding of the Preterm Infant. Blackwell, Oxford, pp 6–9
13. Berger H 1988 (ed) Vitamins and Minerals in Pregnancy and Lactation. Raven Press, New York, pp 1–437
14. Dostálová L, Salmanperä, Václavinková V, Keinz-Erian P, Schüep W 1988 Vitamin concentrations in term milk of European mothers. In: Berger H (ed) Vitamins and Minerals in Pregnancy and Lactation. Raven Press, New York, pp 275–298
15. Sanders TAB, Reddy S 1992 The influence of a vegetarian diet on the fatty acid composition of human milk and the essential fatty acid status of the infant. J Pediatr 120:S71–S77
16. British Nutrition Foundation 1995 Vegetarianism. Briefing Paper: British Nutrition Foundation, London, pp 1–27
17. Specker B 1994 Nutritional concerns of lactating women consuming vegetarian diets. Am J Clin Nutr 59 (suppl):1182S–1186S
18. Wharton PA, Eaton PM, Day KC 1983 Sorrento Asian food tables: food tables, recipes and customs of mothers attending Sorrento Maternity Hospital, Birmingham, England. Human Nutr: Appl Nutr 37A:378–402

19. Eaton PM, Wharton PA, Wharton BA 1984 Nutrient intake of pregnant Asian women at Sorrento Maternity Hospital, Birmingham. Br J Nutr 52:457–468
20. Wharton PA, Eaton PM, Wharton BA 1984 Sub-ethnic variation in the diet of Moslem, Sikh and Hindu pregnant women at Sorrento Maternity Hospital, Birmingham. Br J Nutr 52:469–476
21. Grindulis H, Scott PH, Belton NR, Wharton BA 1985 Combined deficiency of iron and vitamin D in Asian toddlers. Arch Dis Child 61:843–848
22. Walravens PA, Chakar A, Mohni R, Denise J, Lemonnier D 1992 Zinc supplements in breast-fed infants. Lancet 340:683–685
23. Heiskanen K, Siimes MA, Salmanperä L, Perheentupa J 1995 Low vitamin B_6 status associated with slow growth in healthy breast fed infants. Pediatr Res 38:740–746
24. Michaelson KF, Samuelson G, Graham TW, Lönnerdal B 1994 Zinc intake, zinc status and growth in a longitudinal study of healthy Danish infants. Acta Paediatr 83:1115–1121
25. Kent GN, Price RI, Gutteridge DH, Smith M, Allen JR, Bhagat CI, Barnes MP, Hickling CJ, Retallack RW, Wilson SG, Devlin RD, Davies C, St John A 1990 Human lactation: forearm trabecular bone loss, increased bone turnover, and renal conservation of calcium and inorganic phosphate with recovery of bone mass following weaning. J Bone Min Res 5:361–369
26. Wharton BA 1992 Trace element deficiency in man: classification and methods of assessment. Food Chem 43:219–224
27. Keenan WJ, Jewett T, Glueck HI 1971 Role of feeding and vitamin K in hypoprothrombinemia of the newborn. Am J Dis Child 121:271–277
28. Balmer SE, Wharton BA 1989 Diet and faecal flora in the newborn: Breast milk and infant formula. Arch Dis Child 64:1672–1677
29. Balmer SE, Scott PH, Wharton BA 1989 Diet and faecal flora in the newborn: Casein and whey proteins. Arch Dis Child 64:1678–1684
30. Balmer SE, Scott PH, Wharton BA 1989 Diet and faecal flora in the newborn: Lactoferrin. Arch Dis Child 64:1685–1690
31. Balmer SE, Wharton BA 1991 Diet and faecal flora in the newborn: Iron. Arch Dis Child 68: 1390–1394
32. Simhon A, Douglas JR, Drasar BS, Soothill JF 1982 Effect of feeding on infants faecal flora. Arch Dis Child 57:54–58
33. McNinch AW, Tripp JH 1991 Haemorrhagic disease of the newborn in the British Isles: 2 year prospective study. Br Med J 303:1105–1109
34. Lakdawala DR, Widdowson EM 1977 Vitamin D in human milk. Lancet 1:167–168
35. Leerbeck E, Söndergaard H 1980 The total content of vitamin D in human milk and cow's milk. Br J Nutr 44:7–12
36. Cockburn F, Belton NR, Purvis RJ, Giles MM, Brown JK, Turner TL 1980 Maternal vitamin D intake and mineral metabolism in mothers and their newborn infants. Br Med J 281:11–14
37. Heckmatt JZ, Peacock M, Davies AEJ, McMurray J, Isherwood DM 1979 Plasma 25-hydroxyvitamin D in pregnant Asian women and their babies. Lancet 2:546–548
38. Brooke OG, Brown IRF, Bone CD, Carter ND, Cleete HJ, Maxwell JG 1980 Vitamin D supplements in pregnant Asian women: effects on calcium status and fetal growth. Br Med J 280: 751–754
39. Dallman PR 1986 Iron deficiency in the weanling: a nutritional problem on the way to resolution. Acta Paediatr Scand Suppl 323:59–67
40. Barry DMJ, Reeve AW 1977 Increased incidence of gram negative neonatal sepsis with intramuscular iron administration. Pediatrics 60:908–912
41. Van Ashbeck BS, Verbrugh HA, Van Oost BA, Marx JJM, Verhoef J 1982 *Listeria monocytogenes* meningitis and decreased phagocytosis associated with iron overload. Br Med J 284: 542–544
42. Melby K, Slordahl S, Gutteberg TJ, Nordbro SA 1982 Septicaemia due to *Yersinia enterocolitica* after oral overdose of iron. Br Med J 285:467–468
43. Mevissen-Verhage EAE, Marcelis JH, Harmsen-van Amerongen WCM, de Vos NM, Berkel J, Verhoef J 1985 Effect of iron on neonatal gut flora during the first week of life. Eur J Clin Microbiol 4:14–19
44. Mevissen-Verhage EAE, Marcelis JH, Harmsen-van Amerongen WCM, de Vos NM, Verhoef J 1985 Effect of iron on neonatal gut flora during the first three months of life. Eur J Clin Microbiol 4:273–278

45. Wharton BA 1992 Which milk for normal infants? Eur J Clin Nutr 46:S27–S32
46. Lansdown R, Wharton BA 1995 Iron and mental and motor behaviour in children. In: British Nutrition Foundation (ed) Iron: Nutritional and Physiological Significance. Chapman & Hall, London, pp 65–78
47. Wharton BA 1989 Iron nutrition in childhood: the interplay of genes, development and environment. Acta Paediatr Scand 361:5–11

20 Iron deficiency in infants: three topics of current interest

P. R. DALLMAN

ABSTRACT

This discussion will review three somewhat unresolved areas of iron nutrition that are becoming clarified as a result of recent studies. The first is the optimal level of iron fortification for infant formula used by term infants. The studies described favour the use of formulas with intermediate levels of added iron (6–8 mg/l) rather than the high levels (12–13 mg/l) that are widely used in the United States. The second topic of preventing iron deficiency in low birth weight infants emphasizes the importance of starting with premature formulas containing 12–13 mg iron/l or administering 2 mg/kg/day of medicinal iron by about 1 month of age. Very low birth weight infants require as much as 4 mg/kg/day of medicinal iron and have increased iron needs well into their second year. Because of their unusually high risk of developing iron deficiency anaemia, low birth weight infants need to be monitored for anaemia and iron deficiency at regular intervals. Infants breast-fed for more than 6 months are likewise at substantial risk of developing iron deficiency anaemia, even with a diet of iron- and ascorbic-acid-rich foods. Special, iron- and ascorbic-acid-fortified meat and cereal foods used after 4–6 months can be effective. Where these are not available, an alternative is the administration of medicinal iron at a dose of 1 mg/kg/day to a maximum of 15 mg.

INTRODUCTION

Iron deficiency in infancy has deservedly received a great deal of attention in the last two decades. Studies have indicated that it is associated with developmental delay, even when mild (1–3). Furthermore, its prevention in industrialized countries is relatively easy, reliable, and inexpensive using iron-fortified infant foods (4,5). There is a general consensus regarding most matters relating to iron nutrition in infants: first, the iron needs of term infants are greatest after 4–6 months of age; second, formula containing 12 mg added iron/l is effective in preventing iron deficiency, but infants who are breast-fed beyond 6 months are at substantial risk of developing iron deficiency anaemia. Third, unmodified cows' milk should not be used before 12 months of age, and finally, low birth weight infants need additional iron by about 1 month of age.

This discussion will focus on three areas of greater uncertainty. First is the optimal level of iron fortification of infant formula. A second issue is the management of iron nutrition in low birth weight infants. The third topic is the prevention of iron deficiency in infants who are breast fed beyond 6 months of age.

J.G. Bindels, A.C. Goedhart and H.K.A. Visser (eds), Recent Developments in Infant Nutrition, 272–277.

THE OPTIMAL LEVEL OF IRON FORTIFICATION OF INFANT FORMULA

The effectiveness of infant formulas fortified with 12 mg of iron/l (1.8 mg/100 kcal) and with ascorbic acid, first shown in laboratory studies, has been borne out in the USA as a whole. Over the last two decades, the increased use of iron-fortified formula and the decreased use of cows' milk have undoubtedly played a major role in the declining prevalence of anaemia in infants (4–6). In view of such an impressive success, why should this level of iron fortification be modified? An interest in giving no more iron than is needed to prevent iron deficiency is receiving an impetus from a rapidly expanding research literature on the possible risks of modest iron excess (7). There is a theoretical possibility that high intakes of iron could impair the absorption of zinc and copper. Another concern regarding unnecessary iron relates to the role that this element plays in the chemical reactions that lead to oxidative damage to tissues. A third issue involves what has been termed nutritional immunity. The iron-binding proteins transferrin and lactoferrin are believed to protect animals from infection by competitively binding the iron that invading organisms would require for their growth. These are among the factors in support of the lower levels of iron fortification commonly used in Western Europe, about 5–8 mg/l (0.8–1.2 mg/100 kcal).

There have been several indications that iron levels lower than 12 mg/l can effectively prevent iron deficiency in term infants. Gorten and Cross (8) found that this level of fortification was sufficient even to reverse anaemia in premature infants with iron deficiency anaemia, a group with substantially higher iron needs than term infants. A study by Saarinen and Siimes (9) on a relatively small group of infants found no significant difference in the amount of iron that infants absorbed from formula with 6 mg or 12 mg of added iron/l. This finding could be best explained on the basis of the well-established inverse relationship between the dose of iron given and the proportion that is absorbed (10,11). As more iron is given, one reaches a point of diminishing returns.

Two subsequent studies examined the lower levels of iron fortification, closer to or even below the levels of about 7 mg/l, that are commonly used in Western Europe. Haschke and co-workers (12) found that 27 and 24 infants fed 3 or 6 mg iron/l, respectively, of whey-predominant formula did not differ in their iron status and showed little or no evidence of iron deficiency at 9 months of age (there was no comparison group receiving a higher amount of iron). According to estimates of iron intake, iron-fortified solid food (beikost) probably accounted for more dietary iron than did the formula between 6 and 9 months. Consequently, the results apply only to those situations in which there is a substantial intake of iron- and ascorbic-acid-fortified foods other than formula. Nevertheless, the results provide strong support for the efficacy of lower levels of iron fortification.

Lönnerdal and Hernell (13) similarly found no evidence of iron deficiency in the first 6 months of infancy using formula with 4 or 7 mg iron/l. However, the study did not include the period of greatest vulnerability to iron deficiency, after 6 months of age. Interestingly, serum copper and caeruloplasmin concentrations were significantly lower in the group fed the higher iron formula, in accord with other evidence for an interaction between the absorption of the two metals.

For the past several years, Fomon and his group have been studying iron absorption in infants with the use of the stable isotope ^{58}Fe. Recently they studied the absorption of this isotope from infant formula containing either 8 or 12 mg iron/l (Similac, Ross Laboratories) by 46 term infants at about 6 months of age (submitted for publication). The mean incorporation of iron, adjusted for serum ferritin, from the two formulas was virtually identical. They concluded that it is reasonable to use the lower concentration in infant formula.

In Chile at the Institute of Nutrition and Food Technology (INTA), Hertrampf and associates studied the absorption of isotopically labelled iron from various infant formulas administered to menstruating, contraceptive-using women between the ages of 36 and 45 years (manuscript in preparation). Five formulas were investigated and the results were normalized to an absorption of 40% from a reference dose of iron, a percentage that is anticipated in individuals with depleted iron stores but without iron deficiency anaemia. The relevance of the results to infants was supported by a similar percentage of iron absorbed from the same cows' milk in the women as was obtained in infants by the same laboratory in 1978. In the women, the absorption of iron from two formulas with 8 mg iron/l (Similac, Ross Laboratories and Nan, Nestlé) and one with 7 mg iron/l (INTA) approximated 20%, an amount calculated to yield about 1 mg absorbed iron, assuming a consumption of 750 ml of formula. This amount covers the needs of most infants during their period of greatest vulnerability after 6 months of age (14,15). Formulas containing 12 mg or 1.5 mg iron/l would result in the absorption of about twice and less than half this amount respectively. The authors conclude that formulas containing 7–8 mg iron/l would cover the iron needs of term infants during their first postnatal year.

In a second study from Chile by Walter and co-workers, 789 infants were randomized at 6 months of age to one of two formulas (Similac, Ross Laboratories) containing 12 mg or 2.3 mg iron (to meet a label claim of 1.5 mg). The iron status of the infants was evaluated at 12 months of age. The low iron formula was surprisingly effective in pre-venting iron deficiency anaemia. Only 3.8% of the group developed iron deficiency anaemia compared to 2.2% in the group fed the higher iron formula. In the same popu-lation, roughly 30% would have been expected to develop iron deficiency anaemia on a diet based on unmodified cows' milk. The results show what a profound effect small concentrations of iron can have when added to presently marketed formulas. These formulas contain less casein, less calcium and more ascorbic acid than previously, and all of these changes promote iron absorption. However, it is important to point out that the low iron formula did not prevent iron deficiency. A battery of tests of iron status showed 25% to have biochemical evidence of iron deficiency compared with 13% in the high iron group. It is also worth emphasizing that the selection criteria at 6 months, including a birth weight >3.0 kg, would have excluded many of those most vulnerable to iron deficiency.

The results from the studies (Table 1) reviewed above certainly provide a basis for using iron fortification levels well below 12 mg/l. However, as the level of iron fortification is decreased it becomes increasingly important to consider whether other foods provide adequate nutritional iron after 6 months of age. In the USA one of the formula manufacturers has recently started to market an infant formula containing 4 mg iron/l. It would be desirable to determine whether this level is adequate to prevent iron deficiency

Table 1 Studies comparing different levels of iron supplemented to formulas for term infants

	Low	*Intermediate*	*High*
Saarinen and Siimes (9)		6 mg/l	12 mg/l
Haschke et al. (12)	3 mg/l	6 mg/l	
Lönnerdal and Hernell (13)	4 mg/l	7 mg/l	
Fomon et al. (unpublished)		8 mg/l	12 mg/l
Walter et al. (unpublished)	2.3 mg/l		12 mg/l

in term infants particularly after 6 months of age and under current infant feeding practices. It is inappropriate to meet the higher iron needs of low birth weight infants.

LOW BIRTH WEIGHT INFANTS

Low birth weight infants have much greater iron needs than term infants because of their more rapid relative growth and because their neonatal iron stores become depleted at an earlier age. Low birth weight infants who are breast-fed require medicinal iron at a dose of 2 to 3 mg/kg/day to a maximum of 15 mg/day, starting at about one month and continuing for at least one year (16). These needs can also be supplied to most infants by formulas fortified with 12 mg iron/l, but it is doubtful that much lower levels would be adequate. Very low birth weight infants (< 1000 g at birth) require about 4 mg/kg/day (17). These higher levels of iron should probably be started at about 1 month of age. Earlier studies showed that iron deficiency can develop between 2 and 3 months of age in low birth weight infants who are not given supplemental iron or a high iron formula (18). However, starting iron earlier now seems more appropriate in view of the substantial iron losses associated with blood drawn for laboratory studies in many of these infants. Hall and co-workers (19) compared premature formulas containing either 15 or 3 mg iron/l (Similac Special Care, Ross Laboratories) and started before 21 days of age, during the initial hospitalization of 40 infants with a birth weight < 1800 g. After discharge and until 4 months, both groups were fed a formula with 12 mg/l. More infants in the low iron group had evidence of iron deficiency at discharge from the hospital, but by 4 months these differences had largely disappeared. The authors concluded that the higher iron formula was beneficial during the initial hospitalization.

The study of Friel and associates (20) emphasizes the importance of maintaining a high iron intake and monitoring for anaemia well into the second year. The investigators followed an initial group of 81 infants with a birth weight < 1500 g at 3-month intervals until 15 months of age. The infants received a formula containing 13 mg iron/l until at least 8 months of age. Laboratory values were available on a markedly diminishing number during follow up. Nevertheless, it was clear that the proportion with plasma ferritin values < 10 μg/l increased progressively from 0% to 24, 31, 54 and 74% at 3, 6, 9, 12, and 15 months of age, respectively. The results emphasize the importance of regular follow up visits that include monitoring for anaemia and iron deficiency and the prolonged maintenance of high iron formula or iron supplements. How long iron supplements should be maintained into the second year is unclear, but for the present,

18 months seems a reasonable duration for very low birth weight infants. However, low birth weight infants may also need extra iron for as long as 15 months, the duration depending on their birth weight, growth rate, and diet.

Very low birth weight infants may also have increased iron needs when they receive treatment with recombinant human erythropoietin. This form of treatment is increasingly used to minimize the need for transfusions in the first two postnatal months when substantial blood loss for laboratory studies is anticipated. Prevention of severe anaemia requires concurrent administration of relatively large doses of iron. This topic was discussed in a recent review (21) and will not be covered here.

BREAST-FEEDING BEYOND SIX MONTHS

Most term, breast-fed infants can maintain adequate iron nutrition for the first 6 months. Subsequently, they are at potential risk of developing iron deficiency anaemia. Pizarro and associates (6) described a Chilean population of 102 infants with breast-feeding as the sole source of milk and no access to iron-fortified solid foods (beikost). Fifteen percent had iron deficiency anaemia by 9 months of age, compared with <1% of infants fed iron-fortified formula (about 12 mg iron/l).

Calvo and co-workers (22) found eight infants with iron deficiency anaemia among a group of 25 exclusively breast-fed infants (32%) at 9 months of age in Argentina. Recommendations for feeding after 6 months had included freshly prepared orange juice or other ascorbic-acid-rich foods, meat, poultry or beef liver at least once a day, and the avoidance of foods known to inhibit iron absorption, such as tea, eggs and cows' milk. No iron-fortified foods were included.

The potential for preventing iron deficiency among breast-fed infants with iron-fortified foods is indicated by the previously cited study of Haschke and associates (12). Infants were fed commercially prepared meat-containing baby foods fortified with ferrous sulphate and ascorbic acid (2 mg Fe and 10 mg ascorbic acid/100 g) as well as iron-fortified cereal after 6 months of age. A group of 30 infants who were still breast fed had little or no evidence of iron deficiency at 9 months of age. In the earlier study of Siimes and co-workers (23), iron-fortified foods probably had a less prominent role in the diet and there was more evidence of iron deficiency at 9 months.

The overall results on prolonged breast-feeding indicate that infants are increasingly vulnerable to iron deficiency after the age of 4–6 months. The use of well-absorbed forms of fortification iron, such as meat-containing meals with added iron and ascorbic acid, can prevent iron deficiency in most infants. However, this may not be possible with unfortified foods, even if they are selected to enhance iron bioavailability by including meat and ascorbic-acid-rich foods. Where special iron-fortified foods are not available, medicinal iron at a dose of 1 mg/kg/day to a maximum of 15 mg/day may be advisable for infants who are breast fed longer than 4–6 months.

ACKNOWLEDGEMENTS

The author is indebted to Dr Fomon and co-workers (University of Iowa, USA) and Drs Walter and Hertrampf and their groups (INTA, University of Chile) for giving kind permission to outline their most recent unpublished work in this review.

REFERENCES

1. Walter T, De Andraca I, Chadud P, Perales CG 1989 Iron deficiency anemia: adverse effects on infant psychomotor development. Pediatrics 84:7–17
2. Lozoff B, Jimenez E, Wolf AW 1991 Long-term developmental outcome of infants with iron deficiency. N Engl J Med 325:687–694
3. Idjradinata P, Pollitt E 1993 Reversal of developmental delays in iron-deficient anaemic infants treated with iron. Lancet 331:1–4
4. Yip R, Binkin NJ, Fleshood L, Trowbridge FL 1987 Declining prevalence of anemia among low-income children in the United States. JAMA 258:1619–1623
5. Yip R, Walsh KM, Goldfarb MG, Binkin NJ 1987 Declining prevalence of anemia in childhood in a middle-class setting: a pediatric success story? Pediatrics 80:330–334
6. Pizarro F, Yip R, Dallman PR, Olivares M, Hertrampf E, Walter T 1991 Iron status with different infant feeding regimens: relevance to screening and prevention of iron deficiency. J Pediatr 118:687–692
7. Dallman PR 1989 Upper limits of iron in infant formulas. J Nutr 119:1852–1855
8. Gorton MK, Cross ER 1964 Iron metabolism in premature infants. II. Prevention of iron deficiency. J Pediatr 64:509–520
9. Saarinen UM, Siimes MA 1977 Iron absorption from infant milk formula and the optimal level of iron supplementation. Acta Paediatr Scand 66:719–722
10. Heinrich HC, Gabbe EE, Whang DH 1969 Die Dosisabhängigkeit der intestinalen Eisenresorption bei Menschen mit normalen Eisenreserven und Personen mit prälatentem/latentem Eisenmangel. Z Naturforsch 24b:1301–1310
11. Hahn PF, Carothers EL, Darby WJ, Martin M, Sheppard CW, Cannon RO, Beam AS, Densen PM, Peterson JC, McClellan GS 1951 Iron metabolism in human pregnancy as studied with the radioactive isotope, Fe^{59}. Am J Obstet Gynecol 61:477–486
12. Haschke F, Vanura H, Male C, Owen G, Pietschnig B, Schuster E, Kroback E, Huemer C 1993 Iron nutrition and growth of breast- and formula-fed infants during the first 9 months of life. J Pediatr Gastroenterol Nutr 16:151–156
13. Lönnerdal B, Hernell O 1994 Iron, zinc, copper and selenium status of breast-fed infants and infants fed trace element fortified milk-based infant formula. Acta Paediatr 83:367–373
14. Dallman PR 1992 Changing iron needs from birth through adolescence. In: Fomon SJ, Zlotkin S (eds) Nutritional Anemias. Nestlé Nutrition Workshop Series, Vol 30, Nestec Ltd, Vevey/Raven Press Ltd, New York, pp 29–38
15. National Research Council 1989 Recommended Dietary Allowances, 10th edn. Food and Nutrition Board, National Academy Press, Washington, DC
16. Dweck HS, Finberg L, Holmes F, Reynolds JD and American Academy of Pediatrics, Committee on Nutrition 1985 Nutritional needs of low-birth-infants. Pediatrics 75:976–985
17. Siimes MA, Järvenpää AL 1982 Prevention of anemia and iron deficiency in very low-birth-weight infants. J Pediatr 101:277–280
18. Lundström U, Siimes MA, Dallman PR 1977 At what age does iron supplementation become necessary in low-birth-infants? J Pediatr 91:878–883
19. Hall RT, Wheeler RE, Benson F, Harris G, Rippetoe L 1993 Feeding iron-fortified premature formula during initial hospitalization to infants less than 1800 grams birth weight. Pediatrics 92: 409–414
20. Friel JK, Andrews WL, Mathew JD, Long DR, Cornel AM, Cox M, Skinner CT 1990 Iron status of very-low-birth-infants during the first 15 months of infancy. Can Med Assoc J 143:733–737
21. Dallman PR 1993 Anemia of prematurity: the prospects of avoiding blood transfusions by treatment with recombinant human erythropoietin. Adv Pediatr 40:385–403
22. Calvo EB, Galindo AC, Aspres NB 1992 Iron status in exclusively breast-fed infants. Pediatrics 90:375–379
23. Siimes MA, Salmanperä L, Perheentupa J 1984 Exclusively breast-feeding for nine months: risk of iron deficiency. J Pediatr 104:196–199

21 What are the essential amino acids for the preterm and term infant?

P. B. PENCHARZ, J. D. HOUSE, L. J. WYKES and R. O. BALL

ABBREVIATIONS

IAA indispensable amino acid
LBW low birth weight
TPN total parenteral nutrition

INTRODUCTION

The classical definition of essential amino acids is those α-amino acids whose carbon skeleton cannot be synthesized in the human (animal) body (1). Over the past 15 years it has become apparent that the simple division of 20 amino acids for which there exists a t-RNA into essential and non-essential was too simplistic (2). A newer terminology has been created, which will be used in the remainder of this review, namely, indispensable (which is equivalent to essential), conditionally indispensable and dispensable (which is equivalent to non-essential). The concept of conditionally indispensable is particularly important when considering the amino acid needs of infants. Optimal growth of lean tissue requires that an appropriate mixture of all 20 amino acids and sufficient non-protein energy is present to optimize net protein synthesis. Interestingly, Duffy et al. (3) in parenterally fed very low birth weight neonates and Fuller (4) in piglets have shown that an improved balance of amino acid intake results in enhanced net protein deposition by reducing endogenous protein breakdown.

The total demand for dietary nitrogen is generally represented as the requirement for protein. However, amino acids, rather than protein, are required for the maintenance and growth of body tissues. Twenty amino acids are known to be incorporated into body protein. Animals have an ability to synthesize a number of these amino acids to meet their needs for protein synthesis (2,5). These amino acids are classified as nutritionally non-essential or dispensable, since they are not required in the diet, per se, to meet the animals' needs for protein synthesis. Those amino acids for which there is a need for a preformed dietary source are called nutritionally essential or indispensable. The essential nature of certain amino acids results from a complete or partial lack of the enzymatic capacity for their endogenous synthesis. For the purposes of defining a need for amino acids with a well-defined homogeneous population,

J.G. Bindels, A.C. Goedhart and H.K.A. Visser (eds), Recent Developments in Infant Nutrition, 278–296.

Table 1 Classification of amino acids and taurine as indispensable, conditionally indispensable, or dispensable for LBW infants

Dispensable	*Indispensable*	*Conditionally indispensable**
Alanine	Histidine	Arginine
Asparagine	Isoleucine	Cysteine
Aspartate	Leucine	Glutamine
Glutamate	Lysine	Glycine
Serine	Methionine	Proline
	Phenylalanine	Taurine†
	Threonine	Tyrosine
	Tryptophan	
	Valine	

*Insufficient evidence for the classification of these amino acids as either dispensable or indispensable for the growth and development of LBW infants
†Taurine is not synthesized into protein

these two classifications are probably all that is necessary. However, several classification schemes have been proposed for amino acids on the basis of their essentiality or indispensability (5). These schemes include terms such as conditionally indispensable, referring to an amino acid which is normally considered to be dispensable, although a dietary need has been defined during certain metabolic or pathophysiological states. While these terms are useful for comparing requirements across different animal species or ages within species, they are not particularly useful for classifying amino acids within well-defined subgroups. Table 1 outlines the proposed classification of the amino acids for low birth weight (LBW) infants, on the basis of their dietary requirement.

Dispensable amino acids are those for which there exists little evidence of a dietary need for preformed sources. For the LBW infant these amino acids include alanine, asparagine, aspartate, glutamate, and serine (Table 1). While there may not be a specific dietary requirement for these amino acids, they contribute to the non-specific nitrogen pool, from which several amino acids can be synthesized as substrates for protein synthesis.

Indispensable amino acids (IAA) are those which cannot be synthesized endogenously, and they are therefore required preformed in the diet. Initial studies by Rose et al. (6), using amino acid deletion studies, identified eight amino acids (isoleucine, leucine, lysine, methionine, phenylalanine, threonine, tryptophan and valine) as being indispensable for growth of the rat. Rose and colleagues confirmed the indispensability of these amino acids for the adult human in a series of studies which outlined the estimates of amino acid requirements for the adult human male (1). Studies of the amino acid requirements of infants and children also confirmed the indispensability of these eight amino acids, and also documented the indispensability of the amino acid histidine (7). The indispensability of histidine was not shown for adult humans until later (8,9). The supply of these nine indispensable amino acids (Table 1) is therefore imperative in order to provide sufficient substrates for protein synthesis.

The remaining amino acids listed in Table 1 are those for which a definite need of preformed dietary sources is equivocal for preterm and term infants. Arginine and proline are currently not considered to be IAAs for the healthy growing infant (10). However, young rapidly growing piglets have been shown to require a dietary supply of these amino acids (11,12). For LBW infants receiving total parenteral nutrition (TPN), the addition of arginine to an early amino acid solution reduced the incidence of hyperammonaemia (13), but a specific need for this amino acid, as well as proline, for anabolic purposes has not been demonstrated in infants.

In most animals, cysteine is synthesized in vivo from the IAA methionine. However, endogenous production of cysteine from methionine may be inadequate for the LBW infant, due to the low activity of hepatic cystathionase, a key enzyme in the transulfuration pathway (14). Low plasma cysteine concentrations and depressed growth rates have been reported in infants receiving cysteine-free oral formulas (15), although this appears to be the exception rather than the rule (16). The cysteine content of parenteral amino acid solutions is quite low, due to the fact that cysteine is unstable in solution, being oxidized to the insoluble cystine. The delivery of parenteral amino acid solutions containing a cysteine • HCl product resulted in higher plasma cysteine concentrations, although no conclusive effects on nitrogen balance or growth were observed (17). This may be due to the ability of extra-hepatic cystathionase activity to maintain sufficient cysteine for anabolic purposes (18). However, Heird (16) maintains that failures to see beneficial effects of cysteine supplementation to paediatric TPN formulas may be due to other factors, such as the concomitant deficiency of another amino acid (tyrosine) or the conversion of the added cysteine to unusable forms. Further research is warranted, however, to confirm the metabolic need for this amino acid. The same statement can be made for tyrosine, an amino acid whose supply in parenteral solutions is limited by its poor solubility. As with cysteine, tyrosine was shown to be required for the formula-fed infant, as judged by the development of low plasma tyrosine concentrations and reduced rates of nitrogen retention and weight gains (15). Since phenylalanine can be hydroxylated to tyrosine in vivo (19), the demonstration of a need for preformed tyrosine suggests a reduced capacity for phenylalanine hydroxylation in the neonate. Phenylalanine-4-monooxygenase activity is lower in livers of human fetuses and neonates than in adults (20). Furthermore LBW infants receiving amino acid solutions with high phenylalanine and low tyrosine concentrations occasionally exhibit high plasma concentrations of phenylalanine (21,22), supporting the observations of a lower capacity for phenylalanine hydroxylation in the neonate. Despite these data, LBW infants receiving TPN solutions containing high phenylalanine concentrations appear to have a significant capacity for phenylalanine hydroxylation (23–25), as measured using stable isotope tracer techniques. The phenylalanine content of these solutions, however, could result in the presentation of an excess phenylalanine supply which, despite significant hydroxylation, could lead to higher plasma concentrations of phenylalanine. While phenylalanine may be able to meet the endogenous demand for tyrosine, a more effective approach might be to supply tyrosine in a form that is both soluble and biologically effective. Significant interest has been given to testing the biological effectiveness of soluble tyrosine precursors, such as the dipeptide *N*-acetyltyrosine (26,27), and glycyl-L-tyrosine (28). Wykes compared the effectiveness of phenylalanine, *N*-acetyltyrosine, and glycyl-L-tyrosine in supplying tyrosine in vivo, and found that glycyl-L-tyrosine was an effective source of tyrosine, leading to high rates of nitrogen retention in parenterally fed piglets. Evidence from piglet studies (28) supports the use of glycyl-L-tyrosine for providing tyrosine during TPN; however, the tyrosine intakes were in excess, as judged by high plasma tyrosine concentrations and the excretion of tyrosine catabolites in the urine. Studies designed to estimate the correct proportions and overall amount of the aromatic amino acids could dramatically aid in the development of an amino acid profile that results in normal plasma amino acid concentrations.

Taurine, while not synthesized into body proteins, may be required by the LBW infant. Taurine biosynthesis in neonates may be insufficient, due to the low activity of cysteine-sulfinic acid decarboxylase (29). The supply of taurine to neonates is generally accomplished by the feeding of breast milk, which contains high levels of free taurine. However, the feeding of infant formulas lacking taurine results in low plasma taurine concentrations (30). Taurine deficiencies lead to retinal degeneration in cats (31) and depressed growth rates in infant monkeys (32). Retinal abnormalities have also been observed in children receiving taurine-free TPN for more than 6 months (33). For these reasons, taurine is added to virtually all infant formulas and paediatric parenteral amino acid solutions.

The amino acid glutamine is not included in commercially available TPN solutions because of concerns about its stability in aqueous media following heat sterilization (34,35). Although traditionally classified as a dispensable amino acid, growing evidence suggests that endogenous glutamine supply may be insufficient during some metabolic states (36), because of its role as an energy substrate for rapidly dividing cells (37). A dietary source of glutamine may be required for LBW infants receiving TPN, since glutamine synthetase activity is much lower in the fetal and neonatal tissues of the rat (38,39) and lambs (40). Glutamine supply to fetal lambs appears to result from placental synthesis (41). Therefore, we reasoned that a lack of glutamine during parenteral feeding might limit protein synthesis in neonates (42). It is important to note that addition of glutamine to TPN did not improve nitrogen accretion, but in fact had an adverse side effect of increasing the retention of extracellular water (42). Glutamine metabolism was recently studied in children with short-bowel syndrome (43) and it was shown that bowel resection was associated with a reduction in glutamine appearance rate. This was in contrast to leucine appearance, which was unaltered by small bowel resection. This is further evidence that glutamine has a role in small bowel metabolism. This topic is dealt with in more detail in Chapter 23 by Elia.

The appearance of the ^{15}N label in urinary urea was markedly lower than expected during an infusion of [^{15}N]glycine to LBW infants, leading Jackson et al. (44) to speculate that glycine was needed in the diet for the maintenance of protein synthesis and accretion; hence it was not available for transfer from glycine to urea. A similar disappearance of the ^{15}N label in urinary urea (but not in urinary ammonia) has been reported in small for gestational age infants during a period of refeeding (45). Others, however, have not seen this effect (46). A careful comparison of these three papers shows that in the two studies in which the urea labelling disappeared, infants received significantly less protein and hence (probably) less glycine than in the study in which urea labelling was present, but at a lower level than that seen in appropriate for dates infants (46). This hypothesis has not been proven with investigations designed to examine nitrogen balance, growth or other indices of protein status, therefore this observation remains incompletely understood.

In order to promote the accretion of body protein in infants all of the amino acids that are indispensable for growth should be present (5,15,47). An adequate source of non-essential nitrogen, provided by a mixture of the dispensable amino acids, should also be included (15).

NEWER CONCEPTS FOR DETERMINING THE REQUIREMENTS OF INDISPENSABLE AMINO ACIDS AND THEIR APPLICATION TO THE INFANT

Over the past 10 years there has been an active debate regarding the requirement of IAA in adults. Briefly, Young and colleagues championed the use of direct amino acid oxidation as a technique to determine amino acid requirements (48). Using this technique the MIT group have suggested that the requirements of many IAA in adult are underestimated by 2–3 times (5,48). This concept has been challenged on theoretical and experimental grounds (49). Nevertheless our group (50,51) using the completely independent method of indicator amino acid oxidation have confirmed the validity of the direct oxidation estimate of lysine requirement. Indicator amino acid oxidation was developed and validated in growing pigs (52).Table 2 shows the FAO/WHO/UNU estimates of amino acid requirement, from infancy to adults, all of which are derived from nitrogen balance and plasma amino acid level techniques. Also included in this Table are requirement estimates, in adults, derived using amino acid oxidation methods. Currently for practical and ethical reasons there are no amino acid requirement estimates derived by oxidation techniques in either children or infants.

Determination of amino acid needs requires the use of a crystalline amino acid protein source. This source is highly unpalatable, which makes its use in infants and children difficult. Further, the isotope model used to date involves the intravenous infusion of the tracer and sampling of isotope abundance in plasma. We are endeavouring to develop a simpler model which can be used in children by limiting the need for the experimental diet to only the day of study, oral administration of the isotope, using urine to sample plasma amino acid enrichment (66,67) and sampling only breath for oxidation values. At the same time we decided that the best initial approach to this issue in neonates was to use an animal model. A careful review of the literature led us to conclude that the piglet was the most suitable non-primate model (68–72), in part because dietary amino acid requirement, expressed as a percentage of protein, as well as tissue composition of amino acids between humans and piglets were similar (see Table 3).

STUDIES OF AMINO ACID REQUIREMENTS IN A NEONATAL PIGLET MODEL

Partly with a view of later testing the findings from the piglet model in the human neonate we chose to develop a parenterally fed (TPN) piglet model. First we demon-

Table 2 **_Estimates of the indispensable amino acid requirements of humans: Requirements by method* and age (mg/kg body weight/day)_**

	Infants[a] (3–4 months; NB)	*Children*[b] (~2 years; NB)	*Children*[c] (10–12 years; NB)	*Adults*[d] (NB)	*Adults*[e] (PAA)	*Adults*[f] (AAO)	*Adults*[g] (MOL)	*Adults*[h] (IAAO)
Histidine[†]	28	[19]	–	[8–12]	–	–	–	–
Isoleucine	70	31	28	10	–	–	17	–
Leucine	161	73	42	14	30	27	28	–
Lysine	103	64	44	12	32	35	31	36
Methionine plus cysteine	58	27	22	13	–	–	14	–
Phenylalanine plus tyrosine	125	69	22	14	30	30	28	–
Threonine	87	37	28	7	15	15	16	–
Tryptophan	17	12.5	3.3	3.5	3	–	4	4.3
Valine	93	38	25	10	20	17	18	–

*NB, nitrogen balance; PAA, plasma amino acid concentration; AAO, amino acid oxidation; MOL, calculated minimum oxidative losses of indispensable amino acids; IAAO, indicator amino acid oxidation

[†]Values for histidine requirements in brackets were calculated by extrapolation from the FAO/WHO/UNU estimates of requirement (10), Canada Health and Welfare (65)

[a]Based on amino acid concentrations in human milk or cows' milk that supported growth (10,53)

[b]Based on NB supporting ample lean tissue gain (10,54)

[c]Based on upper range of requirement for positive NB (10,55,56)

[d]Based on highest estimate of requirement to achieve NB (reviewed in Ref. 10)

[e]Based on data from Young et al. (57–59), Meredith et al. (60) and Zello et al. (61)

[f]Based on data from Young and colleagues (62,63)

[g]Estimates calculated by equations from minimum oxidation losses by Young et al. (48)

[h]Based on data from Zello et al. (50) and Lazaris-Brunner et al (64)

Table 3 ***Amino acid requirements (g/100g protein) of the piglet and human infant compared with the amino acid composition of various standards***

Amino acid	*Piglet requirements*[a]	*Human requirements*[b]	*Piglet tissue*[c]	*Human fetal tissue*[d]	*Swine milk*[e]	*Human milk*[e]	*Egg protein*[f]
Isoleucine*	3.2	3.2	3.6	3.5	4.0	5.3	5.8
Leucine*	4.2	7.3	6.9	7.5	8.9	10.4	9.0
Valine*	3.4	4.2	4.7	4.7	4.6	5.1	7.2
Lysine*	6.0	4.7	6.7	7.2	7.9	7.1	6.7
Methionine*	} 2.9[g]	2.6[g]	1.6	2.0	2.2	1.6	3.0
Cysteine*			1.5	n.d.	1.6	2.0	2.1
Phenylalanine*	} 4.7[h]	5.7[h]	3.7	4.1	4.3	3.7	5.3
Tyrosine*			2.8	2.9	3.9	4.6	4.3
Threonine*	3.4	4.0	3.9	4.1	3.7	4.4	5.3
Tryptophan*	0.9	0.7	n.d.	n.d.	1.4	1.9	1.8
Histidine*	1.5	1.3	2.5	2.6	2.4	2.3	2.6
Arginine	2.5	n.r.	6.9	7.7	4.4	3.6	6.4
Glycine	n.r.	n.r.	9.7	7.2	3.2	2.2	3.8
Alanine	n.r.	n.r.	6.6	9.0	3.6	4.0	n.d.
Aspartate	n.r.	n.r.	8.0	13.0	7.8	8.6	10.7
Glutamate	n.r.	n.r.	13.8	11.8	20.8	19.0	12.3
Proline	n.r.	n.r.	6.7	8.4	11.7	9.5	4.3
Serine	n.r.	n.r.	4.2	4.1	5.1	6.1	7.7
Total IAA	28.7	33.7	35.4	38.6	44.9	48.4	53.1

*:IAA; n.r.: no requirements established; n.d.: not determined
[a]Calculated from NRC (1988) (73) requirement for swine weighing 1–5kg
[b]Calculated from FAO/WHO (1990) (74) requirement for infants 3–4 months of age
[c]Tissue of piglets 7 days of age (75)
[d]Calculated from Widdowson: (76) tissue of fetuses 2.9–3.4kg
[e]Ref. 77
[f]Ref. 78
[g]Requirement for the sum of methionine and cysteine
[h]Requirement for the sum of phenylalanine and tyrosine

strated that an accretion of protein nitrogen could be achieved which was comparable to that occurring in sow-fed controls (68). Next we used the piglet TPN model to study glutamine requirements (see above and 42), aromatic amino acid metabolism and requirements and lysine requirements using direct and indicator amino acid oxidation techniques.

Aromatic amino acid metabolism

The amino acid requirements of neonates receiving total parenteral nutrition are poorly defined (47). Of particular concern is the concentration of the aromatic amino acids, phenylalanine and tyrosine, due to the relatively poor solubility of crystalline tyrosine. Attempts to meet the parenteral aromatic amino acid requirements of neonates by providing large quantities of phenylalanine (8% by weight of total amino acids) have resulted in extremely elevated plasma phenylalanine concentrations (21,22). However, failure to compensate for tyrosine insolubility can lead to reduced rates of growth and nitrogen retention. In a series of experiments (28,79), we investigated the effectiveness of several tyrosine precursors, as well as two amino acid profiles, in a parenterally fed pig model. Piglets received one of five TPN regimens, differing only in the amino acid profile: (1) Vamin profile; (2) Vaminolact profile; (3) Vaminolact plus phenylalanine; (4) Vaminolact plus glycyl-L-tyrosine; (5) Vaminolact plus *N*-acetyltyrosine. All five regimens contained the same amount of crystalline tyrosine, and the formulations 1, 3, 4 and 5 contained the same amount of total aromatic amino acids (8.8% phenylalanine plus tyrosine). Table 4 shows the complete amino acid profile of the TPN solutions used in infants. The effects of the amino acid profile on nitrogen balance and aromatic amino acid metabolism, over an 8-day study period, are reported in Table 5. Piglets receiving the TPN solution based on the amino acid pattern of Vamin had acceptable nitrogen retention,while those receiving the amino acid pattern of Vaminolact retained less nitrogen. Supplementation of Vaminolact with either phenylalanine or glycyl-L-tyrosine, improved nitrogen retention dramatically. However, the provision of *N*-acetyltyrosine as a source of tyrosine provided no demonstrable benefits when performance criteria of this group were compared to the group given the original Vaminolact-based TPN. Phenylalanine oxidation, measured during a 4 hour primed-continuous infusion of L-[1^{14}C]phenylalanine on study day eight, was higher in the two treatment groups receiving the high phenylalanine intake. The oxidation of phenylalanine (measured as a percentage of the dose oxidized) was more than halved in the three treatment groups receiving the lower level of phenylalanine.

The plasma concentrations of phenylalanine and tyrosine offer significant insights into the metabolism of these amino acids. Delivery of an identical aromatic amino acid pattern, but with other amino acids in a different pattern (Vamin vs Vaminolact + phenylalanine), reduced the plasma phenylalanine concentrations by more than 5-fold. These data suggest that the amino acid pattern of Vamin formulation is deficient in at least one IAA, potentially lysine (79), which is limiting protein synthesis. The highest levels of total aromatic amino acids used in these experiments were patterned after the content in human milk protein but, based on phenylalanine oxidation and

***Table 4* The amino acid profile of commercially available parenteral amino acid solutions (g/100 g total amino acid), compared with the patterns in human and swine milk**

Amino acid	*Human milk*[a]	*Sow's milk*[a]	*Vamin*[b]	*Aminosyn*[c]	*Vaminolact*[b]	*Aminosyn PF*[c]	*Travasol Blend C*[d]	*Trophamine*[e]	*Primene*[d]
Alanine	4.0	3.6	4.3	12.9	9.7	7.0	20.7	5.4	7.9
Arginine	3.6	4.4	4.7	9.9	6.3	12.3	11.2	12.2	8.4
Aspartate	8.6	7.8	5.9	0	6.3	5.3	0	3.2	6.0
Cysteine	2.0	1.6	2.0	0	1.5	0	0	0.1	1.9
Glutamate	19.0	20.8	12.9	0	10.9	8.2	0	5.0	9.9
Glycine	2.2	3.2	3.0	12.9	3.2	3.9	10.3	3.6	4.0
Histidine	2.3	2.4	3.5	3.0	3.2	3.1	4.8	4.8	3.8
Isoleucine	5.3	4.0	5.6	7.3	5.5	7.6	6.0	8.2	6.7
Leucine	10.4	8.9	7.5	9.5	10.8	11.9	7.3	14.0	9.9
Lysine	7.1	7.9	5.5	7.3	8.6	6.8	5.8	8.2	10.9
Methionine	1.6	2.2	2.7	4.0	2.0	1.8	4.0	3.4	2.4
Phenylalanine	3.7	4.3	7.9	4.7	4.2	4.3	5.6	4.8	4.2
Proline	9.5	11.7	11.6	8.7	8.6	8.1	6.8	6.8	3.0
Serine	6.1	5.1	10.7	4.2	5.8	5.0	5.0	3.8	4.0
Threonine	4.4	3.7	4.3	5.2	5.5	5.1	4.2	4.2	3.7
Tryptophan	n.d.	n.d.[f]	1.4	1.6	2.2	1.8	1.8	2.0	2.0
Tyrosine	4.6	3.9	0.7	0.9	0.8	0	0.4	2.3[g]	0.9
Valine	5.1	4.6	6.1	8.1	5.5	6.6	5.8	7.8	7.6
Taurine	–	–	0	0	0.5	0.7	0	0.2	0.6

[a]Data from Ref. 77
Manufacturer: [b]Kabi, [c]Abbott, [d]Clintec, [e]Kendall-McGaw
[f]Tryptophan levels not determined
[g]Approximately 75% of tyrosine present as *N*-acetyltyrosine
n.d.: not determined

***Table 5** The effect of aromatic amino acid profile of neonatal TPN solutions on growth, nitrogen retention, phenylalanine oxidation and plasma phenylalanine and tyrosine concentrations in neonatal piglets*[a]

	Vamin	*VL**	*VL + Phe**	*VL + GT**	*VL + AT**
Phenylalanine intake (g/kg/day)	1.13	0.59	1.13	0.61	0.61
Tyrosine intake (g/kg/day)	0.10	0.11	0.11	0.71[†]	0.71[†]
Nitrogen retention (%)	82.4	70.0	87.3	84.5	74.0
Phenylalanine oxidation (% of dose)	22.8	1.1	24.3	9.0	7.6
Plasma phenylalanine (μmol/l)	2234	156	399	148	118
Plasma tyrosine (μmol/l)	90	12	79	307	27

[a]Adapted from Refs 72 and 79
Data represent arithmetic means of treatment groups
*VL, Vaminolact; VL + Phe, Vaminolact plus phenylalanine; VL + GT, Vaminolact plus glycyl-L-tyrosine; VL + AT, Vaminolact plus *N*-acetyltyrosine
[†]Presented as a tyrosine equivalent assuming 100% availability

plasma phenylalanine concentrations, were probably supplied in excess of requirements.

These data support the ability of glycyl-L-tyrosine to provide tyrosine in vivo, although the rate of inclusion, while consistent with the percentage of tyrosine in human and swine milk proteins (see Table 4), was clearly in excess, based on the high plasma tyrosine concentrations (Table 5) and the presence of tyrosine catabolites in the urine (28). Furthermore, a total aromatic amino acid content of 8.8% of the total amino acids appears to be in considerable excess, supporting the need to specifically determine requirements for phenylalanine and tyrosine in neonates receiving TPN. Phenylalanine oxidation was sensitive to the level of phenylalanine intake, therefore oxidation techniques for the estimation of amino acid requirements should be adapted to the parenterally fed neonate. Amino acid oxidation techniques for the estimation of requirements have previously been confined to the oral feeding animal (for review see Ref. 51). Development of sensitive and minimally invasive oxidation techniques for the estimation of amino acid requirements during TPN would be a significant contribution to clinical research.

The determination of amino acid requirements using direct and indicator oxidation techniques

As part of our research goals, we wished to establish the indicator amino acid oxidation technique (50,52,80) as a method for determining amino acid requirements during TPN. This technique has usually monitored the oxidation of phenylalanine in response to graded additions of another IAA (see Ref. 51 for review).

Phenylalanine oxidation has usually been utilized as the indicator for the adequacy of other IAA. Therefore, before this method could be applied, the phenylalanine requirement during TPN had to be determined. In order to minimize the loss of label from the phenylalanine to the tyrosine pool, tyrosine must be supplied in excess of its requirement. However, an extreme excess of tyrosine must be avoided since both phenylalanine and tyrosine share catabolic pathways. In previous studies (28), we used a level of glycyl-L-tyrosine supplementation at which the tyrosine excess was too great. Therefore, we undertook to measure tyrosine kinetics and to determine the requirement for tyrosine, when supplied as glycyl-L-tyrosine, in parenterally fed piglets. The tyrosine requirement thus determined would be used in future experiments designed to determine amino acid requirements, based on the oxidation of L-[1 ^{14}C]phenylalanine, in neonatal piglets receiving TPN.

Figure 1 depicts the changes in nitrogen retention and tyrosine oxidation in piglets receiving TPN containing graded intakes of tyrosine. Phenylalanine was provided at the concentration in human milk protein (Table 4). Based on a two-phase linear regression cross-over model, a tyrosine intake of 0.31 g/kg/day, equivalent to 2.1 g/100 g amino acids, is estimated to be the minimum tyrosine level necessary to maximize nitrogen retention and minimize tyrosine oxidation, when phenylalanine, total amino acids and available energy are provided at 4.1 g/100 g of amino acids, 15 g/kg/day and 1.1 MJ/kg/day, respectively. However, with the subsequent calculation of the 95% confidence values (51) an adequate supply of tyrosine was concluded to be 2.8% of the total L-amino acids.

This determination of tyrosine requirement provided a level of tyrosine which could then be used in subsequent studies designed to determine the phenylalanine requirement by direct oxidation techniques. Twenty-four piglets received the same TPN for 5.5 days, with phenylalanine provided at 4.2 g/100 g of total amino acids and tyrosine present in excess. Piglets were switched to the test phenylalanine intakes 12 h prior to the initiation of the oxidation study. Phenylalanine oxidation was low and similar for the three lowest intakes, and increased linearly thereafter (Fig. 2). Partitioning the data between two regression equations and determining the cross-over or break-point yielded a result of 2.9 g/100 g of total amino acids. This value represents the requirement for the 50% of the population (51). In order to meet the needs of 95% of the population, the upper 95% confidence value was calculated and taken to represent the safe level of phenylalanine intake, corresponding to 3.3 g/100 g of total amino acids. Therefore, a total aromatic amino acid content of 6.1 g/100 g of total amino acids, with phenylalanine composing at least 54%, is necessary to meet the needs of the parenterally fed neonatal piglet. These data support earlier observations that an aromatic amino acid content of 8.8 g/100 g of the total L-amino acid profile is excessive for the neonate on TPN (79), and may partially explain the high plasma concentrations of phenylalanine and tyrosine observed in some infants receiving TPN with high aromatic amino acid content (21,22).

The use of phenylalanine as an indicator amino acid requires that it be provided at safe intake levels to ensure correct response (51). Therefore it was necessary to estimate safe levels of intake for both phenylalanine and total aromatic amino acids. Having achieved this (Figs 1, 2), it was then possible to adapt the indicator methodology to the case of the neonate receiving TPN. Phenylalanine oxidation was determined in 21 piglets receiving 8.3 g of lysine and 6.1 g total aromatics/100 g of total amino acids with phenylalanine providing 54% of the total (by weight) for 5.5 days. Twelve hours

***Figure 1** Nitrogen retention (▲) and oxidation of L-[1-^{14}C]-tyrosine (+) in neonatal piglets (n = 15) receiving TPN with graded intakes of glycyl-L-tyrosine. Tyrosine intakes were 0.11, 0.31, 0.41, 0.51 and 0.71 g/kg/day. Values are mean ± SE. Values with different superscripts are significantly different ($p < 0.05$)*

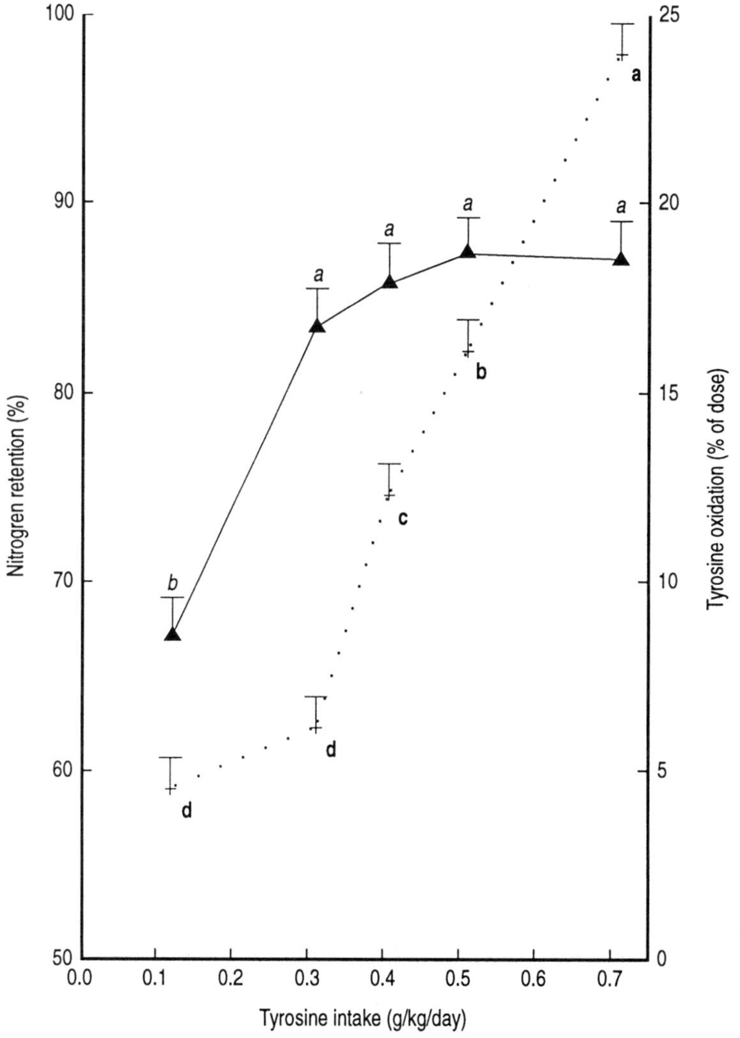

***Figure 2** The oxidation of L-[1[14]C]-phenylalanine in neonatal piglets (n = 24) receiving TPN with graded intakes of phenylalanine. Phenylalanine intakes were 0.2, 0.3, 0.4, 0.5, 0.6, 0.7 or 1.2 g/kg/day. Values are mean ± SE. Values with different superscripts are significantly different (p < 0.05)*

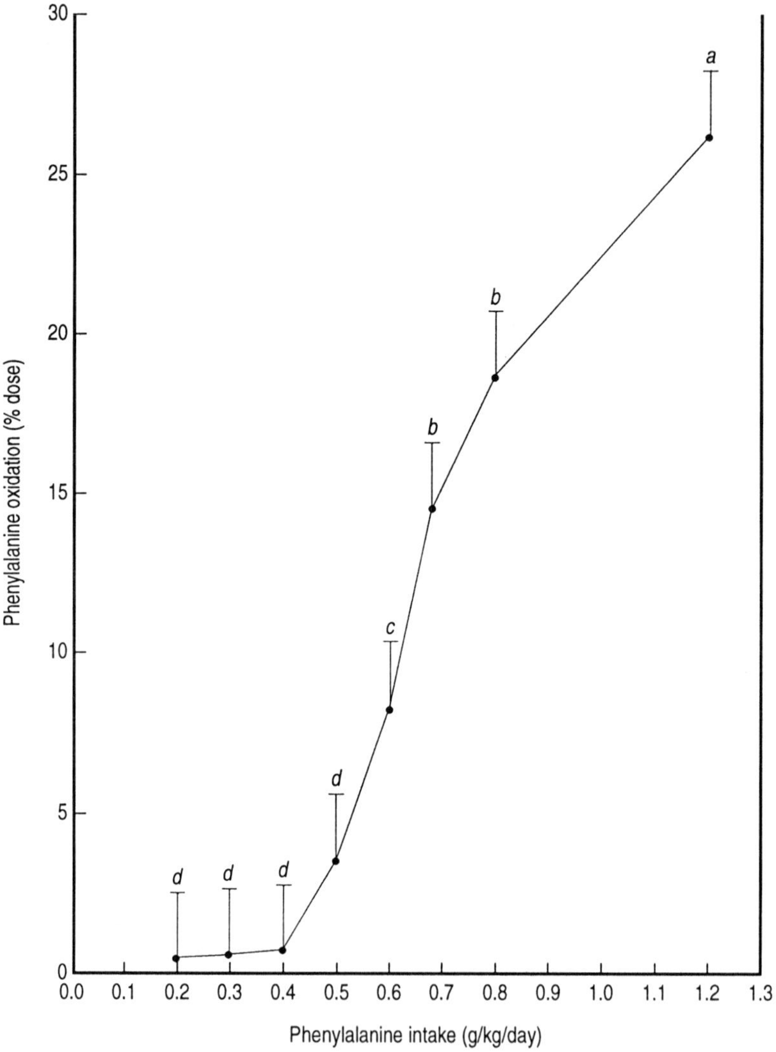

***Figure 3** The oxidation of L-[1^{14}C]-phenylalanine in neonatal piglets (n = 21) receiving TPN with graded intakes of lysine. Lysine intakes were 0.4, 0.55, 0.7, 0.85, 1.0, and 1.3 g/kg/day. Values are mean ± SE. Values with different superscripts are significantly different (p < 0.05)*

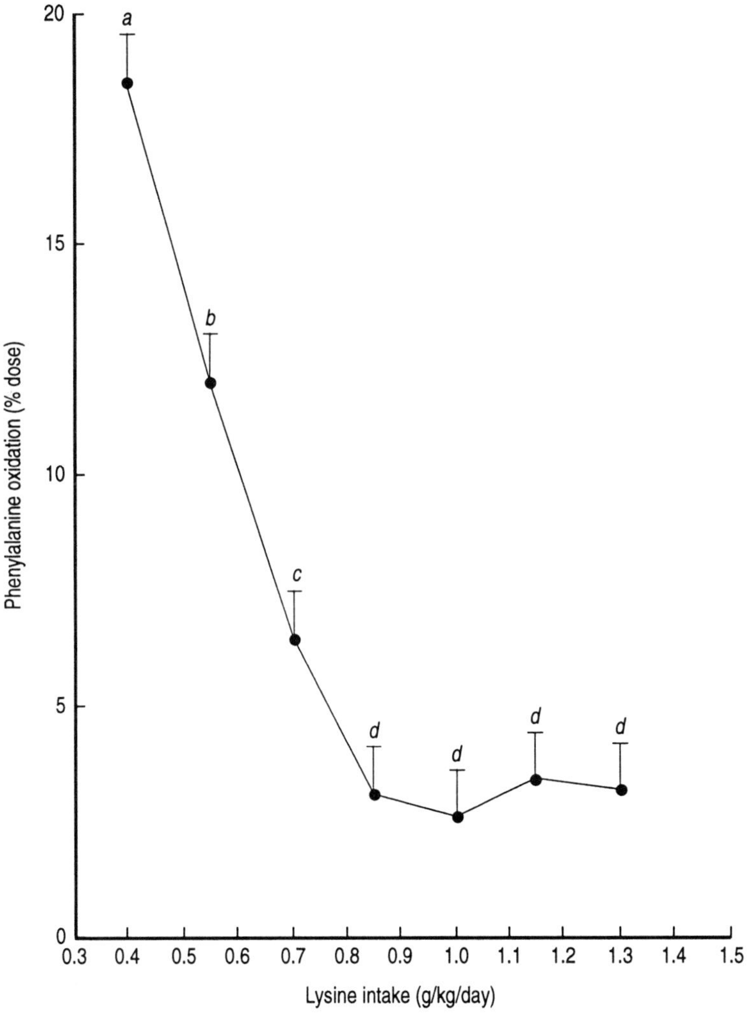

prior to the oxidation studies the piglets received one of seven levels of lysine (Fig. 3). Phenylalanine oxidation decreased in a linear fashion as lysine intakes were increased from 0.4 to 0.85 g/kg/day, after which there was little change in the oxidative response (Fig. 3). A safe level of lysine supply, determined as the upper 95% confidence value of the break-point estimate, would be met by providing a TPN solution containing 5.6 g/100 g of total amino acids as lysine. This value is well below the lysine contents of several parenteral amino acid solutions designed for neonates (Table 1), highlighting the potential pitfalls of the methods used in formulating these solutions and strengthening our standpoint that amino acid requirements need to be specifically defined for neonates receiving TPN. These results also demonstrate that the indicator method for establishing amino acid requirements can be used during TPN.

Advantages of the indicator amino acid oxidation technique

Our study investigating lysine requirements demonstrated that the indicator amino acid oxidation technique could be adapted to estimate amino acid requirements during TPN. This finding is significant due to the fact that all previous estimations of amino acid requirements, using indicator oxidation techniques, have been performed in orally fed subjects (11,50,64,80–83). However, the fact that parenteral nutrients bypass the splanchnic control mechanisms, plus the fact that nutrient delivery is continuous during TPN, could have meant that the oxidation of the indicator amino acid may not have been sensitive to changes in dietary supply during TPN. As our results indicate, this was not the case.

Several factors point to the potential for the indicator oxidation method to improve our knowledge of amino acid requirements of the human neonate. First, unlike nitrogen balance studies, the indicator oxidation technique does not require infants to consume diets that are deficient or excessive in an amino acid for lengthy periods. The current results suggest that 12 h is sufficient, but the choice of this time period was arbitrary, and may be longer than was necessary. Second, the results from our studies indicate that the end-product measure of phenylalanine oxidation (i.e. the percentage of the dose oxidized) was a sensitive variable for the estimation of requirements by breakpoint analysis. This means that studies could be performed that would be relatively non-invasive, requiring only the sampling of breath to measure the oxidative response to a particular dietary amino acid intake. This fact makes this method more attractive for studies involving neonates, due to ethical concerns over the sampling of blood or tissues.

CONCLUSIONS

This review considers the amino acid requirements of the human neonate in the context of recent advances in amino acid oxidation techniques, especially indicator amino acid oxidation. For practical and ethical reasons these techniques are yet to be applied in the human neonate. However, we have developed a model for the human neonate using the metabolically similar newborn piglet. The model achieves the goal of human neonate management: healthy neonates with growth and body composition similar to those fed mother's milk. The model was used to investigate amino acid metabolism and requirements. To our knowledge, our data on tyrosine, phenylalanine and lysine are the first empirical estimates of amino acid requirement during TPN. These data also demonstrate conclusively that the current approaches to providing aromatic amino acids in commercial TPN solutions are inadequate. These data demonstrate that amino acid metabolism during TPN differs from the oral situation and support our hypothesis that the ideal amino acid pattern must be specifically defined for TPN feeding. The piglet model should continue to offer significant insights into various aspects of protein and amino acid metabolism in the neonate. A research programme using a piglet model, when conducted in tandem with a clinical neonate research programme, provides all the aspects of well-designed investigation: identification of the problem, rigorous metabolic investigation, clinical metabolic research and clinical implementation.

ACKNOWLEDGEMENTS

Supported by the Medical Research Council of Canada (Grants #MA10321 and #MT5466), the Natural Sciences and Engineering Research Council of Canada (Grant #OGP0138042) and by The Hospital for Sick Children Foundation. The amino acids used in the piglet TPN studies were kindly supplied by Kabi-Pharmacia-Upjohn.

References

1. Rose WC 1957 The amino acid requirements of adult man. Nutr Abstr Rev 27:631–647
2. Millward DJ, Jackson AA, Price G, Rivers JPW 1989 Human amino acid and protein requirements: current dilemmas and uncertainties. Nutr Res Rev 2:109–132
3. Duffy B, Gunn T, Collinge J, Pencharz P 1981 The effect of varying protein quality and energy intake on the nitrogen metabolism of parenterally fed very low birth weight (< 1600 g) infants. Pediatr Res 15:1040–1044
4. Fuller MF, Reeds PJ, Cadenhead A, Seve B, Preston T 1987 Effects of the amount and quality of dietary protein on nitrogen metabolism and protein turnover of pigs. Br J Nutr 58:287–300
5. Young VR, El-Khoury AE 1995 The notion of the nutritional essentiality of amino acids, revisited, with a note on the indispensable amino acid requirements in adults. In: Cynober L (ed) Amino Acid Metabolism and Therapy in Health and Nutritional Disease. CRC Press, Boca Raton, FL, pp 191–232
6. Rose WC, Oesterling MJ, Womack M 1948 Comparative growth on diets containing ten and nineteen amino acids, with further observations upon the role of glutamic and aspartic acids. J Biol Chem 176:753–762
7. Holt LE Jr, Snyderman SE 1965 Protein and amino acid requirements of infants and children. Nutr Abstr Rev 35:1–13
8. Kopple JD, Swendseid ME 1975 Evidence that histidine is an essential amino acid in normal and uraemic men. J Clin Invest 55:881–890
9. Cho ES, Anderson HL, Wixon RL, Hanson KC, Krause GF 1984 Long-term effects of low histidine intake on men. J Nutr 114:369–384
10. FAO/WHO/UNU Expert Consultation 1985 Energy and protein requirements. WHO Technical Report Series No 724, WHO, Geneva
11. Ball RO, Atkinson JL, Bayley HS 1986 Proline as an essential amino acid for the young pig. Br J Nutr 55:659–668
12. Southern LL, Baker DH 1983 Arginine requirement of the young pig. J Anim Sci 57:402–412
13. Heird WC, Nicholson JF, Driscoll JM Jr, Schullinger JN, Winters RW 1972 Hyperammonemia

resulting from intravenous alimentation using a mixture of synthetic amino acids: a preliminary report. J Pediatr 81:162–165
14. Sturman JA, Gaull G, Raïhä NCR 1970 Absence of cystathionase in human liver: is cysteine essential? Science 169:74–76
15. Snyderman SE 1971 The protein and amino acid requirements of the premature infant. In: Jonxis JHP, Visser HKA, Troelstra JA (eds) Metabolic Processes in the Foetus and Newborn. Stenfert Kroese, Leiden, The Netherlands, pp 128–141
16. Heird WC, Dell RB, Helms RA, Greene HL, Ament ME, Karna P, Storm MC 1988 Amino acid mixture designed to maintain normal plasma amino acid patterns in infants and children requiring parenteral nutrition. Pediatrics 80:401–408
17. Zlotkin SH, Bryan MH, Anderson GH 1981 Cysteine supplementation to cysteine-free intravenous feeding regimens in newborn infants. Am J Clin Nutr 34:914–923
18. Zlotkin SH, Anderson GH 1982 The development of cystathionase activity during the first year of life. Pediatr Res 16:65–68
19. Moss AR, Schoenheimer R 1941 The conversion of phenylalanine to tyrosine in normal rats. J Biol Chem 135:415–429
20. Del Valle JA, Greengard O 1976 Phenylalanine hydroxylase and tyrosine aminotransferase in human fetal and adult liver. Pediatr Res 11:2–5
21. Puntis JWL, Edwards MA, Green A, Morgan I, Booth IW, Ball PA 1986 Hyperphenylalaninaemia in parenterally fed newborn babies. Lancet ii:1105–1106
22. Walker VA, Hall MA, Bulusu S, Allan A 1986 Hyperphenylalanaemia in parenterally fed newborn babies. Lancet ii:1284
23. Castillo L, Ming Y, Marchini S, Chapman TE, Sanchez M, Young VR, Burke JF 1994 Phenylalanine and tyrosine kinetics in critically ill children with sepsis. Pediatr Res 35:580–588
24. Shortland GJ, Walter JH, Fleming PJ, Halliday D 1994 Phenylalanine kinetics in sick preterm neonates with respiratory distress syndrome. Pediatr Res 36:713–718
25. Kilani RA, Cole FS, Bier DM 1995 Phenylalanine hydroxylase activity in preterm infants: is tyrosine a conditionally essential amino acid? Am J Clin Nutr 61:1218–1223
26. Im HA, Meyer PD, Stegink LD 1985 *N*-Acetyltyrosine as a tyrosine source during total parenteral nutrition in adult rats. Pediatr Res 19:514–518
27. Neuhäuser M, Wandira JA, Göttmann U, Bassler KH, Langer K 1985 Utilization of *N*-acetyltyrosine and glycyltyrosine during long term parenteral nutrition in the growing rat. Am J Clin Nutr 42:585–596
28. Wykes LJ, House JD, Ball RO, Pencharz PB 1994 Aromatic amino acid metabolism of neonatal piglets receiving TPN: effect of tyrosine precursors. Am J Physiol 267:E672–E679
29. Sturman JA, Hayes KC (1980) The biology of taurine in nutrition and development. Adv Nutr Res 3:231–299
30. Gaull GE, Rassin DK, Raïhä NCR, Heinonen K 1977 Milk protein quality and quantity in low-birth-weight infants. III. Effects on sulfur amino acids in plasma and urine. J Pediatr 90:348–355
31. Hayes KC, Carey RE, Schmidt SY 1975 Retinal degeneration associated with taurine deficiency in the cat. Science 188:949–951
32. Hayes KC, Stephan ZF, Sturman JA 1980 Growth depression in taurine-depleted infant monkeys. J Nutr 110:2058–2064
33. Geggel HS, Ament ME, Heckenlively JR, Martin DA, Kopple JD 1985 Nutritional requirement for taurine in patients receiving long-term parenteral nutrition. N Engl J Med 312:142–146
34. Tritsch GL, Moore GE 1962 Spontaneous decomposition of glutamine in cell culture media. Exp Cell Res 28:360–364
35. Stehle P, Pfaender P, Furst P 1984 Isotachophoretic analysis of a synthetic dipeptide L-alanyl-L-glutamine: evidence for stability during heat sterilization. J Chromatogr 294:507–512
36. Lacey JM, Wilmore DW 1990 Is glutamine a conditionally essential amino acid? Nutr Rev 48:297–309
37. Windmueller HG, Spaeth AE 1974 Uptake and metabolism of plasma glutamine by the small intestine. J Biol Chem 249:5070–5079
38. Wu C 1963 Glutamine synthetase. I. A comparative study of its distribution in animals and its inhibition by DL-allo-δ-hydroxylysine. Comp Biochem Physiol 8:335–351
39. Knox WE, Kupchik HZ, Liu LP 1971 Glutamine and glutamine synthetase in fetal, adult and neoplastic rat tissues. Enzyme 12:88–98

40. Pell JM, Tooley J, Jeacock MK, Shepherd DAL 1983 Glutamate and glutamine metabolism in tissues of developing lambs. J Agric Sci Camb 101:265–273
41. Pell JM, Jeacock MK, Shepherd DAL 1983 Glutamate and glutamine metabolism in the ovine placenta. J Agric Sci Camb 101:275–281
42. House JD, Pencharz PB, Ball RO 1994 Glutamine supplementation to total parenteral nutrition promotes extracellular fluid expansion in piglets. J Nutr 124:396–405
43. Hankard R, Goulet O, Ricour C, Rongier M, Colomb V, Darmaun D 1994 Glutamine metabolism in children with short-bowel syndrome: a stable isotope study. Pediatr Res 36:202–206
44. Jackson AA, Shaw JCL, Barber A, Golden MHN 1981 Nitrogen metabolism in preterm infants fed human donor breast milk: the possible essentiality of glycine. Pediatr Res 15:1454–1461
45. Van Goudoever JB, Sulkers EJ, Halliday D, Degenhart HJ, Carnielli VP, Wattimena JLD, Sauer PJJ 1995 Whole-body protein turnover in preterm appropriate for gestational age and small for gestational age infants: comparison of [^{15}N]glycine and [1-^{13}C]leucine administered simultaneously. Pediatr Res 37:381–388
46. Pencharz PB, Masson M, Desgranges F, Papageorgiou A 1981 Total body protein turnover in human neonates: effects of birthweight, intrauterine nutritional status and diet. Clin Sci 61: 207–215
47. Ball RO, House JD, Wykes LJ, Pencharz PB 1996 A piglet model for neonatal amino acid metabolism during parenteral nutrition. In: Tumbleson M, Schook L (eds) Advances in Swine Biomedical Research: International symposium. Plenum Press (in press)
48. Young VR, Bier DM, Pellett PL 1989 A theoretical basis for increasing current estimates of the amino acid requirements in adult man, with experimental support. Am J Clin Nutr 50:80–92
49. Fuller MF, Garlick PJ 1994 Human amino acid requirements: can the controversy be resolved? Annu Rev Nutr 14:217–241
50. Zello GA, Pencharz PB, Ball RO 1993 The dietary lysine requirement of young adult males determined by the oxidation of an indicator amino acid, L-[1 ^{13}C]phenylalanine. Am J Physiol 264:E677–E685
51. Zello GA, Wykes LJ, Ball RO, Pencharz PB 1995 Recent advances in methods of assessing dietary amino acid requirements for adult humans. J Nutr 125:2907–2915
52. Ball RO, Bayley HS 1984 Tryptophan requirement of the 2.5-kg piglet determined by the oxidation of an indicator amino acid. J Nutr 114:1741–1746
53. Fomon SJ, Filer LJ Jr 1967 Amino acid requirements for normal growth. In: Nyhan WL (ed) Amino Acid Metabolism and Genetic Variation. McGraw-Hill, New York, pp 391–401
54. Pineda O, Torun B, Viteri FE, Arroyave G 1981 Protein quality in relation to estimate of essential amino acids requirement. In: Bodwell CE, Adkins JS, Hopkins DT (eds) Protein Quality in Humans: Assessment and In Vitro Estimation. AVI Publishing, Westport, CT, pp 29–42
55. Nakagawa I, Takahashi T, Suzuki T, Kobayashi K 1964 Amino acid requirements of children: nitrogen balance at the minimal level of essential amino acids. J Nutr 83:115–118
56. Williams HH, Harper AE, Hegsted DM, Arroyave G, Holt LE 1974 Nitrogen and amino acid requirements. In: Improvement of Protein Nutriture. Report of the Committee on Amino Acids, Food and Nutrition Board. National Academy of Science, Washington, DC
57. Young VR, Hussein MA, Murray E, Scrimshaw NS 1971 Plasma tryptophan response curve and its relation to tryptophan requirements in young adult men. J Nutr 101:45–60
58. Young VR, Tontisirin K, Ozalp I, Lakshmanan F, Scrimshaw NS 1972 Plasma amino acid response curve and amino acid requirements in young men: valine and lysine. J Nutr 102:1159–1169
59. Young VR, Perera WD, Winterer JC, Scrimshaw NS 1976 Protein and amino acid requirements of the elderly. Curr Concept Nutr 4:77–118
60. Meredith CN, Wen Z, Bier DM, Matthews DE, Young VR 1986 Lysine kinetics at graded lysine intakes in young men. Am J Clin Nutr 43:787–794
61. Zello GA, Pencharz PB, Ball RO 1990 Phenylalanine flux, oxidation, and conversion to tyrosine in humans studied with L-[1-^{13}C]phenylalanine. Am J Physiol 259:E835–E843
62. Young VR, Bier DM 1987 Amino acid requirements in the adult human: how well do we know them? J Nutr 117:1484–1487
63. Young VR, Bier DM 1987 A kinetic approach to the determination of human amino acid requirements. Nutr Rev 45:289–298
64. Lazaris-Brunner G, Pencharz PB, Ball RO 1994 Tryptophan requirement of young women determined by indicator amino acid oxidation. FASEB J 8:A462

65. Health and Welfare Canada. Scientific Review Committee 1990 Nutrition Recommendations: The report of the scientific review committee. Canadian Government Publishing Center, Ottawa
66. Zello GA, Tung ASF, Marai L, Pencharz PB, Ball RO 1994 Plasma and urine enrichments following infusion of L-[^{13}C]phenylalanine and L-[*ring*-$^2H_5$]phenylalanine in humans: evidence for an isotope effect in renal tubular reabsorption. Metabolism 43:487–491
67. Wykes LJ, Ball RO, Menendez CE, Maylon Ginther D, Pencharz PB 1992 Glycine, leucine, and phenylalanine flux in low-birth-weight infants during parenteral and enteral feeding. Am J Clin Nutr 55:971–975
68. Wykes LJ, Ball RO, Pencharz PB 1993 The development and validation of a total parenteral nutrition model in the neonatal piglet. J Nutr 123:1248–1259
69. Book SA, Bustad LK 1974 The fetal and neonatal pig in biomedical research. J Anim Sci 38: 997–1002
70. Pond WG, Houpt KA 1978 The Biology of the Pig. Cornell University Press, Ithaca, New York
71. Miller ER, Ullrey DE 1987 The pig as a model for human nutrition. Annu Rev Nutr 7:361–382
72. Moughan PJ, Rowan AM 1989 The pig as a model animal for human nutrition research. Proc Nutr Soc NZ 14:116–123
73. National Research Council 1988 Nutrient Requirements of Swine. 9th edn. National Academy Press, Washington, DC
74. FAO/WHO 1991 Protein Quality Evaluation: report of the joint FAO/WHO expert consultation (FAO Food and Nutrition Paper 51). WHO, Rome
75. Aumaitre A, Dueé PH 1974 Composition en acides amines des protéines corporelles du porcelet entre la naissance et l'âge de huit semaines. Ann Zootech 23:231–236
76. Widdowson EM 1979 Body composition of the fetus and infant. In: Visser HKA (ed) Nutrition and Metabolism of the Fetus and Infant. Martinus Nijhoff, The Hague, p 147–157
77. Davis TA, Nguyen HV, Garcia-Bravo R, Fiorotto ML, Jackson EM, Reeds PJ 1994 Amino acid composition of the milk of some mammalian species changes with stage of lactation. Br J Nutr 72:845–853
78. Ciba Geigy 1970 Scientific Tables, 7th ed. Ciba-Geigy Limited, Basle, Switzerland
79. Wyke LJ, House JD, Ball RO, Pencharz PB 1994 Amino acid profile and aromatic amino acid concentration in total parenteral nutrition: effect on growth, protein metabolism and aromatic amino acid metabolism in the neonatal piglet. Clin Sci 87:75–84
80. Kim KI, Bayley HS 1983 Amino acid oxidation by young pigs receiving diets with varying levels of sulphur amino acids. Br J Nutr 50:383–390
81. Kim KI, Elliott JI, Bayley HS 1983 Oxidation of an indicator amino acid by young pigs receiving diets with varying levels of lysine or threonine, and an assessment of amino acid requirements. Br J Nutr 50:391–399
82. Kim KI, McMillan I, Bayley HS 1983 Determination of amino acid requirements of young pigs using an indicator amino acid. Br J Nutr 50:369–382
83. Duncan AM, Pencharz PB, Ball RO 1995 Lysine requirement of adult males using indicator amino acid oxidation. The effect of a lower protein intake. FASEB J 9:A865

Discussion Session 4

CHAIR: T. STEPHENSON

CHAPTER 18

Koppe In Amsterdam during World War II, we had a famine that started in November 1944 and ended in May 1945. During that period birth weights were reduced by about 300 g. Studies in this cohort at the reproductive age were performed by US colleagues. Right now, in collaboration with Barker and his group we are investigating this cohort for long term effects on the incidence of diseases such as atherosclerosis and diabetes. Our impression after the most recent interim analysis is that in comparing subgroups born before the famine, during it, and afterwards, there are no differences in cholesterol levels. There is, however, a trend towards a higher incidence of diabetes in the group born during the famine.

Hachey Thank you for this information. Your findings on diabetes really confirm what Barker and his colleagues have also observed. However, your impression on cholesterol is somewhat less consistent. We will have to await your final analyses and also accept that information on the mode of feeding during infancy was not recorded.

CHAPTER 19

Donzelli Do you think that we must also pay attention to copper and selenium? Particularly in preterm infants, where the activities of superoxide dismutase and glutathion peroxidase are very low, it might be desirable to provide supplements in order to stimulate the synthesis of these copper-, zinc- and selenium-containing enzymes, which play a key role in the free oxygen radical scavenging mechanism.

Wharton This is indeed another way of looking into the antioxidant story. I mainly focused on the less desired effects of iron in the intestine with respect to the incidence of Gram-negative infections, which has only been demonstrated in parenteral nutrition. Another argument to raise concern is the pro-oxidative capacity of free iron and copper. Perhaps this issue should have also been put forward in the session on LCPUFA, which theoretically are very prone to oxidation. Your remark now on copper and selenium suggests a beneficial contribution to the antioxidant defence mechanism.

Fabris We know that therapeutic use of glucocorticoids in preterm babies may impair normal zinc absorption and storage. Do you think that during corticosteroid administration we should supplement expressed breast milk with zinc?

Wharton I was not aware of the effect of corticosteroids on zinc. However, on the basis of this observation, it is certainly a good suggestion to investigate this on a trial basis.

J.G. Bindels, A.C. Goedhart and H.K.A. Visser (eds), Recent Developments in Infant Nutrition, 297–303.

Böhles During the last several years I have seen a number of infants with hypocalcaemic convulsions. Interestingly, none of the infants were fed on a cows' milk formula but they all received a home-made milk formula from almonds. Such feeding practice is seen in a group with some kind of anthroposophical area of belief. As almonds are very high in phosphorus, it is reasonable to assume that this was causing the problems.

Wharton In our studies we also have faced a high phosphate problem. However, in some of the populations this was covering up a vitamin D problem as well.

Koppe I was interested in your speculation on vitamin K. I have seen those babies, and I am almost sure that there were no problems: no antiseptics were used and I am confident that there is enough vitamin K in breast milk because the babies only need a few nanograms of vitamin K, while breast milk normally contains microgram quantities. However, I do think that there might be something wrong with vitamin K metabolism. Vitamin K is normally continuously recycled within the liver and only if this recycling is impaired will the breast-fed infant run into problems. This seems to me a much more plausible cause for haemorrhagic disease.

Wharton Your theory on the recycling of vitamin K is interesting. However, next to the liver as you suggested, recycling might also take place in the intestine, especially by the microflora.

Marini You said that iron and vitamin D should not be put into a starting formula, but should be added to the follow-on formula. Could you please give your opinion on the timing of introduction of follow-on formulas?

Wharton The answer to this question can only be rather arbitrary. If we use the data on the total iron body stores by Peter Dallman, the dividing point was 4 months. Therefore, let us say at 4 months.

Marini Do you think that this is a good age to change to a follow-on formula in view of the macronutrients like protein? The protein level in follow-on formulas is usually much higher and high protein intakes might not be desired at this age.

Wharton I did not expect to go so much into the discussion about the composition of follow-on formulas, but I see no reason why the protein level has to be that high in follow-on formulas. Indeed, we could go on to a discussion on possibly lowering the protein content of follow-on formulas. The concept which I want to get across is that the formula we use in the first 3 months is a breast milk substitute. The formula we use later on is just a weaning fluid. We may decide to fortify follow-on formulas as we fortify any other weaning food, using it as a particular vehicle. Both types of formulas are quite different approaches and we need to separate them in our minds much more than currently. It was also quite a change for me. I used to think there was no point in follow-on formulations as you may as well use an infant formula throughout the first year of life. I have now become more convinced of the two-pronged approach with its quite different aims.

Pencharz With respect to your plea of supplementing mothers who are insufficiently exposed to sunlight with vitamin D, I would like to point to a supportive study from Saudi Arabia. In this study mothers who gave birth to infants with neonatal rickets obtained vitamin D supplements. The result was that the neonatal rickets could be cured by feeding the mothers.

Stephenson I am intrigued about the final comments on vitamin K, vitamin D and iron in your paper. Do other primates and other mammals produce milk that is uniformly deficient in those, or is it peculiar to the human infant?

Wharton I do not think that iron appears in the milk of any animal and I do not think that vitamin D is present in any large concentration. I am not sure about vitamin K, but compared with human milk it certainly is present in higher concentrations in cows' milk.

CHAPTER 20

Kattamis Is there any relationship between the maternal iron status during pregnancy and the iron body stores of their infants after birth which might be a predictor for iron deficiency later on in infancy?

Dallman A considerable number of studies have addressed this question and there is still disagreement on this topic. However, there is agreement that if there is an effect it is very small and that the haemoglobin level of the newborn is no reliable indicator of its body iron stores. We may therefore assume that while the baby is in utero, it is relatively well protected from iron deficiency. The only restriction on this assumption is the work by Scholl and co-workers (Scholl et al. 1992 Am J Clin Nutr 55:985–988) that suggests that women with iron deficiency anaemia in early pregnancy are at higher risk of giving birth to their infants prematurely or of low birth weight. This may be a response to iron deficiency but in that study the iron deficiency anaemia was also associated with low dietary energy intake and inadequate gestational weight gain.

Hautvast Could you please comment on the new procedures developed by Ray Yip and Fernando Viteri on intermittent iron treatment which has been studied with school children and adults? Might this also be of use with infants?

Dallman On theoretical grounds it is very interesting. It may even work in an experimental situation. However, I am not sure whether it will work in practice because of reasons concerning compliance. It is not easy to come up with a procedure to ensure that a supplement is taken, for instance, once a week.

Donzelli I am puzzled about your conclusion that with respect to iron nutrition, breast-feeding is not appropriate in the second half of infancy. Should we not accept that from an evolutionary point of view it is not detrimental to breast feed for longer periods without giving supplemental iron?

Dallman These evolutionary questions are interesting but they are sometimes hard to resolve. Iron is the fourth most common element in the earth and some studies performed in developing countries show that iron from dirt contamination can be a major source. Whether this is a plausible explanation, or whether the weaning of our ancestors was fairly rich in iron is difficult to say.

Marini Would it not be possible that the status which we now define as mild iron deficiency is in fact a normal physiological condition and that we are overdoing it by supplementing iron?

Dallman I am inclined to agree with you and say that a low iron store is not harmful. But as soon as the production of iron enzymes and haemoglobin, the functional compounds, becomes affected, we do run into problems. In the case of mild anaemia, not only the number of red cells but also their compressibility is reduced so that their mobility through the microcirculation is impaired. This is illustrated in children suffering from congenital heart disease, which are sometimes allowed to become iron deficient to lower their haemoglobin level: they develop more thrombotic problems because of the stiffness of their red cells. I do not think that it is good to lose functional iron compounds but it may not be harmful to have low iron stores.

Stephenson I think that the above comments about where to set the threshold for anaemia underline the need for well-designed intervention studies with well-defined outcome measures. Whatever range we consider as normal, if we show that giving iron as an intervention will result in improved growth and psychomotor development, then this is a very important finding.

Davies Milk is only part of the weaning diet and I wonder whether we are spending too much time thinking about the exact and precise fortification of milk at the expense of considering the importance of young toddlers eating good food. It is the low haemoglobin levels in children at the age of 18 and 24 months which really concerns me.

Dallman One reason why we emphasize on formula so much at the expense of other foods is that, in giving dietary advice, we have control over how much formula is used and for how long. At least in the USA it is the general experience that once solid foods are taken we begin to lose control over what the infant eats. In many cases this will be what the rest of the family eats, not necessarily baby foods. I think the opportunity to intervene is greatest with formula. It becomes less with the baby foods because these foods compete with home-prepared foods.

Wharton We should really change our attitude to follow-on milk. It is not a formula in the ordinary sense, it is a weaning food, a valuable part of the weaning diet. Furthermore, follow-on milk is a very good vehicle for increasing the total iron intake facilitated by vitamin C. A part of this may perhaps also be due to the fact that the introduction of whole cows' milk is postponed and, therefore, the inhibitory effects of calcium and phosphorus on iron uptake.

Cooke We assume that the weaning diet is a uniform thing, but it actually is not. According to our survey, it varies quite remarkably from family to family and I suspect also from city to city. I support the notion that infant formulas fortified with iron may be the type of thing that we should be encouraging families to use, until we have collected good data and developed practical concepts to fundamentally improve the quality of the weaning diet.

Wharton Although I accept that our understanding of the weaning processes is still far from complete, there is already a lot of reasonable data available from the literature. Next to the relative important contribution of follow-on milks to the iron intake, I could voice some concern on the intake of zinc in weanlings and toddlers. Although there is far less evidence for zinc being a problem of the same order as iron, zinc intakes are on the low side compared to the dietary reference values, which, however, need not necessarily be absolutely correct.

CHAPTER 21

Cooke Plasma amino acid patterns are often determined in enteral feeding studies, and conclusions about the adequacy of certain types of enteral formulas are based on differences in amino acid patterns. What is your general view about the usefulness of plasma amino acid patterns as a reflection of protein quality and quantity or even, to a certain extent, toxicity?

Pencharz Plasma amino acid patterns are useful in giving us a clue to what is happening, but the trouble is that any particular level depends on rates of entry and exit from the pool being sampled. It tells us nothing about the kinetics of the separate processes. We therefore largely use plasma amino acid patterns to obtain information on which direction to follow. A good example is the difference in serum threonine levels between breast-fed and whey-dominant-formula-fed infants. Using isotopes, we found that the oxidation of threonine in infants fed human milk is higher than that in formula-fed infants. Interestingly, we found that upon increasing the dietary threonine intake by going from a casein-dominant formula (whey/casein ratio 18:82) via a formula with a whey/casein ratio of 40/60 to a whey-dominant formula (ratio 60/40), the threonine oxidation is not increased. This information enabled the generation of further hypotheses whether it is the dehydratase pathway which is affected or whether an alternative pathway via glycine, the dehydrogenase pathway, is involved. I think that the information we can obtain from plasma amino acid patterns is quite limited and future science has to be increasingly based on isotope studies.

Battaglia Could I ask you to comment about the evidence on different amino acids requirements for specific organs? It seems to me possible that if you look at the whole animal requirements you may be missing what is crucial for particular organ systems.

Pencharz You are absolutely right that amino acid requirements vary from organ to organ, but our approach has been very much at the whole infant or the whole animal level. We are moving into wanting to look at differences whether they are fed enterally versus parenterally. For example, the splanchnic bed, the gut and so on almost certainly have higher needs for glutamine and glycine amongst other amino acids. The body as a whole through the circulation provides attempts to mix up nutrients. There is an important practical issue in neonatology: those who have to be fed intravenously versus those who can be fed enterally. We think that the amino acid patterns should be different in some way and that is why we want to pursue that by looking at splanchnic uptake.

Forsyth It has been traditional practice to withhold amino acids from preterm infants during the first 2–3 days of life and then just gradually reintroduce them. We and others have been concerned that this will be putting them into serious negative nitrogen balances. Is there any significant metabolic reason for withholding amino acids the first days after birth?

Pencharz No. Although I am not a neonatologist, I have performed a lot of neonatal research and in one of our studies (Duffy et al. 1981 Pediatr Res 15:1040–1044) we gave amino acids in the first 24h to babies weighing less than 1500g, some of them 800–900g, which showed no adverse effects. I do not know whether this also applies to babies weighing only about 500–600g. Giving amino acids to preterm infants so early only makes sense if you have the right pattern and that is why we are struggling to find this ideal pattern. You probably may want to start at a lower level and gradually increase the amount, which seems common sense, but to withhold them seems not to make any sense to me.

Uauy We are seeing that nutrition is moving well beyond growth and nitrogen balances into organ function. It is obvious to me that we will have to work towards further fine tuning of our understandings in order to more accurately set nutritional needs. Somehow we will need to face organ growth and development rather than plasma levels, nitrogen balances and oxidation rates. How do you see these new developments in nutrition where growth rates and nitrogen balances are probably not enough?

Pencharz I think you are right. We need to understand more about the metabolism to decide whether we are giving sufficient, and whether an imbalance is producing toxic effects. This is already very complicated with the amino acids and I think it is equally complicated with some of the other nutrients. According to our reasoning so far, we can only resolve these questions by defining individual components and then try and put them together. If, in this context, we also need to take different organs into account, it will become a lot more complicated. We would then need protocols in animals and try to look at specific uptake across an organ. We do not feed that way, however; either we feed by the gut and have to consider the uptake and modulation through the liver, or we might go directly into the systemic circulation.

Elia I wonder, for some specific nutrients, whether we have already reached the point where we can start fine-tuning our concepts about nutrient requirements. If the situation at the moment is such that there are two- to three-fold differences in the estimated requirements for some amino acids as assessed by the direct amino acid oxidation (AAO) method and/or indicator amino acid oxidation (IAAO) method (Table 2) and the nitrogen balance method (the basis for the FAO/WHO/UNU requirements), we still have a huge problem. My question is, that although we accept that there are problems with the nitrogen balance technique, do you think there may also be particular problems with the AAO and IAAO techniques? For example, in adults, and probably to a much lower extent in infants, these techniques may involve disrupting the normal daily pattern of physical activity and the normal pattern of eating. This could be a potential variable that might not necessarily reflect normal physiological 24h events, which you are trying to predict with some of these techniques.

Pencharz I think your question is well taken. First of all, in growing individuals the discrepancy between nitrogen balance and the more novel methods is not that large. Some of the initial IAAO studies were done in older, but growing piglets, and comparisons were carried out with the classical nitrogen balance method; both techniques gave the same answers. We, as well as others have gone back and tried to review the old nitrogen balance studies by Rose, Laberton and others to find an explanation for the discrepancies in adults. We concluded that the experimental conditions in some of those initial nitrogen balance studies were not as rigorous as we would like to have them now and there were problems in terms of giving access of non-protein energy. From the work of Alan Jackson we know that if protein intake is restricted, a considerable amount of recycling will take place and adaptation will take about 10 days. Most of the initial nitrogen balance studies lasted 7–10 days. I do not think they were in true nitrogen balance when they were estimating requirements. In terms of your '24h business', the MIT group and ourselves have both looked at fed and fasting states. We found our estimates of breakpoints in the short-term 4–6h studies to be similar to those in the 24h studies.

Sauer Might it be possible that part of the essentiality of amino acids is a developmental process. When you would repeat the same studies in the same pigs a month later if possible, would you find other levels for requirements?

Pencharz I think this is very possible. Proline, for instance, is often considered as a dispensable amino acid and yet studies in fetal lambs by Battaglia showed that the proline synthesis is limited. My colleague Ron Balder has shown that the glutamate pathway to synthesize proline is not functional in the piglet up to about 3 months of life. After this age that synthetic pathway is open. So yes, there are a lot of metabolic routes which we assume from older animal metabolism studies to be open pathways, which are actually not open, and that is one of the reasons why we talk about conditionally indispensable.

Session 5
Nutrition and the integrity of the gut system

22 Microbial colonization of the intestine of the newborn

S. HUDAULT

ABBREVIATIONS

BbGPF *Bifidobacterium bifidum* growth promoting factor
sASC specific antibody-secreting cells
SCFA short-chain fatty acids

EVOLUTION OF THE INTESTINAL MICROFLORA OF NEONATES

Massive and rapid colonization of the digestive tract of infants by microbes is part of the adaptation to extrauterine life. In utero, the infant is sterile, but following rupture of the membranes there is free access to various body cavities of the infant, which are colonized rapidly by microorganisms. Only a few of the microorganisms encountered in the vaginal canal and in the external environment will become established in the digestive tract (1).

Breast-feeding or formula milk

The first 48 h after birth are characterized by the predominance of a few bacteria, leading to the emergence of two facultative anaerobes *Escherichia coli* and enterococci, which are the first bacteria to colonize the gut (2–6). Some anaerobic bacteria are already present at a low level and their number increases with age. Moreau et al. (3) have found that an equilibrium is reached at the end of the first week of life; this is dependent on the type of feeding (breast milk or infant formula). While 85% of breast-fed infants harboured *Bifidobacterium* as the predominant anaerobic microorganism at 5 days of age, this bacterium was absent in 40% of the formula-fed infants, and, if present, was associated with other anaerobic bacteria such as *Clostridium* or *Bacteroides*. Hall et al. (7) and Roberts et al. (8) agree that feeding human milk leads to a different microflora, which can be seen at 1 month of age. According to these authors, during the first phase of colonization (the first 7–10 days of life) the microflora is not dependent on the nature of the milk: enterobacteria and enterococci are always present in the dominant microflora, associated with Bifidobacteria irregularly. At 1 month of age, 80% of breast-fed infants, compared with only 8% of formula-fed infants have a 'bifidus microflora', defined as a flora in which bifidobacteria outnumber all other components of the fecal flora. Formula-fed infants are colonized by high numbers of bifidobacteria, but these are, nevertheless, less numerous than enterobacteria (Fig. 1). Numerous variations in the colonization rate

J.G. Bindels, A.C. Goedhart and H.K.A. Visser (eds), Recent Developments in Infant Nutrition, 307–317.

***Figure 1** Faecal flora of breast-fed or formula-fed infants. A: total count; B: Bifidobacterium; C: Enterococci; D: Enterobacteria; E: Clostridium; F: Bacteroides sp; circles represent the frequency of isolation of each organism from the faeces. From ref 8*

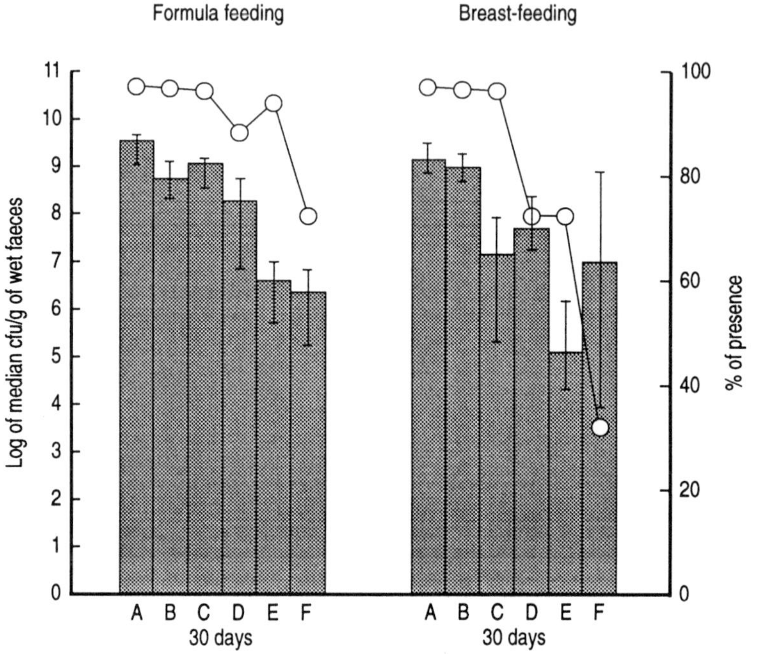

have been observed, for example as a function of the geographic origin and environmental factors at delivery. Simhon et al. (9) found that enterobacteria and enterococci are the predominant bacteria of English breast-fed and formula-fed babies, and they found bifidobacteria in the stools of breast-fed Nigerian babies. Benno et al. (10) found no difference in the number of bifidobacteria in 35 breast-fed and 35 formula-fed infants at 26–48 days of age. They noted that enterobacteria and various anaerobic bacteria (*Bacteroides, Clostridium, Eubacterium, Peptostreptococcus*) were less numerous in breast-fed infants than in formula-fed infants (Table 1).

Hall et al. (7) observed that lactobacilli are as frequent as bifidobacteria in infants at 1 month of age and depend less on the nature of the milk than on the nature of delivery, administration of antibiotics, or nursing in an incubator. However, the high number of lactobacilli found by these authors has not been confirmed by other investigators using identical or different techniques.

The general scheme of the evolution of the microflora of the newborn baby could be as follows. In normal babies, *E. coli* and enterococci become established during the first 24 or 48 h of life, regardless of the diet. At 2–3 days of age, bifidobacteria

***Table 1** Comparison of faecal bacterial populations (mean ± SD) from 35 breast-fed and 35 formula-fed, 28- to 46-day-old Japanese infants*

	Breast-fed		Formula-fed	
Bacterial genera	*Log bacteria/ g stool*	*n*	*Log bacteria/ g stool*	*n*
Bifidobacterium	10.7±0.8	35	10.6±0.5	33
Bacteroides	8.9±1.8	19	9.9±0.6*	23
Eubacterium	7.2±1.9	6	9.7±0.7*	7
Peptococcus	5.7±2.4	2	9.3±0.8*	9
Clostridium	7.2±2.9	16	8.0±2.0	33**
Veillonella	6.6±2.2	15	8.5±2.1	21
Lactobacillus	6.2±1.1	8	8.2±1.1*	13
Enterobacteriaceae	8.2±1.1	34	9.3±0.7**	35
Streptococcus	6.9±1.6	33	9.2±0.8**	35

Data from Benno et al. (10); n = number of positive samples
Significantly different when compared with breast-fed infants:
*$p<0.01$, **$p<0.001$.

are generally one of the first anaerobic bacteria to appear and their population increases at the end of the first week of extrauterine life. In formula-fed infants various other anaerobic bacteria co-exist with bifidobacteria and populations of bifidobacteria rarely outnumber those of enterobacteria. In breast-fed infants, the dominant microflora is generally more simple than that in formula-fed infants. Bifidobacteria are often more numerous than other anaerobic bacteria and enterobacteria. Exceptions to this rule do exist: a breast-fed baby may harbour a microflora without bifidobacteria, while these bacteria can constitute the dominant microflora of a formula-fed baby.

Emergence of a complex microflora

Emergence of a complex microflora similar to that found in adults has been studied and light has been shed on the metabolic role of this microflora. Numerous authors have demonstrated that the presence of the normal intestinal microflora modifies the physiology and the biochemistry of the host. By comparing physiological and biochemical functions and the anatomical structures of the intestine of germ-free animals with their conventional counterparts, some of the effects of the microflora can be evaluated (11). Humans as well as other animals are born germ-free, and the intestine of the newborn consequently lacks microflora-associated characteristics. Colonization begins at birth and subsequently the intestine, as well as other body sites, gradually acquires a functionally active bacterial flora. Different metabolic activities associated with the intestinal microflora of healthy infants have been studied by Midtvedt et al. (11–14): degradation of mucin, regulation of faecal tryptic activity, conversion of bilirubin to urobilinogen and of cholesterol to coprostanol, and production of short chain fatty acids (SCFA). None of these activities were fully established at 6 months of age, at which time the mean urobilinogen level was only one-tenth of that in adults. After birth the relative amounts of acetic acid slowly decreased from nearly 100% to

58%. The relative amounts of all other acids increased during the study but only propionic and butyric acids reached adult levels. The production of total SCFA was increased after exposure to foods other than breast milk between birth and 1 month of life (12). First degradation of mucin was initiated later in breast-fed infants: at 5.5 months of age versus at 2.4 months of age in formula-fed infants (13). The mean age at which children showed complete degradation of mucin pattern was 11.4 months, and at this age no differences between the diet groups were seen. After standardization for the period of exposure to foods other than human milk, a positive correlation between increasing age and increased degradation of mucin was found from 0–1 month to 6–9 months of age. The initiation of the establishment of the conversion of cholesterol to coprostanol was also delayed by breast-feeding (14). At 12 months, by which time all children were weaned, infants who were exclusively breast-fed for more than 4 months or who had been weaned after 6 months had significantly lower conversion rates than the other children. Between 6 and 15 months of age the conversion rate increased with age, but, after exposure to foods other than breast milk, no age factor could be identified. The different bacteria responsible for these different microflora characteristics have not yet been isolated. These results suggest that the colonization sequences are complex.

IMPORTANCE OF THE INTESTINAL MICROFLORA IN THE PHYSIOLOGY OF THE HOST

One of the main roles of the intestinal microflora is protection against the establishment of potential pathogens. In animals, exogenous bacteria inoculated by the oral route are quickly eliminated by the bacteria of the normal resident microflora. This effect, called 'colonization resistance' (15) or 'barrier effect' (16), has also been demonstrated in humans and disappears with antibiotic therapy. In several cases, the bacteria responsible for this antagonistic effect have been isolated. In the newborn such an effect has been observed against a strain of *Lactobacillus casei* (17) and strains of antibiotic-resistant *E. coli* harbouring one or several plasmids (18,19). In some cases, the protective effect is not a result of the number of the bacteria but is due to the modification of their metabolic activity. Such an effect has been observed in germ-free mice in which a strain of bifidobacteria is able to modify the production of cytotoxin by *C. difficile* without modifying the number of the bacteria in situ (20).

Another important role of the microflora for the host is the stimulation of the immune system. IgA plasmocytes are much less numerous in germ-free mice than in their conventional counterparts. Moreau et al. (21) have shown that the establishment of the secretory immune system in conventional mice is related to the sequential colonization by the microflora. Members of the newborn digestive microflora, such as *E. coli, Bacteroides* and *Bifidobacterium*, increase the number and modify the physiology of IgA plasmocytes. IgA, which is abundant in the digestive secretions and maternal milk, inhibits the adhesion of pathogenic bacteria and neutralizes enterotoxins. *Bifidobacterium bifidum* strain B536 is able to stimulate the antibody response against ovalbumin, as does the normal resident microflora of conventional mice (22); *E. coli* has a very weak effect. The intestinal microflora of infants also contributes to the establishment of the suppression of the immune response to dietary proteins, i.e. the establishment of oral tolerance (23). This phenomenon is very important in preventing allergic reactions to dietary proteins. Some bacterial toxins such as the cholera toxin have been shown to profoundly disturb the oral immune tolerance to dietary proteins (24). It is therefore important that the sequential implantation of intestinal microflora occurs in the first weeks of life.

FACTORS INFLUENCING COLONIZATION BY AND THE EQUILIBRIUM OF THE MICROFLORA

Dietary factors

Human milk has a different composition from cows' milk. Numerous studies have attempted to elucidate the nature of the 'beneficial factors' present in human milk, since the health of breast-fed infants is globally better than that of formula-fed infants (25). In 1900 Tissier described the microflora of breast-fed infants as 'blue' because of the presence of a high number of Gram-positive bifidobacteria in the stools. In 1954 György (26) found that a strain of *Bifidobacterium bifidum* var *pennsylvanicus* was dependent for its growth in vitro on factors in human milk, called 'bifidus-factors', which were not detected in cows' milk. These factors were oligosaccharides containing *N*-acetylglucosamine and were supposed to explain the prevalence of bifidobacteria in breast-fed infants. Growth promoting factors for *Bifidobacterium bifidum* were detected later in cows' milk (27,28) and in the faeces of germ-free mice (29). Recently Petchow et al. (30) showed that factors promoting growth of various bifidobacteria are present in both whey and casein fractions of human milk and cows' milk. In two recent studies, we showed that *B. bifidum* growth promoting factors (BbGPF) do not play a role in the colonization of a strain of *Bifidobacterium bifidum* (B536 strain) that requires growth promoting factors from human milk when inoculated into germ-free mice fed cows' milk or human milk (31). The relationship between intestinal colonization by the B536 strain described above and the presence of growth promoting factors in stools or in diet was investigated in 31 healthy newborn infants (32). The infants were inoculated with a single dose of this strain within the first 8 days of life. Spores of a strictly thermophilic *Bacillus* admixed with the *B. bifidum* inoculum were used as transit marker and the faecal populations of both the B536 strain and the transit marker were recorded within 6 days after inoculation. *B. bifidum* strain was found in the predominant flora of six neonates: one breast-fed and five formula-fed. Although the BbGPF amounts were significantly higher in infants fed human milk containing BbGPF than in infants fed milk-based formula without BbGPF, the *B. bifidum* strain became established in only one of the 18 infants fed human milk and in five of the 13 formula-fed infants. During the first 2 weeks of life, no relationship between intestinal colonization by *B. bifidum* and levels of exogenous or endogenous growth promoting factors in stools could be seen.

Human colostrum and mature milk contain several substances which antagonize intestinal bacteria in in vitro conditions. The role of lactoferrin has been studied widely (reviewed in Chapter 25). In an iron-deprived medium, lactoferrin inhibits the growth of *E. coli* and this effect is increased by the addition of immunoglobulins (33). However, in vivo, the role of lactoferrin could not be demonstrated in infants fed formula enriched with bovine lactoferrin and IgG (8,34). Petchow et al. (30) showed that lactoferrin has a growth promoting activity on some species of *Bifidobacterium* but its efficacy in vivo has never been demonstrated.

Prematurity

Because preterm infants are at high risk for infections they are often treated with antibiotics soon after birth. It is therefore difficult to distinguish the effect of prematurity from that of antibiotic therapy on the bacterial colonization of the gut. In a recent study, Hall et al. (7) found that infants of 25–33 weeks' gestation had a lower prevalence of colonization by both lactobacilli and coliforms at 10 days of age and by lactobacilli at 30 days of life compared with term infants. However only 5 infants had not received antibiotics (vs 32 receiving antibiotics) and this precluded any firm conclusions concerning the rate of colonization in preterm infants. The type of feeding (expressed human milk and/or formula) did not influence the pattern of bacterial colonization in these preterm infants.

Inoculation of probiotics

The strong colonization resistance exerted by the normal microflora is supposed to be effective against bacteria from fermented milks and against probiotics. However, even though a bacterial strain is not established in the digestive tract, it can affect the physiology of the host if a sufficient number of bacteria (10^7–10^8 c.f.u./g) reach the gut. To protect preterm infants from nosocomial infections by antibiotic-resistant enterobacteria, Duval et al. (17) and Rastegar et al. (19) supplemented these infants with antibiotic-susceptible, non-pathogenic strains of *E. coli*. Resistant enterobacteria became spontaneously established in the digestive tract of the majority of the unsupplemented infants. The inoculated strains were able to colonize the gut ($>10^7$/g of stool) and significantly reduced the population of antibiotic-resistant enterobacteria. The authors suggested that plasmid-free bacteria have an 'ecological advantage' over antibiotic-resistant bacteria in colonizing the intestine.

To improve bacterial implantation in the newborn intestine and to protect the host against pathogenic bacteria, many studies have looked at the effect of inoculation with lactic acid bacteria, mainly *Lactobacillus* and *Bifidobacterium* strains. We established a strain of *L. casei* in an infant who was maintained germ-free in a sterile bubble because he was at high risk for non-acquired immunodeficiency syndrome (17). The strain remained established as long as it was the only bacterium colonizing the infant. Before he came out of the bubble, the child received a few strains from the microflora of a healthy breast-fed infant of the same age. The *Lactobacillus* strain disappeared as soon as strains of *E. coli* and *Streptococcus* were established in the gut. Several investigators have studied the establishment of the *L. casei* GG strain which produces an antibiotic substance in vitro. This strain was inoculated into full-term infants: it was recovered from nine of 15 and increased the total number of

***Table 2** Effect of ingestion of lactic acid bacteria on mean (SD) duration of diarrhoea, mean number (95% CI) of IgA specific antibody-secreting cells (sASC)/10⁶ cells to rotavirus, and mean levels of serum antibodies to rotavirus during acute and convalescent stage of rotavirus gastro-enteritis*

	LGG	*Lactophilus*	*Yalacta*	*p*
	n=16	*n*=14	*n*=19	
Duration diarrhoea [days]	1.8 (0.8)	2.8 (1.2)	2.6 (1.4)	0.04
sASC response				
Acute	0.1 (0–2.3)	0.2 (0–3.1)	0.5 (0–7.2)	0.70
Convalescent	3.3 (0.9–12.7)	0.1 (0–3.2)	0.1 (0–1.5)	0.01
Serum IgA AB	*n*=13	*n*=11	*n*=12	
Acute	0.3 (0.1–1.7)	0.3 (0–1.6)	0.1 (0–0.3)	0.41
Convalescent	27.5 (18.9–40)	29.6 (21.1–41.4)	6.1 (1.4–26.1)	0.01

49 children aged 6–35 months with rotavirus gastroenteritis randomly received *L. casei* GG (LGG), *L. casei* subsp. *rhamnosus* (Lactophilus) or a combination of *Streptococcus thermophilus* and *L. delbrückii* subsp. *bulgaricus* (Yalacta) twice daily for 5 days
For comparison: duration of diarrhoea in untreated infants ($n=5$) was 2.6 ± 1.3 days, and sASC response (>1 sASC/10^6 cells) was detected in 1/5 control infants at the acute stage
From Ref. 39

lactobacilli compared with that of non-treated infants (35). It remained in eight of the 15 infants after the end of the treatment, but no effect on the number of enterobacteria was observed during or after the end of treatment. Similar results were obtained in preterm infants (36,37).

More recently, the role of *Lactobacillus* has been studied in sick infants. The GG strain was inoculated into children with acute diarrhoea (38), mainly due to rotavirus (82%), and significantly reduced the duration of illness after rehydration and reintake of food. According to Majamaa et al. (39) this effect is specific, since the strain used was more efficient than other lactic acid bacteria studied in parallel in reducing the time for recovery from diarrhoea and increasing the rotavirus-specific immune response (Table 2). This specific effect could be related to immunological properties as suggested by Kaïla et al. (40). Perdigon et al. (41) showed similar results with other *Lactobacillus* strains.

Since bifidobacteria have been found to be more numerous in breast-fed infants, they have been supposed to exert a protective effect against pathogenic bacteria. However, their role has not been clearly shown. Studies (31,42) attempting to regulate the level of enterobacteria by bifido-bacteria in the gut of mice fed human or cows' milk have been unsuccessful (Table 3). A beneficial effect of bifidobacteria has been seen in the recovery from rotavirus diarrhoea in a suckling mice model (43) and in infants (5–24 months) (44) but the mechanisms have not been elucidated. As already

Table 3 Equilibrium between B. bifidum and Bacteroides sp. and K. pneumoniae strains in caeca of gnotobiotic mice

Diets (no. of mice)	*Mice colonized with B. bifidum only*	*Mice colonized with three species*		
		B. bifidum	*Bacteroides*	*Klebsiella*
Autoclaved human milk (4)	9.7 (0.3)	10.0 (0.1)	10.7 (0.1)	10.0 (0.1)
Irradiated human milk (4)	10.0 (0.1)	9.5 (0.2)	10.2 (0.1)	10.1 (0.1)
Autoclaved cows' milk (5)	10.3 (<0.1)	10.1 (0.2)	10.6 (0.1)	10.5 (0.1)

Mean (SEM) $\log^{10}$ bacterial counts/g of caeca; germ-free C3H/He mice were maintained in different sterile isolators and fed sterilized human or cows' milk. Strains of *Bacteroides* and *Klebsiella* were human strains isolated from infants
From Ref. 31

Table 4 Number of children with persistent diarrhoea who presented clinical failure after milk or yoghurt feedings

	Milk (n=24)	*Yoghurt* (n=21)
Duration of diarrhoea >5 days (%)	8 (42%)	3 (14%)
Weight loss >5% in 24 h	2	0
Total number of clinical failures	10	3 ($p<0.05$)

Children aged 3–36 months, with diarrhoea for 13–29 days, receiving formula milk or the same formula fermented by *S. thermophilus* and *L. bulgaricus*
From Ref. 47

mentioned, a strain of *Bifidobacterium* isolated from a healthy breast-fed infant and inoculated into germ-free mice was able to modify the production of cytotoxin by *C. difficile* without modifying the number of the bacteria in situ (20).

Fermented milk nutrition

The importance of fermented milks as a nutritional substitute for milk in infants with persistent diarrhoea has recently been shown in several studies (45–48). Since lactase deficiency and cows' milk intolerance are often identified in children with persistent diarrhoea, replacement of milk with yoghurt could be beneficial. Oral intake of yoghurt significantly decreased the percentage of infants with diarrhoea, shortened the duration of diarrhoea and improved the weight gain (Table 4, Ref. 47). The mechanism underlying this beneficial effect of fermented milk has not yet been elucidated: several hypotheses have been advanced. First, yoghurt has a lower concentration of lactose, which is better absorbed by lactase-deficient adults and children fed fermented milk as shown by breath tests. This is probably due to the intraluminal activity of the microbial β-galactosidase contained in yoghurt. Second, cows' milk proteins are modified in yoghurt. Third, yoghurt stimulates immunity (40). Fourth, the survival of microbial populations in the gut enables a sufficient number of bacteria to be reached to exert biological activity. Although fermented milks are able to decrease the rate of diarrhoea in some cases, one should not conclude that these bacteria become established in the gut of infants, since we have no information on the colonization of the gut after feeding of fermented milk.

CONCLUSION

The digestive microflora colonizes the gut sequentially. This enormous population of microbes plays an important role in the physiology of the host; it is therefore very important to respect the different steps of colonization. A better knowledge of the different populations of the microflora and of their role in the ecosystem is still needed to allow improvement of new types of feeding and to allow adapted inoculation of bacteria in appropriate cases.

REFERENCES

1. Patte C, Tancrède C, Raibaud P, Ducluzeau R 1979 First steps of bacterial colonization of the digestive tract of neonates. Ann Microbiol Inst Pasteur 130A:69–81
2. Cooperstock MS, Zedd AJ 1983 Intestinal flora of infants. In: Hentges DJ (ed) Human Intestinal Microflora in Health and Disease. Academic Press, New York, pp 79–99
3. Moreau MC, Thomasson M, Ducluzeau R, Raibaud P 1986 Cinétique d'établissement de la microflore digestive chez le nouveau-né humain en fonction de la nature du lait. Reprod Nutr Dev 26:745–753
4. Yoshioka H, Iseki K, Fujita K 1983 Development and differences of intestinal flora in the neonatal period in breast-fed and bottle-fed infants. Pediatrics 72:317–321
5. Stark PL, Lee A 1982 The microbial ecology of the large bowel of breast-fed and formula-fed infants during the first year of life. J Med Microbiol 15:189–203
6. Lejeune C, Boussougant Y, de Paillerets F, Ghnassia JC, Gallet JP, Raibaud P, Ducluzeau R 1981 Séquence d'installation de la flore intestinale du nouveau-né. Rev Pédiatr 17:223–243
7. Hall MA, Cole CB, Smith SL, Fuller R, Rolles CJ 1990 Factors influencing the presence of faecal lactobacilli in early infancy. Arch Dis Child 65:185–188
8. Roberts AK, Chierici R, Sawatzki G, Hill MJ, Volpato S, Vigi V 1992 Supplementation of an adapted formula with bovine lactoferrin. I. Effect on the infant faecal flora. Acta Paediatr 81: 119–124
9. Simhon A, Douglas JR, Drasar BS, Soothill JF 1982 Effect of feeding on infants' faecal flora. Arch Dis Child 57:54–58
10. Benno Y, Sawada K, Mitsuoka T 1984 The intestinal microflora of infants: composition of fecal flora in breast-fed and bottle-fed infants. Microbiol Immunol 28:975–986
11. Midtvedt AC, Carlstedt-Duke B, Norin KE, Saxerholt H, Midtvedt T 1988 Development of five metabolic activities associated with the intestinal microflora of healthy infants. J Pediatr Gastroenterol Nutr 7:559–567
12. Midtvedt AC, Midtvedt T 1992 Production of short chain fatty acids by the intestinal microflora during the first two years of human life. J Pediatr Gastroenterol Nutr 15:395–403
13. Midtvedt AC, Midtvedt T 1993 Conversion of cholesterol to coprostanol by the intestinal microflora during the first two years of human life. J Pediatr Gastroenterol Nutr 17:161–168
14. Midtvedt AC, Carlstedt-Duke B, Midtvedt T 1994 Establishment of a mucin-degrading intestinal microflora during the first two years of human life. J Pediatr Gastroenterol Nutr 18:321–326
15. van der Waaij D 1988 Determination of the colonization resistance of the digestive tract of individual mice. J Hyg Camb 72:379–389
16. Ducluzeau R, Bellier M, Raibaud P 1970 Transit digestif de divers inoculums bactériens introduits 'per os' chez des souris axéniques et 'holoxéniques' (conventionnelles): effet antagoniste de la microflore du tractus gastro-intestinal. Zbl Bakt I Orig 213:533–548

17. Hudault S, Ducluzeau R, Dubos F, Raibaud P, Ghnassia JC, Griscelli C 1976 Elimination du tube digestif d'un enfant 'gnotoxénique' d'une souche de *Lactobacillus casei*, issue d'une préparation commerciale: démonstration chez des souris 'gnotoxéniques' du rôle antagoniste d'une souche de *Escherichia coli* d'origine humaine. Ann Microbiol Inst Pasteur 127B:75–82
18. Duval-Iflah Y, Chapuis JP, Ducluzeau R, Raibaud P 1983 Intraspecific interactions between *Escherichia coli* strains in human newborns and gnotobiotic mice and piglets. Prog Food Nutr Sci 7:107–116
19. Rastegar Lari A, God F, Borderon JC, Laugier J, Lafont JP 1990 Implantation and in vivo antagonistic effects of antibiotic-susceptible *Escherichia coli* strains administered to premature newborns. Biol Neonate 58:73–78
20. Corthier G, Dubos F, Raibaud P 1985 Modulation of cytotoxin production by *Clostridium difficile* in the intestinal tracts of gnotobiotic mice inoculated with various human intestinal bacteria. Appl Environ Microbiol 49:250–252
21. Moreau MC, Ducluzeau R, Guy-Grand D, Muller MC 1978 Increase in the population of duodenal immunoglobulin A plasmocytes in axenic mice associated with different living or dead bacterial strains of intestinal origin. Infect Immun 21:532–539
22. Moreau MC, Hudault S, Bridonneau C 1990 Systemic antibody response to ovalbumin in gnotobiotic C3H/HeJ mice associated with *Bifidobacterium bifidum* or *Escherichia coli*. Microecol Ther 20:309–312
23. Moreau MC, Corthier G 1988 Effect of the gastrointestinal microflora on induction and maintenance of oral tolerance to ovalbumin in C3H/HeJ mice. Infect Immun 56:2766–2768
24. Gaboriau-Routhiau V, Moreau MC 1995 Gut flora allows recovery of oral tolerance to ovalbumin in mice after transient breakdown mediated by cholera toxin or *Escherichia coli* heat-labile enterotoxin. Pediatr Res (in press)
25. Feachem RG, Koblinsky MA 1984 Interventions contre les maladies diarrhéiques du jeune enfant: promotion l'allaitement maternel. Bull OMS 62:271–291
26. György P, Norris RF, Rose CS 1954 Bifidus factor. I. Variant of *Lactobacillus bifidus* requiring a special growth factor. Arch Biochem Biophys 48:193–201
27. Petschow BW, Talbott RD 1990 Growth promotion of *Bifidobacterium* species by whey and casein fractions from human and bovine milk. J Clin Microbiol 28:287 –292
28. Kehagias C, Jao YC, Mikolajcik EM, Hansen PMT 1977 Growth response of *Bifidobacterium bifidum* to a hydrolytic product isolated from bovine casein. J Food Sci 42:146–150
29. Romond MB, Romond C, Beerens H, Bourlioux P 1986 Fecal bifidus factors of the germfree mice. In: Dubourguier HC (ed) Biology of Anaerobic Bacteria. Elsevier, Amsterdam, pp 69–74
30. Petschow BR, Talbott RD 1991 Response of *Bifidobacterium* species to growth promoters in human milk and cows' milk. Pediatr Res 29:208–213
31. Hudault S, Bridonneau C, Ducluzeau R, Raibaud P 1991 Effect of bovine or human milk on the establishment of *Bifidobacterium bifidum* and some other bacteria from the human neonatal microflora in gnotobiotic mice. Microbiol Ecol Hum Health Dis 4:1–8
32. Hudault S, Bridonneau C, Raibaud P, Chabanet C, Vial MF 1994 Relationship between intestinal colonization of *Bifidobacterium bifidum* in infants and the presence of exogenous and endogenous growth-promoting factors in their stools. Pediatr Res 35:696–700
33. Bullen JJ, Rogers HJ, Leich L 1972 Iron-binding proteins in milk and resistance to *Escherichia coli* infections in infants. Br Med J 1:69–75
34. Moreau MC, Duval-Iflah Y, Muller MC, Raibaud P, Vial MF, Gabilan JC, Daniel N 1983 Effect of orally administered bovine lactoferrin and bovine IgG on the establishment of *Escherichia coli* in the digestive tract of gnotobiotic mice and human newborn infants. Ann Microbiol (Paris) 134B:429–441
35. Sepp E, Mikelsaar M, Salminen S 1993 Effect of administration of *Lactobacillus casei* strain GG on the gastro-intestinal microbiota of newborns. Microbial Ecol Health Dis 6:309–314
36. Millar MR, Bacon C, Smith SL, Walker V, Hall MA 1993 Enteral feeding of premature infants with *Lactobacillus* GG. Arch Dis Child 69:483–487
37. Stansbridge EM, Walker V, Hall MA, Smith SL, Millar MR, Bacon C, Chen S 1993 Effects of feeding premature infants with lactobacillus GG on the gut fermentation. Arch Dis Child 69: 488–492
38. Isolauri E, Juntunen M, Rautanen T, Sillanaukee P, Koivula T 1991 A human lactobacillus strain (*Lactobacillus casei* sp strain GG) promotes recovery from acute diarrhea in children. Pediatrics 88:90–97

39. Majamaa H, Isolauri E, Saxelin M, Vesikari T 1995 Lactic acid bacteria in the treatment of acute rotavirus gastroenteritis. J Pediatr Gastroenterol Nutr 20:333–338
40. Kaïla M, Isolauri E, Soppi E, Virtanen E, Laine S, Arvilommi H 1992 Enhancement of the circulating antibody secreting cell response in human diarrhea by a human *Lactobacillus* strain. Pediatr Res 32:141–144
41. Perdigon G, Alvarez S, Medici M 1992 Systemic and local augmentation of the immune response in mice by feeding with milk fermented with *Lactobacillus acidophilus* and/or *Lactobacillus casei*. In: Paubert-Braquet M, Dupont Ch, Paoletti R (eds) Foods, Nutrition and Immunity. Dyn Nutr Res Karger, Basel, 1:66–76
42. Hentges D, Marsh WW, Petschow BW, Thal WR, Carter MK 1992 Influence of infant diet on the ecology of the intestinal tract of human flora-associated mice. J Pediatr Gastroenterol Nutr 14: 146–152
43. Duffy LC, Zielezny MA, Riepenhoff-Talty M, Dryja D, Sayahtaheri-Altaie S, Griffiths E, Ruffin D, Barret H, Rossman J, Ogra PL 1994 Effectiveness of *Bifidobacterium bifidum* in mediating the clinical course of murine rotavirus diarrhea. Pediatr Res 35:690–695
44. Saavedra JM, Bauman NA, Oung I, Perman JA, Yolken RH 1994 Feeding of *Bifidobacterium bifidum* and *Streptococcus thermophilus* to infants in hospital for prevention of diarrhoea and shedding of rotavirus. Lancet 344:1046–1049
45. Gonzalez S, Albarracin G, Locascio de Ruiz Pesce M, Male M, Apella MC, Pesce de Ruiz Holgado A, Oliver G 1990 Prevention of infantile diarrhoea by fermented milk. Microbiol Alim Nutr 8:349–354
46. Gendrel D, Richard-Lenoble D, Dupont C, Gendrel C, Nardou M, Chaussain M 1990 Utilisation d'un lait fermenté en poudre chez l'enfant malnutri ou intolerance au lactose. Presse Méd 19: 700–704
47. Boudraa G, Touhami M, Pochart P, Soltana R, Mary JY, Desjeux JF 1990 Effect of feeding yogurt versus milk in children with persistent diarrhea. J Pediatr Gastroenterol Nutr 11:509–512
48. Beau JP, Fontaine O, Garenne M 1990 Management of malnourished children with acute diarrhea and sugar intolerance. J Trop Pediatr 36:86–89

23 Metabolism and nutrition of the gastrointestinal tract

M. ELIA

ABBREVIATIONS

A-PV	arterio-portal venous
BOH	3-hydroxybutyrate
K_{ox}	extracellular concentration of substrate required to achieve half maximum oxidation rate
NEFA	non-esterified fatty acid
SCFA	short-chain fatty acid
V_{max}	maximum velocity
WHO	World Health Organization

INTRODUCTION

Fuel utilization by tissues in vivo is influenced by a variety of factors, including the absolute and relative circulating concentrations of different substrates, the rate of delivery of these substrates to tissues (blood flow), the preference of tissues for specific substrates and metabolic signals such as hormones or cytokines (1). Before these are discussed with respect to the metabolism of the gut, it is necessary to consider some of the unusual features about the substrate supply to this organ.

The gut is essentially unique in that it has dual substrate supply: luminal and vascular. In the fed state the luminal source is likely to be much more important than in the fasted state, particularly for the small intestine. The colonic epithelium may also obtain a substantial proportion of its energy luminally, from short-chain fatty acids (SCFA) derived predominantly from the fermentation of carbohydrates in the colon. This fermentation occurs slowly, which means that SCFA are likely to be produced for considerable periods of time after food ingestion. Indeed SCFA concentrations in the colon show little diurnal variation (2) and other products of fermentation such as hydrogen and methane may peak at various times of the day, including during the night when subjects are asleep (3). It is estimated that in man, 2–4% of the total energy requirements (up to 8% with some diets) (4) is derived from SCFA produced in the colon, while in ruminants SCFA are the major source of energy for the whole body (Table 1). The mucosa of the rumen, and the colonic mucosa of non-ruminants are ideally suited for utilizing SCFA.

Some substrates reach the gut mucosa only from the luminal side. For example, in man butyrate is formed in the lumen of the colon, and is virtually absent from the systemic circulation (14–16). This makes the substrate essentially specific for the

J.G. Bindels, A.C. Goedhart and H.K.A. Visser (eds), Recent Developments in Infant Nutrition, 318–348.

***Table 1** Estimated contribution of short-chain fatty acids (SCFA) from different parts of the gastrointestinal tract to the energy requirement of whole body*

	Contribution (%)	*Reference*
Sheep		
Rumen	70*	5
Hind gut	8†	6
Cattle		
Rumen	63*	7
Total hindgut	9†	7
Beaver		
Large intestine	19††	8
Rabbit		
Total hindgut	30*	9
Pig		
Large intestine	11††	10,11
Human		
Large intestine	2–4§	4

Largely based on Bergman (12)
*Based on in vivo isotope dilution techniques
†Based on amount of fibre and carbohydrate leaving the ileum
††Based on in vitro incubation of gut contents
§May rise to 8% in individuals consuming large amounts of fermentable substrates. A value of 6–10% has been estimated by McNeil (13)

colon and the liver, which extracts most of the butyrate that is absorbed. In contrast, some substrates reach the gut almost entirely from the blood, e.g. ketone bodies, which are absent from the diet.

Arteriovenous catheterization techniques offer a means of studying the metabolism of tissues in vivo, but the inaccessibility of the portal vein in man makes measurement of substrate exchanges across the gut and portal-drained viscera difficult to undertake. As a consequence remarkably little direct information is available regarding fuel selection by the human gastrointestinal tract. However, much information has been gathered indirectly from in vivo studies in animals, in which invasive techniques can be used more readily, or from in vitro studies using a variety of animal and human tissues or epithelial cell preparations (Table 2). These techniques have important limitations, and our concepts about energy and fuel metabolism of the human gut are, therefore, sketchy. In order adequately to interpret the information obtained from various studies, the advantages and disadvantages of various techniques are briefly discussed in the appropriate sections.

O_2 CONSUMPTION AND ENERGY EXPENDITURE OF THE HUMAN GUT

When substrates are oxidized the amount of heat released per litre O_2 consumed varies little with the substrate. Table 3 shows the energy equivalents of O_2 for some substrates that are likely to be used by the gut: the values vary by $< 10\%$, and the ATP gain per litre O_2 consumed generally varies by $<20\%$, even for relatively inefficient fuels such

Table 2 Some methods used to study the nutritional biochemistry of the gastrointestinal tract

1. Arterio-portal venous catheterization techniques with or without blood flow measurements in intact subjects or animals
2. Ligature of isolated segments and exteriorized loops of intestine in vivo with intestinal and/or vascular perfusion
3. In vitro intestinal and/or vascular perfusion
4. Cross-sectional slices of rings of intestine
5. Everted sacs of whole tissue separating two bathing fluids
6. Sheet of mucosal tissue separating bathing fluids
7. Incubation of mucosal biopsy or mucosal scrapings
8. Homogenates of epithelial cell mucosa or whole intestine for measurement of enzyme activity and concentration of metabolites
9. Studies on subcellular fractions, e.g. mitochondrial enzyme activities

Table 3 Heat released during the oxidation of various substrates*

	kJ/mol	*kJ/l O_2*	*kJ/l CO_2*	*RQ*
Glucose	2803	20.84	20.84	1.000
Ketone bodies				
Acetoacetic acid	1775	19.80	19.80	1.000
3-Hydroxybutyric acid	2006	19.89	22.37	0.889
Acetic acid	874.5	19.51	19.51	1.000
Propionic acid	1527	19.47	22.71	0.857
Butyric acid	2183.5	19.48	24.35	0.800
Long-chain fatty acid	10787	19.51	27.75	0.703
Glutamine**	2170	21.51	19.36	1.111

*Based on Elia and Livesey (17) and Livesey and Elia (18)
**With ammonia as the nitrogenous end product

as acetate (Table 4). Therefore O_2 consumption gives an approximate estimate of energy expenditure and ATP gain, but particularly energy expenditure. This conclusion also applies generally to substrates that are incompletely oxidized. Although considerable information is available about the O_2 consumption of human splanchnic tissues (25–30% of whole body O_2 consumption, or 60–80 ml/min in an individual consuming 240 ml O_2/min) there is little information about the O_2 consumption of the constituent tissues, e.g. the relative contribution of the liver and the gut plus portal-drained viscera (19). However, arteriovenous catheterization studies in various mammals, including rodents, carnivores and herbivores, suggest that the gut and portal-drained viscera account for 25–50% of the total splanchnic O_2 consumption. Even in ruminants, which have a large mass of gut relative to the liver (2–3:1 in sheep and cows (20) compared with a value of 0.67 in man; Ref. 21, and data in Table 5), the gut and portal-drained viscera frequently account for 35–50% of total splanchnic production (23,24), or even slightly more than 50% (25). If the value of 25–50% is extrapolated to a 70 kg man who consumes 240 ml O_2/min, then about 17–34 ml O_2 will be consumed by the gut and portal-drained viscera.

Table 4 ATP gain/l oxygen and ATP gain/J associated with the direct oxidation of short-chain fatty acids and glucose, assuming no cost for absorption of either glucose or short-chain fatty acids

	ATP/l O_2	Efficiency relative to glucose	ATP/J	Efficiency relative to glucose
Glucose	0.273	1.00	13.0	1.00
Cytoplasmic activation†				
Acetic acid	0.216	0.79	11.1	0.85
Propionic acid	0.220	0.81	11.3	0.87
Butyric acid	0.235	0.86	12.0	0.92
Mitochondrial activation†				
Acetic acid	0.229	0.84	11.8	0.91
Propionic acid	0.229	0.84	11.8	0.91
Butyric acid	0.241	0.88	12.6	0.97

†There is uncertainty about the proportion of cytoplasmic and mitochondrial activation based on Livesey and Elia (4)

Table 5 O_2 consumption and blood flow through different human tissues*

	Mass (kg)	Blood flow (l/min)	Blood flow/kg tissue (l/kg/min)	O_2 consumption ml/min	% whole body	ml/kg tissue per min	ml/l blood flow
Brain	1.4	0.9	0.64	50	20	36	56
Heart	0.33	0.25	0.76	22.5	9	68	90
Kidney	0.31	1.0	3.22	20	8	65	20
Liver†	1.8	1.4††	0.78	40†	17	22	29
Gut†	1.2	1††	0.83	12.5†	5	10†	12.5
Muscle	28	1	0.036	55	22	2	55
Adipose tissue	15	0.45	0.03	10	4	0.7	22
Other tissues (by difference from whole body)§	21.96	1.0	0.05	35	15	1.6	35
Whole body	70	6	0.086	240	100	3.4	40

*Based on Elia (19) and Wade et al. (22)

†Tentative values for O_2 consumption – see text. The mass of the 'gut' in reference mass (21) includes the mass of the small intestine (0.64 kg), large intestine (0.37 kg) and stomach (0.15 kg)

††The liver has a dual blood supply from hepatic artery and portal vein which drains the gut (~1 l/min) and portal-drained viscera

§These tissues include bone, endocrine glands and skin. The blood flow to skin (estimated ~0.6 ml/min at rest, Ref. 22) may increase considerably in a warm environment

The O_2 consumption of the portal-drained viscera (e.g. pancreas and spleen) is unknown. Although their mass (pancreas, 100 g in the male, 85 g in the female; spleen, 180 g in the male, 150 g in the female (21)) is considerably less than that of the small and large intestine (~1.2 kg), it is possible that their energy expenditure and

O_2 consumption are relatively high. If, for example, the O_2 consumption of the spleen and pancreas in the resting post-absorptive state is similar to that of other organs (Table 5), such as the brain (35 ml/kg/min), kidney and heart (65 ml/kg/min), and liver (probably 20–30 kg/ml/min), then the oxygen consumption of the human gut is likely to be about 10–15 ml/min (or 5–20 ml/min with less stringent assumptions). This corresponds to only 4–6% of the resting O_2 consumption of the whole body. Although this tentative value may appear small both in absolute terms (8–13 mg/kg of intestine) and in relation to other organs (Table 5), it is higher than that of the whole body (3.4 ml/kg/min). This is because the overall value for the body is influenced by the low metabolic rate of resting muscle and adipose tissue, which are the two largest tissues of the body (Table 5). There is even greater uncertainty about the contribution of the mucosal and non-mucosal layers of the gut to total O_2 consumption, but studies using tissue slices of the rat jejunum suggest that muscle contributes 7–14% of the total O_2 consumption (26,27). Although muscle contraction will increase the contribution of the muscular component, in vitro results clearly suggest that most of the O_2 consumption in the gut arises from the mucosa.

Fuel selection and metabolite exchanges across the gut and portal-drained viscera

There is considerable uncertainty about the proportion of energy derived from the vascular route compared with the luminal route, but one estimate based on animal studies of the vascularly perfused gut in vivo suggests that in the fed state about 45% of the CO_2 produced by the jejunum arises from the oxidation of three luminal amino acids (glutamine, which accounts for 30%, aspartate, glutamate) and luminal glucose (6%) (28). Oxidation of other luminal substrates will increase the value even higher. There are no comparable studies for the large intestine, but Roediger (29) suggested that the large bowel may derive as much as 70% of its energy from luminal substrates (short chain fatty acids) on the basis of in vitro studies of fuel selection by colonocytes.

Due to the inaccessibility of the portal vein, few studies have assessed the exchange of circulating substrates across the gut and portal-drained viscera in humans. Most of these studies (Table 6) have been carried out in patients or situations which cannot be described as 'normal': during surgery (at cholecystectomy); 3–11 days after abdominal surgery; 2–5 days after severe accidental injury; in patients with cirrhosis and in 'normal' subjects requiring diagnostic cardiac catheterization. The most consistent finding (Table 6) is the net uptake of glutamine. If the arterio-portal concentration of glutamine is 40–50 μmol/l and the portal blood flow is 1 l/min, for example after an overnight fast, then the rate of net uptake is 40–50 μmol/min or 58–72 mmol/day (8.4–10.5 g/day), which is more than the dietary intake of glutamine (estimated to be about 4 g/day in adults ingesting a mixed diet). Table 7 puts these figures into perspective, by considering the net inter-organ flux of glutamine between various human tissues. It appears that the splanchnic region is a major site of glutamine utilization, with the gut plus portal-drained viscera making a particularly large contribution. Furthermore, if all the vascular glutamine ($C_5O_3N_2H_{10}$) taken up by the gut

Table 6 Arterio-portal venous (A-PV) concentration differences of various metabolites in man (μmol/l) †

	Overnight fast (30) (*n*=8)†	*Overnight fast cirrhotic* (31) (*n*=16)††	*3-day fast cirrhotic* (31) (*n*=9)††	*3-day fast normal* (31) (*n*=13)††	*Day 2 after trauma* (32) (*n*=8)§	*Day 5 after trauma* (32) (*n*=8)§	*Overnight fast; during surgery* (33) (*n*=5)‡	*Fast ≥ 1 day 3–11 days after surgery* (34) (*n*=5)‡‡
Glucose	−10	+0±30	+60±0.04	+40±30				+120±20*
Lactate	−10	−4±21	−24±74	−36±22				−10±20
Pyruvate	+1	−5±4	−11±8	−19±6				−5±2*
Glycerol	−2	+4±10	+23±11	+20±8				+10±8
NEFA		+9±36	0±31	+238±94				
Triglycerides		−10±8	5±19	−55±44				
AcAc	−1	+6	−7±16	+12±8				
BOH	−13	+27*	+24±18	+43±13*				+29±8*
Total ketone bodies	−14	+33*	+17±20	+35±19				
Alanine	−1	−20±8*	−17±2*	−21±11	10	+7	−54±13*	+27±12
Aspartate					0	−2	−8±6	
Citrulline					−4*	−7*	−19±4*	−18±5*
Glutamate		0±2	+6±5	−1±2	+4	−1		
Glutamine		+55±14*	+52±13*	+14±6	+33*	+37*	+38±16*$	98±29*
Glycine					−4	+2	−30±11*	−17±6*
Isoleucine					−2	+9	−4±2	−2±3
Leucine					+1	+7	−10±2*	−4±3
Methionine					−1	+2	−2±1*	−1±1

Table 6 continued …

Table 6 ... continued

	Overnight fast (30) (*n*=8)[†]	*Overnight fast cirrhotic* (31) (*n*=16)[††]	*3-day fast cirrhotic* (31) (*n*=9)[††]	*3-day fast normal* (31) (*n*=13)[††]	*Day 2 after trauma* (32) (*n*=8)[§]	*Day 5 after trauma* (32) (*n*=8)[§]	*Overnight fast; during surgery* (33) (*n*=5)[‡]	*Fast ≥ 1 day 3–11 days after surgery* (34) (*n*=5)[‡‡]
Phenylalanine					0	0	−3±1	−2±2
Serine					+3	+9	−3±7*	3±2
Taurine					+1	−1	−17±4*	0±1
Threonine					−1	+3	−10±1*	−1±3
Tyrosine					+1	−3	−2±2	−1±2
Valine					+12	33	−9±3*	0±4

Results are expressed as mean or mean ± SEM. Negative values indicate release; positive values indicate uptake. Metabolite concentrations were measured in blood (whole blood) except in the studies of McAnena et al. (32) and Bjorkman et al. (amino acid measurements only (34)) where measurements were made in plasma. Red blood cells are involved in the exchange but with glutamine most of the exchange involves plasma (33)

*Statistically significant exchange

[†]5–6 days after cholecystectomy (no blood flow measurements)

[††]Blood flow measurements were made in these studies. In the normal subjects who fasted 3 days only total hepatic flow was reported (1.52 ± 0.16 l/min)

[§]Injury severity score 6–36 (no blood flow measurements)

[‡]At cholecystectomy (no blood flow measurements). The A-PV concentration difference for glucose in 5 patients undergoing cholecystectomy has been reported to be 80 ± 30 μmol/l (35)

[‡‡]3–11 days after abdominal surgery for cholecystectomy or malignancy

[$]Includes asparagine

Table 7 Net inter-organ flux of glutamine (g/day) in man after an overnight fast*

Net uptake		*Net release*	
Splanchnic tissues†	15.3†	Muscle	8.3
Brain	1.5	Adipose tissue	2.1
Kidney	6.0	Other non-muscular tissues	≥12.1
Other	?		
Total	≥23.5	Total	≥23.5

*Based on Elia (36)

†The gut and portal-drained viscera are estimated to account for more than half the splanchnic uptake (see text), e.g. about 2/3 in the study of Felig et al. (33). Plasma arterio-venous concentration differences (μmol/l) for glutamine have been reported to be 139±21 across the luminal jejunum, 42±9 across the ileum, and 32±9 across the colon (measurements in patients with abdominal malignancy undergoing surgery without an assessment of blood flow (37))

and portal-drained viscera (e.g. 40–50 μmol/min) was completely oxidized to CO_2, H_2O and ammonia, according to the following equation (17):

$$C_5O_3N_2H_{10} + 4.5O_2 = 5CO_2 + 2H_2O + 2NH_3 + 2170\,\text{kJ (519 kcal)}$$

then the amount of O_2 required for this oxidation (4–5 ml/min) would make a substantial contribution to the total O_2 consumption of the gut (~10–15 ml/min; see above and Table 5). However, this contribution to the oxidative metabolism of the gut is probably an overestimate for two reasons: first, the gut is not the only tissue that drains into the portal vein (the human spleen in vivo may utilize glutamine) and second, not all the glutamine taken up by the gut is fully oxidized. In the in vivo perfused rat small intestine, only 50–55% of the glutamine carbon utilized by the gut was oxidized to CO_2. The remaining carbon was converted to citrulline, protein, other amino acids, lactic and citric acids, non-volatile organic acids and perchloric acid-insoluble substances (38). Furthermore, isolated pig enterocytes oxidize only about 10% of the glutamine, glucose and galactose that they metabolize (39). There is also some evidence from in vivo catheterization studies in pigs that after a meal amino acids are used to synthesize a labile protein pool which is located in the gut and portal-drained visera (40).

A few other points emerge from measurement of the arterio-portal concentration differences in humans. First, although the ketone body exchange is variable (perhaps partly due to the formation of ketone bodies by the colonic mucosa; see below) a number of studies report a significant uptake of 3-hydroxybutyric acid. If the uptake of 3-hydroxybutyric acid ($C_4H_8O_3$) occurred at a rate of about 25 μmol/min (Table 6) and all of it was oxidized [$C_4H_8O_3 + 4.5O_2 = 4CO_2 + 4H_2O + 2006\,\text{kJ}$ (479 kcal)] (17) there would be a requirement for 2.5 ml O_2/min, which is less than the corresponding value for glutamine (see above). However, during starvation and diabetes, when the circulating concentration of ketone bodies increases (20-fold or more) the concentration of glutamine decreases, ketone bodies could become more important fuels for the gut than glutamine. Animal studies in vivo support this view (see below). Second, the exchange of non-esterified fatty acids (NEFA; triacyl-glycerol, glucose and lactate) is variable and almost always not significant. This may be partly due to the confounding

Table 8 Contribution of arterial substrates to respiratory CO_2 in vascularly perfused segments of jejunum of fed and fasted rats in vivo*

	% of total	
	Fed	*Fasted*
Glutamine	46	35
3-hydroxybutyrate	3	26
Acetoacetate	11	24
Glucose	10	7
Other	30	8**
Total	100	100

*Based on Windmueller and Spaeth (28,42), and Windmueller (38)
**In the fasted state, NEFA accounted for only 3% of the CO_2 production

effects of exchange across adipose tissue (e.g. omental fat), and partly to the variable fate of the substrates. Studies in the rat suggest that NEFA are not as important a source of energy for the gut as originally thought (41) because they can be esterified instead of oxidized (42). The lack of significant exchange across the gut may also be due in part to the small arterio-venous concentration differences for some metabolites, which can make accurate estimates of net exchange extremely difficult. For example, the percentage uptake of glucose in the studies indicated in Table 6 ranged from 0.0 to 2.6%, which is close to the coefficient variation for the measurement of glucose. This small arterio-portal venous (A-PV) difference reflects the high blood flow to the gut relative to its mass and rate of oxidative metabolism (Table 5). Therefore if the blood glucose concentration was 5 mmol/l and the gut took up only 1% of the glucose presented to it (~50 μmol/min), there would be a requirement for 6.7 ml O_2 for complete oxidation [$C_6H_{12}O_6+6O_2=6CO_2+6H_2O+2803$ kJ (17)], or an amount that could account for about half of the estimated O_2 consumption of the gut. However, glucose measurements are not sensitive enough to assess accurately the true glucose exchange. The mean percentage extraction of a variety of metabolites (A-PV concentration differences × 100/arterial concentration) have been calculated for each study indicated in Table 6, and are presented below as a range for comparative purposes: glucose, 0–2.6%; lactate, −0.5 to 6.5%, pyruvate, 2.1–31.1%; 3-hydroxybutyrate, −12.4 to 17.9%; acetoacetate, 1.2–4.3%; glutamine, 3.5–21.8%.

It is clear that a more sensitive and more specific approach is necessary to assess accurately the contribution of some substrates to the oxidative and non-oxidative metabolism of the human gut in vivo. The use of tracers and the measurement of production of labelled CO_2 and products of metabolism in venous blood draining the vascularly perfused gut in vivo is one such technique which has provided important concepts about the fuel selection by the gut of rodents (38). Such a technique has identified glutamine and ketone bodies as important vascular substrates for the small intestine of the rat (see Table 8), which is consistent with results obtained in human catheterization studies (Table 6). The labelled studies have also been used to assess

Table 9 Metabolic fate of the carbon and nitrogen of glutamine by the rat small intestine in vivo

	Metabolic fate of carbon†		*Metabolic fate of nitrogen*††	
	Arterial administration	*Luminal administration*	*Arterial administration*	
CO_2	55.5	53.4	Ammonia	37.9
Lactate	8.5	14.8	Citrulline	27.6
Citrate	1.6	1.7	Alanine	24.4
Other organic acids	1.1	2.1	Proline	7.2
Citrulline	4.7	4.2	Glutamate	1.8
Proline	5.2	5.5	Ornithine	1.1
Alanine	3.3	3.8	Total	100
Other amino acids	2.0	6.6		
Glucose	4.7	2.4		
Incorporated into tissue	13.4	5.5		
Total	100	100		

Based on Windmueller (38). Results are expressed as percentage of total
†Assessed by tracing the fate of L-[U-^{14}C] glutamine
††Assessed by arteriovenous concentration differences (jejunum)

the metabolic fate in vivo of important substrates such as glutamine (Table 9). Finally, the human arterio-portal catheterization technique can be used to assess the validity of in vitro studies. For example, if the human gut contains 200 g of enterocytes [the villous mass of the human small intestine is ~230 g of which 86% is due to enterocytes (21)], and these utilize 14 μmol glucose/g dry weight/min (corresponding to about 2.1 μmol/g wet weight/min), as suggested by human enterocyte studies in vitro (43), then the amount of glucose utilized will correspond to 420 μmol/min (0.6 mol/day). This rate of utilization, which does not even take into account glucose utilization by other parts of the small intestine, or by the large intestine and portal-drained viscera, is a little more than half the glucose utilized by the body (~1 mol/day). Such a rate of utilization should produce an arterio-portal (A-PV) concentration difference of about 0.42 mmol (~10% extraction when the blood glucose concentration is 4.0–4.5 mmol/l). Clearly the in vitro observations do not apply in vivo. Similarly, if the rate of lactate production suggested by studies with isolated human enterocytes occurred in vivo [~20 μmol/g dry weight/min (43)], the PV concentration of lactate would be almost double the arterial concentration. There is again serious concern about the significance of the in vitro observations, since most human catheterization studies report no significant release of lactate by the gut of individuals fasted either overnight or for 3 days (Table 6; Ref. 37). Concerns have also been raised by Windmueller and Spaeth (28,42) about the significance of the high in vitro glycolytic flux, which has been reported to be up to 10-fold greater than that observed in their in vivo rat perfusion studies. They suggest that the high glycolytic rates in vitro may result from inadequate oxygenation (28). In vascular perfusion studies a high O_2 gradient can be created by incorporating red blood cells in the vascular circuit. When the haematocrit increases above 15% there is a significant reduction in the proportion of glucose converted to

Table 10 Effect of substrate infusions on the arterial concentration and on the arterio-portal concentration difference and rate of substrate exchange in vivo

	Arterial concentration (μmol/l)	*A-PV* (μmol/l)	*Uptake* (μmol/kg/min)
Dogs (48)			
Glutamine			
Before infusion	577	4	1.43
During infusion	1060*	154*	4.6*
3-Hydroxybutyrate			
Before infusion	46	10	0.27
During infusion	748*	85*	4.33*
Glucose			
Before infusion	4200	144	4.11
During infusion	7550*	83	2.68
Rats (49)			
Glutamine			
Before infusion	418	113	2.88
During infusion	775*	229*	6.39*

*Significant change

lactate (44). Furthermore, Windmueller and Spaeth (42) have demonstrated that the rat gut in vivo tends to take up lactate when the arterial lactate concentration is low and to release lactate when the arterial lactate concentration is high. Since enterocytes/colonocytes are usually incubated in lactate-free media, this may favour lactate release.

Substrate availability

Substrate concentration

Fuel selection by a variety of human tissues can be influenced by the availability of substrates in the circulation. A classic example concerns the utilization by the human brain of ketone bodies during starvation. They become the dominant fuel for the brain, largely because of their mass action effect (their circulating concentration increases 20–40-fold to reach levels >5 mmol/l) (45), whilst glucose becomes less important as its concentration decreases.

Similar principles apply to the gut. Windmueller and Spaeth (46) demonstrated that the uptake of glutamine in the isolated vascularly perfused small intestine of the rat increased as its concentration increased between 200 and 700 μmol/l; above 700 μmol/l there was no further increase in uptake, suggesting 'saturation'. Hanson and Parsons (47) also found that glutamine utilization by the vascularly perfused small intestine of the rat in situ was concentration-dependent, and was unaffected by the presence of glucose. A variety of studies in vivo have also demonstrated that acute elevations in the circulating concentrations of ketone bodies and glutamine result in increased substrate utilization by the gut and portal-drained viscera (Table 10).

Other workers have reached similar conclusions about the mass action effects of substrates by comparing the results obtained in different groups of animals. Watford et al. (50) infused glutamine into 40-day-old diabetic rats and raised the arterial glutamine concentration several-fold. This resulted in an increase in the A-PV concentration difference (360 μmol/l compared with 50 μmol/l in animals infused with saline). Another example is the positive relationship between the arterial glutamine load and the rate of glutamine uptake by the gut of chronically acidotic rats (51). Yet another possible example concerns diabetes (50). In diabetic rats with a circulating ketone body concentration of 4.4 mmol/l, the A-PV difference of ketone bodies was 131 μmol/l, while in control animals with a ketone body concentration of ~0.2 mmol/l the exchange was undetectable.

Other workers have used colonocytes or enterocytes to study the effect of substrate concentration on oxidative and non-oxidative metabolism. For most substrates the extracellular concentration needed for near maximal oxidation is 5 mmol/l, although in the case of butyrate and colonocytes it is ~0.5 mmol/l (Table 11). Several points can be made about these observations. First, the concentration of glucose in arterial blood (5 mmol/l) is similar to that required for maximum saturation in vitro, but this is not the case for glutamine (see Table 11; K_{ox} ~0.45 mmol/l) which has a circulating concentration of about 0.5 mmol/l and a reported luminal concentration of 1.2 mmol/l after a protein-rich meal in humans (55). In vitro studies of enterocytes/colonocytes have used concentrations of glutamine ranging from <1 mmol/l (a non-saturating concentration) to 5 mmol/l (close to saturating concentration). The circulating concentration of acetate, propionate and butyrate in vivo is well below the K_{ox} (extracellular concentration required for half maximal oxidation in colonocytes in vitro), and considerably less than that required for near maximal oxidation. In contrast the colonic concentrations of SCFA in rats and adult humans are normally well in excess of the saturating concentrations.

Second, the colonic luminal concentrations of butyrate in young infants, especially those who are breast fed, are low (Table 12; close to the K_{ox} for butyrate, observed in colonocyte preparations in vitro – see Table 11). This is partly because in young infants the overall SCFA concentrations are low, and partly because the contribution of butyrate to the total SCFA concentration is considerably lower than in adults (56,57). The lower luminal concentration of SCFA in breast-fed infants compared with those fed formula milk or cows' milk (Table 12) could be due to multiple factors, including the quantity and type of fermentable substrate reaching the colon, the extent of fermentation, the type of bacteria responsible for fermentation, and the rate of absorption and utilization of butyrate by the colon. It is possible that the low luminal butyrate concentrations in young breast-fed infants may limit butyrate oxidation in the colon (by a reduced mass action effect), assuming that colonocytes from infants respond to substrate concentrations in a similar way as colonocytes from adults.

Third, these studies have not assessed the oxidative characteristics of ketone bodies, which are potentially important fuels for epithelial cells of the gut.

Fourth, the results in Table 11 are based on studies involving incubations with single substrates, although it is known that other substrates and hormones can influence the oxidation (53,59,60). Clausen and Mortensen (53) found that both V_{max} and K_{ox} of

Table 11 ***Oxidative metabolism of substrates by enterocytes and colonocytes and approximate vascular and luminal concentration of substrates in vivo***

	V_{max}†	K_{ox}†† (mmol/l)	*Approx. conc. required for near maximal oxidation* (mmol/l)	*Approx. arterial conc.* (mmol/l)	*Approx. luminal conc.* (mmol/l)
Rat					
Enterocytes (52)					
Glucose	4.39[a]	0.40	5	5	<5–>10
Glutamine	6.98[a]	0.45	5	0.5	<0.5–5
Colonocytes (52)					
Glucose	7.17[a]	0.83	5	5	<0.5
Glutamine	6.67[a]	0.80	5	0.5	<0.5
Acetate	6.95[a]	0.88	5	<0.2	~70
Butyrate	11.15[a]	0.33	0.5	≤0.01	~13
Colonocytes (53)					
Glucose	0.34[c]	0.78	5	5	<0.5
Acetate	1.11[c]	0.49	5	≤0.2	~70
Propionate	0.99[c]	0.34	3	≤0.01	~30
Butyrate	1.01[c]	0.18	0.5	<0.03	~13
Human					
Colonocytes (normal) (54)					
Glucose	0.07[b]	0.61	5	5	<0.5
Acetate	0.29[b]	0.57	5	<0.2	~60
Propionate	0.11[b]	0.37	3	≤0.01	~20
Butyrate	0.19[b]	0.08	0.5	≤0.01	~20
Colonocytes (ulcerative colitis) (54)					
Glucose	0.08[b]	0.71	5	5	<0.5
Acetate	0.36[b]	0.62	5	<0.2	~60
Propionate	0.27[b]	0.53	5	<0.01	~20
Butyrate	0.22[b]	0.10	0.5	<0.01	~20

†V_{max} is the maximum rate of CO_2 production with single substrate incubation under the conditions of study

††K_{ox} is the extracellular concentration of substrate required to achieve half maximum oxidation rate

*Production of CO_2 (derived from single labelled carbon) is multiplied by the number of carbon atoms per molecule

**Production of CO_2 (derived from single labelled carbon) is not multiplied by the number of carbon atoms per molecule

[a]μmol CO_2/g dry weight/min*

[b]μmol CO_2/g wet weight/min**

[c]μmol CO_2/g wet weight/min**

specific substrates oxidized by rat colonocytes could be altered by substrates, e.g. butyrate (5 mM) reduced the V_{max} of acetate four-fold (strong, non-competitive inhibition), and this was associated with a reduction in the K_{ox}. With other substrates, no inhibition of acetate oxidation was observed. Furthermore, a few examples were found in which the K_{ox} of a substrate was increased by the presence of another substrate,

Table 12 **Concentration (mmol/l) of short-chain fatty acids (SCFA) in faeces† or meconium+**

	Infants						
		Breast-fed		*Formula-fed*		*Fed cows' milk*	
	Birth†	*1 month*	*2–5 months*	*1 month*	*2–5 months*	*2–5 months*	*Adults*‡
Total SCFA	5.7	20.3	44.1	51.5	71.1	63.6	104.0
Acetic acid	>5.5	18.8	38.5	36.3	52.2	47.7	58.8
Propionic acid	<0.1	0.5	1.1	8.6	12.0	9.2	17.3
Butyric acid	<0.1	0.2	0.4	3.2	2.8	2.4	19.5
Others	<0.1	0.2	0.2	2.3	2.2	1.7	8.4

+All data except those on meconium are based on Ogawa et al. (56)
†Concentration in meconium which rapidly becomes colonized by bacteria after birth; based on Midtvert and Midtvert (57)
‡Meijer-Severs and van Santen (58)

e.g. in rat colonocytes propionate increased the K_{ox} of butyrate but had no effect on its V_{max} (53).

Fifth, the K_{ox} for glutamine by rat enterocytes in vitro (~0.45 mmol/l) is similar to the glutamine concentration needed for half-maximal utilization of the glutamine by the isolated vascularly perfused gut of the rat (46). These values are close to the physiological concentrations, and therefore changes in circulating concentrations are likely to influence the extent of glutamine metabolism in vivo (see also Table 10).

Sixth, the gut may utilize substrates and deal with them in a different way when they are presented to it from the vascular side rather than the luminal side, e.g. glucose (44). Oxygen also seems to penetrate epithelial cells much more readily when delivered from the vascular bed than from the lumen (44).

Seventh, some of the differences between studies involving epithelial cell and those involving gut perfusion may be due to the metabolic activity of non-epithelial cells (e.g. lymphocytes, macrophages, muscle cells etc), which are components of the gut.

Eighth, a general issue concerning enterocyte and colonocyte preparations concerns their orientation: in vivo, one side faces the lumen, the other to the vascular supply, and the sides are 'attached' to each other. This orientation can be associated with widely different concentrations of substrates on either side of epithelial cells. For example, the SCFA concentration in the lumen of the colon is 100- to 1000-fold greater than in portal blood and considerably more than in arterial blood. The loss of cellular orientation (so that all sides of the cells are exposed to the same concentration of substrates), and the loss of the lateral cell-to-cell contact have unknown effects on metabolism.

Blood flow

Since the blood flow to the gut is high relative to its oxidative metabolic activity (Table 5) it may seem unlikely that substrate utilization is limited by blood flow – changes in blood flow are more likely to affect the utilization of substrates with a high fractional extraction (1). However, in pathological conditions that produce hypotension, blood flow to the gut is frequently reduced to a greater extent than blood flow to a number of other major organs, and there may be shunting of blood across the base of the villi, which disrupts the normal countercurrent flow dynamics (61). In critical illness the oxygen tension in the mucosa of the gut can become markedly reduced and as a result the mucosa may become disrupted and ulcerated.

Competition between substrates

A number of in vivo and in vitro studies have demonstrated that the utilization or oxidation of one substrate is inhibited by the presence of another substrate. Table 13 illustrates such an inhibition in rat colonocytes. However, not all studies have demonstrated this phenomenon. Ashy and Ardawi (43) found that the utilization of glucose (10 mM) by human enterocytes was not significantly affected by the presence of other substrates such as glutamine (5 mM) or acetoacetate (5 mM), which were utilized in substantial quantities at the same time as glucose. Similarly, Fleming et al. (63) found that oxidation of butyrate (5 mM) by rat colonocytes was not affected by the addition

Table 13 Inhibition of substrate utilization by addition of another substrate to rat colonocytes

Substrate oxidized or utilized*	*Concentration (mM)*	*Inhibitory effect of added substrate*	*Concentration (mM)*	*Reference*
glucose	10	glutamine	5	(62)
	10	acetoacetate	5	(62)
	10	butyrate	10	(62)
	0.125–2*	butyrate	5	(53)
	5*	acetate	5	(63)
	5*	butyrate	5	(63)
	5*	glutamine	5	(63)
glutamine	5	acetoacetate	5	(64)
	5*	glucose	5	(63)
	5*	acetate	5	(63)
	5*	butyrate	5	(63)
acetate	0.125–2*	butyrate	5	(53)
	0.125–2*	propionate	5	(53)
	0.125–2*	glucose	5	(53)
butyrate	10*	acetate	20	(65)
	0.125–2*	glucose	2	(53)
	0.125–2*	acetate	5	(53)
	5*	3-hydroxybutyrate	5	(63)
propionate	0.125–2*	butyrate	5	(53)

*Indicates oxidation was assessed using carbon labelled substrates. The other studies measured utilization

of glucose (5 mM) or glutamine to the incubation medium, despite substantial utilization of the added substrate. Furthermore, in the vascularly perfused rat intestine in vivo, acute elevations of ketone body concentrations produce an increase in the rate of ketone body uptake without affecting the uptake of glutamine (50). These observations raise important questions about the metabolic fate of the substrates. At least three major possibilities exist. First, the substrates may replace the oxidative metabolism of endogenous fuels. In the absence of substrates, colonocytes or enterocytes depend entirely on endogenous fuels for energy. However, oxidation of endogenous fuels can be replaced by oxidation of exogenous fuels. In the presence of glucose (10 mM) isolated human colonocytes from the descending colon were estimated to derive only about a quarter of their energy from glucose (66,67). When the cells were incubated in the presence of both glucose (10 mM) and butyrate (10 mM) there was a small non-significant change in the rate of glucose utilization, and butyrate became the dominant energy substrate, apparently replacing the oxidative metabolism of virtually all the endogenous fuels. Some tissues in vivo (e.g. muscle) increase their energy (triacyl-glycerol and glycogen) and protein (amino acid) content after a meal (68,69) and utilize it during fasting or exercise (70). The extent to which fuels are stored in the intestine after a meal and released for oxidative or non-oxidative purposes during the post-absorptive period is unknown, although catheterization studies in pigs suggest that this may occur (40). Such considerations raise the possibility of important interactions between the metabolism of endogenous intestinal fuels and exogenous fuels derived from the lumen of the gut or the blood.

Second, there may be an increase in oxidative metabolism and energy expenditure. The O_2 consumption of the gut or of enterocytes or colonocytes may increase when they are exposed to substrates, analogous to the way in which food ingestion increases energy expenditure of the whole body (dietary induced thermogenesis). In man, the splanchnic bed appears to be a major site of dietary induced thermogenesis (71,72), accounting for up to half the total increase in O_2 consumption after a mixed meal. The separate effects of the human gut and liver are unknown, but in sheep and cattle in the fed state the liver is more responsive than the gut (24,25). In an in vivo jejunal perfusion model, luminal feeding was reported to produce only a small increase in O_2 consumption (~4%) (28). A variable increase (0–36%) in O_2 consumption of the ileum in vitro was observed in various animals after exposure to substrates such as acetate, palmitate, glutamine and glutamate (73). Similarly, isolated enterocytes and colonocytes show a variable increase in O_2 consumption in response to added substrates. In one study the O_2 consumption of isolated human colonocytes increased by only 5–41% (66,67) after exposure to various substrates either alone or in combination (5–10 mM). In another study involving rat colonocytes (62) the increase was greater (27–80%), and in yet another study (43) the O_2 consumption of human enterocytes almost trebled (280%) in the presence of glutamine (5 mM) and glucose (10 mM) in combination. Watford et al (74) also noted that glutamine (5 mM) and glucose (10 mM) in combination could more than double the O_2 consumption of rat enterocytes.

Third, not all the substrates utilized by the gut are used for oxidative metabolism, and the extent to which non-oxidative metabolism occurs can be affected by the presence of other fuels. For example, although the amount of glucose (7.5 mM) utilized by the vascularly perfused small intestine of fasted rats may be unaffected by the addition to the perfusate of 3-hydroxybutyrate (2 mM) and acetoacetate (1 mM), the percentage of glucose converted to lactate has been reported to increase from 46% when glucose alone was perfused to essentially 100% when ketone bodies were added to the perfusion medium (75). The effect of ketone bodies in inhibiting glucose oxidation probably involves inhibition of the pyruvate dehydrogenase complex. Another example of substrate interaction concerns the effect of adding butyrate to rat colonocytes, which not only reduces the rate of glucose utilization, but also increases the rate of formation of lactate from glucose (62) and thus increases the proportion of glucose used in non-oxidative metabolism. This again may be due to inhibition of the pyruvate dehydrogenase complex, by an increase in the ratio of acetyl CoA (formed from butyrate) to coenzyme A.

In isolated enterocyte and colonocyte preparations a very large proportion of glucose is converted to lactate (≥90% in some colonocyte preparations, irrespective of whether the cells are incubated with glucose alone or with additional substrates). It is therefore likely that glucose oxidation in these colonocytes accounts for only a small proportion of energy expenditure (i.e. addition of glucose is likely to replace oxidation of endogenous fuels to only a small extent). A variety of studies involving enterocytes and colonocytes have reached a similar conclusion (42,44,50,66,67,75). On the other hand glucose may have indirect effects on the metabolism of other exogenous fuels: colonocytes incubated with butyrate release acetoacetate, but this release is decreased in the presence of glucose (66), implying more complete utilization of the butyrate carbon. In some cases the change in non-oxidative metabolism can be subtle: glucose increases the yield of alanine and aspartate from glutamine in rat colonocytes, although this is compensated by a decrease in glutamate production (62).

There is also the potential for some anabolic processes to occur in the gut, such as the conversion of NEFA to triacylglycerol, amino acids to protein and glucose to glycogen. The large uptake of NEFA observed in some studies of the isolated perfused small intestine of the rat (41) may largely reflect incorporation of NEFA into triacylglycerol of mesenteric adipose tissue (42) rather than oxidation, as previously thought (41). Furthermore, some amino acids may be used for the synthesis of secretory and non-secretory proteins rather than for oxidative purposes. In a study of protein turnover in rat enterocytes about half of the newly synthesized protein was secreted into the incubation medium (76); data from other rat studies (77,78) suggest that 30% of leucine taken up from the circulation is used to form apolipoproteins and other proteins synthesized by the jejunum and released into the mesenteric lymph.

Competition between luminal and vascular substrates

Using the vascularly perfused small intestine in vivo, Windmueller and Spaeth (28) clearly demonstrated that luminal substrates can replace vascular fuels or endogenous intestinal fuels as oxidative fuels. There is also evidence that luminal glutamine or glutamate can decrease utilization of glutamine from the blood. High concentrations of luminal glutamate can decrease the uptake and metabolism of arterial glutamine in rat jejunal segments in situ by about 25% (79). A 40% reduction in the rate of metabolism can also be produced by perfusing glutamine (6 mM) through the intestinal lumen. Hanson and Parsons (47) found that the net uptake of glutamine (1.5 mM) from the vascular side of an in vitro perfused segment of rat small intestine was reduced three-fold following luminal perfusion with glutamine (4.5 mM). However, without tracers it is not possible to assess flux rates to and from the vascular side.

A hierarchy of fuel preference

A number of authors have attempted to establish a hierarchy for the preferential utilization of oxidation of substrates by the gut, by incubating isolated colonocytes or enterocytes with similar concentrations of substrates either alone or in combination (Tables 14 and 15). Although the uptake of metabolism in vivo depends on a variety of factors that do not operate in vitro (e.g. variable concentrations of multiple hormones and substrates, and variable short term changes in perfusion rates) the enterocyte/colonocyte data (Table 14) are interesting in at least three ways. First, they identify glutamine as a preferred fuel for enterocytes, which is consistent with the A-PV concentration differences in vivo in man and results from vascularly perfused rat intestine in vivo. Second, they identify butyrate as a preferred fuel for colonocytes.

Table 14 Hierarchy of fuel oxidation suggested from in vitro studies of enterocytes, colonocytes or mucosal biopsies of the colon*

	Substrate concentration (mM)	*Reference*
Rat		
Colonocytes		
Butyrate>acetate>propionate>glucose and glutamine	All substrates 5	63
Butyrate>propionate>acetate>glucose	All substrates 0.2	53
Butyrate>propionate>acetate and glucose	All substrates 2	53
Butyrate>acetoacetate>glutamine>glucose	Glutamine and aceto-acetate 5; butyrate and glucose 10	65
Butyrate and acetoacetate>glutamine>glucose	All substrates 5 except glucose, 10	62
Enterocytes		
Glutamine>glucose>acetate, propionate and butyrate	All substrates 5	63
Human		
Colonocytes		
Butyrate>acetate>propionate>glucose	All substrates 0.5	54
Butyrate>acetate>propionate>glucose	All substrates 2	54
Butyrate>glucose	Both substrates 10	66,67
Colonic biopsies		
Butyrate>glutamine>glucose	Butyrate and glutamine 5; glucose 11	80

*The hierarchy is established by calculating the relative rates of CO_2 production. The rate of $^{14}CO_2$ production from the singly labelled carbon (data provided by authors) is multiplied by the number of carbons in the molecule

Unfortunately substrate balance studies across the large bowel in vivo and in vitro are lacking. In a human post-mortem study (16) the ratio of butyrate to acetate and propionate was lower in portal blood than in the lumen of the bowel, which is consistent with preferential utilization of butyrate by the colon. However, it is difficult to be certain about this interpretation because the study did not provide information about either the rate of SCFA production, which is influenced by the rate and type of substrate fermented, or the arterial concentration of SCFA, which influences portal concentration (especially acetate, which has a higher circulating concentration than the other SCFA). Third, studies with isolated pig enterocytes suggest that galactose can be utilized and oxidized shortly after birth, but the rate rapidly decreases by the time of weaning, and continues to do so thereafter. These studies also suggest that galactose is quantitatively a much less preferred substrate for pig enterocytes than are glutamine and glucose (39).

Table 15 Preferential sites of oxidation of substrates as suggested by rat enterocyte and colonocyte studies in vitro

	Reference
Glucose	
Colon>caecum=jejunum	63
Colon>jejunum	52
Glutamine	
Jejunum=caecum=colon	63
Ileum>colon (elemental diet)	81
Ileum=colon (animals on fibre-containing diet)	81
SCFA (butyrate, propionate, acetate)	
Colon>small intestine	63,81
Butyrate	
Colon>jejunum	52
Colon>distal small intestine	52
Acetate	
Colon>distal small intestine	81
Colon>distal jejunum	81
Propionate	
Colon>small intestine	52

Metabolism in different parts of the gastrointestinal tract

There is virtually no information about fuel metabolism in the oesophagus and stomach, but there is sufficient information from studies on other parts of the gastrointestinal tract to suggest important regional differences in fuel selection and oxidation. For example, SCFA are preferentially utilized by colonocytes compared to enterocytes. In one study, colonocytes oxidized butyrate at a rate nine times greater than enterocytes (81). Furthermore, ingestion of fibre increased the oxidation of SCFA and other substrates by colonocytes but not by enterocytes. Some, but not all, studies suggest that enterocytes oxidize glutamine at a faster rate than colonocytes.

Some metabolites are also preferentially formed in certain parts of the gut. For example, ketone bodies are formed as a result of SCFA utilization by the large bowel, while enterocytes, which generally have little capacity to utilize SCFA, produce few or no ketone bodies (rat, rabbit and probably man). Another regional difference concerns the metabolism of glutamine, which is converted to aspartate, alanine and ammonia by isolated rat colonocytes, and to alanine instead of aspartate by rat enterocytes (62). There may be species differences as well, since ketone body utilization is reported to be lower in human enterocytes than rat enterocytes (43).

Hormones

There has been no systematic examination of the effect of hormones on substrate metabolism of the gut. However, glucagon administration to chronically catheterized dogs (fasted 18–26 h) was reported to produce a 2- to 3-fold increase in the uptake of glutamine by the gut and portal-drained viscera, and reduce the circulating concentration of glutamine by 20–30% (82). Dexamethasone has also been reported to increase glutamine uptake in animals (rats and dogs) in vivo (83,84) and to increase the activity of glutaminase which metabolizes glutamine (85). In contrast, dexamethasone reduced the rate of glucose uptake by the gut and portal-drained viscera, and reduced the activities of glycolytic enzymes as well as the activity of oxoglutarate dehydrogenase, which is a key enzyme of the citric acid cycle (84). Uptake of glutamine by the inverted jejunum of dexamethasone-treated rats also increased glutamine utilization, and this may again be mediated by an increase in glutaminase activity. Recent catheterization studies in pigs suggest that growth hormone treatment before trauma can double glutamine utilization by the gut and portal-drained viscera (86). Finally, there is evidence that insulin can acutely regulate glucose absorption, transport and metabolism in the rat small intestine in vivo (59).

Combination of factors

In a specific situation different factors may affect intestinal metabolism in opposing ways. For example, rats with long-standing streptozotocin-induced diabetes showed increased glutaminase activity in the small intestine (50,87,88) but not in the colon (88). There was an overall increase in gut glutaminase activity, which would be expected to increase the rate of glutamine utilization. However, arterio-portal catheterization studies in vivo suggested reduced (88) or no net uptake of plasma glutamine (50,87,89). These changes can be explained by a marked reduction in the arterial glutamine concentration (almost two-fold in some studies of long-term diabetes) and an increase in both the circulating concentration and uptake of ketone bodies by the gut, which replaced circulating glutamine as a fuel. Some of the changes in fuel selection in diabetes may also be mediated by changes in hormones and altered activities of key enzymes. However, as indicated above, acute elevation of ketone bodies may not necessarily decrease glutamine utilization in vivo (50), possibly because infusion of salts of ketone bodies can produce an alkalosis (cf. hyperketonaemic states which are associated with acidosis), and possibly as a result of long-term effects of hormones or substrates on key enzymes involved in glutamine and ketone body metabolism.

Finally, the circulating concentration of substrates is influenced by the metabolic activity of many tissues, particularly those that are active in their production or utilization. Thus it is possible for flux generation steps in one tissue (e.g. glutamine production in muscle or ketone body production in the liver) to influence metabolism in the gut. In the case of SCFA the flux generating step is colonic fermentation, which depends on ingestion of fibre, resistant starch or certain oligosaccharides, such as those present in milk (90).

The effects of disease

In vivo studies in dogs and rats (48,51,83,89,91–94) have demonstrated that surgery, bacteraemia and steroid administration increase glutamine uptake by the gut and portal-drained viscera, whereas acidosis, starvation, diabetes mellitus and malignancy decrease glutamine utilization. Surgery does not appear to increase glucose utilization in vivo (92) and corticosteroids, which mediate some of the effects of surgery, decrease glucose utilization in vitro and in vivo (64). Studies of isolated rat colonocytes (95) suggest that malnutrition can decrease the utilization of most substrates and produces different substrate interactions. For example, the conversion of glucose to lactate in the presence of ketone bodies appears to be greater in colonocytes obtained from undernourished animals compared with that in normal animals (95).

Roediger (66) was the first to report that colonocytes from patients with colitis exhibited reduced oxidation of butyrate. Support for this observation came from further studies on human colonocytes (96) and biopsies (80), and led to the suggestion that a metabolic block in butyrate oxidation may be the cause of the colitis. The induction of colitis in the rat by 2-bromo-octanoate, which inhibits β-oxidation, supported this concept (97). However, other studies with human colonocytes (54) and colonic mucosal biopsies (98) have not found evidence of reduced butyrate oxidation in colitis. In interpreting these discrepant observations it is necessary to take into account the presence in colitic bowel of necrotic tissue which contributes to the weight and protein content of the mucosa, but has little or no effect on the overall rate of oxygen consumption, resulting in a reduction in the rate of O_2 consumed per g wet tissue or per g protein. In addition, colitis may produce important changes in the proportion of different 'colonocytes' which, according to Roediger (99), refers to all cells constituting the colonic epithelium: columnar, goblet, crypt and endocrine cells. Other potential explanations include possible differences in the methods of preparation of cells or biopsies, and difficulties in assessing and interpreting oxidation rates. Most studies have not measured total O_2 consumption or CO_2 production of the cell or biopsy preparations, and not all studies have used labelled substrates. Furthermore, some studies have assessed the oxidation rates of only one or two substrates. Finally, it is possible that differences in nutritional state, dietary fibre intake and drug treatment may influence the results: animal studies suggest that malnutrition may impair substrate oxidation rates by isolated colonocytes (95), whilst fibre increases oxidation rates of SCFA and other substrates. In addition steroids, which are used in the treatment of colitis, increase the activity of colonic glutaminase which could affect glutamine utilization and its interaction with other substrates.

Some clinical implications

The above discussion suggests that butyrate is likely to be an important fuel for the large bowel, while glutamine is important for the small bowel. Both of these substrates have a variety of other important metabolic, haemodynamic and other effects (Table 16): it is difficult to assess the extent to which these effects depend on each other, and particularly the extent to which they are linked to energy metabolism. However, because of their multiple effects, these two substrates have generated considerable clinical interest, especially with respect to their therapeutic potential in disorders of the gastrointestinal tract.

Table 16 Multiple effects of glutamine and butyrate on the intestinal mucosa

1. *Energy substrate*
   - Butyrate (large bowel), glutamine (small and large bowel)
   - Luminal nutrition (butyrate) or luminal and vascular (glutamine)
   - Oxidation of substrate <6 carbons (glutamine 5C, butyrate 4C)
   - Non-glycolytic pathway (glutamine and butyrate)
   - Energy and ATP equivalent of O_2 for glutamine and butyrate are similar

2. *Other effects*
   - Increase in gut blood flow (glutamine, small intestine (100); butyrate, large intestine (101))
   - Sodium co-transport (glutamine, small intestine (102); butyrate, large intestine (103))
   - Trophic effects/cell division: effect of glutamine (104) and butyrate (105) on small and large bowel; effect on cell differentiation (mainly butyrate)
   - Nucleic acids. Glutamine is essential for formation (amino groups) of nucleic acids; butyrate is involved in acetylation and methylation of DNA (106)
   - Signalling, e.g. glutamine influences action of epidermal growth factor (107); butyrate has effects on urokinase and urokinase inhibitors (108) as well as on enteroglucagon and peptide PYY
   - Gene expression. Butyrate increases mRNA expression of several proteins (109) and gangliosides (110). Glutamine increases mRNA expression of glutaminase, which is involved in glutamine metabolism
   - Mucus formation. Glutamine donates amino groups to the sugars present in mucus. Butyrate affects secretion of colonic fluid/mucus
   - Effect on other cells, e.g. leucocyte and phagocyte function, is thought to be influenced by glutamine (111,112), which is an important substrate for these cells (113), and by butyrate (114)
   - Mucosal barrier function believed to be improved by glutamine and butyrate by multiple mechanisms

Butyrate

Since butyrate is formed from the fermentation of fibre, the long-term administration of fibre-free feeds may prevent butyrate from serving as an important fuel for the colon, and at the same time prevent it from exerting multiple colonic effects (Table 16). In addition, long-term administration of fibre-free feeds is unphysiological and may lead to constipation. In one study 75% of patients on long-term home enteral tube feeding had constipation which required the use of laxatives and enemas (115), although the underlying disease may also have contributed.

Human milk is unusual in being a rich source of oligosaccharides (up to 18 g/l). The chemical structure of these oligosaccharides suggests that they are likely to escape digestion in the small intestine and therefore act as possible substrates for fermentation in the large bowel. More than 130 oligosaccharides (3–11 carbons per molecule) had been characterized by 1991 (90) and the list has expanded several-fold since then. The concentrations of different oligosaccharides in human milk decrease at different rates over the first few months of lactation, with a rapid decline of sialicacid-containing oligosaccharides, but little or no significant change in those containing fucose (116). The level of oligosaccharides in cows' milk and formula milk is considerably lower than in human milk, but the significance of this with respect to the metabolism and function of the colon remains uncertain, especially since the luminal concentration of SCFA in infants fed human milk is lower than in infants fed cows' milk or formula milk (Table 12). Nevertheless, a case has been made for providing sufficient amounts of fermentable substrate to the colon in a variety of conditions (29,67,117) including recovery from malnutrition. Without it, SCFA are unable to exert their putative beneficial effects on the colon (Table 16).

Butyrate has effects on gene expression in the colonic mucosa, which is of particular interest with respect to cancer of the colon, the incidence of which is related to dietary fibre intake. Butyrate also has effects on other cell lines that are not normally exposed to butyrate: human erythroid cell lines exposed to butyrate show increased expression of fetal and embryonic haemoglobin. Systemic administration of arginine-butyrate (initially 500 mg/kg/day) to individuals with thalassaemia and sickle cell disease (118) for a typical period of 2–3 weeks led to an increase in fetal Hb synthesis (6–45%), an increase in the percentage of fetal reticulocytes (6–25%) and globin mRNA (×2–6) and an increase in Hb concentration. In one patient who was given butyrate infusions intermittently for 7 weeks the Hb concentration rose from 4.7 to 10.2 g/dl. These observations emphasize that the administration of large doses of butyrate via an unphysiological route may have quite different effects to those seen under normal circumstances.

The apparent inhibition of butyrate oxidation in colitis has led to the administration of butyrate (or SCFA) enemas in patients with colitis. Patients receiving butyrate enemas had a significantly better response than control patients with respect to endoscopic and histological scores (119). Senagore et al. (120) compared the effect of corticosteroid (hydrocortisone 6 mg, mesalazine 4 mg/day) and SCFA enemas in a randomized prospective study of patients with proctosigmoiditis and found that SCFA were as effective as the standard drug treatment. Harig et al. (121) used SCFA enemas for 4–6 weeks in patients with diversion colitis. There was an improvement in endoscopic appearance of the mucosa within 2 weeks, whereas the control group given saline showed no improvement. Histological changes paralleled the endoscopic changes. Finnie et al. (98), who observed no inhibition of butyrate oxidation in colonic biopsies obtained from patients with colitis, suggested that the condition may be a 'deficiency state' since SCFA concentrations in faeces have been reported to be low in patients with colitis (122). Therefore a possible alternative explanation for the beneficial effects of butyrate in colitis is that the increased concentration of SCFA resulting from the enemas might produce 'mass action' effects (54). However, the faecal concentration of SCFA in patients with colitis is variable: some reports suggest decreased concentrations (122), others a normal concentration (123,124) and yet others increased concentrations (125). Furthermore, the extracellular concentration of butyrate required for maximal or near maximal oxidation by isolated colonocytes (Table 11) appears to be much lower than the colonic butyrate concentrations. It is therefore possible that some of the clinical benefits produced by butyrate are related to its other effects on the colon (Table 16).

Glutamine

Because of the multiple effects of glutamine (an energy source, trophic effects on the gut, effects on lymphocyte/macrophage function etc.), and remarkable studies in which morbidity and mortality were reported to be reduced in animals suffering from mucositis due to cytotoxic drugs and radiation treatment (126), the therapeutic potential of glutamine in gastrointestinal disorders was investigated. The first human trial with glutamine involved treatment of peptic ulcers (127). Its beneficial effect was confirmed experimentally in rats which developed aspirin-induced peptic ulcers (128). Not all subsequent studies of glutamine supplementation in humans with gastrointestinal disorders have shown benefit. In a randomized double blind crossover study in patients with carcinoma of the colon, who developed mucositis of the mouth and oesophagus as a result of receiving 5-fluorouracil (5 mg/day), glutamine had no significant effect over the placebo with respect to mouth comfort, ease of eating or the WHO mucositis score (129). In another randomized study (glutamine vs placebo) in patients receiving aggressive cytotoxic therapy following bone marrow transplantation, both the patient and observer-rated mucositis scores were not significantly affected by glutamine administration (130). In contrast, encouraging results have been obtained in preliminary studies of patients with pouchitis (inflammation of the surgically formed ileal pouch which serves as a faecal reservoir following colectomy) and possibly in Crohn's disease, both of which are small-intestinal conditions. Administration of intravenous glutamine to patients receiving aggressive cytotoxic therapy following bone marrow transplantation has been reported to reduce infection rates and hospital stay (131). Preliminary analysis of another randomized prospective study involving a similar group of patients suggests that glutamine-treated patients had significantly fewer pyrexial episodes and lower circulating endotoxin levels than the control group, who received an isonitrogenous parenteral nutrition regimen (T. Maugham, C. Pointon, S. Jebb, M. Elia, unpublished data). Presumably glutamine had an effect on the gut barrier function of the small and/or large bowel, which normally prevents the translocation of bacteria and endotoxin into the circulation. However, improved clearance of the translocated material is also a possibility.

There are several possible explanations for the negative effects of glutamine in mucositis of the upper gastrointestinal tract (mouth and oesophagus) and its positive effects on the small intestine. One of these concerns the distribution of enzymes involved in the glutamine metabolism. The activity of glutamine synthetase, which forms glutamine, is weak throughout the mucosa of the human gastrointestinal tract (132), with the possible exception of the upper stomach. This means that the gut depends on other sources for its supply of glutamine (diet and the circulation). Glutaminase, which catalyses the initial metabolism of glutamine in a flux generating reaction and is therefore a key enzyme involved in the metabolism of glutamine, has its highest activity in the small intestine, intermediate in the colon, and lowest in the oesophageal mucosa (132).

CONCLUSION

Concepts about nutrient utilization by the human gut have been slow to develop for at least three reasons: difficulties in assessing fuel selection from two sources (luminal and vascular) which vary in importance during different parts of the feeding–fasting cycle, the inaccessibility of the vessels draining the gut, and the high blood flow to the gut relative to its oxidative metabolism, which means that the arterio-portal venous concentration difference for some metabolites is very small. Despite this, there is evidence to suggest a special role for glutamine, ketone bodies (especially in starvation when ketone body concentration increases) and short chain fatty acids (mainly butyrate) as important fuels for the gastrointestinal tract. Butyrate is essentially entirely derived from the lumen of the colon, while ketone bodies are essentially derived from the blood. Glutamine is derived from both the lumen and the blood, but most is derived from the blood. The colon probably derives very little glutamine from the lumen, because most of it is already absorbed by the small intestine. Butyrate and glutamine have multiple effects on the integrity, structure and function of the bowel and these are probably partly linked to their energy metabolism. Interest in these effects has led to important investigations into the therapeutic use of these substrates in the developing field of nutritional pharmacology (133).

REFERENCES

1. Elia M 1995 General integration and regulation of metabolism at the organ level. Proc Nutr Soc 54:213–232
2. Parker DS, McMillan RJ 1976 The determination of volatile fatty acids in the caecum of the conscious rabbit. Br J Nutr 35:365–371
3. Poppitt SD, Livesey G, Faulks RM, Roe M, Prentice AM, Elia M 1996 Circadian patterns and 24h total hydrogen and methane excretion in man in relation to the intake and fermentation of non-starch polysaccharide (NSP) diets, including implications for indirect calorimetry and $D_2{}^{18}O$. Eur J Clin Nutr (in press)
4. Livesey G, Elia M 1995 Short-chain fatty acids as an energy source in the colon: metabolism and clinical implications. In: Cummings JH, Rombeau JL, Sakata T (eds) Physiological and Clinical Aspects of Short-Chain Fatty Acids, Cambridge University Press, pp 427–482
5. Gray FV, Weller RA, Pilgrim AF, Jones GB 1967 Rates of production of volatile fatty acids in the rumen. Aust J Agric Res 18:625–634
6. Hoover WH 1978 Digestion and absorption in the hindgut of ruminants. J Anim Sci 46: 1789–1799
7. Siciliano-Jones J, Murphy MR 1989 Production of volatile fatty acids in the rumen and caecum-colon of steers as affected by forage: concentrate and forage physical form. J Dairy Sci 72: 485–492
8. Hoover WH, Clarke SD 1972 Fiber digestion in the beaver. J Nutr 102:9–15
9. Marty J, Vernay M 1984 Absorption and metabolism of the volatile fatty acids in the hindgut of the rabbit. Br J Nutr 51:265–277
10. Imoto S, Namioka S 1978 VFA production in the pig large intestine. J Anim Sci 47:467–478
11. Kim KI, Benevenga NG, Grimmer WH 1978 Lactose activity and VFA production in the caecum and colon of pigs fed a corn-soy or 40% whey diet. J Anim Sci 46:1648–1657
12. Bergman EN 1990 Energy contributions of volatile fatty acids from the gastrointestinal tract in various species. Physiol Rev 70:567–590
13. McNeil NI 1984 The contribution of the large intestine to energy supplies in man. Am J Clin Nutr 39:338–342
14. Bridges SR, Anderson JW, Deakins DD, Dillon DW, Wood CL 1992 Oat bran increases serum acetate of hypercholesterolaemic men. Am J Clin Nutr 56:455–459

15. Dankert J, Zijlstra JB, Wolthers BG 1981 Volatile fatty acids in human peripheral and portal blood: quantitative determination by vacuum distillation and gas chromatography. Clin Chim Acta 110:301–307
16. Cummings JH, Pomare EW, Branch WJ, Naylor CPE, MacFarlane GT 1987 Short-chain fatty acids in human large intestine, portal, hepatic and venous blood. Gut 28:1221–1227
17. Elia M, Livesey G 1992 Energy expenditure and fuel selection in biological systems: the theory and practice of calculations based on indirect calorimetry and tracer methods. World Rev Nutr Diet 70:68–131
18. Livesey G, Elia M 1988 Estimation of energy expenditure, net carbohydrate utilization and net fat oxidation and synthesis by indirect calorimetry. Evaluation of errors with special reference to the detailed composition of fuels. Am J Clin Nutr 47:608–628
19. Elia M 1992 Organ and tissue contribution to metabolic rate. In: Kinney JM, Tucker HN (eds) Energy Metabolism, Tissue Determinants and Cellular Corollaries. Raven Press, New York, pp 61–80
20. Brody S 1945 Bioenergetics of growth. Reinhold Press, New York, pp 575–662
21. Snyder WS, Cook MJ, Nassett Karhausen LR, Parry-Howells G, Tipton IH 1974 Report of the taskforce on reference man. International Commission on Radiological Protection No. 23. Pergamon Press, Oxford
22. Wade OL, Bishop JM, Donald KW 1962 Cardiac output and regional blood flow. Blackwell Scientific Publications, Oxford
23. Huntington GB 1990 Energy metabolism in the digestive tract and liver of cattle: influence of physiological state and nutrition. Reprod Nutr Dev 30:35–47
24. Burrin DG, Ferrell CL, Eisemann JH, Britton RA, Nienaber JA 1989 Effect of level of nutrition on splanchnic blood flow and oxygen consumption in sheep. Br J Nutr 62:23–34
25. Reynolds CK, Tyrrell HF, Reynolds PJ 1991 Effect of diet forage-to-concentrate ratio and intake on energy metabolism in growing beef heifers: whole body energy and nitrogen balance and visceral meat production. J Nutr 121:984–1003
26. Sherratt HSA 1968 The metabolism of the small intestine. Oxygen uptake and L-lactate production along the length of the small intestine of the rat and guinea pig. Comp Biochem Physiol 24:745–761
27. Bronk JR, Parsons DS 1965 The polarographic determination of the respiration of the small intestine of the rat. Biochim Biophys Acta 107:397–404
28. Windmueller HG, Spaeth AE 1980 Respiratory fuels and nitrogen metabolism in vivo in small intestine of fed rats. J Biol Chem 255:107–112
29. Roediger WEW 1986 Metabolic basis of starvation diarrhoea: implications for treatment. Lancet i:1082–1084
30. Dietze G, Wicklmayr M, Hepp KD, Bogner W, Mehnert H, Czempiel H, Henftling HG 1976 On gluconeogenesis of human liver. Diabetologia 12:555–561
31. Owen OE, Reichle FA, Mozzoli MA, Kreulen T, Patel MS, El Ehnbein IB 1981 Hepatic, gut and renal substrate flux in patients with hepatic cirrhosis. J Clin Invest 68:240–252
32. McAnena OJ, Moore FA, Moore EE, Jones TN, Parsons P 1991 Selective uptake of glutamine in the gastrointestinal tract: confirmation in a human study. Br J Surg 78:480–482
33. Felig P, Wahren J, Raf L 1973 Evidence of inter-organ amino acid transport by blood cells in humans. Proc Natl Acad Sci 70:1775–1779
34. Bjorkman O, Eriksson LS, Nyberg B, Wahren J 1990 Gut exchange of glucose and lactate in basal state and after oral glucose ingestion in post-operative patients. Diabetes 39:747–781
35. Wahren J, Felig P, Ahlborg G, Jorfeldt L 1971 Glucose metabolism during leg exercise in man. J Clin Invest 50:2715–2725
36. Elia M 1993 Glutamine metabolism in adipose tissue in vivo. Clin Nutr 12:51–53
37. van der Hulst RRW, Deutz NEP, Soeters PB, von Meyenfeldt M 1993 Glutamine uptake at different sites of the human gut. Clin Nutr 12 (suppl): 29, 10–11
38. Windmueller HG 1982 Glutamine utilization by the small intestine. Adv Enzymol 53:202–237
39. Darcy-Vrillon B, Posho L, Morel MT, Bernard F, Blachier F, Meslin JC, Duee PH 1994 Glucose, galactose and glutamine metabolism in pig isolated enterocytes during development. Pediatr Res 36:175–181
40. Deutz NEP, TenHave GAM, Soeters PB, Moughan PJ 1995 Increased intestinal amino acid retention from the addition of carbohydrate to a meal. Clin Nutr 14:354–364

41. Hulsmann WC 1971 Preferential oxidation of fatty acids by rat small intestine. FEBS Lett 17: 35–38
42. Windmueller HG, Spaeth E 1978 Identification of ketone bodies and glutamine as the major respiratory fuels in vivo for postabsorptive rat small intestine. J Biol Chem 253:69–76
43. Ashy AA, Ardawi MSM, Salleh M 1988 Glucose, glutamine and ketone-body metabolism in human enterocytes. Metabolism 37:602–609
44. Porteous JW 1978 Glucose as a fuel for small intestine. Biochem Soc Trans 6:534–539
45. Hasselbach SG, Knudsen GM, Jakobsen I, Hageman LP, Holm S, Paulson OB 1995 Blood-brain barrier permeability of glucose and ketone bodies during short-term starvation in humans. Am J Physiol 268:E1161–E1166
46. Windmueller HG, Spaeth AE 1974 Uptake and metabolism of plasma glutamine by the small intestine. J Biol Chem 249:5070–5079
47. Hanson PJ, Parsons DS 1977 Metabolism and transport of glutamine and glucose in vascularly perfused small intestine rat. Biochem J 166:509–519
48. Souba WW, Scott TE, Wilmore DW 1985 Intestinal consumption of intravenously administered fuels. JPEN 9:18–22
49. Buttrose M, McKellar D, Welbourne TC 1987 Gut–liver interaction in glutamine homeostasis: portal ammonia role in uptake and metabolism. Am J Physiol 252:E746–E750
50. Watford M, Erbelding EJ, Smith EM 1987 The regulation of glutamine and ketone body metabolism in the small intestine of the long-term (40 day) streptozotocin-diabetic rat. Biochem J 242:61–68
51. Welbourne TC, Childress D, Givens G 1986 Renal regulation of interorgan glutamine flow in metabolic acidosis. Am J Physiol 251:R859–R866
52. Kight CE, Fleming SE 1993 Nutrient oxidation by rat intestinal epithelial cells is concentration dependent. J Nutr 123:876–882
53. Clausen MR, Mortensen PB 1994 Kinetic studies on the metabolism of short-chain fatty acid and glucose by isolated rat colonocytes. Gastroenterology 106:423–432
54. Clausen MR, Mortensen PB 1995 Kinetic studies on colonocyte metabolism of short-chain fatty acids and glucose in ulcerative colitis. Gut 37:684–689
55. Adibi SA, Mercer DW 1973 Protein digestion in human intestine as reflected in luminal mucosal and plasma amino acid concentrations after meals. J Clin Invest 52:1586–1594
56. Ogawa K, Ben RA, Pons S, de Paolo MIL, Fernandez LB 1992 Volatile fatty acids, lactic acid and pH in the stools of breast-fed and bottle-fed infants. J Pediatr Gastroenterol Nutr 15: 248–252
57. Midtvedt C, Midtvedt J 1992 Production of short chain fatty acids by the intestinal microflora during the first 2 years of human life. J Pediatr Gastroenterol Nutr 15:395–403
58. Meijer-Severs GL, van Santen E 1989 Short-chain fatty acids and organic acid concentrations in feces of 10 human volunteers and their correlation with anaerobe cultural counts over a 15 month period. Scand J Gastroenterol 24:1276–1280
59. Kellett GL, Jamal A, Robertson JP, Wollen N 1984 The acute regulation of glucose absorption, transport, and metabolism of the rat in vivo. Biochem J 219:1027–1035
60. Elia M, Folmer P, Schlatmann A, Goren A, Austin S 1988 Carbohydrate, fat and protein metabolism in muscle and in the whole body of man after mixed meal ingestion. Metabolism 37:542–547
61. Haglund V 1986 Gastrointestinal, hepatic and renal complications of shock and trauma. In: Little RA, Frayn KN (eds) The Scientific Basis for the Case of the Critically Ill. Manchester University Press, pp 293–306
62. Ardawi MSM, Newsholme EA 1985 Fuel utilization in colonocytes of the rat. Biochem J 231:713–719
63. Fleming SE, Fitch MD, DeVries S, Liu ML, Kight C 1991 Nutrient utilization by cells isolated from rat jejunum, cecum and colon. J Nutr 121:869–878
64. Ardawi MSM 1988 Glutamine and ketone body metabolism in the gut of streptozotocin-diabetic rats. Biochem J 249:565–572
65. Roediger WEW 1982 Utilization of nutrients by isolated epithelial cells of the rat colon. Gastroenterology 83:424–429
66. Roediger WEW 1980 The colonic epithelium in ulcerative colitis: an energy-deficiency disease? Lancet 2:712–715

67. Roediger WEW 1980 Role of anaerobic bacteria in the metabolic welfare of the colonic mucosa of man. Gut 21:793–798
69. Elia M, Folmer P, Schlatmann A, Goren A, Austin S 1989 Amino acid metabolism in muscle and in the whole body of man before and after ingestion of a single meal. Am J Clin Nutr 49: 1203–1210
70. Romyn JA, Coyle EF, Sidossis LS, Gastaldelli A, Horowitz JF, Endert E, Wolfe RR 1993 Regulation of endogenous fat and carbohydrate metabolism in relation to exercise intensity and duration. Am J Physiol 265:E380–E391
71. Brundin T, Wahren J 1991 Influence of a mixed meal on splanchnic and inter-scapular energy expenditure in humans. Am J Physiol 260:E232–E237
72. Brundin T, Thorne A, Wahren J 1992 Heat leakage across the abdominal wall and meal induced thermogenesis in normal weight and obese subjects. Metabolism 41:49–55
73. Neptune EM 1965 Respiration and oxidation of various substrates by ileum in vitro. Am J Physiol 209:329–332
74. Watford M, Lund P, Krebs HA 1979 Isolation and metabolic characteristics of rat and chicken enterocytes. Biochem J 178:589–596
75. Hanson PJ, Parsons DS 1978 Factors affecting the utilization of ketone bodies and other substrates by rat jejunum: effects of fasting and of diabetes. J Physiol 278:55–67
76. Higashiguchi T, Hasslgren PO, Wagner K, Fischer J 1993 Effect of glutamine on protein synthesis in isolated intestinal epithelial cells. JPEN 17:307–314
77. Sigurdsson G, Noel S-P, Havel RJ 1979 Quantification of the hepatic contribution to the metabolism of high density lipoproteins in rats. J Lipid Res 20:316–324
78. Wu A-L, Windmueller HG 1979 Relative contributions by liver and intestine to individual plasma apolipoproteins in the rat. J Biol Chem 254:7316–7322
79. Windmueller HG, Spaeth AE 1975 Intestinal metabolism of glutamine and glutamate from the lumen as compared to glutamine from blood. Arch Biochem Biophys 171:662–672
80. Chapman MAS, Grahn MF, Boyle MA, Hutton M, Rogers J, Williams NS 1994 Butyrate oxidation is impaired in the colonic mucosa of sufferers of quiescent ulcerative colitis. Gut 35: 73–76
81. Marsman KE, McBurney MI 1995 Dietary fibre increases oxidative metabolism in colonocytes but not in distal small intestine enterocytes isolated from rats. J Nutr 125:273–282
82. Geer RJ, William PE, Lairmore T, Abumrad AN 1987 Glucagon: an important stimulator of gut and hepatic glutamine metabolism. Surg Forum 38:27–29
83. Souba WW, Smith RJ, Wilmore DW 1985 The effects of glucocorticoids on metabolism in visceral organs. Metabolism 34:450–456
84. Salleh M, Ardawi MSM, Majzoub MF, Newsholme EA 1988 Effect of glucocorticoid treatment on glucose and glutamine metabolism by the small intestine of the rat. Clin Sci 75:93–100
85. Fox AD, Kripke SA, Berman JM, McGitney RM, Settle RG, Rombeau JL 1988 Dexamethasone administration induces increased glutaminase specific activity in the jejunum and colon. J Surg Res 44:391–396
86. Mjålland M, Unneberg K, Jenssen TG, Revhaug A 1995 Experimental study to show that growth hormone treatment before trauma increases glutamine uptake in the gastrointestinal tract. Br J Surg 82:1076–1079
87. Watford M, Smith EM, Erbelding EJ 1984 The regulation of phosphate-activated glutaminase activity and glutamine metabolism in the streptozotocin-diabetic rat. Biochem J 224:207–214
88. Ardawi MSM 1987 The maximal activity of phosphate-dependent glutaminase and glutamine metabolism in the colon and small intestine of the streptozotocin-diabetic rats. Diabetologia 30:109–114
89. Brosnan ME 1983 Interorgan metabolism of amino acids in streptozotocin-diabetic ketoacidotic rats. Am J Physiol 244:E151–E158
90. Sabharwal H, Sjoblad S, Lundblad A 1991 Affinity chromatographic identification and quantitation of blood group A-active oligosaccharides in human milk and feces of breast fed infants. J Pediatr Gastroenterol Nutr 12:274–279
91. Souba WW, Wilmore DW 1983 Postoperative alteration of arteriovenous exchange of amino acids across the gastrointestinal tract. Surgery 94:342–350
92. Souba WW 1991 Glutamine: a key substrate for the splanchnic bed. Annu Rev Nutr 11:285–308
93. Souba WW, Herkowitz K, Klimberg SV, Sallöm RM, Plumley DA, Flynn T, Copeland EM 1990

The effect of sepsis and endotoxaemia on gut glutamine metabolism. Ann Surg 211:543–551
94. Souba WW, Klimberg S, Plumley DA, Sallöm RM, Flynn TC, Bland KI, Copeland EM 1990 Role of glutamine in maintaining a healthy gut and supporting the metabolic response to injury and infection. J Surg Res 48:383–391
95. Firmansyah A, Penn D, Lebenthal E 1989 Isolated colonocyte metabolism of glucose, glutamine, n-butyrate and β-hydroxybutyrate in malnutrition. Gastroenterology 97:622–629
96. Ireland A, Jewell DP 1990 5-aminosalicylic acid has no effect on butyrate metabolism in human colonic epithelial cells. Gastroenterology 98 (Suppl) A176
97. Roediger WEW, Nance S 1986 Metabolic induction of experimental colitis by inhibition of fatty acid oxidation. Br J Exp Pathol 67:773–782
98. Finnie IA, Taylor BA, Rhodes JM 1993 Ileal and colonic epithelial metabolism in quiescent ulcerative colitis: increased glutamine metabolism in distal colon but no defect in butyrate metabolism. Gut 34:1552–1558
99. Roediger WEW, Truelove SC 1979 Method of preparing isolated colonic epithelial cells (colonocytes) for metabolic studies. Gut 20:484–488
100. Houdijk APJ, van Leeuwen PAM, Boermeester MA, van Lambalgen T, Teerlink T, Flinkerbusch EL, Sauerwein HP, Wesdorp RI 1994 Glutamine-enriched enteral diet increases splanchnic blood flow in the rat. Am J Physiol 267:G1035–G1040
101. Mortensen FV, Nielsen 1995 In vivo and in vitro effects of short-chain fatty acids on intestinal circulation. In: Cummings JH, Rombeau JL, Sakata T (eds) Physiological and Clinical Aspects of Short-chain Fatty Acids. Cambridge University Press, Cambridge, pp 341–400
102. Bradford NM, McGivan JD 1982 The transport of alanine and glutamine into isolated rat epithelial cells. Biochim Biophys Acta 689:55–62
103. Engelhardt W 1995 Absorption of short-chain fatty acids from the large intestine. In: Cummings JH, Rombeau JL, Sakata T (eds) Physiological and Clinical Aspects of Short-chain Fatty Acids. Cambridge University Press, Cambridge, pp 149–170
104. Yoshida S, Leskiw MJ, Schluter MD, Bush KT, Nagele RG, Lanza-Jacoby S, Stein JP 1992 Effect of total parenteral nutrition, systemic sepsis and glutamine on gut mucosa in rats. Am J Physiol 263:E368–E373
105. Sakata T 1995 Effects of short-chain fatty acids on the proliferation of gut epithelial cells in vivo. In: Cummings JH, Rombeau JL, Sakata T (eds) Physiological and Clinical Aspects of Short-chain Fatty Acids. Cambridge University Press, Cambridge, pp 391–400
106. Young GP, Gibson PR 1995 Butyrate and the human cancer cell. In: Cummings JH, Rombeau JL, Sakata T (eds) Physiological and Clinical Aspects of Short-chain Fatty Acids. Cambridge University Press, Cambridge, pp 391–400
107. Ko TC, Beauchamp RD, Townsend CM, Thompson JC 1993 Glutamine is essential for epidermal growth factor stimulated intestinal cell proliferation. Surgery 114:147–154
108. Gibson PR, Rosella O, Young GP, Rosella G 1994 Butyrate is a potent inhibitor of urokinase secretion by normal colonic epithelium in vitro. Gastroenterology 107:410–419
109. Kruh J, Defer N, Tichonicky L 1995 Effects of butyrate on cell proliferation and gene expression. In: Cummings JH, Rombeau JL, Sakata T (eds) Physiological and Clinical Aspects of Short-chain Fatty Acids. Cambridge University Press, Cambridge, pp 275–288
110. Berensen CS, Patterson MA, Migdadi JA, Lance P 1995 N-butyrate mediation of ganglioside expression of human and murine cancer cells demonstrates relative cell specificity. Clin Sci 88:491–499
111. Newsholme EA, Parry-Billings M 1990 Properties of glutamine release from muscle and its importance for the immune system. JPEN 14 (suppl): 63S–67S
112. Caldwell MD 1989 Local glutamine metabolism in wounds and inflammation. Metabolism 38 (suppl 1):34–39
113. Newsholme EA, Newsholme P, Cuci R, Challoner E, Salleh M, Ardawi M 1988 A role for muscle in the immune system and its importance in surgery, trauma, sepsis and burns. Nutrition 4:261–268
114. Brisseau GF, Rotstein OD 1995 The effect of short-chain fatty acids on phagocytic function. In: Cummings JH, Rombeau JL, Sakata T (eds) Physiological and Clinical Aspects of Short-chain Fatty Acids. Cambridge University Press, Cambridge, pp 391–400
115. Parker T, Neale G, Elia M 1996 Home enteral tube feeding in East Anglia. Eur J Clin Nutr (in press)

116. Miller JB, Bull S, Miller J, McVeagh P 1994 The oligosaccharide composition of human milk: temporal and individual variations in monosaccharide components. J Pediatr Gastroenterol Nutr 19:371–376
117. Roediger WEW 1990 Starved colon: diminished mucosal nutrition, diminished absorption and colitis. Dis Colon Rectum 33:858–862
118. Perrine SP, Ginder GD, Fuller DV, Dover GH, Ikuta T, Witkrowska HE, Cai S-P, Vichinsky EP, Olivieri N 1993 A short-term trial of butyrate to stimulate fetal-globin-gene expression the β-globin disorders. N Engl J Med 328:81–86
119. Scheppach W, Sommer H, Kirchner T, Paganelli G-M, Bartram P, Christl S, Richter F, Dusel G, Kasper H 1992 Effect of butyrate enemas on the colonic mucosa in distal ulcerative colitis. Gastroenterology 103:51–56
120. Senagore AJ, MacKeigan JM, Schneider M, Ebrom S 1992 Short-chain fatty acid enemas: a cost effective alternative in the treatment of non-specific proctosigmoiditis. Dis Colon Rectum 35:923–927
121. Harig JM, Soergel KH, Komorowski RA, Wood CM 1989 Treatment of diversion colitis with short-chain fatty acid irrigation. New Engl J Med 320:23–28
122. Vernia P, Caprilli R, Latella G, Barbetti F, Magliocca FM, Cittadini M 1988 Fecal lactate and ulcerative colitis. Gastroenterology 95:1564–1568
123. Hove H, Mortensen PB 1995 Influence of intestinal inflammation (IBD) and small and large bowel length on fecal short-chain fatty acids and lactate. Dig Dis Sci 40:1372–1380
124. Holtug K, Rasmussen HS, Mortensen PB 1988 Short-chain fatty acids in inflammatory bowel disease. Scand J Clin Lab Invest 48:667–671
125. Roediger WEW, Heyworth M, Willoughby P, Piris J, Moore A, Truelove SC 1982 Luminal ions and short-chain fatty acids as markers of functional activity of the mucosa in ulcerative colitis. J Clin Pathol 35:323–326
126. Elia M 1992 Glutamine in parenteral nutrition. Int J Food Sci Nutr 43:47–59
127. Shire W, Snider RN, DuBilier B, Rude JC, Clarke GE, Ravel JO 1957 Glutamine treatment of peptic ulcer. Texas State J Med, Nov, 840–843
128. Okabe S, Honda K, Takeuchi K, Takagi K 1975 Inhibitory effect of L-glutamine on gastric irritation. Dig Dis 20:626–631
129. Jebb SA, Osborne RJ, Maughan TS, Mohideen N, Mack P, Mort D, Shelley MD, Elia M 1994 5-fluorouracil and folinic acid-induced mucositis: no effect of oral glutamine supplementation. Br J Cancer 70:732–735
130. Jebb SA, Marcus SA, Elia M 1995 Pilot study of oral glutamine supplementation in patients receiving bone marrow transplants. Clin Nutr 14:162–165
131. Zeigler TR, Young LS, Benfell K, Schetlinga M, Hortos K, Bye R, Morrow FB, Jacobs DO, Smith RJ, Antin JH, Wilmore DW 1992 Clinical and metabolic efficacy of glutamine-supplemented parenteral nutrition after bone marrow transplantation. A randomized double-blind trial. Finn Int Med 116:821–828
132. James LA, Lunn PG, Elia M, Hunter JO, Neale G, Middleton S 1996 Glutaminase and glutamine synthetase activities in the gastrointestinal tract of humans. Proc Nutr Soc (in press)
133. Elia M 1995 Changing concepts of nutrient requirements in disease: Implications for artificial nutritional support. Lancet 345:1279–1284

24 Development of the mucosal immune system in humans

P. BRANDTZAEG

ABBREVIATIONS

APC	antigen-presenting cells
BALT	bronchus-associated lymphoid tissue
CD	cluster of differentiation
DTH	delayed-type hypersensitivity
GALT	gut-associated lymphoid tissue
HLA	human leucocyte antigen
IEL	intraepithelial lymphocytes
IFN-γ	interferon-γ
IL	interleukin
KLH	keyhole limpet haemocyanin
MΦ	macrophage
MALT	mucosa-associated lymphoid tissue
M cells	membrane cells
NALT	nasal-associated lymphoid tissue
pIg	polymeric immunoglobulin
RAG	recombinase-activating gene
SC	secretory component
SIgA	secretory immunoglobulin A
SIgM	secretory immunoglobulin M
TCR	T-cell receptor
TdT	terminal deoxyribonucleotidyltransferase
TGF-β	transforming growth factor-β

INTRODUCTION

Mucosal epithelia cover a surface area that is at least two hundred times that of the skin, perhaps amounting to some 400 m^2 in an adult. This extensive and generally quite vulnerable epithelial barrier is protected by numerous chemical and physical innate mechanisms that cooperate intimately with a local adaptive immune system. The main humoral mediators of specific mucosal immunity are secretory IgA and IgM (SIgA and SIgM); the former class of antibodies constitute the largest non-inflammatory defence system of the body (1–3).

Successful interactions between innate and specific adaptive immune protection is

J.G. Bindels, A.C. Goedhart and H.K.A. Visser (eds), Recent Developments in Infant Nutrition, 349–376.

a prerequisite for health, because the various mucosae are favoured as portals of entry by most infectious agents, allergens and carcinogens. The neonatal period is particularly critical in this respect, as the newborn is immediately exposed to numerous micro-organisms, foreign proteins and chemicals. This review will summarize salient features of postnatal mucosal immunity and discuss cellular and molecular interactions of putative importance for adaptive surface protection and for the development of oral tolerance.

The evolution of immunogenetics

Much of the genetic information instrumental to successful immunological handling of environmental confrontations was established several million years before our ancestors (*homo sapiens*) started migration through Africa and to other continents, some 100000 years ago, reaching the region of present Europe approximately 35000 B.C. Evolutionary gene modifications occurred as an adaptation to the microbial pressure and to the dietary impact of hunting, fishing, gathering, agriculture and husbandry. Over time, the mucosal immune system apparently generated two arms of defence: immune exclusion to combat penetration of noxious antigens and colonization of pathogens; and hyporesponsiveness to innocuous antigens bombarding the extensive mucosal surfaces (Fig. 1). The latter phenomenon is called oral tolerance when it is induced in the gut by dietary antigens, and it exerts down-regulatory immune effects both locally and in peripheral tissues (4).

In the region of present Europe, farming started only some 12000 years ago (5). It is believed that before domestication, approximately 70% of the energy intake came from plants, and the remainder from meat, bone marrow and brain. Subsequently, grain (mainly wild wheat) and milk from cows and goats probably provided almost 90% of the energy. Immunoregulatory mechanisms underlying oral tolerance have therefore had a relatively short time to evolve genetically in response to the changing impact of dietary antigens, and it should be no big surprise that adverse immunological reactions to certain recently introduced food proteins such as gluten can occur. Although the association of gluten-sensitive enteropathies (coeliac disease and dermatitis herpetiformis) with defined HLA-DQ molecules is well established, other genetic variables of the immune response to gliadin peptides remain obscure (6). The immunogenetics of cows' milk allergy is even less understood.

Mucosal surface defence and oral tolerance

Considering the fact that more than a ton of nutrients may pass through the gastro-intestinal tract of an adult every year, oral tolerance mechanisms seem to have developed rather successfully; the immunological challenge represented by the large variety of dietary immunogenic proteins causes no hypersensitivity in most individuals, despite substantial absorption of intact antigens even in the healthy state. Paganelli et al. (7) found up to 3 ng/ml of β-lactoglobulin in peripheral blood after an intake of 1.2 litre of bovine milk. Husby et al. (8) detected up to 10 ng/ml of ovalbumin, which

***Figure 1** Two major defence mechanisms induced in the gut. Immune exclusion limits epithelial colonization of microorganisms and inhibits penetration of foreign material. This first-line defence is principally mediated by secretory antibodies of the IgA (and IgM) class in cooperation with various non-specific innate protective factors. Secretory immunity is mainly stimulated by live microbial and particulate antigens. Soluble protein antigens and haptens, on the other hand, primarily induce suppression of proinflammatory humoral immune responses (IgG and IgE antibodies) as well as delayed-type hypersensitivity (DTH) mediated by activated helper T cells ($CD4^+$) of the Th1 subset. This poorly defined phenomenon induced by dietary antigens is called oral tolerance; it may operate both locally and in the periphery*

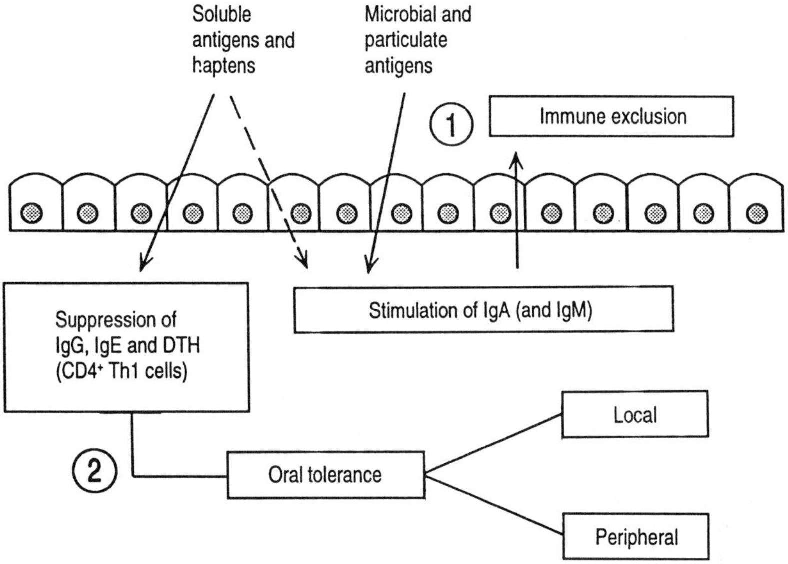

corresponded to about 10^{-5} of the amount consumed. Most of the antigenic material appeared in the circulation after 2–5 h and was partly present in immune complexes.

Nevertheless, mucosal defence is by no means patent, as highlighted in relation to infections, particularly in the developing countries. This situation could have been significantly improved by re-implementation of traditional breast-feeding (9). Mucosal pathogens are now a major killer of children below the age of 5 years, being responsible for more than 14 million deaths of children annually. Diarrhoeal disease alone claims a toll of 5 million children per year or about 500 deaths every hour. These figures document the need for mucosal vaccines to enhance surface defence against common infectious agents in addition to advocating breast-feeding. Convincing epidemiological documentation suggests that the risk of dying from diarrhoea is reduced 14–24 times in nursed children (9,10).

Importance of breast-feeding

There is an increasing consensus that controlling population growth in developing countries not only requires an improvement in economic and educational status, but also requires parents to have a realistic expectation that their children will survive to adulthood. Promotion of breast-feeding is important to this end, not only because of the protective value of this natural immunological 'substitution therapy' but also because of its nutritional and contraceptive properties (9). It is also important to emphasize the psychological advantages of nursing (for the mother as well as for the child), and to consider that the nutritional value of breast milk beyond 6 months seems to be underestimated (11). Finally, breast milk contains a number of immunoregulatory factors that may have significant effects on the development of the infant's mucosal immune system, thereby potentially enhancing in an indirect way the subsequent health of the individual.

PROTECTIVE AND IMMUNOREGULATORY COMPONENTS OF BREAST MILK

Antibodies

The initial breast milk, or colostrum, is much richer in antibodies than other secretions (12) because of its remarkably high concentration of SIgA (~12 g/litre) and SIgM (~0.6 g/litre). Individual variations are large, however, and the level decreases approximately four-fold after 2 weeks, then remains fairly stable throughout lactation (13). Antibodies of these two classes are locally produced by plasma cells as polymeric immunoglobulins (pIg), mainly dimeric IgA and pentameric IgM (14). These J chain-containing polymers are transported selectively through the secretory epithelium by the pIg receptor (Fig. 2), which is a unique 100-kDa glycoprotein expressed basolaterally on glandular cells throughout the body. The receptor is also called the transmembrane secretory component (SC), previously named 'secretory piece' or 'transport piece' (4).

When the secretory antibodies are released apically from the epithelial cell to the lumen, the pIg receptor is cleaved and an 80-kDa fragment called bound SC remains non-covalently attached to SIgA and SIgM (Fig. 2). Bound SC is in addition covalently linked to SIgA (but not to SIgM) by disulphide bridges, thereby endowing considerable stability on this class of antibody (2), which contributes to its remarkable survival when passing through the gastrointestinal tract (15). Unoccupied pIg receptor molecules are also cleaved and released to the lumen (14). Such 80-kDa free SC is present at quite high levels (~2 g/litre) in colostrum (12); its exerts a stabilizing effect on SIgM (16) and may by itself contribute to intestinal defence through inhibiting adhesion of *Escherichia coli* (17) and by activating mucosal eosinophils (18).

Because the lactating mammary glands are part of the integrated mucosal immune system (Fig. 3), milk antibodies reflect antigenic stimulation of mucosa-associated lymphoid tissue (MALT) in the gut as well as in the airways (see below). This has been documented by showing that SIgA from breast milk exhibits specificity for an

***Figure 2** Model for epithelial transport of J chain-containing dimeric IgA and pentameric IgM by transmembrane secretory component (SC or pIg receptor) expressed basolaterally on glandular cells. The resulting secretory immunoglobulins act in a first line of defence by performing immune exclusion of antigens in the mucus layer at the epithelial surface (to the right). Serum-derived or locally produced IgG is not subjected to active external transport. Free SC is generated when unoccupied pIg receptor (at the top) is cleaved at the apical face of the epithelial cell*

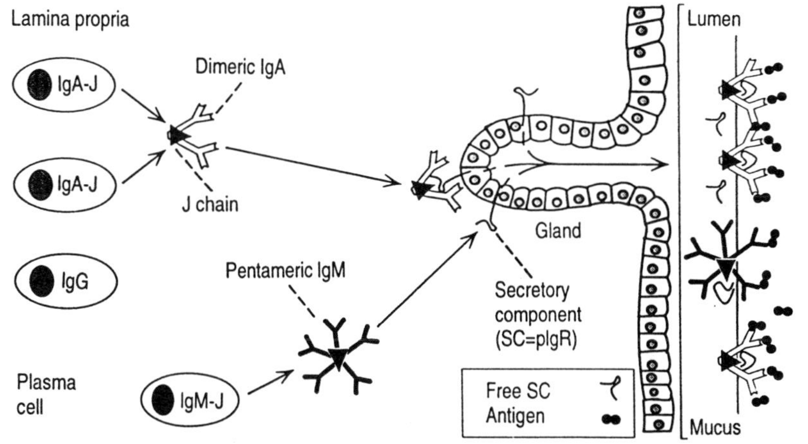

array of common intestinal and respiratory pathogens (19). The secretory antibodies are thus highly targeted against infectious agents in the mother's environment, which are those likely to be encountered by the baby during the first weeks of life. Breast-feeding therefore represents an ingenious immunological integration of mother and child (Fig. 4). Although the protection provided by this defence mechanism is most readily demonstrable in populations that live in poor sanitary conditions, a beneficial clinical effect is also apparent in the industrialized world (20), even in relation to relatively common diseases such as otitis media and acute lower respiratory tract infections (21,22).

Antibodies to various dietary antigens such as cow's milk proteins (23) and gluten (24) are also present in breast milk. The possible role of such antibodies in the clinical presentation of immune-mediated food intolerance will be discussed below.

Non-specific antimicrobial factors

Numerous constituents of breast milk are thought to exert innate defence functions. These include lysozyme, lactoferrin (see Chapter 25, this volume), peroxidase, complex oligosaccharides (receptor analogues), fatty acids (lipids) and mucins (19–21). A

***Figure 3** Schematic depiction of the human mucosal immune system. Inductive sites are constituted by the Peyer's patches with their B-cell (B) follicles and M-cell (M)-containing follicle-associated epithelium through which luminal antigens are actively transported, solitary lymphoid follicles scattered in the intestinal mucosa, the appendix, and the pharyngeal lymphoid tissue. Memory B and T lymphocytes migrate from these sites to lymph and peripheral blood for subsequent extravasation at mucosal effector sites. As an example, the intestinal lamina propria is depicted with its various immune cells including B lymphocytes, J chain-positive IgA- and IgM-producing plasma cells, $CD4^+$ T cells (CD4), macrophages ($M\phi$) as well as other types of antigen-presenting cells (APC), eosinophils and natural killer cells (NK). Additional features are the generation of secretory IgA (SIgA) and secretory IgM (SIgM) as well as leakage of smaller amounts of mainly serum-derived IgG antibodies (IgG) into the lumen. The composition of intraepithelial lymphocytes (mainly $TCR\alpha/\beta^+CD8^+$ and some $TCR\gamma/\delta^+$ T cells) is also illustrated*

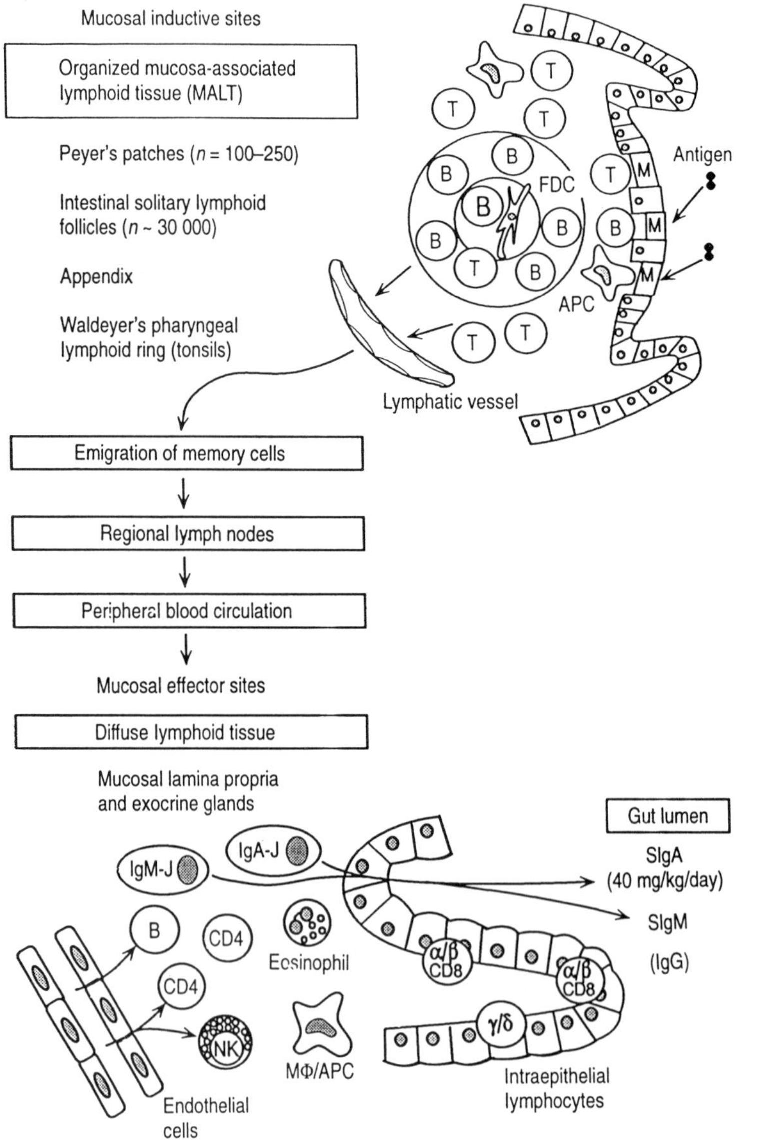

***Figure 4** Integration of mucosal immunity between mother and newborn child, with emphasis on migration of primed B (and probably T) cells from Peyer's patch via lymph and peripheral blood to the lactating mammary gland. Such distribution (arrows) beyond the gut of precursors for IgA plasma cells is crucial for local glandular production and subsequent occurrence in breast milk of secretory antibodies (SIgA and SIgM) specific for enteric antigens (microorganisms and food proteins). By this mechanism, the breast-fed baby will receive relevant secretory antibodies directed against the microbiota colonizing its mucosae (initially reflecting the microflora of the mother) and hence be better protected both in the gut and in the upper airways (shaded areas) in the same way as the mother's gut mucosa is protected by similar antibodies (shaded areas)*

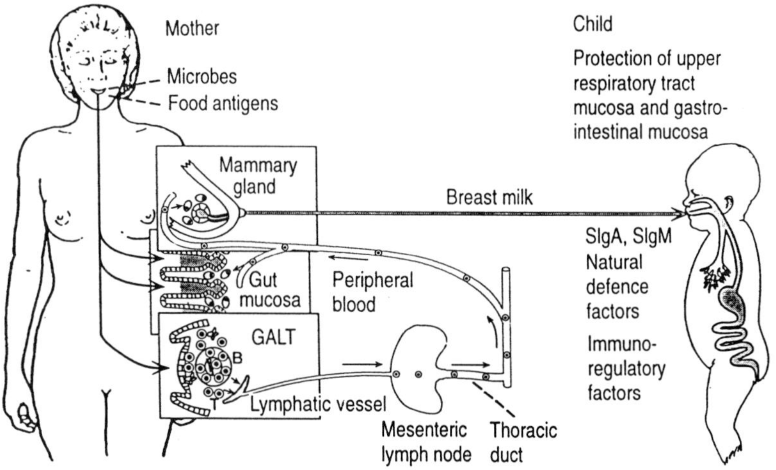

further discussion of their potential protective effects is beyond the scope of this chapter.

Immune cells

A variety of leucocytes occur in colostrum ($\sim 4\times10^6$/ml) and in later milk ($\sim 10^5$/ml). Macrophages (55–60%) and neutrophilic granulocytes (30–40%) dominate over lymphocytes (5–10%), the latter being mainly (75–80%) T lymphocytes (19,20). Oral administration of macrophages in newborn mice showed survival of these cells for several hours in the gut and even some mucosal uptake (25). A series of subsequent experiments with milk leucocytes in newborn rats, calves and lambs likewise demonstrated transepithelial migration, with lymphocytes as the predominant cell type (26). The distribution of labelled human colostral leucocytes after enteral administration to premature baboons suggested epithelial adherence in the gut as well as mucosal uptake and persistence for more than 60 h, along with some peripheral migration (27). The contribution of milk lymphocytes to the infant's intestinal immune system therefore remains an intriguing possibility.

Milk leucocytes generally express markers of previous priming and respond readily to restimulation. The macrophages contain engulfed SIgA that they may release on contact with bacteria in the gut (26); they may also secrete an array of immunologically important cytokines (20). It is furthermore of interest that, compared with peripheral blood, breast milk is relatively enriched in T lymphocytes with the alternative T-cell receptor (TCR) γ/δ; this subset mainly employs the Vδ1/Jδ1-encoded receptor (28) as do γ/δ cells among intraepithelial lymphocytes (IEL) in the gut (see below). T lymphocytes of this phenotype with specificity for *Mycobacterium tuberculosis* appear in colostrum of tuberculin-positive nursing mothers (29). This observation is particularly intriguing in view of the direct antimicrobial activity mediated by various T-cell phenotypes upon interaction with microbial targets (30). In this respect, mucosal defence may especially engage γ/δ cells; their supply from breast milk may be immunologically important before the baby's IEL population has developed numerically as well as in terms of phenotypes and TCR repertoire (see below).

Putative immunoregulatory factors

Glycoproteins from human colostrum have an enhancing effect on T-cell proliferation at low concentrations but an inhibitory effect at high concentrations (31). The biological significance of this in vitro phenomenon is obscure as long as the active factors remain unidentified. One inhibitory mechanism of breast milk was suggested to be down-regulation of interleukin-2 (IL-2) production (32), and multimeric colostral α-lact-albumin was shown to induce apoptosis in lymphocytes (33). Conversely, the stimulatory effect of milk on T cells was tentatively ascribed to lactoferrin (31), but great confusion exists about the possible immunoregulatory properties of this protein (34).

Several studies have reported that unfractionated supernatants of breast milk cell cultures preferentially stimulate IgA production by peripheral blood lymphocytes (26). This effect may be due to the various cytokines that are secreted by stimulated milk macrophages (20). The same soluble cytokines are found in breast milk (19,20), and the presence of transforming growth factor-β (TGF-β), IL-6 and IL-10 is of particular interest for the development and differentiation of IgA-producing cells (14). Direct evidence to this end has been provided for colostral IL-6 (35). Even if these cytokines are unable to survive the passage through the gastrointestinal tract, they may be locally released from milk macrophages stimulated in the neonatal gut and thereby enhance the development of mucosal immunity.

INDUCTION AND DISSEMINATION OF MUCOSAL IMMUNITY

Gut-associated lymphoid tissue

Primary mucosal B-cell responses that give rise to secretory antibodies (SIgA and SIgM) appear to be elicited mainly in lymphoepithelial structures where antigens are sampled from the lumen of the aeroalimentary tract (1,14). The organized gut-associated lymphoid tissue (GALT), that includes scattered as well as aggregated (Peyer's patches) secondary lymphoid B-cell follicles, represents the best-studied inductive sites of the mucosal immune system (Fig. 3). In humans, Peyer's patches are mainly found in the distal ileum, whereas most of the approximately 30 000 solitary follicles occur in the appendix and the large bowel, particularly in the rectum. All these components of GALT appear functionally similar; they contain a characteristic follicle-associated epithelium with membrane (M) cells capable of transporting live and dead particulate antigens into the underlying lymphoid tissue (36). Because of this antigen-sampling function, the M cells are vulnerable sites in the gut epithelium that may allow entry of certain enteropathogens such as *Salmonella typhi, Yersinia enterocolitica* and reovirus (37).

B and T cells as well as macrophages and other antigen-presenting cells (APC) of the dendritic variety occur immediately underneath the follicle-associated epithelium, even in the basal pockets of the M cells (28). Antigen that is transported into GALT will furthermore be presented to memory B cells as immune complexes retained on follicular dendritic cells within the germinal centres of the activated lymphoid follicles (Fig. 3). However, little is known about the preferential portal of entry for soluble luminal antigens; such molecules, including dietary proteins, may mainly be taken up through the extensive epithelial surfaces covering the diffuse lymphoid tissue of the intestinal mucosa. Their fate and possible immunological effects will depend on how they are handled locally and on whether they reach lymph or portal blood (1,4,39).

Heterogeneity in mucosa-associated lymphoid tissue

Although GALT constitutes the major part of MALT, mucosal immunity also appears to depend on induction of immune responses in palatine tonsils and other lymphoepithelial structures of Waldeyer's pharyngeal ring, including nasal-associated lymphoid tissue (NALT) such as the adenoids (40,41). This notion has recently been supported by the preferential appearance of IgA antibodies in parotid secretions of rabbits after tonsillar antigen exposure (42). Accumulating evidence suggests regionalization or subcompartmentalization in the mucosal immune system, especially dichotomy between the gut and the upper aeroalimentary tract with regard to homing properties and terminal differentiation of B cells (43). This disparity may be explained by microenvironmental differences in the antigenic repertoire as well as in the lymphoid and vascular adhesion molecules involved in local leucocyte extravasation (see below). It seems that primed immune cells preferentially home to effector sites corresponding to the inductive sites where they initially responded to antigen. Because bronchus-associated lymphoid tissue (BALT) is lacking in normal lungs of newborns and adults (44,45), the Waldeyer's ring may represent a particularly important component of MALT in humans, especially in the postnatal period (46).

Pre- and postnatal development of MALT

In humans, Peyer's patches and other GALT structures are well developed early in fetal life (47–49), discrete T- and B-cell areas being apparent at 19 weeks' gestation

(50). The primary lymphoid follicles seem to form around follicular dendritic cells. In these initial intestinal lymphocyte accumulations, the B cells have been reported to contain cytoplasmic IgM and express CD5 (51); they may therefore represent a transitional phenotype between pre-B cells and naive B cells. The latter phenotype is identified by the coexpression of surface IgM and IgD as seen in the mantle zones of postnatal Peyer's patches (38,50).

Germinal centres signifying B-cell activation do not appear in the intestine until shortly after birth (51–53), which reflects their dependence on antigenic stimulation. Animal studies have also shown failure of secondary follicles to develop in Peyer's patches of germ-free mice (54). The B cells of germinal centres express small amounts of surface IgA along with less IgM or IgG (55). This skewing probably reflects isotype switching in the course of preferential clonal differentiation to precursors for IgA-producing cells, which is the hallmark of Peyer's patches.

The fetal development of Waldeyer's lymphoid ring also seems to be antigen independent, apparently starting somewhat earlier than the Peyer's patch anlage, primary B-cell follicles being present at 16 weeks' gestation (56). Plasma cells appear in the palatine tonsils about 2 weeks after birth, or at about the same time as in the intestinal lamina propria (see below).

Migration of primed lymphoid cells from MALT

After antigen-induced activation, proliferation and partial differentiation in MALT, memory B and T cells migrate rapidly to regional lymph nodes and further to the peripheral blood circulation (Fig. 3). Very few B cells with IgA-secreting capacity are present in the blood of newborns ($<8\times10^{-6}$ lymphocytes), but after 1 month this number is remarkably increased ($\sim600\times10^{-6}$ lymphocytes), reflecting the progressive intestinal antigen stimulation (57,58). Initial elevation of positive cells may be seen in preterm infants, especially in those who have contracted intrauterine infections, but IgM production dominates in these cases (58). Such observations suggest that the mucosal immune system is competent even before birth, at least during the final trimester. This is further supported by the finding that fetal lamina propria T cells can be activated by mitogens or bacterial superantigens in vitro (59).

Memory B and T cells derived from GALT express adhesion molecules or 'homing receptors' specific for corresponding counter-receptors or ligands ('addressins') on endothelial cells of venules in organized as well as in diffuse mucosal lymphoid tissue (60); they therefore home preferentially to effector sites such as the gut lamina propria and exocrine glands (Fig. 3). Most B and T cells stimulated in GALT extravasate along the intestinal mucosa; an important 'traffic signal' to this end appears to be the prominent expression of $\alpha E\beta 7$ integrin on such memory cells (60). Certain T-cell subsets, particularly those whose integrin $\alpha E\beta 7$ expression is induced locally (probably by TGF-β), tend to migrate into the epithelium (especially in the small intestine) where they are retained basolaterally as IELs (see below), apparently by the binding of $\alpha E\beta 7$ to epithelial E-cadherin (62,63).

An unknown fraction of GALT-derived B cells (and probably T cells) end up in diffuse mucosal lymphoid tissues and exocrine glands outside the gut (64). The presence of SIgA antibodies to bovine β-lactoglobulin in neonatal breast milk ('witch milk') from infants fed cows' milk formula (65) documents early homing of primed B-cells from gut to mammary glands. Such a 'common' or integrated mucosal immune system is the rationale for many desired oral vaccines (66). Conversely, migration to the gut lamina propria of B cells induced in Waldeyer's ring appears to be negligible (40,43), perhaps because of a disparity in homing receptors or endothelial adhesion molecules, as alluded to above. There is good evidence that mucosal endothelial determinants are shared among the venules of Peyer's patches, mesenteric lymph nodes, and the intestinal lamina propria (60); another set of adhesion molecules may be shared between the inductive sites and the exocrine effector tissues of the upper aeroalimentary tract (40,43). Such a putative homing heterogeneity in the mucosal immune system must be further documented and eventually taken into account to develop effective vaccines for local protection (64,66).

DEVELOPMENT OF HUMORAL MUCOSAL IMMUNITY

Passively acquired antibodies in secretions

Most full-term babies growing up under privileged conditions exhibit satisfactory mucosal resistance, provided a normal development of the innate defence mechanisms has occurred. Adequate systemic antibody protection of the mucosae is provided by circulating maternal IgG, of which 50–60% is distributed extravascularly. Selective placental transfer of IgG in the fetus is unique for primates (67), and postnatal uptake of breast-milk-derived macromolecules seems to be of minor importance in supporting systemic immunity in humans (68,69), except perhaps in preterm babies (39,70).

Immediately after birth only occasional traces of SIgA and SIgM are found in human exocrine fluids, whereas some IgG is usually present as a result of passive external diffusion or epithelial 'leakage'; its source is primarily interstitial tissue fluid, which after 13–17 weeks' gestation contains readily detectable maternal IgG, particularly in the highly vascularized mucosae (68). Some intraluminal IgG and SIgA may be derived from amniotic fluid taken up by the fetus (39,71), and it is possible that maternal anti-idiotypic antibodies of these isotypes constitute a low-grade antigenic stimulus of MALT before birth (68). In addition, SIgA from breast milk, either as anti-idiotypic antibodies or bound to luminal antigen, might provide an immunological stimulus in the suckling baby because promiscuous Fc receptors seem to exist on the M cells (72). It may hence be visualized that antibodies from the mother in various ways exert a guidance function in early priming of the newborn's mucosal immune system, but the importance of this possibility is unknown.

Complement-activating IgG antibodies against dietary proteins and infectious agents may, in theory, adversely affect mucosal penetrability for a variety of macromolecules and hence act as a two-edged sword while contributing to local protection. Such a possibility has been suggested by in vitro (73) and in vivo (74) test models in which IgG antibodies against one antigen were shown to enhance mucosal penetration of bystander proteins. Mucosal integrity can apparently be damaged by lysosomal enzymes released from polymorphonuclear granulocytes that are attracted when immune complexes are formed locally. This may explain why abrupt introduction of native cow's milk proteins in infants often causes gastrointestinal bleeding (75).

The proinflammatory potential of maternal IgG or locally produced IgM and IgG antibodies is probably less important in infants who are breast-fed because milk SIgA antibodies, which have little or no complement-activating capacity (3), probably exert a non-inflammatory blocking effect. It is also noteworthy in this context that breast milk contains large amounts of the soluble complement inhibitor CD59 or protectin (76). This and other complement regulatory proteins are, moreover, produced by the epithelium both in the gut (77) and in the airways (78).

Development of local Ig production

Scattered B and T lymphocytes are seen in fetal intestinal lamina propria from 14 weeks' gestation (59,68). Occasional IgM- and IgG-producing plasma cells have been reported to appear somewhat later and they remain few until birth, whereas IgA-producing cells are either absent or extremely rare even until 10 days of age. This is true also for the appendix (68). By contrast, human fetal salivary glands sometimes contain scattered IgA-producing immunocytes in addition to the IgM and IgG isotypes, especially after 30 weeks' gestation; most of these IgA plasma cells (~90%) synthesize the IgA1 subclass and virtually all co-express the J chain (79). This apparent difference between the salivary glands and the gut mucosa with regard to low-grade fetal B-cell activation is intriguing. Maternal anti-idiotypic antibodies, cytokines and perhaps also foreign protein antigens might be present in amniotic fluid (68), to which the oral cavity is continuously exposed during fetal life. In particular the minor salivary glands, which have rather short ducts, could be stimulated by retrograde passage of such factors. Tonsillar immunostimulation may also be of special importance for immunity in the upper aeroalimentary tract during the perinatal period, as alluded to above.

Numbers of IgM- and IgA-producing cells in the intestinal lamina propria and salivary glands start to rise rapidly 2–4 weeks after birth, the latter class becoming predominant after 1–2 months (68). At an average age of 15 months the number of IgA immunocytes in the salivary glands has approached the lower normal adult range, and the subsequent increase throughout early childhood seems to be small. The increase of IgA immunocytes normally appears to be somewhat more rapid in the gut and may peak at about 12 months, while the numbers of IgM-producing cells continue to decrease (68).

Effect of antigen exposure and nutrition on local IgA production

The above observations have been made in industrialized countries; a much faster development of the local IgA system is usually seen in developing countries (68). This appears to be true even in malnourished children (80), which shows that mucosal immunity is highly adaptable to the antigenic load of the environment. *Bacteroides* spp. and *E. coli* seem to be particularly stimulatory for the development of intestinal

IgA immunocytes (68). Even artificial colonization of infants with a non-enteropathogenic strain of *E. coli* was found to enhance formation of intestinal IgA (81). Reduced amounts of microbial and dietary antigens apparently explain why the numbers of colonic IgA- and IgM-producing cells were decreased by about 50% after 2–11 months in children subjected to defunctioning colostomies (82). Postnatal and prolonged observations on defunctioned ileal segments in lambs revealed still more strikingly a reduced number of Ig-producing cells in the lamina propria; this result was explained by decreased local accumulation of B-cell blasts and might have involved both hampered precursor homing to the mucosa and reduced local proliferation and differentiation of B cells (83). Antigenic constituents of food may exert a substantial stimulatory effect, as suggested by a less active intestinal B-cell system in mice fed on hydrolysed milk proteins (84) as well as in parenterally fed babies (85).

An immunohistochemical study of children with protein–energy malnutrition revealed selective reduction of intestinal IgA-producing immunocytes (86); experiments in mice supported a similar negative effect of prolonged and severe malnutrition (87). Hampered homing of IgA precursor cells to the mucosal effector sites might explain this adverse development (88). Conversely, it has been reported that undernourished children respond to bacterial overgrowth in the gut with enhanced IgA synthesis and secretion (89). It is also of great clinical importance that although IgA as well as epithelial SC might be adversely affected by severe malnutrition, these changes have been shown to be reversible with renutrition (90).

Development of secretory antibodies

As discussed above, the transmembrane SC or pIg receptor is imperative for active external transport of SIgA and SIgM (Fig. 2). Constitutive intestinal SC expression is quite weak in the crypt epithelium before the 29th week in utero; thereafter it increases up to 32 weeks' gestation, and adult SC levels are reached 1–2 weeks after birth (68). A comparable development is seen in fetal salivary glands where SC first appears in a few acini and small ducts, followed by increased expression between 20 and 30 weeks' gestation. Thereafter, apical staining for IgA in the same epithelial elements has been observed, which shows that some external transport of salivary SIgA indeed may take place before birth (68). This observation accords with the presence of occasional J chain-positive IgA immunocytes in the glandular stroma of some fetuses as described above.

Expression of SC in the salivary glands is remarkably enhanced between 2 weeks and 5 months after birth; this might reflect a temporary stimulation by interferon-γ (IFN-γ), tumour necrosis factor-α and IL-1 present particularly in early breast milk (19,20). These cytokines are known to stimulate SC expression in secretory epithelial cells (91), and the concurrently enhanced epithelial HLA-DR expression (presumably IFN-γ-dependent) supports this notion (68). Moreover, the salivary levels of IgA (92) and free SC (93,94) have been found to be relatively high during the same postnatal period.

Total SIgM as well as antibodies of this class have often been demonstrated to be relatively prominent in saliva after the first few postnatal months (68). Several studies have reported a marked increase of salivary IgA during the same period, usually peaking between 1 and 2 months, whereas very little increase seems to take place over the following few years (68). Most studies agree that adult salivary IgA levels are reached much later on in childhood (68).

It takes up to 3 months after birth before the IgA1 to IgA2 immunocyte ratio observed in salivary glands approaches that of the normal adult, with approximately 35% IgA2-producing cells (43). This progressive shift from local IgA1 to IgA2 is reflected in the development of the respective salivary SIgA subclass levels (95,96) and may depend on an increased postnatal dissemination of IgA2 precursor cells from the Peyer's patches, where this subclass dominates (97). The striking neonatal predominance of IgA1 in the upper aeroalimentary tract might be of relevance for the development of allergy; it has recently been reported that atopic infants early in life have an increased proportion of pharyngeal bacteria capable of cleaving selectively IgA1 (98). This may temporarily jeopardize their mucosal surface defence.

We have been unable to see signs of IgA transport through the crypt epithelium in fetal intestinal mucosa. The SIgA that has been reported to occur in the fetal gut lumen may therefore be derived both from amniotic fluid (71) and from fetal saliva. In fact, SIgA and SIgM antibodies to poliovirus as well as to *E. coli* have been detected by sensitive immunoassay in meconium and saliva of some neonates as early as the first day of life (99). The report showing that adult levels of IgA are attained in faeces as early as 1–2 months after birth (100) is at variance with the fact that the average rectal number of IgA-producing cells around 2 months is only about 20% of that seen after 2 years (101). This discrepancy probably reflects the difficulties encountered in obtaining a reliable measure of actual IgA secretion in the gut. However, quantitation of Ig levels in jejunal fluid showed no significant increase of SIgA after the age of 1 year whereas IgM tended to decrease (102), in accordance with the intestinal immunocyte development described above.

Effect of breast-feeding on infant SIgA levels

The effect of breast-feeding on salivary IgA in early infancy has been a matter of much dispute. The influence of colostral cytokines and other growth factors, contamination of saliva with milk IgA, shielding of the mucosal immune system of breast-fed infants by maternal SIgA antibodies, and altered growth of the baby's intestinal microbial flora have been offered as alternative explanations for the discrepant observations (68).

Breast-feeding may promote the postnatal development of secretory immunity (103), apparently even in the urinary tract (21); it further enhances secretory and systemic immune responses to oral as well as to parenteral vaccines (104,105). Nevertheless, several prospective studies have reported that the early increase of salivary IgA (and IgM) is more prominent in formula-fed than in solely breast-fed infants (68,103,106), but the difference apparently disappears after weaning (107). Observations in infants likewise suggest that breast-feeding, in comparison with formula-feeding, reduces the salivary IgA antibody levels to cow's milk proteins (108). This decrease was seen after a nursing period of only 3 weeks and also appeared in infants receiving mixed feeding (106,108). Levels of salivary IgG, which is serum-derived rather than locally produced, likewise tended to be lower for several weeks in the breast-fed infants; this suggested earlier 'closure' or reduced permeability of their oral epithelia (106). The finding of elevated salivary IgA antibody levels to casein in saliva of infants at high risk of allergy might have reflected retarded 'gut closure', perhaps because of extended postnatal mucosal permeability or reduced amounts of SIgA antibodies to cows' milk proteins in the breast milk of their mothers (23,109).

Although breast-feeding initially may reduce the salivary SIgA antibody levels, recent reports have suggested that later on in infancy (up to 8 months) it boosts the development of SIgA (96,110). An early shielding effect of breast milk antibodies has likewise been documented in suckling mice, which show a delayed intestinal IgA response against their natural microbial flora (111).

INTRAEPITHELIAL LYMPHOCYTES (IEL)

Development of IEL in the gut

Electron-microscopic studies first demonstrated IEL in human intestinal epithelium as early as at 11 weeks' gestation (112). As in adults, fetal IEL occur mainly in the villi of the small intestine and are dominated by $CD3^+$ T cells. Their numbers increase throughout the gestational period (113), which suggests that migration into the epithelium is to some extent antigen-independent and perhaps reflects the development of Peyer's patches, where the IEL precursors may largely be generated (114–116). However, luminal factors clearly determine the magnitude of the accumulation of IEL as shown by their rapid postnatal increase – up to 10-fold by the age of 1–2 years (117,118). Moreover, germ-free animals have only few IEL; and intestinal colonization experiments in germ-free mice (119) as well as in germ-free rats (120) have demonstrated a marked stimulatory effect of the microbiota and also its apparent impact on the TCR repertoire of IEL (120). As mentioned above, almost all IEL are positive for the integrin $\alpha E\beta 7$, which exhibits affinity to epithelial E-cadherin (63). Because the expression of $\alpha E\beta 7$ is enhanced by TGF-β (62), this cytokine from colostrum or colostral macrophages may be an additional luminal factor contributing to the accumulation of IEL in breast-fed infants.

Small intestinal IEL are mainly of the CD8 phenotype (80–90%) both in children and adults (117); this dominance over the CD4 phenotype is somewhat less in the fetus (113). Also, the presumably immature (pre-T-cell?) $CD3^-CD7^+$ IEL subset that amounts to ~10% in adults is totally absent in fetal life (50). $CD3^+CD4^-CD8^-$ ('double-negative') cells constitute ~6% of postnatal $CD3^+$ IEL, whereas this fraction accounts for 35–70% in the fetal gut (50). In adults, the latter subset is included in the small proportion (2–10%) of $CD3^+$ IEL that express TCRγ/δ rather than TCRα/β (121). Because the TCRγ/δ^+ fraction amounts to only about 20% of the fetal $CD3^+$ IEL, many double-negative TCRα/β^+ IEL must exist in the fetus (50). Moreover, while TCRγ/δ^+ IEL in adults are mainly (~70%) reactive with the monoclonal antibody δTCS1 that detects the Vδ1/Jδ1-encoded receptor (121,122), they are all negative for this epitope in the fetal gut and rather use the Vδ2 gene segment (50). The remarkably preferential postnatal use of the Vδ1 gene segment may reflect some antigen specificity in the local accumulation or intraepithelial expansion of γ/δ IEL, but molecular analysis has nevertheless revealed considerable diversity of their TCR usage (123).

There is a continuing debate whether a fraction of human IEL are thymus-independent in their development. A marker of such a subset in mice appears to be the expression of CD8α/α homodimers rather than CD8α/β heterodimers (122). In this context it is interesting that almost half of the human fetal lamina propria and intraepithelial $CD8^+$ T cells express the homodimer, in contrast to the almost 95% dominance of the heterodimer on T cells within fetal Peyer's patches (123). Only 5% of adult jejunal IEL bear the CDα/α homodimer (122).

A requirement for intraepithelial development of T cells in the gut would be that IEL express the recombinase-activating gene-1 and gene-2 (RAG-1 and RAG-2) as well as terminal deoxyribonucleotidyltransferase (TdT) that are involved in the shaping of TCR, the latter enzyme being particularly required for generation of N-region diversity. RAG expression has been demonstrated for adult jejunal IEL in two investigations (125,126), particularly for a small fraction with an immature or thymocyte-like phenotype ($CD2^+TCR/CD3^-$ or $CD3^+/TCR^-$) but rarely for the ordinary TCRα/β^+ IEL (126). Using a more quantitative molecular method, it was recently shown that very few adult IEL (less than one in 10^7 cells) express TdT (127). Intra-epithelial selection and maturation of IEL is therefore unlikely to be a significant pathway for T-cell development in adults. However, despite the low or absent TdT expression also shown by IEL in the fetus, fetal IEL apparently undergo considerable TCR gene rearrangement in situ but nevertheless show limited N-region diversity (128). Because such diversity is extensive for TCR of adult IEL (129), fetal IEL do not seem to contribute substantially to the postnatal IEL population.

Putative function of IEL in the gut

Some evidence exists for an epithelium-specific cytotoxic activity of human TCR$\alpha/\beta^+$$CD8^+$ IEL (129), but their possible contribution to oral tolerance is also of great interest (see below). This poorly defined functional suppressor phenomenon refers to the experimental observation that rodents fed soluble proteins develop immunological hyporesponsiveness by down-regulation of systemic (and probably local) immunity, particularly proinflammatory IgG and IgE antibody responses as well as delayed-type hypersensitivity mediated by activated $CD4^+$ T cells and macrophages (4,130). As discussed in the introduction, oral tolerance probably also exists in humans and is a likely explanation for the fact that most individuals do not show hypersensitivity to dietary proteins, despite some absorption of such antigens in an intact form (7,8).

The major TCR$\alpha/\beta^+$$CD8^+$ IEL population in adults is oligoclonal as deemed from a restricted TCR usage (129); this suggests specificity for a limited number of antigens and may speak against a putative involvement in oral tolerance towards dietary antigens. Epithelial HLA class I or class I-like (class Ib or CD1) molecules are believed to perform antigen presentation to T cells, and some evidence indicates that nonpolymorphic CD1d molecules act as recognition element for α/β as well as for γ/δ human IEL (129). Although epithelial cells in the small intestine express HLA class II both apically and basolaterally (91), epithelial class II molecules (which have affinity to CD4) show an unusual conformation, at least in the mouse, and may therefore be inadequate for efficient antigen presentation (131). This could contribute to a preferential engagement of $CD8^+$ T cells by the gut epithelium.

It has been speculated that the γ/δ T cells have a special role in the immunological surveillance of mucosal surfaces, perhaps by eliminating infected or damaged epithelial cells through interaction with heat-shock proteins (132). In this context it is especially intriguing that γ/δ cells can recognize unprocessed antigens directly on the surface of microorganisms without involving APC and HLA molecules. Such a mechanism could endow γ/δ cells with unique defence properties in the gut epithelium, although their antigenic repertoire remains obscure. It furthermore appears that γ/δ IEL support the integrity and immunobiology of the epithelium by producing growth-promoting factors such as keratinocyte growth factor that enhances repair and function of the epithelial layer (133,134).

A disproportionately increased $TCR\gamma/\delta^+$ jejunal IEL population is present in patients with coeliac disease, although its phenotypic composition remains fairly normal (121,122). The putative immunopathological role of this IEL subset is intriguing in view of the suggestion that murine $TCR\gamma/\delta^+$ IEL may act as contrasuppressor cells, thereby being able to abrogate tolerance (135). If this is true also for the human counterparts, the expansion of intraepithelial γ/δ cells in the coeliac lesion could contribute to the striking increase of Ig-producing immunocytes and activated ($CD25^+$) lamina propria $CD4^+$ T cells seen in untreated patients, apparently reflecting a break of oral tolerance to gluten peptides (121,136). However, the existence of contrasuppressor cells is quite controversial, and the increase of $TCR\gamma/\delta^+$ IEL in coeliac disease might rather reflect enhanced cytotoxic clearance of damaged epithelial cells in an attempt to maintain the integrity of the surface epithelium. Moreover, experimental induction of tolerance in the respiratory tract of rats has suggested an important role of γ/δ cells in suppressing T-cell-dependent IgE production (137). It has to be admitted that the function of both $TCR\alpha/\beta^+CD8^+$ and $TCR\gamma/\delta^+$ IEL in mucosal immunity remains enigmatic.

DEVELOPMENT OF ORAL TOLERANCE

Mechanisms for tolerance induction

Recent experiments in CD8 knock-out mice led to the conclusion that separate mechanisms may operate in mucosal induction of local and peripheral tolerance. Thus, $CD8^+$ T cells were crucial for the down-regulation of orally elicited mucosal immunity but not for oral suppression of systemic antibody responses (138). Moreover, the read-out system could have a marked impact on the interpretation of experimental results; that is, the conclusion may depend on whether one examines humoral immunity, cell-mediated immunity, or both (139). Altogether, information obtained from animal studies shows that oral tolerance probably is an extremely complex phenomenon (4).

A central role of the gut epithelium in tolerance induction is supported by the observation that this phenomenon depends on preserved integrity of the mucosal barrier (140,141). Moreover, recent studies have provided considerable information about the immunoregulatory function of nonprofessional APC such as keratinocytes (139). As reviewed elsewhere (142), the chief effect obtained when intestinal epithelial cells are used in APC in various test systems is stimulation of $CD8^+$ T cells with suppressor function. Although enterocytes from the small intestine express abundant HLA class II antigens (91), recent evidence suggests that their interaction with $CD8^+$ T cells is rather mediated by the non-classical class I molecule CD1d (143). In addition, human enterocytes apparently express an additional ligand (gp180) that by interaction with the α chain of CD8 may rapidly activate the tyrosin kinase $p56^{lck}$ and thereby trigger preferentially $CD8^+$ T cells (144). Antigen presentation by enterocytes in this context could theoretically leave IEL in a state of anergy. Such unresponsive cells might inhibit the IL-2 receptor on mucosal $CD4^+$ helper cells, a possibility suggested by the effect of $CD8^+$ suppressor cell lines (145). Perhaps CD8 molecules released from stimulated IEL function as inhibitory ligands affecting naive T cells during immunological priming (146).

It was mentioned above that oral tolerance undoubtedly operates in humans because the normal intestinal mucosa exhibits no substantial IgG response (1,142) and contains very few T cells with markers of recent activation such as the IL-2 receptor or CD25 (121). Moreover, the systemic IgG response to dietary antigens tends to decrease in humans with increasing age (147,148), and direct evidence for a hyporesponsive state to bovine serum albumin was obtained by intradermal testing of adults with this antigen (149). Quite recently, experimental feeding of keyhole limpet haemocyanin (KLH) to healthy adults resulted in down-regulation of the peripheral T cell response but stimulation of local as well as systemic humoral immunity (150). Conversely, intranasal application of KLH tended to suppress both cell-mediated and humoral peripheral immunity to this antigen (151).

As alluded to above, coeliac disease apparently represents an interesting 'model disease' in which a break of oral tolerance to gluten induces hyper-responsiveness of both humoral and T-cell-mediated immunity, including a proliferative response of IEL (152). In an experimental model of food-sensitive enteropathy in mice, specific activation of T cells observed in the lamina propria as well as in the epithelium was likewise assumed to explain the associated mucosal damage (153). It is interesting in this context that the normal microbial flora of the gut seems to play an important role in the recovery of suppressive mechanisms after toxin-mediated abrogation of oral tolerance as well as in prolongation of a hyporesponsive state in terms of systemic (IgG and IgE) humoral immunity in mice (154). Such observations add to the overwhelming complexity of oral tolerance.

Effect of breast-feeding

As discussed above, experiments in mice have demonstrated that SIgA antibodies in breast milk affect the normal microbial flora in the suckling newborn by retarding its contact with the developing GALT (111). The presence of TGF-β in breast milk (26) may further contribute to its tolerogenic properties because this cytokine exerts a pronounced immunosuppressive effect on GALT (155) and supports the epithelial barrier function (156). When the host's mucosal immune response subsequently is successfully elicited, GALT will be further shielded by the SIgA antibodies produced; local immunostimulation is thereby attenuated despite the continued presence of microorganisms (157). This may partly explain the hyporesponsiveness that exists towards members of the indigenous gut bacteria (158), resembling some sort of oral tolerance phenomenon both in rodents (159) and in humans (160). There is evidence to suggest that in patients with inflammatory bowel disease, oral tolerance to their own normal intestinal flora is broken (160), although it is not clear whether this is a primary or secondary event in the pathogenesis (158). Perhaps better immunological tolerance towards intestinal bacteria explains the apparent beneficial effect of breast-feeding in terms of a significantly decreased risk of later development of Crohn's disease (19).

The shielding effect on GALT of SIgA in breast milk may likewise contribute to the establishment of oral tolerance towards gluten. Antibodies to alcohol-extractable proteins of gluten (gliadin) are present in breast milk, and breast-feeding has been shown to protect significantly against the development of coeliac disease in children, unrelated to the time of solid food introduction (161–163). On the basis of such studies, it can be tentatively concluded that mixed feeding rather than abrupt weaning seems to promote tolerance to food proteins. This notion is supported by the report suggesting that cow's milk allergy is more likely to develop in infants whose mothers have relatively low levels of SIgA antibodies to cow's milk proteins in their breast milk (23). Although still a quite controversial issue, the balance of epidemiological studies altogether appears to support the view that breast-feeding protects against food allergy (26,164). The reported beneficial effect with regard to development of type I diabetes may also be related to less uptake of cows' milk proteins in breast-fed infants (19). A protective effect of SIgA antibodies is further supported by the increased incidence of atopic disease seen in children with a delayed development of their IgA-producing capacity as reflected in serum (165,166) as well as in intestinal mucosa (167,168) and external secretions (169).

CODA

Several more or less well-defined variables influence the development of mucosal immunity as well as oral tolerance, and some of these components are reciprocally modulated by the immune system (Fig. 5). Epithelial permeability or penetrability for luminal antigens is likely an important primary or secondary event in the pathogenesis of several mucosal diseases; it is determined by the individual's age (e.g. preterm vs term infant), interactions between mast cells, nerves and neuropeptides (170), concurrent infection or atopic allergy, and the shielding effect of SIgA provided by breast milk or produced by the infant itself (39,59). The consequences will depend on how fast 'closure' of the epithelial barrier can be attained or re-established, which is influenced both by the age of the infant and by the successful mounting of a specific SIgA response as well as generation of oral tolerance towards innocuous antigens

***Figure 5** Summary of variables influencing the development of mucosal immunity and oral tolerance. Several of the components are reciprocally modulated*

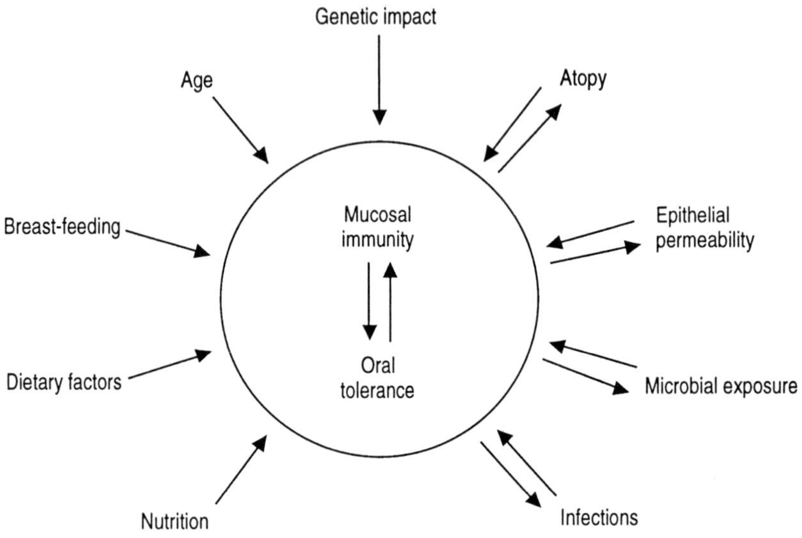

from the diet and from the normal microbial flora. The temporary immaturity of systemic immunocompetence in the newborn period also probably affects the mucosal immune system (59). Over and above all these variables is the immunogenetics of the infant, as revealed not only by possible atopic predisposition but also by other types of immune-mediated food intolerance such as coeliac disease. Clinical problems with certain dietary antigens are not surprising, however, in view of the relatively short time that has elapsed since the introduction of husbandry and agriculture. As a curiosity, it may be interesting that the Lundehund ('puffin-dog'), a small Norwegian spitz breed, also presents with an apparently food-related and genetically determined enteropathy that shows certain similarities to coeliac disease (171); this particular dog has probably been domesticated for only some 2000 years and is still adapting.

CONCLUSIONS

The mucosal immune system is under complex genetic control, and much of the information that modulates adaptive immune responses was encoded several million years ago. Subsequent evolutionary gene modifications have been introduced by the impact of the environment, both in terms of microorganisms and food antigens. The latter were determined by the way of life, and were different for the numerous generations of hunter-gatherers from those encountered during the relatively recent practice of agriculture and husbandry. Facing these various challenges, the mucosal immune system has developed two arms: immune exclusion performed by secretory antibodies, which acts mainly to inhibit colonization by pathogenic microorganisms and penetration of harmful antigens; and down-regulatory mechanisms to avoid local and peripheral overreaction (hypersensitivity) towards innocuous substances bombarding the mucosal surfaces. In the gut, the latter phenomenon is called oral tolerance and apparently explains why most individuals show no adverse immune reactions to persistent contact with food proteins. For a variable period after birth, both immune exclusion and oral tolerance are poorly developed; appropriate generation of the underlying network of regulatory mechanisms depends on both the establishment of a normal microbial flora and the introduction of dietary antigens (timing and dose). Breast-feeding is immunologically important during this vulnerable period, not only as a natural 'substitution therapy' for the infant's lacking secretory antibodies but likely also by its immune-modulating properties. Much remains to be learned about how mucosal immunoregulatory mechanisms adapt postnatally to avoid adverse reactions to the normal microbiota as well as to food, while providing an efficient immunological first-line defence against pathogens.

ACKNOWLEDGEMENTS

Studies in the author's laboratory are supported by the Norwegian Cancer Society, the Research Council of Norway, Anders Jahre's foundation, and Rakel and Otto Kr. Bruun's Legacy. Hege E. Svendsen and Erik K. Hagen are thanked for excellent secretarial assistance. The useful discussions with Dr Med. Vet. Bergljot Børresen, concerning domestication in Europe and the particular features of the Lundehund, are gratefully acknowledged.

REFERENCES

1. Brandtzaeg P, Baklien K, Bjerke K, Rognum TO, Scott H, Valnes K 1987 Nature and properties of the human gastrointestinal immune system. In: Miller K, Nicklin S (eds) Immunology of the Gastrointestinal Tract. CRC Press, Boca Raton, FL, pp 1–85
2. Hanson LÅ, Brandtzaeg P 1989 The mucosal defense system. In: Stiehm ER (ed) Immunologic Disorders in Infants and Children, 3rd edn. WB Saunders Co, Philadelphia, pp 116–155
3. Kilian M, Russell MW 1994 Function of mucosal immunoglobulins. In: Ogra PL, Mestecky J, Lamm ME, Strober W, McGhee JR, Bienenstock J (eds) Handbook of Mucosal Immunology. Academic Press, Orlando, FL, pp 127–137
4. Brandtzaeg P 1996 History of oral tolerance and mucosal immunity. Ann NY Acad Sci 778: 1–27
5. Harlan JR 1986 Plant domestication: diffuse origins and diffusions. In: Barigozzi C (ed) The Origin and Domestication of Cultivated Plants. Elsevier, Amsterdam, pp 21–34
6. Thorsby E 1995 HLA-associated disease susceptibility – which genes are primarily involved? Immunologist 3:51–58
7. Paganelli R, Atherton DJ, Levinsky RJ 1983 Differences between normal and milk allergic subjects in their immune responses after milk ingestion. Arch Dis Child 58:201–206
8. Husby S, Jensenius JC, Svehag S-E 1985 Passage of undegraded dietary antigen into the blood of healthy adults. Quantification, estimation of size distribution, and relation of uptake to levels of specific antibodies. Scand J Immunol 22:83–92
9. Hanson LÅ, Ashraf R, Carlsson B, Jalil F, Karlberg J, Lindblad BS, Khan SR, Zaman S 1993

Child health and the population increase. In: Hanson LÅ, Köhler L (eds) Peace, Health and Development 4:31–38
10. Anonymous 1994 A warm chain for breastfeeding (Editorial). Lancet 344:1239–1241
11. Børresen HC 1995 Rethinking current recommendations to introduce solid food between four and six months to exclusively breastfeeding infants. J Hum Lact 11:201–204
12. Brandtzaeg P 1973 Structure, synthesis and external transfer of mucosal immunoglobulins. Ann Inst Pasteur Immunol 124C:417–438
13. Haneberg B 1974 Human milk immunoglobulins and agglutinins to rabbit erythrocytes. Int Arch Allergy 47:716–729
14. Brandtzaeg P 1995 Molecular and cellular aspects of the secretory immunoglobulin system. APMIS 103:1–19
15. Haneberg B 1974 Immunoglobulins in feces from infants fed human or bovine milk. Scand J Immunol 3:191–197
16. Brandtzaeg P 1975 Human secretory immunoglobulin M. An immunochemical and immunohistochemical study. Immunology 29:559–670
17. Giugliano LG, Ribeiro STG, Vainstein MH, Ulhoa CJ 1995 Free secretory component and lactoferrin of human milk inhibit the adhesion of enterotoxigenic *Escherichia coli*. J Med Microbiol 42:3–9
18. Lamkhioued B, Gounni AS, Gruart V, Pierce A, Capron A, Capron M 1995 Human eosinophils express a receptor for secretory component. Role in secretory IgA-dependent activation. Eur J Immunol 25:117–125
19. Goldman AS 1993 The immune system of human milk: antimicrobial, antiinflammatory and immunomodulating properties. Pediatr Infect Dis J 12:664–671
20. Wold AE, Hanson LÅ 1994 Defense factors in human milk. Curr Opin Gastroenterol 10: 652–658
21. Newman J 1995 How breast milk protects newborns. Sci Am 273:58–61
22. Pisacana A, Graziano L, Zona G, Granata G, Dolezalova H, Cafiero M, Coppola A, Scarpellino B, Ummarino M, Mazzarella G 1994 Breast feeding and acute lower respiratory infection. Acta Paediatr 83:714–718
23. Savilahti E, Tainio V-M, Salmenperä L, Arjomaa P, Kallio M, Perheentupa J, Siimes M-A 1991 Low colostral IgA associated with cow's milk allergy. Acta Paediatr Scand 80:1207–1213
24. Juto P, Holm S 1992 Gliadin-specific and cow's milk protein-specific IgA in human milk. J Pediatr Gastroenterol Nutr 15:159–162
25. Hughes A, Brock JH, Parrott DMV, Cockburn F 1988 The interaction of infant formula with macrophages: effect on phagocytic activity, relationship to expression of class II MHC antigen and survival of orally administered macrophages in the neonatal gut. Immunology 64:213–218
26. Slade HB, Schwarz SA 1987 Mucosal immunity: the immunology of breast milk. J Allergy Clin Immunol 80:346–356
27. Jain L, Vidyasagar D, Xanthou M, Ghai V, Shimada S, Blend M 1989 In vivo distribution of human milk leucocytes after ingestion by newborn baboons. Arch Dis Child 64:930–933
28. Bertotto A, Gerli R, Castellucci G, Scalise F, Vaccaro R 1991 Human milk lymphocytes bearing the γ/δ T-cell receptor are mostly δTCS1-positive cells. Immunology 74:360–361
29. Bertotto A, Gerli R, Castellucci G, Crupi S, Scalise F, Spinozzi F, Fabietti G, Forenza N, Vaccaro R 1993 Mycobacteria-reactive γ/δ T cells are present in human colostrum from tuberculin-positive, but not tuberculin-negative nursing mothers. AJRI 29:131–134
30. Levitz SM, Mathews HL, Murphy JW 1995 Direct antimicrobial activity of T cells. Immunol Today 16:387–391
31. Mincheva-Nilsson L, Hammarström M-L, Juto P, Hammarström S 1990 Human milk contains proteins that stimulate and suppress T lymphocyte proliferation. Clin Exp Immunol 79:463–469
32. Hooton JWL, Pabst HF, Spady DW, Paetkau V 1991 Human colostrum contains an activity that inhibits the production of IL-2. Clin Exp Immunol 86:520–524
33. Håkansson A, Zhivotovsky B, Orrenius S, Sabharwal H, Svanborg C 1995 Apoptosis induced by a human milk protein. Proc Natl Acad Sci USA 92:8064–8068
34. Brock J 1995 Lactoferrin: a multifunctional immunoregulatory protein? Immunol Today 16: 417–419
35. Saito S, Maruyama M, Kato Y, Moriyama I, Ichijo M 1991 Detection of IL-6 in human milk and its involvement in IgA production. J Reprod Immunol 20:267–276

36. Kato T, Owen RL 1994 Structure and function of intestinal mucosal epithelium. In: Ogra PL, Mestecky J, Lamm ME, Strober W, McGhee JR, Bienenstock J (eds) Handbook of Mucosal Immunology. Academic Press, Orlando, FL, pp 11–26
37. Owen RL 1994 M cells – entryways of opportunity for enteropathogens. J Exp Med 180:7–9
38. Farstad IN, Halstensen TS, Fausa O, Brandtzaeg P 1994 Heterogeneity of M-cell-associated B and T cells in human Peyer's patches. Immunology 83:457–464
39. Sanderson IR, Walker WA 1993 Uptake and transport of macromolecules by the intestine: possible role in clinical disorders (an update). Gastroenterology 104:622–639
40. Brandtzaeg P, Halstensen TS 1992 Immunology and immunopathology of tonsils. Adv Oto-Rhino-Laryngol 47:64–75
41. Kuper CF, Koornstra PJ, Hameleers DMH et al. 1992 The role of nasopharyngeal lymphoid tissue. Immunol Today 13:219–224
42. Fukuizumi T, Inoue H, Anzai Y, Tsujisawa T, Uchiyama C 1995 Sheep red blood cell instillation at palatine tonsil effectively induces specific IgA class antibody in saliva in rabbits. Microbiol Immunol 39:351–359
43. Brandtzaeg P 1992 Humoral immune response patterns of human mucosae: induction and relation to bacterial respiratory tract infections. J Infect Dis 165 (suppl 1):S167–S176
44. Pabst R, Gehrke I 1990 Is the bronchus-associated lymphoid tissue (BALT) an integral structure of the lung in normal mammals, including humans? Am J Resp Cell Mol Biol 3:131–135
45. Tschernig T, Kleemann WJ, Pabst R 1995 Bronchus-associated lymphoid tissue (BALT) in the lungs of children who had died from sudden infant death syndrome and other causes. Thorax 50: 658–660
46. Brandtzaeg P 1996 The B-cell development in tonsillar lymphoid follicles. Acta Otolaryngol Suppl 523:55–59
47. Cornes JS 1965 Number, size, and distribution of Peyer's patches in the human small intestine. Gut 6:225–233
48. Husband AJ, Gleeson M 1990 Developmental aspects of gut-associated immunity: a comparative review. In: MacDonald TT (ed) Ontogeny of the Immune System of the Gut. CRC Press, Boca Raton, FL, pp 83–116
49. O'Leary AD, Sweeney EC 1986 Lymphoglandular complexes of the colon: structure and distribution. Histopathology 10:267–283
50. Spencer J, MacDonald TT 1990 Ontogeny of human mucosal immunity. In: MacDonald TT (ed) Ontogeny of the Immune System of the Gut. CRC Press, Boca Raton, FL, pp 23–50
51. Spencer J, MacDonald TT, Finn T, Isaacson PG 1986 The development of gut associated lymphoid tissue in the terminal ileum of fetal human intestine. Clin Exp Immunol 64:536–543
52. Bridges RA, Condie RM, Zak SJ, Good RA 1959 The morphologic basis of antibody formation. Development during the neonatal period. J Lab Clin Med 53:331–357
53. Gebbers J-O, Laissue JA 1990 Postnatal immunomorphology of the gut. In: Hadziselimovic F, Herzog B, Bürgin-Wolff A (eds) Inflammatory Bowel Disease and Coeliac Disease in Children. Kluwer Academic Publishers, Dordrecht, pp 3–44
54. Parrott DMV 1976 The gut-associated lymphoid tissue and gastrointestinal immunity. In: Ferguson A, MacSween NRM (eds) Immunological Aspects of the Liver and Gastrointestinal Tract. MTP Press, Lancaster, pp 1–32
55. Butcher EC, Rouse RV, Coffman RL, Nottenburg CN, Hardy RR, Weissman IL 1982 Surface phenotype of Peyer's patch germinal center cells: implications for the role of germinal centers in B cell differentiation. J Immunol 129:2698–2707
56. von Gaudecker B, Müller-Hermelink HK 1982 The development of the human tonsilla palatina. Cell Tissue Res 224:579–600
57. Nahmias A, Stoll B, Hale E, Ibegbu C, Keyserling H, Innis-Whitehouse W, Holmes R, Spira T, Czerkinsky C, Lee F 1991 IgA-secreting cells in the blood of premature and term infants: normal development and effect of intrauterine infections. Adv Exp Med Biol 310:59–69
58. Stoll BJ, Lee FK, Hale E, Schwartz D, Holmes R, Ashby R, Czerkinsky C, Nahmias AJ 1993 Immunoglobulin secretion by the normal and the infected newborn infant. J Pediatr 122: 780–786
59. MacDonald TT, Spencer J 1993 Development of gastrointestinal immune function and its relationship to intestinal disease. Current Opin Gastroenterol 9:946–952
60. Picker LJ 1994 Control of lymphocyte homing. Curr Opin Immunol 6:394–406

61. Hamann A, Andrew DP, Jablonski-Westrich D, Holzmann B, Butcher EC 1994 Role of α_4-integrins in lymphocyte homing to mucosal tissues in vivo. J Immunol 152:3282–3293
62. Parker CM, Cepek KL, Russell GJ, Shaw SK, Posnett DN, Schwarting R, Brenner MB 1992 A family of β_7 integrins on human mucosal lymphocytes. Proc Natl Acad Sci USA 89:1924–1928
63. Cepek KL, Shaw SK, Parker CM, Russell GJ, Morrow JS, Rimm DL, Brenner MB 1994 Adhesion between epithelial cells and T lymphocytes mediated by E-cadherin and the $\alpha^E\beta 7$ integrin. Nature 372:190–193
64. Mestecky J, Abraham R, Ogra PL 1994 Common mucosal immune system and strategies for the development of vaccines effective at the mucosal surfaces. In: Ogra PL, Mestecky J, Lamm ME, Strober W, McGhee JR, Bienenstock J (eds) Handbook of Mucosal Immunology. Academic Press, Orlando, FL, pp 357–372
65. Roberton DM, Forrest PJ, Frangoulis E, Jones CL, Mermelstein N 1986 Early induction of secretory immunity in infancy: specific antibody in neonatal breast milk. Arch Dis Child 61:489–494
66. Staats HF, Jackson RJ, Marinaro M, Takahashi I, Kiyono H, McGhee JR 1994 Mucosal immunity to infection with implications for vaccine development. Curr Opin Immunol 6:572–583
67. Mackenzie NM 1990 Transport of maternally derived immunoglobulin across the intestinal epithelium. In: MacDonald TT (ed) Ontogeny of the Immune System of the Gut. CRC Press, Boca Raton, FL, pp 69–81
68. Brandtzaeg P, Nilssen DE, Rognum TO, Thrane PS 1991 Ontogeny of the mucosal immune system and IgA deficiency. Gastroenterol Clin North Am 20:397–439
69. Klemola T, Savilahti E, Leinikki P 1986 Mumps IgA antibodies are not absorbed from human milk. Acta Pediatr Scand 75:230–232
70. Weaver LT, Wadd N, Taylor CE, Greenwell J, Toms GL 1991 The ontogeny of serum IgA in the newborn. Pediatr Allergy Immunol 2:72–75
71. Cleveland MG, Bakos M-A, Pyron DL, Rajaraman S, Goldblum RM 1991 Characterization of secretory component in amniotic fluid. Identification of new forms of secretory IgA. J Immunol 147:181–188
72. Weltzin R, Lucia-Jandris P, Michetti P, Fields BN, Kraehenbuhl JP, Neutra MR 1989 Binding and transepithelial transport of immunoglobulins by intestinal M cells: demonstration using monoclonal IgA antibodies against enteric viral proteins. J Cell Biol 108:1673–1685
73. Brandtzaeg P, Tolo K 1977 Mucosal penetrability enhanced by serum-derived antibodies. Nature 266:262–263
74. Lim PL, Rowley D 1982 The effect of antibody on the intestinal absorption of macromolecules and on intestinal permeability in adult mice. Int Arch Allergy Appl Immunol 68:41–46
75. Ziegler EE, Fomon SJ, Nelson SE, Rebouche CJ, Edwards BB, Rogers RR, Lehman LJ 1990 Cow milk feeding in infancy: further observations on blood loss from the gastrointestinal tract. J Pediatr 116:11–18
76. Bjørge L, Jensen TS, Vedeler CA, Ulvestad E, Kristoffersen EK, Matre R 1993 Soluble CD59 in pregnancy and infancy. Immunol Lett 36:233–234
77. Bjørge L, Vedeler CA, Ulvestad E, Matre R 1994 Expression and function of CD59 on colonic adenocarcinoma cells. Eur J Immunol 24:1597–1603
78. Varsano S, Frolkis I, Ophir D 1995 Expression and distribution of cell-membrane complement regulatory glycoproteins along the human respiratory tract. Am J Resp Crit Care Med 152: 1087–1093
79. Thrane PS, Rognum TO, Brandtzaeg P 1991 Ontogenesis of the secretory immune system and innate defence factors in human parotid glands. Clin Exp Immunol 86:342–348
80. Nagao AT, Pilagallo MIDS, Pereira AB 1993 Quantitation of salivary, urinary and faecal sIgA in children living in different conditions of antigenic exposure. J Trop Pediatr 39:278–283
81. Lodinova R, Jouja V, Wagner V 1973 Serum immunoglobulins and coproantibody formation in infants after artificial intestinal colonization with *Escherichia coli* 083 and oral lysozyme administration. Pediatr Res 7:659–669
82. Wijesinha SS, Steer HW 1982 Studies of the immunoglobulin-producing cells of the human intestine: the defunctioned bowel. Gut 23:211–214
83. Reynolds JD, Morris B 1984 The effect of antigen on the development of Peyer's patches in sheep. Eur J Immunol 14:1–6
84. Sagie E, Tarabulus J, Maeir DM, Freier S 1974 Diet and development of intestinal IgA in the mouse. Isr J Med Sci 10:532–534

85. Knox WF 1986 Restricted feeding and human intestinal plasma cell development. Arch Dis Child 61:744–749
86. Green F, Heyworth B 1980 Immunoglobulin-containing cells in jejunal mucosa of children with protein-energy malnutrition and gastroenteritis. Arch Dis Child 55:380–383
87. Wade S, Lemonnier D, Alexiu A, Bocquet L 1982 Effect of early postnatal under- and overnutrition on the development of IgA plasma cells in mouse gut. J Nutr 112:1047–1051
88. McDermott MR, Mark DA, Befus AD, Baliga BS, Suskind RM, Bienenstock J 1982 Impaired intestinal localization of mesenteric lymphoblasts associated with vitamin A deficiency and protein-calorie malnutrition. Immunology 45:1–5
89. Beatty DW, Napier B, Sinclair-Smith CC, McCabe K, Hughes EJ 1983 Secretory IgA synthesis in kwashiorkor. J Clin Lab Immunol 12:31–36
90. Watson RR, McMurray DN, Martin P, Reyes MA 1985 Effect of age, malnutrition and renutrition on free secretory component and IgA in secretion. Am J Clin Nutr 42:281–288
91. Brandtzaeg P, Halstensen TS, Huitfeldt HS, Krajči K, Kvale D, Scott H, Thrane PS 1992 Epithelial expression of HLA, secretory component (poly-Ig receptor), and adhesion molecules in the human alimentary tract. Ann NY Acad Sci 664:157–179
92. Kuitunen M, Savilahti E 1995 Mucosal IgA, mucosal cow's milk antibodies, serum cow's milk antibodies and gastrointestinal permeability in infants. Pediatr Allergy Immunol 6:30–35
93. Burgio GR, Lanzavecchia A, Plebani A, Jayakar S, Ugazio AG 1980 Ontogeny of secretory immunity: levels of secretory IgA and natural antibodies in saliva. Pediatr Res 14:1111–1114
94. Roberts SA, Wincup G, Harries DA 1980 Mucosal receptor for IgA in the breast-fed neonate. Early Hum Dev 4:161–166
95. Smith DJ, King WF, Taubman MA 1989 Isotype, subclass and molecular size of immunoglobulins in salivas from young infants. Clin Exp Immunol 76:97–102
96. Fitzsimmons SP, Evans MK, Pearce CL, Sheridan MJ, Wientzen R, Cole MF 1994 Immunoglobulin A subclasses in infants' saliva and in saliva and milk from their mothers. J Pediatr 124:566–573
97. Bjerke K, Brandtzaeg P 1990 Terminally differentiated human intestinal B cells. IgA and IgG subclass-producing immunocytes in the distal ileum, including Peyer's patches, compared with lymph nodes and palatine tonsils. Scand J Immunol 32:61–67
98. Kilian M, Husby S, Høst A, Halken S 1995 Increased proportions of bacteria capable of cleaving IgA1 in the pharynx of infants with atopic disease. Pediatr Res 38:182–186
99. Mellander L, Carlsson B, Hanson LA 1986 Secretory IgA and IgM antibodies to *E. coli* O and poliovirus type I antigens occur in amniotic fluid, meconium and saliva from newborns. A neonatal immune response without antigenic exposure: a result of anti-idiotypic induction? Clin Exp Immunol 63:555–561
100. Haneberg B, Aarskog D 1975 Human faecal immunoglobulins in healthy infants and children, and in some with diseases affecting the intestinal tract or the immune system. Clin Exp Immunol 22:210–222
101. Perkkio M, Svailahti E 1980 Time of appearance of immunoglobulin-containing cells in the mucosa of the neonatal intestine. Pediatr Res 14:953–955
102. Hjelt K, Sørensen CH, Nielsen OH, Krasilnikoff PA 1988 Concentrations of IgA, secretory IgA, IgM, secretory IgM, IgD, and IgG in the upper jejunum of children without gastrointestinal disorders. J Pediatr Gastroenterol Nutr 7:867–871
103. Stephens S 1986 Development of secretory immunity in breast fed and bottle fed infants. Arch Dis Child 61:263–269
104. Hahn-Zoric M, Fulconis F, Minoli I, Moro G, Carlsson B, Bottiger M, Raiha N, Hanson LA 1990 Antibody responses to parenteral and oral vaccines are impaired by conventional and low protein formulas as compared to breast-feeding. Acta Paediatr Scand 79:1137–1142
105. Pabst HF, Spady DW 1990 Effect of breast-feeding on antibody response to conjugate vaccine. Lancet 336:269–270
106. Gleeson M, Cripps AW, Clancy RL, Hensley MJ, Dobson AJ, Firman DW 1986 Breast feeding conditions a differential developmental pattern of mucosal immunity. Clin Exp Immunol 66: 216–222
107. Tappuni AR, Challacombe SJ 1994 A comparison of salivary immunoglobulin A (IgA) and IgA subclass concentrations in predentate and dentate children and adults. Oral Microbiol Immunol 9:142–145

108. Renz H, Brehler C, Petzoldt S, Prinz H, Rieger CH 1991 Breast feeding modifies production of SIgA cow's milk-antibodies in infants. Acta Paediatr Scand 80:149–154
109. Renz H, Vestner R, Petzoldt S, Brehler C, Prinz H, Rieger CHL 1990 Elevated concentrations of salivary secretory immunoglobulin A anti-cow's milk protein in newborns at risk of allergy. Int Arch Allergy Appl Immunol 92:247–253
110. Avanzini MA, Plebani A, Monafo V, Pasinetti G, Teani M, Colombo A, Mellander L, Carlsson B, Hanson LÅ, Ugazio AG, Burgio GR 1992 A comparison of secretory antibodies in breast-fed and formula-fed infants over the first six months of life. Acta Paediatr 81:296–301
111. Kramer DR, Cebra JJ 1995 Early appearance of 'natural' mucosal IgA responses and germinal centers in suckling mice developing in the absence of maternal antibodies. J Immunol 154: 2051–2062
112. Orlic D, Lev R 1977 An electron microscopic study of intraepithelial lymphocytes in human fetal small intestine. Lab Invest 37:554–561
113. Spencer J, Dillon SB, Isaacson PG, MacDonald TT 1986 T cell subclasses in fetal human ileum. Clin Exp Immunol 65:553–558
114. Dunkley ML, Husband AJ 1987 Distribution and functional characteristics of antigen-specific helper T cells arising after Peyer's patch immunization. Immunology 61:475–482
115. Guy-Grand D, Griscelli C, Vassalli P 1978 The mouse gut T lymphocyte, a novel type of T cell. Nature, origin, and traffic in mice in normal and graft-versus-host conditions. J Exp Med 148: 1661–1677
116. Cuff CF, Cebra CK, Rubin DH, Cebra JJ 1993 Developmental relationship between cytotoxic α/β T cell receptor-positive intraepithelial lymphocytes and Peyer's patch lymphocytes. Eur J Immunol 23:1333–1339
117. Cerf-Bensussan N, Guy-Grand D 1991 Intestinal intraepithelial lymphocytes. Gastroenterol Clin North Am 20:549–576
118. Machado CSM, Rodrigues MAM, Maffei HVL 1994 Gut intraepithelial lymphocyte counts in neonates, infants and children. Acta Paediatr 83:1264–1267
119. Umesaki Y, Okada Y, Matsumoto S, Imaoka A, Setoyama H 1995 Segmented filamentous bacteria are indigenous intestinal bacteria that activate intraepithelial lymphocytes and induce MHC class II molecules and focosyl asialo GM1 glycolipids on the small intestinal epithelial cells in the ex-germ-free mouse. Microbiol Immunol 39:555–562
120. Helgeland L, Vaage JT, Rolstad B, Midtvedt T, Brandtzaeg P 1996 Microbial colonization influences the T cell receptor Vβ repertoire of intraepithelial lymphocytes in rat intestine: differential Vβ skewing in the CD8$\alpha\alpha$ and CD8$\alpha\beta$ subsets. Submitted
121. Halstensen TS, Brandtzaeg P 1994 Phenotypic characteristics of human intraepithelial lymphocytes. In: Kiyono H, McGhee JR (eds) Mucosal Immunology: Intraepithelial Lymphocytes, Advances in Host Defense Mechanisms, Vol. 9. Raven Press, pp 147–161
122. Cerf-Bensussan N, Cerf M, Guy-Grand D 1993 Gut intraepithelial lymphocytes and gastrointestinal diseases. Curr Opin Gastroenterol 9:953–961
123. Giachino C, Rocci MP, De Libero G, Oderda G, Ansaldi N, Migone N 1994 An alternative approach to the assessment of $\gamma\delta$ T-cell clonality in celiac disease intestinal lesions through cDNA heteroduplex analysis of T-cell receptor VJ junctions. Human Immunol 40:303–311
124. Latthe M, Terry L, MacDonald TT 1994 High frequency of CD8$\alpha\alpha$ homodimer-bearing T cells in human fetal intestine. Eur J Immunol 24:1703–1705
125. Lynch S, Kelleher D, McManus R, O'Farrelly C 1995 RAG1 and RAG2 expression in human intestinal epithelium: evidence of extrathymic T cell differentiation. Eur J Immunol 25:1143–1147
126. Lundqvist C, Baranov V, Hammarström S, Athlin L, Hammarström M-L 1995 Intra-epithelial lymphocytes. Evidence for regional specialization and extrathymic T cell maturation in the human gut epithelium. Int Immunol 7:1473–1487
127. Taplin M-E, Frantz ME, Canning C, Ritz J, Blumberg RS, Balk SP 1996 Evidence against T-cell development in the adult human intestinal mucosa based upon lack of terminal deoxynucleotidyltransferase expression. Immunology 87:402–407
128. Koningsberger JA, Chott A, Logtenberg T, Blumberg RS, Balk SP 1995 The TCR$\alpha\beta$ repertoire in human intestinal fetal mucosa during gestation. Gastroenterology 108:A851
129. Blumberg RS, Balk SP 1994 Intraepithelial lymphocytes and their recognition of non-classical MHC molecules. Int Rev Immunol 11:15–30
130. Mowat AM 1994 Oral tolerance and regulation of immunity to dietary antigens. In: Ogra PL,

Mestecky J, Lamm ME, Strober W, McGhee J, Bienenstock J (eds) Handbook of Mucosal Immunology. Academic Press, Orlando, FL, pp 185–201
131. Vidal K, Samarut C, Magaud J-P, Revillard J-P, Kaiserlian D 1993 Unexpected lack of reactivity of allogeneic anti-Ia monoclonal antibodies with MHC class II molecules expressed by mouse intestinal epithelial cells. J Immunol 151:4642–4650
132. Born W, Happ MP, Dallas A, Reardon C, Kubo R, Shinnick T, Brennan P, O'Brien R 1990 Recognition of heat shock proteins and $\gamma\delta$ cell function. Immunol Today 11:40–43
133. Erle DJ 1995 Scratching the surface. Curr Biol 5:252–254
134. Komano H, Fujiura Y, Kawaguchi M, Matsumoto S, Hashimoto Y, Obana S, Mombaerts P, Tonegawa S, Yamamoto H, Itohara S, Nanno M, Ishikawa H 1995 Homeostatic regulation of intestinal epithelia by intraepithelial $\gamma\delta$ T cells. Proc Natl Acad Sci USA 92:6147–6151
135. Kiyono H, McGhee JR 1994 T helper cells for mucosal immune responses. In: Ogra PL, Mestecky J, Lamm ME, Strober W, McGhee JR, Bienenstock J (eds) Handbook of Mucosal Immunology. Academic Press, Orlando, FL, pp 263–274
136. Brandtzaeg P, Halstensen TS, Hvatum M, Kvale D, Scott H 1993 The serologic and mucosal immunologic basis of celiac disease. In: Walker WA, Harmatz PR, Wershil BK (eds) Immunophysiology of the Gut. Bristol-Myers Squibb/Mead Johnson Nutrition Symposia, Vol 11. Academic Press, London, pp 295–333
137. McMenamin C, Pimm C, McKersey M, Holt PG 1994 Regulation of IgE responses to inhaled antigen in mice by antigen-specific $\gamma\delta$ T cells. Science 265:1869–1871
138. Hörnquist E, Grdič D, Mak T, Lycke N 1996 CD8-deficient mice exhibit augmented mucosal immune responses and intact adjuvant effects to cholera toxin. Immunology 87:220–229
139. Nickoloff BJ, Turka LA 1994 Immunological functions of non-professional antigen-presenting cells: new insights from studies of T-cell interactions with keratinocytes. Immunol Today 15: 464–469
140. Nicklin S, Miller K 1983 Local and systemic immune responses to intestinally presented antigen. Int Arch Allergy Appl Immunol 72:87–90
141. Strobel S, Mowat AM, Drummond HE, Pickering MG, Ferguson A 1983 Immunological responses to fed protein antigens in mice. II. Oral tolerance for CMI is due to activation of cyclophosphamide-sensitive cells by gut-processed antigen. Immunology 49:451–456
142. Brandtzaeg P, Halstensen TS, Kett K, Krajči P, Kvale D, Rognum TO, Scott H, Sollid LM 1989 Immunobiology and immunopathology of human gut mucosa: humoral immunity and intraepithelial lymphocytes. Gastroenterology 97:1562–1584
143. Panja A, Blumberg RS, Balk SP, Mayer L 1993 CD1d is involved in T cell–intestinal epithelial cell interactions. J Exp Med 178:1115–1119
144. Li Y, Yio XY, Mayer L 1995 Human intestinal epithelial cell-induced $CD8^+$ T cell activation is mediated through CD8 and the activation of CD8-associated $p56^{lck}$. J Exp Med 182:1079–1088
145. Aune TM, Pogue SL 1989 Generation and characterization of continuous lines of $CD8^+$ suppressor T lymphocytes. J Immunol 142:3731–3739
146. Hambor JE, Kaplan DR, Tykocinski ML 1990 CD8 functions as an inhibitory ligand in mediating the immunoregulatory activity of $CD8^+$ cells. J Immunol 145:1646–1652
147. Rothberg RM, Farr RS 1965 Anti-bovine serum albumin and anti-alpha lactalbumin in the serum of children and adults. Pediatrics 35:571–588
148. Scott H, Rognum TO, Midtvedt T, Brandtzaeg P 1985 Age-related changes of human serum antibodies to dietary and colonic bacterial antigens measured by an enzyme-linked immunosorbent assay. Acta Pathol Microbiol Immunol Scand [C] 93:65–70
149. Korenblat PE, Rothberg RM, Minden P, Farr RS 1968 Immune responses of human adults after oral and parenteral exposure to bovine serum albumin. J Allergy 41:226–235
150. Husby S, Mestecky J, Moldoveanu Z, Holland S, Elson CO 1994 Oral tolerance in humans. T cell but not B cell tolerance after antigen feeding. J Immunol 152:4663–4670
151. Waldo FB, W Van den Wall Bake A, Mestecky J, Husby S 1994 Suppression of the immune response by nasal immunization. Clin Immunol Immunopathol 72:30–34
152. Halstensen TS, Brandtzaeg P 1993 Activated T lymphocytes in the celiac lesion: non-proliferative activation (CD25) of $CD4^+$ α/β cells in the lamina propria but proliferation (Ki-67) of α/β and γ/δ cells in the epithelium. Eur J Immunol 23:505–510
153. Ohtsuka Y, Yamashiro Y, Maeda M, Oguchi S, Shimizu T, Nagata S, Yagita H, Yabuta K, Okumura K 1996 Food antigen activates intraepithelial lymphocytes and lamina propria lymphocytes in food-sensitive enteropathy in mice. Pediatr Res 39:862–866

154. Gaboriau-Routhiau V, Moreau M-C 1996 Gut flora allows recovery of oral tolerance to ovalbumin in mice after transient breakdown mediated by cholera toxin or *Escherichia coli* heat-labile enterotoxin. Pediatr Res 39:625–629
155. Ishizaka S, Kimoto M, Tsujii T, Saito S 1994 Antibody production system modulated by oral administration of human milk and TGF-β. Cell Immunol 159:77–84
156. Planchon SM, Martins CAP, Guerrant RL, Roche JK 1994 Regulation of intestinal epithelial barrier function by TGF-β1. Evidence for its role in abrogating the effect of a T cell cytokine. J Immunol 153:5730–5739
157. Shroff KE, Meslin K, Cebra JJ 1995 Commensal enteric bacteria engender a self-limiting humoral mucosal immune response while permanently colonizing the gut. Infect Immun 36: 3904–3913
158. MacDonald TT 1995 Breakdown of tolerance to the intestinal bacterial flora in inflammatory bowel disease (IBD). Clin Exp Immunol 102:445–447
159. Foo MC, Lee A 1972 Immunological response of mice to members of the autochthonous intestinal microflora. Infect Immun 6:525–532
160. Duchmann R, Kaiser I, Hermann E, Mayet W, Ewe K, Meyer Zum Büschenfelde K-H 1995 Tolerance exists towards resident intestinal flora but is broken in active inflammatory bowel disease (IBD). Clin Exp Immunol 102:448–455
161. Greco L, Auricchio S, Mayer M, Grimaldi M 1988 Case control study on nutritional risk factors in celiac disease. J Pediatr Gastroenterol Nutr 7:395–399
162. Ansaldi N, Faussone D, Tavassoli K, Balocco L, Dell'Olio D, Oderda G, Bramante L 1991 Clinical data on celiac disease with an early or late onset (Italian). Minerva Pediatr 43:377–381
163. Cataldo F, Traverso G, Maltese I, Albeggiani A, Paternostro D 1991 Celiac disease and dietary habits in the 1st year of life (Italian). Minerva Pediatr 43:7–10
164. Polanco I 1992 Current status of digestive intolerance to food protein. J Pediatr 121:S108–S110
165. Taylor B, Norman AP, Orgel HA, Stokes CR, Turner MW, Soothill JF 1973 Transient IgA deficiency and pathogenesis of infantile atopy. Lancet ii:111–113
166. Ludviksson BR, Eiriksson TH, Ardal B, Sigfusson AB, Valdimarsson H 1992 Correlation between serum immunoglobulin A concentrations and allergic manifestations in infants. J Pediatr 121: 23–27
167. Perkkiö M 1980 Immunohistochemical study of intestinal biopsies from children with atopic eczema due to food allergy. Allergy 35:573–580
168. Sloper KS, Brook CGD, Kingston D, Pearson JR, Shiner M 1981 Eczema and atropy in early childhood: low IgA plasma cell counts in the jejunal mucosa. Arch Dis Child 56:939–942
169. Van Asperen PP, Gleeson M, Kemp AS, Cripps AW, Geraghty SB, Mellis CM, Clancy RL 1985 The relationship between atopy and salivary IgA deficiency in infancy. Clin Exp Immunol 62:753–757
170. Marshall JS, Bienenstock J 1994 The role of mast cells in inflammatory reactions of the airways, skin and intestine. Curr Opin Immunol 6:853–859
171. Flesjå K, Yri T 1977 Protein-losing enteropathy in the Lundehund. J Small Anim Pract 18: 11–23

25 The role of lactoferrin in the gastrointestinal tract

O.M. CONNEELY, P.P. WARD, X. ZHOU and S. WAGNER

ABBREVIATIONS

LPS lipopolysaccharides
nLF native human lactoferrin
rLF recombinant human lactoferrin

INTRODUCTION

Lactoferrin is an iron-binding glycoprotein that consists of a single polypeptide chain with a molecular weight of 78 kDa. It is the second most abundant protein in human milk (~1 g/l) (1,2) and is found in most exocrine secretions including tears, nasal secretions, saliva, intestinal mucus and genital secretions (3–5). The protein also is expressed and secreted by the secondary granules of polymorphonuclear neutrophils (6). The polypeptide structure of lactoferrin comprises two homologous domains that appear to have arisen by intragenic duplication (7). The crystal structure of the protein has been resolved (8,9) and it has been shown that each domain binds one ferric and one carbonate anion. In addition, each domain contains one glycosylated site to which *N*-linked glycan residues are attached (10).

Lactoferrin is a member of the transferrin family of non-haem iron-binding proteins (11), which includes transferrin, the major iron transporting protein in blood (12,13), the egg-white protein ovotransferrin (14) and the melanocyte protein melanotransferrin (15). Together, these proteins are responsible for controlling the levels of free iron in biological fluids. They are distinguished from other iron-binding proteins by their unique anion requirement for binding of iron. With regard to its iron-binding properties, lactoferrin differs from serum transferrin in its higher affinity for iron and in its unique ability to retain iron over a broad pH range (16,17). These differences, together with its differential tissue distribution relative to other transferrins, contribute to the unique functional properties of lactoferrin.

Lactoferrin is a multifunctional protein to which several physiological functions have been ascribed (reviewed in 18,19). These include regulation of iron absorption in the intestine (20,21), promotion of intestinal cell growth (22), protection against microbial infection (23–25), regulation of myelopoiesis (26,27) and systemic immune responses (28,29). However, definitive evidence in vivo for most of these functions is not yet available. Here, we summarize the results of studies by other investigators and ourselves that address the role of lactoferrin in regulation of iron homeostasis and in protection against infection.

J.G. Bindels, A.C. Goedhart and H.K.A. Visser (eds), Recent Developments in Infant Nutrition, 377–386.

REGULATION OF IRON ABSORPTION IN THE INTESTINE BY LACTOFERRIN

The mechanism of iron uptake in the intestine is still a subject of debate. A role for lactoferrin in this process is supported by the observation that iron is absorbed better by infants who consume human breast milk rather than cows'-milk-based infant formula that contains low levels of inactivated bovine lactoferrin (20) and by the demonstration that lactoferrin, but not transferrin, can deliver its bound iron to human duodenal cells (30). Specific receptors for lactoferrin have been identified and partially purified from enterocytes of a number of species (19,31,32), although the role of these receptors in mediating uptake of lactoferrin-bound iron has not been established. Lactoferrin is very stable in the gastrointestinal tract (33) and this stability, together with the high affinity and pH stability of iron binding, suggest that the protein also may be involved in sequestration and detoxification of free iron.

ANTIBACTERIAL PROPERTIES OF LACTOFERRIN

The antibacterial functions of lactoferrin are perhaps the most clearly substantiated by both in vitro (34–36) and in vivo (37,38) evidence. It is likely that three different mechanisms involving at least two and possibly three separate domains of the protein contribute to its antimicrobial functions. The first mechanism is a bacteriostatic effect related to the high iron-binding affinity of the protein that deprives iron-requiring bacteria of this essential growth nutrient (34,39,40). Since the bacteriostatic properties of lactoferrin are due to its iron-binding ability, the protein is capable of retarding the growth of a broad range of microorganisms including a variety of Gram-negative and Gram-positive bacteria and certain yeasts (39,41). However, bacteriostasis is often temporary because some Gram-negative bacteria adapt to iron-restrictive conditions by synthesizing low molecular weight iron chelators (siderophores) that can remove iron from lactoferrin. The efficacy of bacteriostasis is dependent on the iron status of lactoferrin and is overcome by saturation of lactoferrin with iron (42). In this regard, it should be noted that lactoferrin expressed in mothers' milk is only 5–8% saturated with iron and therefore would be expected to function as an efficient bacteriostatic agent.

The second antibacterial property of lactoferrin is due to a direct bactericidal function within the protein. Lactoferrin has a direct bactericidal effect against some Gram-negative and Gram-positive bacteria that cannot be attributed to simple iron deprivation. Apolactoferrin can cause a rapid loss of bacterial viability that cannot be reversed by the addition of exogenous iron to the growth medium (34,43). At physiological concentrations, apolactoferrin directly damages the outer membrane of Gram-negative bacteria by causing the release of lipopolysaccharides (LPS) (44). Recent results suggest that a cationic domain at the N-terminus of lactoferrin is responsible for the bactericidal properties. Synthetic peptides containing this cationic domain (amino acid residues 18–40) of human lactoferrin have a more potent bactericidal effect (45) and lead to a greater release of LPS (46) than intact lactoferrin (45). The N-terminal amino acids responsible for the bactericidal effect are distinct from those involved in iron binding, indicating that the bactericidal effect of lactoferrin is exerted by a mechanism distinct from metal chelation. However, the contribution of the bactericidal peptide to the antibacterial properties of the native lactoferrin protein is not known. Although no sequence similarities exist between the isolated peptide and other known antimicrobial peptides, the asymmetrical clustering of basic residues observed int his region of lactoferrin has been identified in a variety of naturally occurring antimicrobial/anti-endotoxin peptides which disrupt bacterial cells, including magainins, cecropins and defensins (47). These proteins and peptides have been the focus of considerable investigation in recent years for the treatment of infectious disease because, in addition to their microbicidal function, they possess an important anti-endotoxin activity that, unlike synthetic antibiotics, allows them to bind to and inactivate LPS released from bacteria and thereby diminish or inactivate the potentially toxic inflammatory properties of LPS.

The third mechanism by which lactoferrin exerts its antimicrobial activity is more complex and perhaps the most speculative, and relates to the ability of the protein to regulate immune responses (28,29). Lactoferrin is expressed in neutrophil secondary granules together with several additional antimicrobial proteins (6). Lactoferrin has been reported to have both positive (48) and negative (26,27,49,50) regulatory effects on myelopoiesis. A rare condition in humans has been reported in which neutrophil granules are reduced in quantity and almost devoid of lactoferrin (51,52). The deficiency is accompanied by a lack of response to chemotactic signals and diminished adhesiveness of neutrophils to particle surfaces, leading to pyogenic infections and in particular deep-seated skin abscesses (51). Whether lactoferrin deficiency alone is the primary cause of this syndrome has not been confirmed. Systemic infection with bacteria is accompanied by a rapid rise in serum levels of lactoferrin secreted from granulocytes (53) and a concomitant decrease in serum iron levels (hypoferraemia). Intravenous or intraperitoneal administration of lactoferrin before experimental infection with a lethal dose of *E. coli* can protect mice against the infection (38). Both in vitro and in vivo studies suggest that this effect of lactoferrin involves inhibition of production of several cytokines, including tumour necrosis factor α (TNFα) and interleukin-1 (IL-1), which are key mediators of the inflammatory response leading to death from toxic shock (28,29) in response to Gram-negative LPS endotoxins. The ability of lactoferrin to interact with the conserved lipid A moiety of lipopolysaccharides that is required for induction of these cytokines suggests that lactoferrin may have a broad spectrum anti-endotoxin role with regard to protection against inflammation (54). High affinity receptors for lactoferrin also have been identified on monocytes (55), macrophages (56), platelets (57) and lymphocytes (58). However, the role of these receptors in mediating the immune regulatory and/or anti-inflammatory functions of lactoferrin remains to be established. While the regulation of cytokine production by lactoferrin clearly has exciting implications with regard to the use of lactoferrin as an immune regulatory/anti-inflammatory agent when administered alone or in combination with synthetic antibiotics for the treatment of infectious disease, more in vivo studies are needed to investigate this aspect of lactoferrin activity.

***Figure 1** Iron saturation of native and recombinant human lactoferrin. Native (nLF, ▲) or recombinant (rLF, ●) human lactoferrin (500 μg) that had been exposed to pH 2.0 for several hours and then neutralized was incubated with increasing concentrations of ^{59}Fe for 30 min at 22°C. Free ^{59}Fe was removed by passage of the samples over a NAP10 column. ^{59}Fe-bound lactoferrin was quantified by liquid scintillation counting. For these and the studies described below, nLF was purchased from Sigma and rLF was produced and purified as described previously (59) by expression of the human lactoferrin cDNA in Aspergillus awamori*

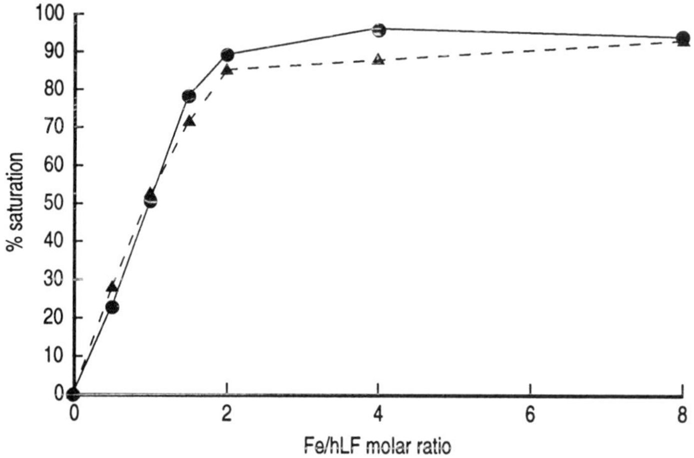

ANTIBACTERIAL MECHANISMS OF LACTOFERRIN: IMPLICATIONS WITH REGARD TO INTESTINAL INFECTION

The bacteriostatic, bactericidal and immune regulatory mechanisms described above may function synergistically to provide an antibacterial activity that is unique to lactoferrin. However, as indicated above, the contribution of each of these mechanisms to the antibacterial activity of lactoferrin has not been established. We have recently begun to investigate the mechanisms that may contribute to these antibacterial properties. First, using iron and receptor binding assays, we have demonstrated that both natural and recombinant human lactoferrin that have been exposed to low pH (pH 2.0) and then neutralized are still capable of binding ferric iron and intestinal receptors in a specific and saturable manner (Figs 1,2). These data suggest that human lactoferrin that has passed through the stomach should retain the ability to bind iron and enteric receptor sites when it reaches the intestine if the protein remains intact. Second, in vitro antibacterial assays have demonstrated that the ability of lactoferrin to inhibit the growth of the enteropathogenic strains of Gram-negative bacteria *Shigella flexneri* and *Klebsiella pneumoniae* is enhanced by prior exposure of the protein to pH 2.0 followed by neutralization (Fig. 3A,B), a process that occurs naturally in vivo after exposure of the protein to the human digestive tract. Third, using human enterocyte cells grown in tissue culture, we have demonstrated that binding of lactoferrin to

***Figure 2** Binding of recombinant human lactoferrin (rLF) to human enterocyte cells. The human colon carcinoma cell line Caco2 was grown in tissue culture for 10 days until the cells reached confluency and differentiated into human enterocytes. Biotinylated rLF was saturated with iron and incubated (0.05–5.0 μM) with solubilized Caco2 membranes (200 ng) in the presence or absence of a 20-fold molar excess of unlabelled ligand. The assay was carried out for 4 h at 4°C. Binding of biotinylated rLF to the Caco2 membranes was determined using a biotin/avidin microtitre assay*

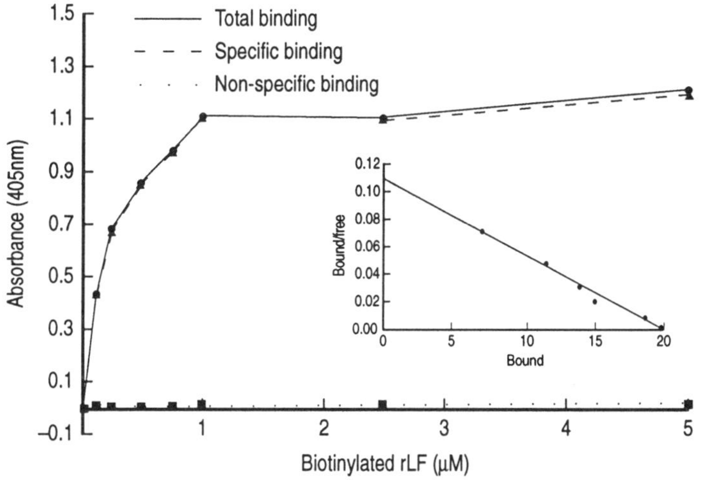

human enterocyte receptors plays an important role in mediating the antibacterial function of the protein (Fig. 4A,B). In these studies, we inoculated tissue culture media with *Klebsiella pneumoniae* in the presence or absence of human enterocytes containing receptors for lactoferrin and examined the ability of recombinant lactoferrin (1 μM) to inhibit bacterial growth under both conditions. The results of these experiments indicated that inhibition of bacterial growth in the tissue culture medium by lactoferrin was significantly enhanced in the presence of enterocytes. Most interesting, however, was the observation that lactoferrin decreased the number of bacteria attaching to the enterocyte cell surface by 90%. Analysis of the dose-dependent antibacterial efficacy of lactoferrin in this assay (data not shown) indicated that the maximal effective concentration correlated well with the kinetics of receptor binding by lactoferrin ($K_d = 0.2\,\mu M$), as expected for a receptor-mediated antibacterial activity. Thus, both exposure to low pH and interaction with intestinal receptors have the capacity to bring about some functional (presumably conformational) change in the protein that results in an activation or enhancement of antibacterial activity.

Finally, we have established that the iron-binding function of lactoferrin plays an essential role in the antibacterial activity of the protein when it is bound to enterocyte receptors. In these studies, we used a genetically engineered mutant of human lactoferrin in which the two essential tyrosine residues in both halves of the protein

Figure 3 ***The in vitro antibacterial activity of nLF and hLF against the Gram-negative enteropathogens Shigella flexneri (A) and Klebsiella pneumoniae (B) is enhanced by prior exposure of the proteins to pH 2.0 and neutralization. 1×10^6 c.f.u/ml of each bacterial strain was incubated for 4 h in bactopeptone broth in the absence or presence of nLF or rLF treated as indicated in the figure. The bacterial cultures were then serially diluted and plated on McConkey (S. flexneri) or plate count (K. pneumoniae) agar plates for bacterial enumeration***

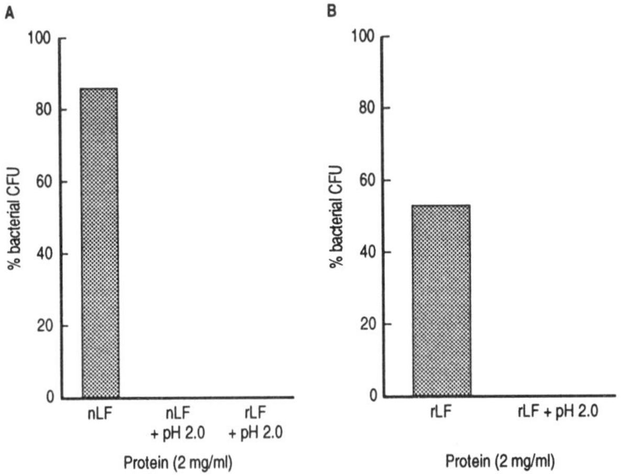

Figure 4 ***Contribution of enteric receptor binding to the antibacterial activity of lactoferrin. Caco2 cells in MEM media (final volume 400 μl) were incubated with a standard inoculum of Klebsiella pneumoniae (1×10^6 c.f.u/ml) in the presence or absence of rLF (1 μM) (A). After 4 h, the extracellular media and cell-associated bacteria were harvested, serially diluted and plated on plate count agar for enumeration.***

The antibacterial activity of lactoferrin in the absence of Caco2 cells (tissue culture medium alone) was determined in parallel using a microtitre plate assay (B). Bacteria (1×10^6 c.f.u/ml) were incubated in MEM media (100 μl total volume) in the presence or absence of lactoferrin (1 μM). After 4 h the medium was harvested, serially diluted and plated on plate count agar for bacterial enumeration

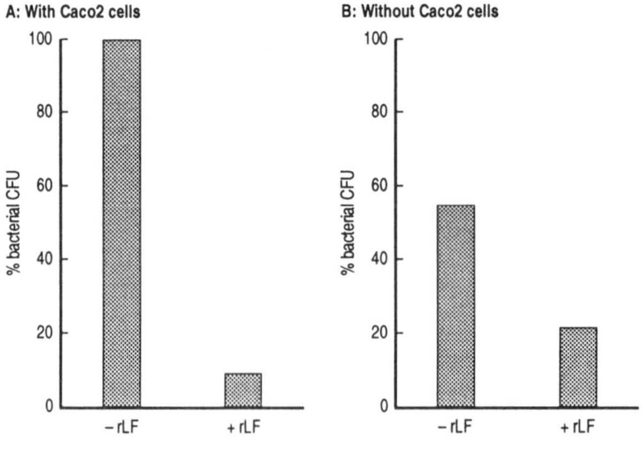

***Figure 5** Contribution of the iron binding function of lactoferrin to the growth inhibition of Klebsiella by the protein. An iron-binding-defective mutant of lactoferrin was generated by site-directed mutagenesis of the human lactoferrin cDNA and the mutant cDNA was expressed in Aspergillus awamori as previously described (59). The resulting mutant lactoferrin protein contained alanine residues in place of the two tyrosine residues that are essential for iron binding in both halves of lactoferrin (Tyr93, Tyr193, Tyr436 and Tyr529) and the mutant was unable to bind iron when tested in an iron binding assay. The mutant (M4-Y) then was tested in the ex vivo antibacterial assay described in the legend of Figure 4*

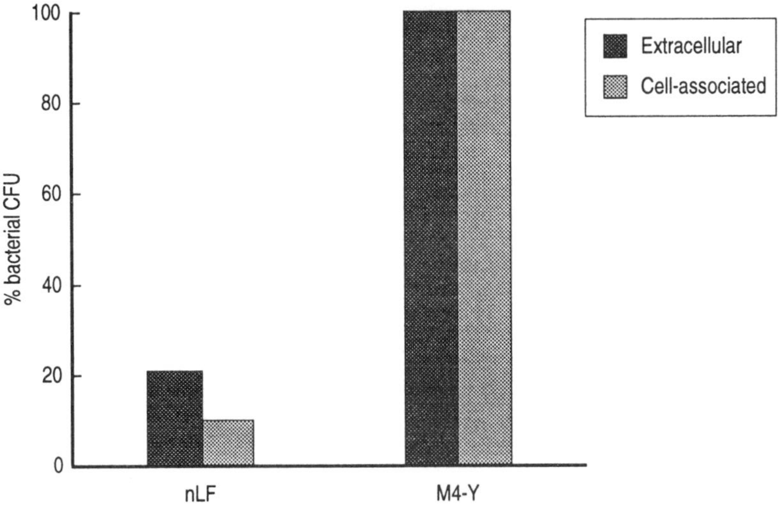

that are essential for iron binding were replaced by alanine residues. The resulting mutant was unable to bind iron but retained its enterocyte receptor binding properties. However, when tested in the enterocyte-cell-culture-based ex vivo antibacterial assay, the mutant was unable to inhibit the growth of *Klebsiella pneumoniae* (Fig. 5), indicating that loss of the iron-binding activity of the protein resulted in a loss of antibacterial function. Thus both the human intestinal receptor-binding and iron-binding activities of lactoferrin can play an important and perhaps synergistic role in mediating the antibacterial activity of the protein.

CONCLUSIONS

The results of our ex vivo antibacterial assays indicate that lactoferrin can significantly inhibit the growth and association of *Klebsiella pneumoniae* with human enteric cells. Further, the antibacterial activity of lactoferrin against this Gram-negative pathogen appears to be bacteriostatic in nature and is dependent on the iron-binding function of the protein. Most importantly, the antibacterial activity of lactoferrin is also facilitated by binding of the protein to specific enterocyte receptors. These results indicate that specific lactoferrin–host interactions play an important role in mediating the antibiotic activity of the protein in the gastrointestinal tract. While the assay system used in these studies provides a valuable model system to address the contribution of human enterocyte interactions to the activity of lactoferrin, the assay is limited in its ability to address all of the antibacterial functions of lactoferrin. For example, this system does not address the potential lipopolysaccharide-inactivating/anti-endotoxin function of the protein or the effect of lactoferrin on the regulation of host immune responses associated with the gut associated lymphatic system (GALT). These properties are likely to play an important role in the ability of lactoferrin to protect the host against bacteria. Thus, the elucidation of the multifactorial antibacterial properties of lactoferrin will require the use of in vivo animal models that take into account the contribution of species-specific receptor interactions and the anti-endotoxin and immune regulatory properties of the protein. In this regard, the availability of recombinant DNA methods by which various species of lactoferrins can be produced for use in animal models and the ability to produce mutant lactoferrins to allow identification of the functional domains of the protein will greatly facilitate our understanding of the multifunctional antibacterial properties of lactoferrin as well as aid in the rational design of trials to establish the efficacy of this protein as a natural broad-spectrum antibiotic in humans.

REFERENCES

1. Masson PL, Heremans JF 1971 Lactoferrin in milk from different species. Comp Biochem Physiol 39:119–129
2. Hennart PF, Brasseur DJ, Delogne-Desnoeck JB, Dramaix MM, Robyn CE 1991 Lysozyme, lactoferrin and secretory immunoglobin A content in breast milk: Influence of duration of lactation, nutrition status, prolactin status, and parity of mother. Am J Clin Nutr 53:32–39
3. Masson PL, Heremans JF, Dive C 1966 An iron-binding protein common to many external secretions. Clin Chim Acta 14:735–739
4. Pentecost BT, Teng CT 1987 Lactotransferrin is the major estrogen inducible protein of mouse uterine secretions. J Biol Chem 262:10134–10139
5. Yu L-C, Chen Y-H 1993 The developmental profile of lactoferrin in mouse epididymis. Biochem J 296:107–111
6. Masson PL, Heremans JF, Schonne E 1969 Lactoferrin, an iron-binding protein in neutrophilic leukocytes. J Exp Med 130:643–658
7. Metz-Boutigue M-H, Jolles J, Mazurier J, Schoentgen F, Legrand D, Spik G, Montreuil J, Jolles P 1984 Human lactotransferrin: amino acid sequence and structural comparisons with other transferrins. Eur J Biochem 145:659–676
8. Anderson BF, Baker HM, Norris GE, Rice DW, Baker EN 1989 Structure of human lactoferrin: Crystallographic structure analysis and refinement at 2.8Å resolution. J Mol Biol 209:711–734
9. Anderson BF, Baker HM, Norris GE, Rumball SV, Baker EN 1990 Apolactoferrin structure demonstrates ligand-induced conformational change in transferrins. Nature 344:784–787
10. Spik G, Coddenville B, Montreuil J 1988 Comparative study of the primary structures of sero-, lacto- and ovotransferrin glycans from different species. Biochemistry 70:1459–1469
11. Aisen P, Listowsky I 1980 Iron transport and storage proteins. Annu Rev Biochem 49:357–393
12. MacGillivray RTA, Mendez E, Shewale JG, Sinha SK, Lineback-Zins J, Brew J 1983 The primary structure of serum transferrin. J Biol Chem 258:3543–3553
13. Yang F, Lum JB, McGill JR, Moore CM, Naylor SL, Van Bragt PH, Baldwin WD, Bowman BH 1984 Human transferrin: cDNA characterization and chromosomal location. Proc Natl Acad Sci USA 81:2752–2756
14. Jeltsch J-M, Chambon P 1982 The complete nucleotide sequence of the chick ovotransferrin mRNA. Eur J Biochem 122:291–295

15. Rose TM, Plowman GD, Teplow DB, Dreyer WJ, Hellström K-E, Brown JP 1986 Primary structure of the human melanoma-associated antigen p97 (melanotransferrin) deduced from mRNA sequence. Proc Natl Acad Sci USA 83:1261–1265
16. Aisen P, Leibman A 1972 Lactoferrin and transferrin: A comparative study. Biochim Biophys Acta 257:314–323
17. Mazurier J, Spik G 1989 Comparative study of the iron-binding properties of human transferrins. I. Complete and sequential iron saturation and desaturation of the lactotransferrin. Biochim Biophys Acta 629:399–408
18. Sanchez L, Calvo M, Brock JH 1992 Biological role of lactoferrin. Arch Dis Child 67:657–661
19. Iyer S, Lönnerdal B 1993 Lactoferrin, lactoferrin receptors and iron metabolism. Eur J Clin Nutr 47:232–241
20. Fransson G-B, Lönnerdal B 1980 Iron in human milk. J Pediatr 96:380–384
21. Davidson LA, Lönnerdal B 1989 Fe-saturation and proteolysis of human lactoferrin: effect on brush-border receptor-mediated uptake of Fe and Mn. Am J Physiol 257:G930–G934
22. Nichols BL, McKee KS, Huebers HA 1990 Iron is not required in the lactoferrin stimulation of thymidine incorporation into the DNA of rat crypt enterocytes. Pediatr Res 27:525–528
23. Nemet K, Simonovits I 1985 The biological role of lactoferrin. Haematologia 18:3–12
24. Weinberg ED 1978 Iron and infection. Microbiol Rev 42:45–66
25. Bullen JJ, Rogers HJ, Leigh L 1971 Iron-binding proteins in milk and resistance to *Escherichia coli* infections in infants. Br Med J 1:69–75
26. Zucali JR, Broxmeyer HE, Levy D, Morse C 1989 Lactoferrin decreases monocyte-induced fibroblast production of myeloid colony-stimulating activity by suppressing monocyte release of interleukin-1. Blood 74:1531–1536
27. Hangoc G, Falkenburg JHF, Broxmeyer HE 1991 Influence of T-lymphocytes and lactoferrin on the survival promoting effects of IL-1 and IL-6 on human bone marrow granulocyte–macrophage and erythroid progenitor cells. Exp Hematol 19:697–703
28. Machnicki M, Zimecki M, Zagulski T 1993 Lactoferrin regulates the release of tumour necrosis factor alpha and interleukin 6 in vivo. Int J Exp Pathol 74:433–439
29. Crouch SPM, Slater KJ, Fletcher J 1992 Regulation of cytokine release from mononuclear cells by the iron-binding protein lactoferrin. Blood 80:235–240
30. Cox TM, Mazurier J, Spik G, Montreuil J, Peters TJ 1979 Iron binding proteins and influx of iron across the duodenal brush border. Evidence for specific lactotransferrin receptors in the human intestine. Biochem Biophys Acta 558:120–128
31. Hu WL, Mazurier J, Sawatzki G, Montreuil J, Spik G 1988 Lactotransferrin receptor of mouse small-intestinal brush border: binding characteristics of membrane-bound and Triton X-100-solubilized forms. Biochem J 249:435–441
32. Hu WL, Mazurier J, Spik G 1990 Isolation and partial characterization of a lactotransferrin receptor from mouse intestinal brush border. Biochemistry 29:535–541
33. Spik G, Strecker G, Fournet B, Bouquelet S, Montreuil J, Dorland L, van Halbeek H, Vliegenthart JFG 1982 Primary structure of the glycans from human lactotransferrin. Eur J Biochem 121: 413–419
34. Arnold RR, Cole MF, McGhee JR 1977 A bactericidal effect for human lactoferrin. Science 197: 263–265
35. Oram JD, Reiter B 1968 Inhibition of bacteria by lactoferrin and other iron-chelating agents. Biochim Biophys Acta 170:351–365
36. Arnold RR, Brewer M, Gauthier JJ 1980 Bactericidal activity of human lactoferrin: Sensitivity of a variety of microorganisms. Infect Immun 28:893–898
37. Trümpler U, Straub PW, Rosenmund A 1989 Antibacterial prophylaxis with lactoferrin in neutropenic patients. Eur J Clin Microbiol Infect Dis 8:310–313
38. Zagulski T, Lipinski P, Zagulska A, Broniek S, Jarzabek Z 1989 Lactoferrin can protect mice against a lethal dose of *Escherichia coli* in experimental infection in vivo. Br J Exp Pathol 70:697–704
39. Bullen JJ, Rogers HJ, Griffiths E 1978 Role of iron in bacterial infection. Curr Topics Microbiol Immunol 80:1–35
40. Reiter B 1978 Review of non-specific antimicrobial factors in milk. Ann Rech Vet 9:205–224
41. Reiter B, Brock JH, Steel ED 1975 Inhibition of *E. coli* by bovine colostrum and post-colostral milk. II. The bacteristatic effect of LF on a serum susceptible and serum resistant strain of *E. coli*. Immunology 28:83–95

42. Griffiths E, Humphreys J 1977 Bacteriostatic effect of human milk and bovine colostrum on *Escherichia coli*: Importance of bicarbonate. Infect Immun 15:396–401
43. Arnold RR, Russell JE, Champion WJ, Gauthier JJ 1981 Bactericidal activity of human lactoferrin: Influence of physical conditions and metabolic conditions of the target microorganism. Infect Immun 32:655–660
44. Ellison RT, Giehl T-J, La Force FM 1988 Damage of the outer membrane of enteric Gram-negative bacteria by lactoferrin and transferrin. Infect Immun 56:2774–2781
45. Bellamy W, Takase M, Yamauchi K, Wakabayashi H, Kawase K, Tomita M 1992 Identification of the bactericidal domain of lactoferrin. Biochim Biophys Acta 1121:130–136
46. Yamauchi K, Tomita M, Giehl TJ, Ellison RT 1993 Antibacterial activity of lactoferrin and a pepsin-derived lactoferrin peptide fragment. Infect Immun 61:719–728
47. Zasloff M 1987 Magainins, a class of antimicrobial peptides from *Xenopus* skin: isolation, characterization of two active forms and partial cDNA sequence of a precursor. Proc Natl Acad Sci USA 84:5449–5453
48. Sawatzki G, Rich C 1989 Lactoferrin stimulates colony stimulating factor production in vitro and in vivo. Blood Cells 15:371–375
49. Broxmeyer HE, Williams DE, Hangoc G, Cooper S, Gentile P, Shen RN, Ralph P, Gillis S, Bicknell DC 1987 The opposing actions in vivo on murine myelopoiesis of purified preparations of lactoferrin and the colony stimulating factors. Blood Cells 13:31–48
50. Zucali JR, Broxmeyer HE, Ulatowski JA 1979 Specificity of lactoferrin as an inhibitor of granulocyte–macrophage colony-stimulating activity production from fetal mouse liver cells. Blood 54:951–954
51. Breton-Gorius J, Mason DY, Buriot D, Vilde JL, Griscelli C 1980 Lactoferrin deficiency as a consequence of a lack of specific granules in neutrophils from a patient with recurrent infections. Am J Pathol 99:413–419
52. Boxer LA, Coates TD, Haak RA, Wolach JB, Hoffstein S, Baehner RL 1982 Lactoferrin deficiency associated with altered granulocyte function. New Engl J Med 307:404–409
53. Gutteberg TJ, Rokke O, Andersen O, Jørgensen T 1989 Early fall of circulating iron and rapid rise of lactoferrin in septicemia and endotoxemia: An early defence mechanism. Scand J Infect Dis 21:709–715
54. Wang D, Pabst KM, Aida Y, Pabst MJ 1995 Lipopolysaccharide-inactivating activity of neutrophils is due to lactoferrin. J Leukoc Biol 57:865–874
55. Birgens HS, Hansen NE, Karle H, Kristensen LO 1983 Receptor binding of lactoferrin by human monocytes. Br J Haematol 54:383–391
56. van Snick JL, Masson PL 1976 Binding of human lactoferrin to mouse peritoneal cells. J Exp Med 144:1568–1580
57. Leveugle B, Mazurier J, Legrand D, Mazurier C, Montreuil J, Spik G 1993 Lactotransferrin binding to its platelet receptor inhibits platelet aggregation. Eur J Biochem 213:1205–1211
58. Mazurier J, Legrand D, Hu WL, Montreuil J, Spik G 1989 Expression of human lactotransferrin receptors in phytohemagglutinin stimulated human peripheral blood lymphocytes: Isolation of the receptors by antiligand-affinity chromatography. Eur J Biochem 179:481–487
59. Ward PP, Piddington CS, Cunningham GA, Zhou X, Wyatt RO, Conneely OM 1995 A system for production of commercial quantities of human lactoferrin: a broad spectrum natural antibiotic. Biotechnology 13:498–503

Discussion Session 5

CHAIR: J.G. BINDELS

CHAPTER 23

Desjeux Could you, from your calculation of oxygen consumption, estimate the use of ATP and from that the amount of nutrients transported by the gut? Second, is it possible to manipulate the diet in such a way that, at least at the intestinal level, certain nutrients are spared for specific organs? For instance, could we by using glutamine spare glucose for the brain?

Elia That is precisely the situation that has been suggested to occur after surgery. I think that glutamine utilization can indeed spare glucose for tissues. However, the amount of glucose utilized by the gut is actually very small, compared with the total glucose utilization within the body. This highlights where enterocyte studies can be misleading: measuring the glucose utilization in enterocytes and then extrapolating the results to the total number of enterocytes and colonocytes in the human body would suggest that about 60% of the total glucose utilization of the body occurs in the gut. That is clearly not the case. Competition between substrates could indeed spare substrate for other tissues, but the effect is probably quantitatively not important, particularly as far as the brain is concerned, since glucose is not a major substrate for the gut. As far as ATP is concerned, one can calculate the ATP gain relative to oxygen consumption,which varies with the type of fuel. However, the ATP gain per litre O_2 consumed varies by less than 20%, even for relatively inefficient fuels such as acetate. The energy costs of absorption of nutrients have been difficult to define. Use of classical stochiochemistry would suggest that the amount used for absorption is not very high in proportion to the total oxygen consumption. Some of the in vivo studies in the rat suggest that there is only very little change in oxygen consumption by the gut in the fed state, when there is active absorption, compared with the fasted state. If nutrient absorption processes would significantly increase energy expenditure of the whole body, one would expect a much larger increase in oxygen consumption.

Marini Do you have any information about the influence of the type of diet on intestinal metabolism? Secondly, could you tell us something about the importance of short-chain fatty acids for colostomy patients?

Elia I do not think there are any data on the influence of the type of diet on intestinal metabolism in man. I would like to make one point related to this issue. Malnourished animals show decreased oxidative metabolism and decreased oxidative capacity in their epithelial cells. The components of the diet which are responsible for this effect are not known. With respect to your second question about the importance of short-chain fatty acids for patients with a colostomy, I would like to remind the audience

J.G. Bindels, A.C. Goedhart and H.K.A. Visser (eds), Recent Developments in Infant Nutrition, 387–390.

that short-chain fatty acids, particularly butyrate and propionate, are essentially non-existent in arterial blood. Short-chain fatty acids are metabolized either by the colonic mucosa, especially butyrate, which is an important fuel for the large intestine, or by the liver. Most of the epithelial cells of the colon are ideally suited for using short-chain fatty acids. There is little potential for the small intestine to use short-chain fatty acids as fuel, although there might be some adaptation if there is bacterial overgrowth. There are some very interesting effects of fibre on the proliferation of enterocytes, and the mechanism of it is not entirely clear. In colostomy patients, colonic fermentation, and therefore production of short-chain fatty acids, is lacking. Since the colon has been removed, there are no colonocytes to utilize the short-chain fatty acids. Therefore, I am not sure if I can fully answer your question.

Marini You have measured the nutritional biochemistry of the gastrointestinal tract using arterio-portal catheterization techniques. Have you also looked at the lymphatic drainage, especially of short-chain fatty acids?

Elia About 95% of short-chain fatty acids are absorbed. Most short-chain fatty acids will be taken up in the portal system and transported to the liver. However, we should not dismiss the lymphatic system because, for example, some studies in the rat suggest that up to 30% of the labelled leucine taken up may find its way into lymph as the protein part of a lipoprotein. There is no reason to suspect that human lymph does not drain some proteins synthesized by the epithelial cells of the gut.

CHAPTER 24

Pietrzyk You mentioned the cytokines which are produced by macrophages. Recently, a paper was published in Pediatric Research showing that ductal cells contain mRNA for macrophage colony-stimulating factor. Do you have information about other cytokines which can be produced by ductal cells?

Brandtzaeg I guess that like most epithelial cells they will produce the pro-inflammatory cytokines IL-6, IL-8 and perhaps TNFα.

Pietrzyk Do you know whether human milk contains interleukin-12, the cytokine which would be responsible for the selection of T_{h1} or T_{h2}-cells?

Brandtzaeg I have no information on the content of IL-12 in breast milk. This cytokine is commonly produced by macrophages. Therefore I assume that it is present in human milk, with its high content of macrophages.

CHAPTER 25

Dallman To what extent does food iron saturate lactoferrin in the intestine under the conditions that would prevail in the intestine of the infant? Could the iron content of the diet influence the activity of lactoferrin?

Conneely About 30% of the iron in human milk is lactoferrin-bound, and yet lactoferrin is 8–20% saturated with iron. Thus, there is some partitioning of iron in the milk that favours lactoferrin in the iron-free form. To answer your question on how iron supplementation in the diet may affect the activity of lactoferrin in the intestine is much more complex; it is very likely that again there is a partitioning. How that partitioning works and what factors contribute to it are unknown. It is known that lactoferrin prefers ferric iron, and that there are certain sets of conditions in the intestine, such as the presence of bicarbonate, that favour uptake of iron by lactoferrin. Further studies are required to evaluate the effect of food iron on the antibacterial activity of lactoferrin. The best way to study this would be to compare the antibacterial activity of iron-free and iron-saturated lactoferrin, both supplemented in a diet.

Dallman Absorption studies suggest that non-haem iron forms a single pool in the intestine. You might find out that it is not what happens.

Conneely That is right.

Elia Could you tell us something about the extent to which lactoferrin is digested by intestinal enzymes? If it does escape digestion in the small intestine, could lactoferrin have effects in the large bowel as well? Have there been any studies with colonocytes?

Conneely Geneviève Spik and her group in France and other investigators have looked at the fate of ingested lactoferrin. They have shown that a significant portion of human lactoferrin can be isolated intact in faeces, and that this faecal lactoferrin is still able to bind iron. This observation suggests that a significant portion of lactoferrin is resistant to proteolytic enzymes and low pH in the gastrointestinal tract. Secondly, remember that in adults lactoferrin is also produced in situ in the intestine and in the stomach by goblet cells. I presume that lactoferrin has been made to be resistant both in structure and in function to these circumstances, as are many of the regulatory milk proteins such as growth factors.

Plaut What enterocyte cells did you use, a cell line or transformed cells?

Conneely We used a standard cell line, the human colon carcinoma cell line, Caco2. These cells are grown for 7–10 days on tissue culture, when they reach confluency and differentiate into human enterocytes.

Plaut Does lactoferrin only bind to the apical side of the enterocyte, or does it also bind to the lateral side?

Conneely These Caco2 cells can be grown in a directional matter on matrices, forming an apical and a lateral surface. It appears that lactoferrin only binds to the apical side of the enterocyte, while transferrin binds to the basolateral side.

Berger In newborn babies, both preterm and term, plasma ferritin is highly loaded with iron, and a relatively large percentage of the iron present in plasma is non-protein-bound. It has been speculated that this free iron might play a role in the pathogenesis of diseases such as respiratory distress syndrome, bronchopulmonary dysplasia and necrotizing enterocolitis. A couple of years ago Professor Widdowson suggested that unsaturated lactoferrin could reduce the content of body iron by transferring it from plasma into the lumen. Could you comment on this?

Conneely I am not aware of any direct studies that have addressed this issue. The clearance of iron by lactoferrin that has been reported occurs in plasma. There is some evidence of iron transfer from the extracellular fluid to intracellular ferritin in macrophages by lactoferrin, but I have not seen any direct evidence of iron transfer back into the gut. It has been reported, however, that lactoferrin may be involved in sequestration and detoxification of free iron in the intestine, preventing the absorption of excess iron. In that way lactoferrin may lower plasma levels of iron by preventing the transfer from the lumen to the circulation.

Index